Form Follows Nature

For Zéza

Rudolf Finsterwalder (Hg./Ed.)

FORM FOLLOWS NATURE

Zweite, überarbeitete Auflage / 2nd revised Edition

EINE GESCHICHTE DER NATUR ALS MODELL FÜR FORMFINDUNG IN INGENIEURBAU, ARCHITEKTUR UND KUNST

A HISTORY OF NATURE AS MODEL FOR DESIGN IN ENGINEERING, ARCHITECTURE AND ART.

Mit einem Vorwort von / With a preface by *Kristin Feireiss*

Und einem unveröffentlichten Interview mit / And an unpublished interview with *Frei Otto*

Birkhäuser
Basel

Inhalt / Content

Kristin Feireiss

Vorwort / Preface

»Ich bin Frei, ich bin Frei Otto. Ich bin ein freier Mensch.« Das sagte der große Architekt und Menschenfreund in einem Video-Interview am 13. Februar 2015 in seinem Haus in Warmbronn nur wenige Wochen vor seinem Tod. Und er fügte hinzu: »Auch wenn ich jetzt blind bin, meine Gedanken über Architektur sind klarer als jemals zuvor. (...) Ich habe es immer als meine Pflicht angesehen, neue Gebäudetypen zu entwickeln, um Menschen zu helfen zu überleben: nach Erdbeben, Überschwemmungen und von Menschen verursachten Katastrophen.« | Dieses letzte Interview* Frei Ottos ist ebenso wie sein gesamtes Lebenswerk das humanitäre Vermächtnis eines der bedeutendsten Architekten, Visionäre, Forscher und Erfinder des 20. Jahrhunderts — und es richtet sich ganz besonders an eine junge Architektengeneration. || Viel ist inzwischen über Frei Otto gesagt und geschrieben worden, der trotz seines hohen Alters für alle unerwartet am 9. März dieses Jahres verstorben ist, aber noch zu Lebzeiten mit dem Pritzker Preis 2015 ausgezeichnet worden war. Seine Reaktion auf diese große Ehrung: »Was könnte mir Schöneres passieren!« Die Zeremonie selbst hat er nicht mehr erlebt. || Es ist mir eine große Freude, in dieser neuen Auflage des Buches *Form Follows Nature* von Rudolf Finsterwalder über Frei Otto und sein Werk zu schreiben, da ich ihm über Jahrzehnte hinweg verbunden war. Ich habe ihn und seine Frau Ingrid in ihrem Wohnhaus in Warmbronn bei Stuttgart besucht, bei dem Frei Otto seine architektonische Haltung innovativ, ja fast spielerisch in Raum umgesetzt hat. Ich hatte die

* Interview Edward Lifson, *The Pritzker Prize*. Kamera: Moritz Dirks

Abb. 1 Kristin Feireiss, Frei Otto und Shigeru Ban

Fig. 1 Kristin Feireiss, Frei Otto and Shigeru Ban

"I'm Frei, Frei Otto. I am a free man." These words were spoken by the great architect and philanthropist in a video interview on February 13, 2015, at his house in Warmbronn only a few weeks before his death. He went on to add, "Even though I'm now blind, my thoughts on architecture are clearer than they've ever been (...) I have always considered it my duty to develop new building types that can help people to survive in the wake of major catastrophes such as earthquakes, floods and man-made disasters." | Frei Otto's final interview*, much like his entire life's work, is the humanitarian legacy of one of the 20th century's most important architects, visionaries, researchers and inventors – and his words are specifically addressed to a new generation of architects. || Much has been said and written about Frei Otto since his death on March 9th of this year, which came as a surprise to everyone despite his advanced age. Shortly before his death, he was awarded the 2015 Pritzker Architecture Prize. His reaction on the occasion of this great honor: "I can't imagine anything better than this!" However, he wouldn't live to attend the ceremony itself. || It gives me great pleasure to write about Frei Otto and his work in this new edition of the book *Form Follows Nature* by Rudolf Finsterwalder. I was associated with Frei Otto for decades and visited him and his wife Ingrid at their house in Warmbronn close to Stuttgart where he had realized his architectural vision in an innovative and even playful fashion. Several years ago I had the pleasure of rummaging around two garages close to his house, which were stacked to the ceiling with his architectural and engineering models, all the while Frei Otto explained every single model to me. I felt like Alice in Wonderland. This would result years later in the exhibition *Form Follows Nature* curated by Rudolf and Maria Finsterwalder at our Aedes Architecture Forum in Berlin. The exhibition featured several of these "garage models", which had subsequently been painstakingly restored. || As a jury member for the Pritzker Prize, I had to give a personal statement regarding my decision, from which I would like to quote: "Frei Otto's innovations, his humanitarian concepts revolving around social and ecological development and his belief in the importance of continual basic research made him one of the most inspirational and influential personalities worldwide in the modern history of architecture and, thanks to his holistic approach that was always ahead of its time, he will remain so long after his death." || Despite his cult status as a star architect, Frei Otto always emphasized teamwork and cooperation between equally empowered experts across multiple disciplines. || Today more than ever this approach serves as a model for new generations of architects seeking to manage building projects of increasing complexity that require shared responsibility. || Frei Otto's pioneering drive to develop a democratic architecture was shaped early on by his experiences as a fighter pilot in Nazi Germany. This period would further reinforce his rejection of all things monumental and

* Interview Edward Lifson, *The Pritzker Prize*. Camera: Moritz Dirks

Freude, vor etlichen Jahren in zwei Garagen in der Nähe seines Hauses, in denen sich damals seine Architektur- und Konstruktionsmodelle bis zur Decke stapelten, zu stöbern, während Frei Otto mir jedes einzelne Modell erklärte – eine Situation, in der ich mich fühlte wie Alice im Wunderland. Jahre später folgte in unserem Aedes Architekturforum in Berlin die Ausstellung *Form Follows Nature*, kuratiert von Rudolf und Maria Finsterwalder, in der wir viele der »Garagen-Modelle« von einst, inzwischen sorgfältig restauriert, zeigen konnten. || Als Jurymitglied des Pritzker Preises habe ich, wie alle anderen Juroren, ein persönliches Statement zu unserer Entscheidung gegeben, aus dem ich hier zitieren möchte: »Frei Otto war mit seinen Erfindungen, seinen humanistischen, auf gesellschaftlichen und ökologischen Entwicklungen basierenden Konzepten und seiner Überzeugung von der Bedeutung kontinuierlicher Grundlagenforschung einer der inspirierendsten und weltweit einflussreichsten Persönlichkeiten der modernen Architekturgeschichte. Er wird dies noch weit über seinen Tod hinaus bleiben, da er mit seinem holistischen Ansatz seiner Zeit immer einen Schritt voraus war.« | Frei Otto hat – dem Stararchitektenkult zum Trotz – zeitlebens auf die Zusammenarbeit im Team gesetzt, auf das gleichberechtigte Miteinander disziplinübergreifender Experten, und er ist mit dieser gelebten Haltung gerade heute ein Vorbild für eine junge Architektengeneration bei Bewältigung und Lösung immer komplexer werdender Bauaufgaben, die sich nur in gemeinsamer Verantwortung bewältigen lassen. || Der Antrieb für die Entwicklung einer demokratischen Architektur, deren Vorreiter Frei Otto war, wurde schon früh durch seine Erfahrungen im Nationalsozialismus, den er als Kampfflieger erlebt und überlebt hat, geprägt, einer Zeit, die seine Ablehnung gegenüber allem Monumentalen, Unmenschlichen noch verstärkt hat. | Frei Ottos Sehnsucht nach sozialer Gerechtigkeit spiegelt sich auch in seiner Vorstellung vom Paradies wider, über die er in seinem denkwürdigen Interview gesprochen hat. Da es in englischer Sprache geführt ist, möchte ich es diesmal auch so wiedergeben: »I have only one dream: I would like to give paradise to everybody with all its beauty. I learned that you cannot make beauty, but you can live beauty.« || Was kann man Frei Ottos eigenen Worten noch hinzufügen? Und was gibt es, was in den zahlreichen Würdigungen und Nachrufen noch nicht zur Sprache gekommen ist? Vielleicht ist nicht allen bewusst, dass Frei Otto einer der Ersten war – wenn nicht der Erste –, der sich in Europa mit nachhaltigem Bauen auseinandergesetzt hat, als noch kaum einer diesen Begriff auf die Architektur angewandt hat; dass sich Frei Otto immer klar von der Haltung der Moderne und ihrem exponiertesten Vertreter Le Corbusier abgegrenzt hat, der die Natur als »Unordnung« und damit als Gegensatz zur Architektur verstanden hat; dass Frei Otto als einer der Ersten die Beziehung zwischen Natur und Architektur in den Mittelpunkt seiner Forschung gestellt und Naturprozesse

inhumane. | His quest for social justice is reflected in his idea of paradise, which he touched upon in his memorable interview, given in English: "I have only one dream: I would like to give paradise to everybody with all its beauty. I learned that you cannot make beauty, but you can live beauty." || What could possibly add to Frei Otto's own words? And what if anything has not yet been said in the countless tributes and obituaries? Perhaps not everyone knows that Frei Otto was one of the first – if not *the* first – in Europe to take up the subject of sustainability in construction, long before the concept was commonly applied to architecture; that Frei Otto always clearly distanced himself from the mindset of modernity and its most prominent representative Le Corbusier, who saw nature as "disorder" and thus opposed to architecture; that Frei Otto was one of the first to focus his research on the relationship between nature and architecture and to apply natural processes to the development and implementation of architectural structures and forms. || Much has also been written about his style of lightweight construction which revolutionized architecture worldwide. Rarely mentioned, however, is the fact that he not only implemented his concepts in his own architecture but also deliberately did *not* place them under patent protection. He considered it his duty to make his innovations available to all for the collective advancement of architecture. Architects such as Zaha Hadid, Richard Rogers and Shigeru Ban have continued to make use of his innovations. || One last thing in closing: in the mid 90s, when I was a member of the Zollverein Coal Mine Industrial Complex, I proposed the construction of a "Frei Otto Museum for Architecture and Engineering" on the site of the coal mine with a building designed by Shigeru Ban. The project had already been presented to the public in a press conference. Ultimately, nothing became of it and Frei Otto was thoroughly disappointed. | Perhaps the momentum provided by international attention and appreciation for this great architect and visionary will breathe new life into the idea of a Frei Otto Museum.

auf die Entwicklung und Anwendung architektonischer Strukturen und Formen übertragen hat. || Auch über die von ihm entwickelte Leichtbauweise, mit der er die Architektur weltweit revolutioniert hat, ist viel geschrieben worden. Nur selten erwähnt wird aber, dass Otto seine zahlreichen Erfindungen nicht nur bei seinen eigenen Bauten umgesetzt hat, sondern auch ganz bewusst *nicht* unter Patentschutz gestellt hat. Er sah es als seine Verpflichtung an, seine Erfindungen als Teil des kollektiven Wissens allen verfügbar zu machen, um so die Entwicklung der Architektur voranzutreiben. Architekten wie Zaha Hadid, Richard Rogers oder Shigeru Ban haben immer wieder auf seine Erfindungen zurückgegriffen. || Zum Schluss noch eine Anmerkung: Mitte der 90er Jahre, als ich dem Beirat der Zeche Zollverein angehörte, habe ich den Vorschlag eingebracht, auf der Zeche ein »Frei Otto Museum für Architektur und Ingenieurwesen« zu installieren, wofür Shigeru Ban das Gebäude entworfen hat. In einer Pressekonferenz wurde das Projekt bereits der Öffentlichkeit vorgestellt. Es ist nichts daraus geworden. Frei Ottos Enttäuschung war groß. | Vielleicht gibt das Momentum der internationalen Aufmerksamkeit und Würdigung dieses großen Architekten und Visionärs der Idee eines Frei Otto Museums neuen Auftrieb.

Rudolf Finsterwalder

Einführung / Introduction
Form Follows Nature

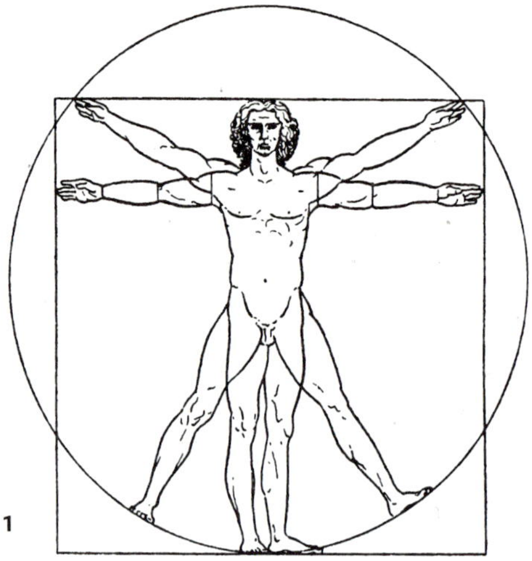

1

Seit jeher beschäftigt die Menschen die Natur als Vorbild, als Ziel ihrer Bemühungen, als Herausforderung in ihrem Streben. In vielerlei Hinsicht ist die Natur ein Fundus für den schaffenden Menschen, aber auch ein Kontrapunkt zur eigenen Arbeit. | Naturwissenschaftler, Ingenieure, Architekten und Künstler schöpfen aus dem reichen Fundus der Natur, lernen daraus und lassen sich inspirieren. | Das vorliegende Buch will einen Abriss der Geschichte der Auseinandersetzung des Menschen mit der Natur liefern und einen Ausblick auf weitere Möglichkeiten des Lernens von der Natur bieten. Nicht der Anspruch auf Vollständigkeit soll im Vordergrund stehen, sondern die Vielfalt. Vertreter verschiedener Disziplinen zeigen auf ihre Weise ihre Auseinandersetzung mit den Formen und Strukturen der Natur. »Denn wahrhaftig steckt die Kunst in der Natur, wer sie heraus kann reißen, der hat sie.«[1] (Abb.1) | Als Symbol der Übereinstimmung von Geist und Natur ermutigt die *divina proportio* Vitruvs den neuzeitichen Menschen, sich eine harmonische Welt nach eigenen Maßvorstellungen zu entwerfen. | Aus Vitruvs Proportionsverständnis sprechen die pythagoreische Zahlenlehre und die Vorstellung von der Gottebenbildlichkeit des Menschen als dem Maß aller Dinge. In den einfachen Zahlen offenbart sich nach Pythagoras das Walten der göttlichen Vernunft. | »Ferner ist natürlicherweise der Mittelpunkt des Körpers der Nabel. Liegt nämlich ein Mensch mit gespreizten Armen und Beinen auf dem Rücken und setzt man die Zirkelspitze an der Stelle des Nabels ein und schlägt einen

1 Albrecht Dürer: *Vier Bücher von menschlicher Proportion*, Nürnberg 1528

Abb.1 Vitruvs Mensch im Kreis

Fig.1 Vitruvian Man inscribed in circle and square

Since time immemorial, man has engaged with nature as a paradigm, as a target for their efforts, as a challenge in their quests. In many ways nature is a resource for the creative man, but also a counterpoint to their own work. Natural scientists, engineers, architects and artists draw from nature's rich pool, to learn from and get inspired. | This book aims to outline the history of man's examination with nature, while also offering a view to further possibilities of learning from nature. Not to be exhaustive, but rather to offer an overview. Representatives of different disciplines show their ways of examination of natural forms and structures. "Because truly the art is in nature, who can tear it out, got it."[1] (Fig. 1) | As a symbol for conformity between spirit and nature the *divina proportio* by Vitruvius encourages the modern man to design his own harmonic world according to his own ideas of proportions. | In Vitruvius, understanding the proportions of the Pythagorean doctrine of numbers is inherent, while also the notion of the divine image of man as the measure for everything. In Pythagoras' idea, the divine rule is revealed in the simple numbers. | "Further, naturally, the navel is the center of the body. If namely a man lies on his back with his arms and legs outstretched, and a circle is inscribed around the center point of the navel, than the tips of both hands and toes are touched by this circle. Along with the circle formed, the figure of a square can be found. If a measure is taken from the soles of the feet to the top of the head and compared with a measure of the outstretched arms, it will be the same width and height as are defined according to the square protractor."[2] | The relation between nature and culture is based on the Pythagorean belief in a universal validity of simple harmonic numerical ratios, which the master builder of the Universe took as a base for his plan of the world as a unitary creation. "The book of nature is written with mathematical symbols."[3]

1 Albrecht Dürer: *Four books of human proportion*, Nürnberg 1528

2 August Rode, trans., *Baukunst*, by Vitruvius, from Roman original (Third ed., first chapter), Leipzig 1796

3 Galileo Galilei, shortform of a quotation in rossaepfeltheorie.de from *Saggiatore* 1623

Frei Otto Frei Otto was the most prominent authority that masterminded the concept of *Form Follows Nature*. His pursuit of real understanding and his clear academic approach served as both an example and a benchmark. It was developed by us over the course of many discussions on the topic and the concept of this book. Following his advice, the title was supposed to end with a question mark to emphasize the debate surrounding the topic and to prevent it from being misunderstood as a simplistic slogan. Frei, who had often been misunderstood in life both intentionally and unintentionally, wanted to avoid that. In much the same way as with Sullivan, whose concept of *Form Follows Nature* was misinterpreted and whose quotation has often been misused, *Form Follows Nature* is not a demand to adopt or imitate forms from nature, but rather to explore the laws of nature and thus develop a deeper understanding of them. This is modeled especially on the organic development of growth and forms, resulting in a conclusively

2

Kreis, dann werden von dem Kreis die Fingerspitzen beider Hände und die Zehenspitzen berührt. Ebenso, wie sich am Körper ein Kreis ergibt, wird sich auch die Figur des Quadrats an ihm finden. Wenn man nämlich von den Fußsohlen bis zum Scheitel Maß nimmt und wendet dieses Maß auf die ausgestreckten Hände an, so wird sich die gleiche Breite und Höhe ergeben wie bei Flächen, die nach dem Winkelmaß quadratisch angelegt sind.«[2] | Die Beziehung zwischen Natur und Kultur beruht auf dem pythagoreischen Glauben an die universale Gültigkeit einfacher harmonischer Zahlenverhältnisse, die der Baumeister des Universums seinem Plan für die Welt als einheitliche Schöpfung zugrunde gelegt hat. »Das Buch der *Natur* ist mit mathematischen Symbolen geschrieben.«[3]

Frei Otto Frei Otto war die wichtigste Instanz für das Entwickeln von *Form Follows Nature*. Sein Streben nach wirklichem Verständnis und sein klarer wissenschaftlicher Weg waren Vorbild und Maßstab. In vielen Gesprächen zum Thema und zum Konzept des Buches haben wir es entwickelt. Auf seinen Rat hin sollte hinter dem Titel ein Fragezeichen stehen, um die Diskussion zum Thema zu betonen und um zu vermeiden, dass er als billiger Slogan missverstanden wird. Er, der in seinem Leben so oft missverstanden wurde, absichtlich und unabsichtlich, wollte das vermeiden. Wie Sullivan, der mit seinem Spruch »form follows function« missverstanden wurde und dessen Zitat oft missbraucht wurde. »Form follows

2 Vitruv: *Baukunst*, aus der römischen Urschrift übersetzt von August Rode, Leipzig 1796, Drittes Buch, Erstes Kapitel

3 Galileo Galilei: Kurzform eines Zitats. rossaepfeltheorie.de aus dem *Saggiatore* von 1623, Abschnitt 25 (engl. Übersetzung)

Abb. 2 Frei Otto und Rudolf Finsterwalder während des Interviews

Fig. 2 Frei Otto and Rudolf Finsterwalder during the interview

3

evolving interpretation. Regardless of whether a musician relies on algorithms borrowed from nature or an architect is creating buildings, what matters is that when you draw on inspiration from nature, you develop an understanding for it and adapt it to your own craft. Adapting it true to life rarely serves that purpose and won't produce groundbreaking results. | The second edition of this book features my interview with Frei on the topic of fluid wood. He describes in great detail the possibility of creating organic forms with new materials whose only limitations are determined by the formwork. The outlined process resembles the stages of growth encountered in nature. This lays the foundation for creating structures that evoke the shapes of living organisms. The only prerequisite for achieving this is a radically different approach than that applied to traditionally built reinforced concrete structures. Steel and concrete or timber form a unit that can be molded into form. Iron does not need to be worked into shape, but rather everything is molded in one step. Frei Otto's outstanding achievements have been to question everything, think differently and propose new solutions. Unfortunately, he died much too soon. His head was full of ideas, some of which remain forever lost for they have not been documented. Fortunately, some have been preserved; however they are still largely unexplored and untapped. This remains an endeavor for the future and the present book will hopefully serve as an inspiration.

Abb.3 Experimentelle Modelle in Frei Ottos Depot

Fig.3 Experimental models in the storage of Frei Otto

nature« ist nicht die Aufforderung, Formen der Natur zu übernehmen oder nachzuahmen. Es ist vielmehr die Aufforderung, sich mit den Gesetzmäßigkeiten der Natur auseinanderzusetzen und ein Verständnis dafür zu entwickeln. Vorbild soll vor allem das organische Entwickeln von Wachstum und Formen sein, eine sich schlüssig entwickelnde Interpretation. Egal ob man als Musiker mit Algorithmen aus der Natur arbeitet oder als Architekt an Konstruktionen, wichtig ist es, wenn man sich die Inspiration aus der Natur holt, diese zu verstehen und dann für die eigene Arbeit zu adaptieren. Ein Übernehmen 1:1 ist selten zielführend und wegweisend. | In der vorliegenden 2. Auflage ist mein Interview mit Frei zu flüssigem Holz abgedruckt. Sehr detailliert beschreibt er darin die Möglichkeit, mit einem neuen Material organische Formen herzustellen, deren einzige Einschränkung die begrenzten Möglichkeiten des Schalungsbaus sind. Der beschriebene Prozess ähnelt sehr einem Wachstumsprozess in der Natur. Es ist eine Möglichkeit, Konstruktionen zu schaffen, die an lebende Organismen erinnern. Voraussetzung dafür ist ein radikal anderer Ansatz, als in der klassischen Weise Stahlbetonkonstruktionen zu bauen. Stahl und Beton beziehungsweise Holz bilden eine Einheit, die gegossen werden kann. Es müssen nicht Eisen geflochten werden und in Form gebracht werden, sondern alles passiert in einem einzigen Prozess, in einem Guss. Es sind die einzigartigen Leistungen Frei Ottos, Dinge zu hinterfragen, anders zu denken und neue Lösungen zu finden. Leider ist er viel zu früh verstorben. Vieles war in seinem Kopf, was nun verloren ist, da es nicht dokumentiert wurde. Einiges ist zum Glück erhalten, aber zu einem großen Teil immer noch nicht entdeckt und nutzbar gemacht. Das ist eine Aufgabe für die Zukunft, das vorliegende Buch soll dazu anregen.

Form Die Formenviefalt in der Natur erscheint unendlich. Aus einer begrenzten Anzahl von mathematischen Gesetzen entsteht eine scheinbar endlose Anzahl von Formen und Mustern. Muster entstehen aus mathematischen Regeln. Die Schönheit der Muster und Formen der Natur beruht auf den zugrunde liegenden mathematischen Regeln, das Regelhafte und Gleichmäßige, die Symmetrie sind das, was der Mensch als harmonisch empfindet. Eine innere Ordnung der Dinge wirkt auf das menschliche Auge und den Geist als angenehm. | D'Arcy Wentworth Thompson sagte: »Denn die Harmonie der Welt kommt in Form und Zahl zum Ausdruck, und das Herz und die Seele wie alle Poesie der Naturphilosophie sind im Begriff der mathematischen Schönheit vereint.«[4] (Abb. 4) | Ein Beispiel für diese versteckten Regeln ist die auffallend gleiche Anzahl von Blütenblättern bei Pflanzen. Bei den meisten entsprechen sie Zahlen aus der Reihe 3, 5, 8, 13, 21, 34, 55, 89 und so weiter. Diese Reihe, bei der jede Zahl die Summe der beiden vorhergehenden ist, nennt man Fibonacci-Zahlen, sie ist nicht nur bei Blüten, sondern auch bei ande-

4 *On Growth and Form.* 1917, Nachwort, S. 448

4

Form The variety of forms in nature seems endless. Out of a limited quantity of mathematical rules a seemingly endless quantity of forms and patterns arise. Patterns derive from mathematical rules. The beauty of patterns and forms in nature based on the mathematical rules, the regularity and uniformity, and symmetry are what man feels as harmonious. An inherent order in things appeals to human eye and spirit. | D'Arcy Wentworth Thompson said: "For the harmony of the world is made manifest in Form and Number, and the heart and soul and all poetry of Natural Philosophy are embodied in the concept of mathematical beauty."[4] (Fig. 4) | An example for these hidden rules is the striking quantity of plant petals. In most of the plants they correspond to numbers in the series 3, 5, 8, 13, 21, 34, 55, 89 and so on. This row, in which every number is the sum of the two numbers before is called the Fibonacci sequence. It can be found very often not only with petals but also with other parts of plants. In the sunflower's head there are small flowers that later on become seeds, which are positioned in two intersecting spiral families. Of these spirals some kinds have 34 or 55, others 55 and 89 flowers, all numbers of the Fibonacci series | This mathematical regularity is what we find so attractive in nature. In 1951 Le Corbusier published the *Modulor*, a theory to develop a sequence of harmonic numbers, stemming from proportions and measurements of the human body. He started with a standard size of man and developed from there a sequence of numbers in red, which are related to each

4 Afterword to *On Growth and Form* (1917), p. 448

Abb. 4 Sonnenblume

Fig. 4 Sunflower

5
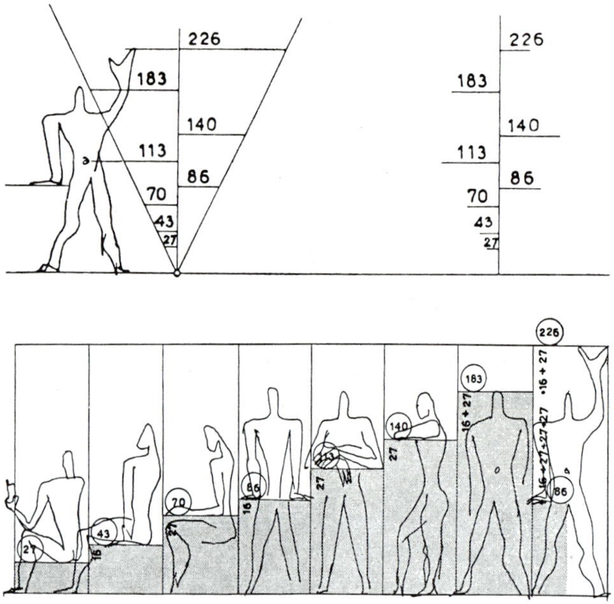

6

ren Merkmalen von Pflanzen außergewöhnlich oft zu finden. In den Köpfen von Sonnenblumen sind kleine Blüten, die später zu Samen werden, in zwei sich überschneidenden Spiralfamilien angeordnet. Von diesen Spiralen haben einige Arten 34 bzw. 55, andere 55 und 89, alles Zahlen der Fibonacci-Reihe. | Diese mathematische Regelhaftigkeit ist es, was wir an der Natur so anziehend empfinden. | 1951 erschien der *Modulor*, eine Theorie Le Corbusiers, der versuchte eine Reihe harmonischer Zahlen zu entwickeln, die sich aus den Proportionen und Maßen des menschlichen Körpers ableitete. Er ging von einer Standardgröße des Menschen aus und entwickelte daraus eine rote Reihe von Zahlen, die untereinander im Verhältnis des goldenen Schnittes in Beziehung stehen. Aus der Verdoppelung der Zahlen entsteht die blaue Reihe. | Eine solche Reihe, die Fibonacci-Reihe oder Le Corbusiers Modulor bieten die Möglichkeit, Dinge nach gewissen Regeln zu gestalten. Entscheidend ist, dass keine Zufallsmaße entstehen, sondern die Maße immer wieder auftauchen und in Beziehung zueinander stehen, diese Harmonie ist erlebbar und spürbar. Auch ohne das Wissen um die Regelhaftigkeit und mathematische Präzision empfindet der Mensch dies als schön und ästhetisch. Ein Garant für ästhetische Formen oder Schönheit ist die Verwendung solcher Zahlenreihen natürlich nicht, sondern vielmehr ein Werkzeug (Abb.5). | Seit den späten 20er Jahren entwickelte Richard Buckminster Fuller material- und energieffiziente Konstruktionen, unter ihnen die geodätische Kuppel, eine Konstruktion, die aus

Abb.5 / Fig.5
Le Corbusier, Modulor

Abb.6 Geodätische Kuppel
(Buckminster Fuller)

Fig.6 Geodesic dome
(Buckminster Fuller)

7

other through the ratio of the golden section. The doubling of the numbers creates the sequence of blue numbers. | Such a sequence, the Fibonacci sequence or Le Corbusier's modulor, offer the possibility to design according to certain rules. It is crucial that there are no random measurements; instead, repeating dimensions create a relationship among each other. This harmony can be felt. Even without knowledge about the observance of rules or mathematical precision, one can still feel the underlining beauty. Of course, the use of such sequences of numbers is no guarantee for aesthetical forms or beauty, only a tool. (Fig. 5). | Since the late 20th Richard Buckminster Fuller developed constructions efficient in material use and energy, among them the geodetic dome, a construction developed with forms from nature. Octahedrons, which allowed a package as dense as possible, were used as the base. A super lightweight construction was the result. In 1954 the geodetic dome was patented, and since then has been built approximately 300,000 times worldwide. His biographer Lloyd Steven Sieden wrote: "Buckminster Fuller believed that natural creations such as the human hand represented technology at its finest and believed that the best human created technology was that which most likely mirrored nature's creations."[5] (Fig. 6)

Culture-Nature Today in our western culture, we normally call nature that which hasn't been created by man. | Whether man itself belongs to nature or

5 Lloyd Steven Sieden: *Buckminster Fuller's Universe.* Perseus Publishing, Cambridge 1998, ed. Philip Jodidio from *Architecture: Nature,* Prestel 2006, p. 42

Abb. 7 Schwimmbad (Alvaro Siza)

Fig. 7 Swimming pool (Alvaro Siza)

Formen der Natur entwickelt wurde. Oktaeder, die eine dichtestmögliche Packung erlauben, wurden als Grundlage verwendet, eine sehr leichte und stabile Konstruktion war das Ergebnis. 1954 wurde die geodätische Kuppel patentiert, die seitdem weltweit ca. 300.000 Mal gebaut wurde. Sein Biograph Lloyd Steven Sieden schrieb: »Buckminster Fuller believed that natural creations such as the human hand represented technology at its finest and believed that the best human created technology was that which most likely mirrored nature's creations.«[5] (Abb.6)

5 Lloyd Steven Sieden: *Buckminster Fuller's universe*. Perseus Publishing, Cambridge 1998. in *architecture: nature*, S. 42

Kultur – Natur Unter Natur wird heute in westlichen Kulturkreisen im Allgemeinen das bezeichnet, was nicht vom Menschen geschaffen wurde. | Ob der Mensch selbst zur Natur gehört oder nicht, ist bereits nicht mehr gesellschaftlicher Konsens. Natur ist nach einer verbreiteten Auffassung das Gegenteil von Kultur. Dieses Spannungsfeld zwischen Natur und Kultur ist höchst interessant, wo beginnt Kultur, wo endet Natur? Welche Wechselbeziehungen gibt es? | Natur wird wegen ihres Ebenmaßes und ihrer Perfektion vom Menschen als schön empfunden, manchmal jedoch als allzu perfekt. Der Mensch empfindet oftmals Spannungen und Störungen als wichtig, als ästhetisch reizvoll. Ein perfekt gestaltetes Objekt kann durch eine Irritation, einen Bruch interessanter werden. Álvaro Siza beschreibt das Phänomen so: »What is made by man is not natural. More and more I think that there must be a certain distance between the natural and the manmade. But there must be also a dialogue between the two. Architecture comes from natural forms, but it also transforms nature. If you look at the small restaurant (*boa nova* in Leca, Portugal) I designed, immediately before the swimming pool (Piscina in Leca, Portugal), there is a profile of rocks, and the building almost follows that outline. Of course, the swimming pool involved a longer, more horizontal space, but I tried to do a critique of my own approach in the restaurant. I thought that the reaturant was to mimetical – it seems to be a part of the rocks in a way. In the swimming pool, I designed five or six walls that relate to some of the rocks. The important thing is how this geometry encounters the natural elements, and how the landscape is transformed. I did not touch the rocks, I only added something recognizeable as not natural. I put those things together, that is the way I see the relation between nature and architecture. If you look at the work of F. L. Wright, in his first phase he works with nature, but later on he works with the landscape. He transforms the natural environment. There is an affirmation of the built without ignoring the basic aspects of topography or the typology of the landscape.«[6] (Abb.7) | Interessant ist das Konzept, die Natur in ihrer Eigenart bestehen zu lassen und durch einen bewusst gesetzten Kontrast in ihrer Wirkung zu steigern, was ein profundes Verständnis von Natur voraussetzt. | Einen anderen Ansatz, sich dem Verständnis von Natur und ihrer Erscheinungsformen zu

6 *Interview with Álvaro Siza*. Porto: Mai 1998; in *architecture: nature*. Philip Jodidio, Prestel 2006, S. 48

not is not a common consensus. This area of conflict between nature and culture is very interesting, where culture begins, where nature ends. What kind of interchanges exists? | Man sees nature as beautiful because of its regularity and perfection, but sometimes this seems too perfect. Man quite often feels tensions and interferences as important, aesthetically attractive. A perfectly designed object can become more interesting with an irritation or interference. Álvaro Siza describes that phenomenon this way: "What is made by man is not natural. More and more I think that there must be a certain distance between the natural and the manmade. But there must be also a dialogue between the two. Architecture comes from natural forms, but it also transforms nature. If you look at the small restaurant (*boa nova* in Leca, Portugal) I designed, immediately before the swimming pool (Piscina in Leca, Portugal), there is a profile of rocks, and the building almost follows that outline. Of course, the swimming pool involved a longer, more horizontal space, but I tried to do a critique of my own approach in the restaurant. I thought that the reaturant was to mimetical – it seems to be a part of the rocks in a way. In the swimming pool, I designed five or six walls that relate to some of the rocks. The important thing is how this geometry encounters the natural elements, and how the landscape is transformed. I did not touch the rocks, I only added something recognizeable as not natural. I put those things together, that is the way I see the relation between nature and architecture. If you look at the work of F. L. Wright, in his first phase he works with nature, but later on he works with the landscape. He transforms the natural environment. There is an affirmation of the built without ignoring the basic aspects of topography or the typology of the landscape."[6] (Fig. 7) | An interesting concept is to leave nature in its original form and to enhance it with a well-considered contrast in its impact, which assumes a profound understanding of nature. | Another approach that comes closer to an understanding of nature and its form of expression is shown by modern mathematics and physics. | A phenomenon that is contrary to the principle of determinism is the chaos-principle, the non-linear dynamic. According to modern physics, there are rules in the smallest units of space and time that govern the so-called accidental. Chaotic behaviour follows determinstic rules, but is so unregular that it looks like coincidental. The important point is the perception that simple rules can create complex impacts. Sometimes simple rules create predictable patterns, but other times very complex and casually structures. The so-called butterfly effect tells us that the smallest interferences in a dynamic process can cause big impacts. | Edward N. Lorenz, the discoverer of the phenomena: "Can the flap of a butterfly in Brazil cause a tornado in Texas?"[7] (Fig. 8) | So the contradiction between nature and culture would be called into question. The decisive question could be: Which rules are behind certain phenomenas, or from the perspective of the creator, which

6 *Interview with Álvaro Siza*. Porto: May 1998 from *architecture: nature*. Philip Jodidio, Prestel 2006, p. 48

7 Title of his speech 1972 during the centenary of the American Association for the Advancement of Science; according to Science 320, 2008, p. 431.

nähern, liefert die moderne Mathematik und Physik. | Ein Phänomen, das dem Determinismus entgegensteht, ist das Chaos-Prinzip, die nichtlineare Dynamik. Nach den Erkenntnissen der modernen Physik regiert in den kleinsten Einheiten von Raum und Zeit der Zufall. Chaotisches Verhalten gehorcht deterministischen Gesetzen, ist aber so unregelmäßig, dass es zufällig erscheint. Das Entscheidende ist die Erkenntnis, dass einfache Regeln komplexe Wirkungen erzeugen können. Manchmal erzeugen einfache Regeln, vorhersagbare Muster, manchmal jedoch sehr komplexe, zufällig anmutende Strukturen. Der so genannte Schmetterlingseffekt besagt, dass kleinste Störungen in einem dynamischen Prozess sich zu großen Wirkungen entwickeln können. | Edward N. Lorenz, der Entdecker des Phänomens: »Kann der Flügelschlag eines Schmetterlings in Brasilien einen Tornado in Texas auslösen?«[7] (Abb. 8) | So würde sich der Widerspruch zwischen Natur und Kultur relativieren, die entscheidende Frage könnte so lauten: Welche Regel steckt hinter gewissen Phänomenen bzw., aus der umgekehrten Sicht des Schaffenden, welche Regel wende ich an und wie steuere ich den Prozess der Form- und Gestaltfindung. Können durch bewusst eingeführte Störungen interessante Ergebnisse erzielt werden? | Wie kann dieser Prozess kontrolliert werden? Es können in der Natur vorhandene Muster adaptiert oder transformiert werden, es können aber auch neue generiert werden. Mit einem tiefen Verständnis von Natur und deren mathematischen Grundlagen ist es auch vorstellbar, dass neue synthetische Formen und Strukturen in diesem Duktus entwickelt werden.

7 Titel seines Vortrags im Jahr 1972 während der Jahrestagung der American Association for the Advancement of Science; laut *Science* 320, 2008, S. 431.

Effizienz Die Natur funktioniert nach klaren Regeln und Gesetzen, da Effizienz in der Natur eine Frage des Überlebens ist. Nach dieser Maxime der energetischen Effizienz entsteht Wachstum in der Natur. Eine Lebensform, die nicht optimal funktioniert, ist auf Dauer nicht überlebensfähig. | Eine Eierschale ist ein Beispiel für sparsamen Umgang mit Material und Energie. Die dünnstmögliche Schale genügt für die Aufgabe, die zu erfüllen ist, ein sehr effizienter Einsatz von Material. Das Verhältnis der Schalenstärke zum Radius der Schale beträgt bei einer Eierschale ca. 1:40. Dieses Verhältnis betrug bei der Schale für das Zeissplanetarium in Jena, die bereits 1922 von Walter Bauersfeld und Ulrich Finsterwalder realisiert wurde, 1:267 und konnte bei einer größeren Kuppel für die Firma Schott in Jena auf 1:472 verbessert werden. | So gesehen ist die Eierschale ein Konstrukt, das einen Kompromiss zwischen Festigkeit und Zerbrechlichkeit schafft, was für ihren Zweck – die Schale darf während des Legens nicht brechen, muss aber später von den schlüpfenden Jungen durchbrochen werden können – auch notwendig ist. | Ein anderes Beispiel sind Seifenblasen, die z. B. Frei Otto untersucht hat. Seifenblasen bilden immer eine optimierte Fläche, die kleinstmögliche Fläche,

rules do I use and how can I control the process of form and design finding? Can deliberately introduced disturbances produce interesting results? | How can this process be controlled? Existing patterns in nature can be adapted or transformed, while also new ones can be created. With a deep understanding of nature and its mathematical base its even conceivable that new synthetic forms and structures can be developed based on the same characteristics.

Efficency Nature follows clear rules and laws, because efficency in nature is a question of survival. Following this maxim, the energetic efficiency of growth occurs in nature. A life form that is not functioning properly cannot survive in the long run. An eggshell is a good example for economical handling of material and energy. The thinnest possible shell is enough to fullfill the purpose, a very efficient use of material. The ratio from the shell's thickness to its radius, in the case of the eggshell, is approximately 1:40. The ratio used in the gridshell of the Zeissplanetarium in Jena, which was realised by Walter Bauersfeld and Ulrich Finsterwalder in 1922, was 1:267 and could be improved in the larger gridshell for Schott in Jena to 1:472. | The eggshell is a construction that creates a compromise between strength and fragility. What is necessary because of its duty; the shell should resist while the egg is being laid and must be breakable by the hatching chick. | Another example are foam bubbles, that, for instance, Frei Otto had investigated. Foam bubbles always form an optimized surface, the smallest possible surface, which not only saves material but also because of its form shows the biggest possible stability (Fig. 9, Fig. 10). | Frei Otto used the knowledge of these experiments for the development of bearing structure, which corresponds to his experiments. But a principle cannot be simply transmitted from one scale to another one entirely. There are other rules that have to be considered, so that certain principles only work out in one particular scale. | Ian Stewart, a natural scientist and mathematican, shows an analogy between a flea and an elephant in his book "The secret of the snowflake". If an elephant would have a jump force relatively to a flea, it could jump over the Eiffel-Tower. But this is not possible. | Critical for the jump force is the width of the muscles, which grows with the 2nd potency, while the mass is growing with the 3rd potency, so that the relative jump force decreases with increasing size and mass. Meaning the elephant cannot jump over the tower. Even the percentage of the skeleton is rising with the increasing size, since also in this case the mass that has to be carried is increasing with the 3rd potency. || For all kinds of systems, however, an appropriate structure has to be found. Nature can only give a suggestion, but cannot simply be copied 1:1. A critical reflection of the study is essential, so as to use the experience in another context. || For architects

8

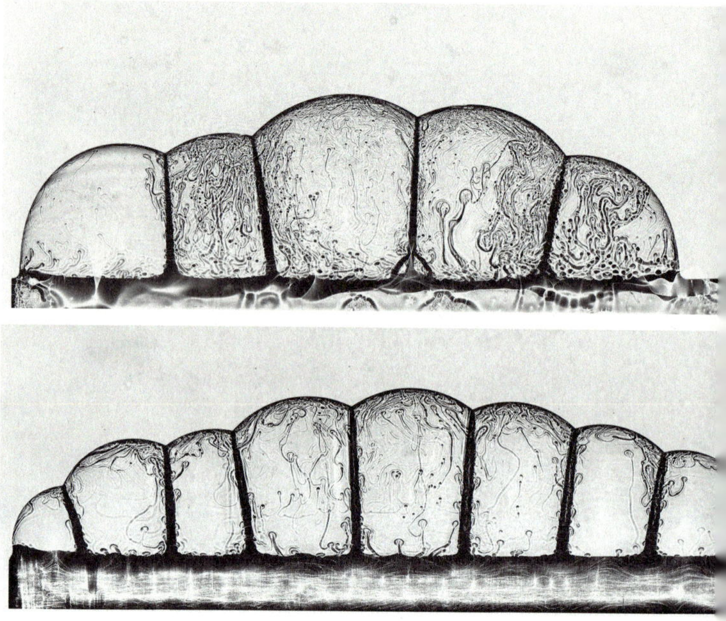

9

die nicht nur Material spart, sondern auch aufgrund ihrer Form die größtmögliche Stabilität aufweist. (Abb. 9, Abb. 10) | Die Erkenntnisse dieser Untersuchungen nutzte Frei Otto zur Entwicklung von Tragwerken, die analog zu seinen Versuchen funktionieren können. Ein Prinzip kann jedoch nicht von einem Maßtab in einen gänzlich anderen übertragen werden. Es spielen auch andere Gesetze eine Rolle, so dass manche Prinzipien nur in einem bestimmten Maßtab funktionieren. | Ian Stewart, ein Naturwissenschaftler und Mathematiker, zeigt in seinem Buch »Das Rätsel der Schneeflocke« eine Analogie zwischen Floh und Elefant. Ein Elefant, der eine relative Sprungkraft hätte wie ein Floh, könnte über den Eiffelturm springen. Das ist jedoch nicht möglich, da für die Sprungkraft der Querschnitt der Muskeln entscheidend ist, die mit der 2. Potenz wächst, die Masse wächst jedoch mit der 3. Potenz, so dass die relative Sprungkraft mit zunehmender Größe und Masse abnimmt, d.h der Elefant kann nicht über den Turm springen. | Das gleiche Phänomen hat auch Auswirkungen auf die Skelettbildung bei Tier und Mensch. Der Anteil des Skelettes am Körper nimmt bei zunehmender Größe auch zu, da auch hier die zu tragende Masse mit der 3. Potenz zunimmt. || Für alle Systeme gilt jedoch, dass eine jeweils passende Struktur gefunden werden muss. | Die Natur kann jedoch nur eine Anregung geben, aber nicht simpel 1 : 1 kopiert werden. Eine kritische Reflektion der Beobachtung ist notwendig, um die Erfahrungen in einem anderen Zusammenhang zu verwenden. || Für Architekten und Ingenieure

Abb. 8 Selbstinverse fraktale Flicken (Benoît B. Mandelbrot)

Fig. 8 Self-inverse fractal patch (Benoît B. Mandelbrot)

Abb. 9 Seifenblasen (Frei Otto)
Fig. 9 Soap bubbles (Frei Otto)

10

11

and engineers understanding of the forces is critical. Quite often, with the help of models, an optimized form can be developed. A graphic sample is the chain model by the Spanish architect Antonio Gaudi, who found the ideal form with hanging chains (Fig. 11). | The tensile forces in the hanging chains corespond to the compressive forces in the upside down construction, this way bending forces could be avoided. The chains hang according to the ideal line, which also applies for the inverted construction. The forces could have been visualised in a simple way.

Ecology Ecology (from Greek *oikos*, "house, household", and *logos*, "doctrine", so the "doctrine of household") means a branch of biology, which is concerned about the interactions of organisms with their abiotic environment. | In the 20th century the term became more and more political and today mainly referres to themes of environmental protection and sustainability. | Nature develops forms along the rules of energy efficiency; this applies both to the growth, as well as for the care of creatures. The skeleton is an example of improved support structures, each bone is an example that illustrates the forces that shape the actual course accordingly. | For example, a human thighbone consists of a foamlike structure in the interior, which corresponds very closely to the distribution of forces. The natural growth is developing as efficiently as possible. Today foams are used as ultra lightweight constructions in different fields, such as lightweight boards or in car

Abb. 10 Gitterschale für Jena
Fig. 10 Grid shell for Jena

Abb. 11 Kettenmodell Gaudi
Fig. 11 Chain model Gaudi

ist das Verständnis des Kräfteverlaufs entscheidend. Oftmals kann anhand von Modellen geklärt werden, wie die optimale Form aussieht. Ein plastisches Beispiel ist das Kettenmodell des spanischen Architekten Antonio Gaudi, der mit hängenden Ketten die ideale Form für seinen Bau fand. (Abb.11) | Die Zugkräfte in den nach unten hängenden Ketten entsprechen den Druckkräften in der auf den Kopf gestellten Konstruktion, Biegemomente wurden so vermieden. Die Ketten hängen der Ideallinie entsprechend, die eben auch für die umgedrehte Konstruktion gilt. Die Kräfte konnten so auf einfachem Wege sichtbar gemacht werden.

Ökologie Unter Ökologie (von griechisch *oikos*, »Haus, Haushalt«, und *logos*, »Lehre«, also »Lehre vom Haushalt «) versteht man ursprünglich ein Teilgebiet der Biologie, welches sich mit den Wechselbeziehungen der Organismen untereinander und mit ihrer abiotischen Umwelt beschäftigt. | Im 20. Jhdt. wurde der Begriff mehr und mehr politisch aufgeladen und bezeichnet heute vor allem die Themen Umweltschutz und nachhaltiges Handeln. | Die Natur entwickelt Formen nach dem Gesetz des geringsten Energieverbrauchs, dies gilt sowohl für das Wachstum wie auch für die Versorgung von Lebewesen. Der Skelettbau ist ein Beispiel für optimierte Tragstrukturen, jeder Knochen ist ein Beispiel, das die Formung dem tatsächlichen Kräfteverlauf entsprechend zeigt. | So zeigt z.B. ein menschlicher Oberschenkelknochen eine schaumartige Struktur im Inneren, die sehr genau dem Kräfteverlauf entspricht. Das natürliche Wachstum entwickelt sich so effizient wie möglich. Heute werden Schäume als ultraleichte Konstruktionen in den verschiedensten Bereichen eingesetzt, z.B. als Leichtbauplatten am Bau oder in der Autoindustrie. (Abb.12) | Viele Formen der Natur können eine Inspiration für Konstruktionen sein. Die Natur denkt nicht im rechten Winkel, sie sucht sich den effizientesten Weg, die optimale Form. Das Problem bei der Übertragung in andere Bereiche ist, dass die Fertigungsweisen nicht mit diesen organischen Formen kompatibel sind. Nicht alles, was effizient erscheint, ist auch effizient zu bauen. Es müssen zuerst entsprechende Fertigungstechnologien entwickelt werden, die dies ermöglichen. Die zunehmende Material- und Energieknappheit macht ein Umdenken in diese Richtung unumgänglich. (Abb.13) || Buckminster Fuller prägte den Begriff *Ephemerisierung*, mit dem das Bestreben nach immer leichteren und effizienteren Strukturen und Technologien beschrieben wird. Als Beispiel verglich er einen Satelliten, der einen Bruchteil des Gewichtes von Transatlantikkabeln besitzt und damit sehr viel ressourcenschonender ist und deren Aufgabe übernimmt. || Es wird mehr und mehr ein drängendes Problem für Architekten und Ingenieure, sich mit nachhaltigem Bauen auseinanderzusetzen. Das heißt zum einen den Verbrauch an Material zu verringern und zum anderen den Energieverbrauch im Betrieb zu reduzieren. Auch in dem Bereich des Energie-

industry (Fig. 12). | Many forms of nature can be an inspiration for construction. Nature is not thinking in a right angle, it is searching the most efficient way, the optimum form. The problem with the transmission in other fields is that the manufacturing methods are not compatible with these organic forms. Not everything that appears efficient is also efficient to build. Firstly, appropriate ways of production have to be developed that allow this. The increasing lack of material and energy makes a rethinking in the direction unavoidable (Fig. 13). || Buckminster Fuller formed the term *ephemerisation*, that describes the search for lighter and more efficient structures and technologies. For example he compared a satellite, that weights a small percentage of a transatlantic cable, and so is a much more resource-efficient and takes its job. || It is becoming a more and more urgent problem for architects and engineers to deal with a sustainable construction. On one hand this means to reduce the consumption of material and on the other hand to reduce the consumption of energy during operation. Even in the field of energy consumption one can learn from nature. | An example from the fauna shows the energy consumption in terms of enhanced physique. The emperor penguin lives in much colder regions than his relatives. Because of his bigger body, he has a better ratio of volume to surface, which reduces the relative energy consumption. A principle that is also a decisive factor for building planning. Even here it is evident that a compact building uses less energy than a building with a larger envelope surface.

Bionic Bionic is concerned with the transmission of priciples and techniques from the field of living nature to technology. Bionics is an interdisciplinary area in which natural scientists as well as members of other disciplines like, for example, architects and engineers work together. | The term derives from the military field and is composed of the words *biology* and *technics*. | Bionics is the systematic learning from nature and thus stands in contrast to the pure nature of inspiration. | For example, the Wright brothers developed their first aircrafts from their studies of bird flight. They studied the wings of large birds, as well as the construction of the wings and consequently developed their technique. In contrast, for the balloon flight there was no example in living nature. Thanks to human spirit it was discovered that something lighter than the air could fly. | Even in the field of bionics, the decisive developments didn't arise from the direct transmission in the technology, but through an understanding of the natural processes, which was then translated in the field of technology. | A well-known example is the so-called lotus effect, which was discovered at the lotus plant, almost all watersolvent substances roll off of its leaves. This effect could be used for paints and coatings. Another example is the shark's skin; it consists of small, closely placed scales. On theses scales are sharp-edged, fine channels, that are positioned parallel to the

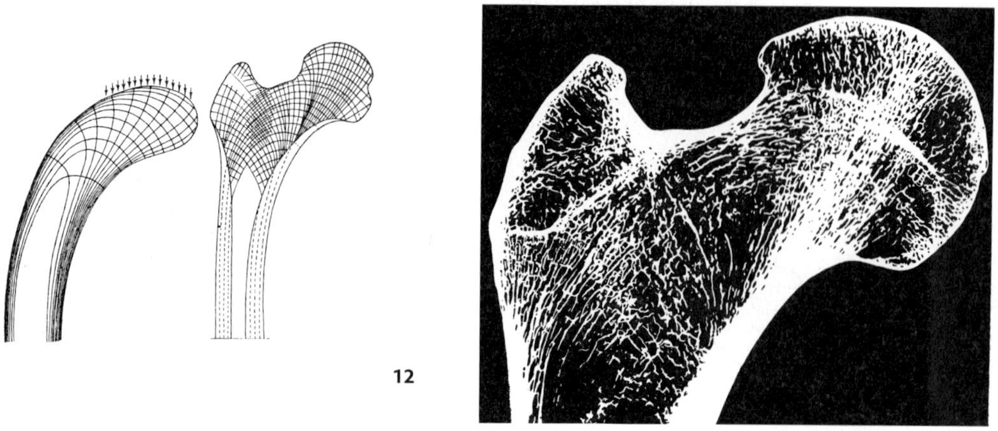

12

verbrauchs lässt sich aus der Natur lernen. | Ein treffendes Beispiel aus der Tierwelt zeigt den in Hinsicht auf Energieverbrauch optimierten Körperbau. Der Kaiserpinguin lebt in kälteren Regionen als seine Verwandten. Durch seine größere Figur hat er ein besseres Verhältnis von Volumen zu Hüllfläche, was den relativen Energieverbrauch reduziert. Ein Prinzip, das auch in der Planung von Gebäuden ein entscheidender Faktor ist. Auch hier gilt, ein kompaktes Gebäude verbraucht weniger Energie als ein Bau mit vergleichsweise großer Hüllfläche.

Bionik Die Bionik beschäftigt sich mit der Umsetzung von Prinzipien und Techniken aus dem Bereich der lebenden Natur in Technik. Die Bionik ist ein interdisziplinärer Bereich, in dem Naturwissenschaftler sowie auch Vertreter anderer Disziplinen wie etwa Architekten und Ingenieure zusammenarbeiten. | Der Begriff stammt ursprünglich aus dem Bereich des Militärs und setzt sich aus den Wörtern *biology* und *technic* zusammen. | Bionik ist systematisches Lernen von der Natur und steht damit in Abgrenzung zur reinen Naturinspiration. | So haben z.B. die Gebrüder Wright aus dem Studium des Vogelfluges ihre ersten Flugapparate entwickelt. Sie untersuchten den Flügelschlag großer Vögel sowie die Konstruktion der Flügel und entwickelten daraus ihre Technik. Wohingegen es für den Ballonflug kein Beispiel in der lebenden Natur gibt. Dem menschlichen Geist ist die Erkenntnis zu verdanken, dass etwas, was leichter ist als Luft, flie-

Abb.12 Oberschenkelknochen und Kräftediagramm (Thompson)

Fig.12 Thighbone and force diagram (Thompson)

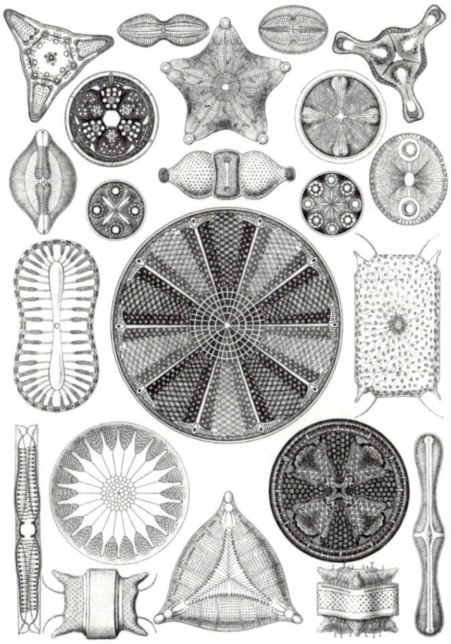

13

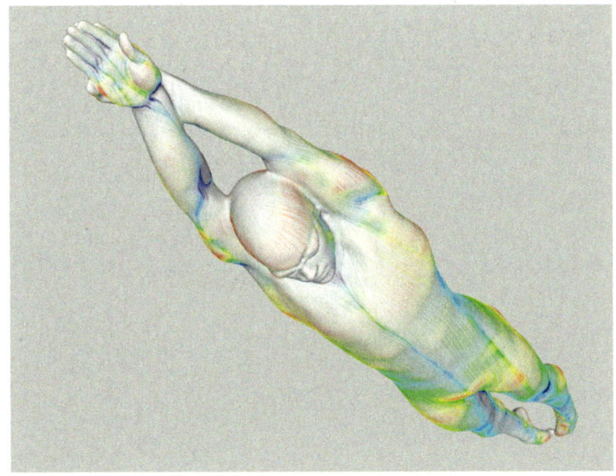

14

flow. These microscopic small channels cause a reduction of frictional resistance. This skin frictioning effect is effective in all kinds of flows, also in the air. Airplanes can be covered with a special coating, which on its surface has a similar structure and so decreases the air friction of the aircraft. The scientific bases derive from the investigations of fossil sharks and their scales. A similar application is used on modern swimsuits. (Fig.14)

Living and not living nature and technic One differenciates between *living nature* (plants, animals) and *not living nature* (stones, liquids, gases). | Technology is often considered as a counterpoint to nature. | Technology means the process and ability for the practical use of natural science and for the production of industrial, handcraft and artistic products, whereas the Greek origin of the term did not distinguish between today's categories of art and technics. | Technique is basically the use of special methods, principles, single or in combination to achieve a certain effect. Thus technology is work-saving work. | Technique can be understood as the ability of man to use or transform, in a senseible way, natural laws, forces and raw materials to safeguard their livelihood or for the satisfaction of his need to self-fulfillment. Aside the materialistic needs (food, clothes, housing) also cultural needs are granted through technique. || It is questionable whether technology is the antithesis of nature or a man-made part of it. There

Abb.13 Diatomea. Triceratium. (Ernst Haeckl)

Fig.13 Sea-hedgehog skeleton (Ernst Haeckl)

Abb.14 Bionik: Schwimmanzug (Speedo)
Fig.14 Bionics: swimsuit (Speedo)

gen kann. | Auch im Bereich der Bionik sind die entscheidenden Entwicklungen nicht aus einem direkten Übertragen in die Technik entstanden, sondern durch ein Verstehen der natürlichen Vorgänge, was dann übersetzt wurde in den Bereich der Technik. | Ein bekanntes Beispiel ist der so genannte Lotuseffekt, der an der Lotuspflanze entdeckt wurde, an ihren Blättern perlen fast alle wasserlöslichen Substanzen ab. Dieser Effekt konnte für Farben und Beschichtungen genutzt werden. | Ein anderes Beispiel ist die Haifischhaut, die aus kleinen, dicht aneinander liegenden Schuppen besteht. Auf diesen Schuppen befinden sich scharfkantige feine Rillen, die parallel zur Strömung ausgerichtet sind. Diese mikroskopisch kleinen Rillen bewirken eine Verminderung des Reibungswiderstandes. Dieser widerstandsvermindernde Effekt ist in allen turbulenten Strömungen, also auch in Luft wirksam. Flugzeuge können mit einer speziellen Folie beklebt werden, die auf ihrer Oberseite über eine ähnliche Struktur verfügt und so den Luftwiderstand des Flugzeugs senkt. Die wissenschaftliche Grundlage entstammt Untersuchungen an fossilen Haien und deren Schuppen. Ein ähnlicher Anwendungsbereich sind moderne Schwimmanzüge. (Abb.14)

Lebende und nicht lebende Natur und Technik Man unterscheidet zwischen lebender Natur (Pflanzen, Tiere) und nicht lebender Natur (z.B. Flüssigkeiten, Steine, Gase) | Als Gegenpol zur Natur wird oftmals der Begriff Technik gesehen. | Unter Technik versteht man Verfahren und Fähigkeiten zur praktischen Anwendung der Naturwissenschaften und zur Produktion industrieller, handwerklicher oder künstlerischer Erzeugnisse, wobei der griechische Begriff zwischen den heutigen Kategorien Kunst und Technik nicht unterschied. | Technik ist grundsätzlich die Anwendung von besonderen Methoden, Prinzipien, einzeln oder in Kombination, um bestimmte Wirkungen zu erzielen. Somit ist Technik Arbeit, um Arbeit zu sparen. | Technik kann als die Fähigkeit des Menschen verstanden werden, Naturgesetze, Kräfte und Rohstoffe zur Sicherung seiner Existenzgrundlage oder zur Befriedigung seines Bedürfnisses zur Selbstverwirklichung sinnvoll einzusetzen oder umzuwandeln. Neben den materiellen Bedürfnissen (Nahrung, Kleidung, Wohnen) werden auch kulturelle Bedürfnisse durch die Technik gesichert. | Fraglich ist, ob die Technik ein Gegenpol zur Natur ist oder als vom Menschen gemacht ein Teil von ihr. Es gibt auch Beispiele von Technik im Bereich der Tierwelt, wie zb hexagonale Bienenwaben, Biberdämme, Vogelnester, Termitenbauten und Spinnennetze. (Abb.15 - 18)

Inspiration Die Beziehung des Menschen zur Natur ist ein Thema, das sich durch die menschliche Entwicklungsgeschichte zieht. Sowohl als Antipode als auch als Vorbild ist sie stets ein Thema zur Auseinandersetzung. Verschiedene

15

16

are examples of technology in the animal kingdom, such as the hexagonal honeycombs, beaver dam, birds nest, termits nest and spider web. (Fig. 15 - 18)

Inspiration The relationship between man and nature is a theme that runs through the history of human development. Both as an antipode, as an ideal, it is always a topic for debate. Different contributions show how different approaches have come closer to the subject. The fascinating thing is that through different disciplines, scientific as well as cultural, there is an interest for nature and the results are quite often similar. Crucial is a well-considered argument, which does not want to imitate, but to understand and to learn. If one approaches, from a deep understanding, the living and not living nature debate can become an endless fountain of inspiration for forms, technics, structures ... | Plato's understanding of nature was: "Nature is a letter from God to mankind."[8] | The question is how to read this letter and what to learn from it. | The selection of the following contributions from different disciplines is not intended to be complete. It wants to stimulate the endless debate of nature. Whether one is trying to imitate nature, or take it as a counterpoint to one's own ideas and projects, whether as a fountain of inspiration or as a challenge. Nature remains the defining issue for manhood. | Follow it!

8 *Lebensweisheiten berühmter Philosophen. 4000 Zitate von Aristoteles bis Wittgenstein*, p. 12; Edition 6. Humboldt Verlag, 2007

Abb.15 Bienenwabe
Fig.15 Honeycomb

Abb.16 Vogelnest
Fig.16 Bird nest

17

18

Beiträge zeigen unterschiedliche Ansätze, sich dem Thema zu nähern. Das Faszinierende daran ist, dass es durch verschiedene Disziplinen, naturwissenschaftliche wie kulturelle, dieses Interesse an der Natur gibt und dass die Ansätze oft sehr ähnlich sind. Entscheidend ist eine reflektierte Auseinandersetzung, die nicht nachmachen will, sondern verstehen will. Wenn man von einem tiefen Verständnis ausgeht, kann die lebende und nicht lebende Natur eine nicht endenwollende Inspirationsquelle sein für Formen, Techniken, Strukturen… | Platons Naturverständnis drückt sich so aus: »Die Natur ist ein Brief Gottes an die Menschheit.«[8] | Die Frage ist, wie dieser Brief gelesen wird und was daraus zu lernen ist. | Die Auswahl der folgenden Beiträge aus den verschiedensten Disziplinen erhebt keinen Anspruch auf Vollständigkeit. Sie sollen anregen, sich mit dem unendlichen Thema Natur auseinanderzusetzen. Egal, ob man versucht, die Natur zu imitieren oder in ihr einen Kontrapunkt zu den eigenen Ideen und Projekten findet, ob als Inspirationsquelle oder als Herausforderung für das eigene Schaffen. Die Natur bleibt das bestimmende Thema für die Menschheit. | Follow it!

8 *Lebensweisheiten berühmter Philosophen. 4000 Zitate von Aristoteles bis Wittgenstein*, S. 12; Edition 6. Humboldt Verlag, 2007

Abb. 17 Termitenbau
Fig. 17 Termite-build

Abb. 18 Spinnennetz
Fig. 18 Spider web

1

Flüssiges Holz

Gespräch mit Frei Otto am 8. August 2013 in Warmbronn

Liquid wood

Conversation with Frei Otto, August 8, 2013, in Warmbronn

Rudolf Finsterwalder

R. Finsterwalder Du hast uns in einem Gespräch im AUT/ Innsbruck von deiner Zukunftsvision erzählt, dem flüssigen Holz. Was stellst du dir unter flüssigem Holz vor?

F. Otto Ich habe ja verschiedene Studien mit Materialien gemacht, die vielleicht in der Zukunft der Architektur einsetzbar wären. Die Frage war, ob man Faserbeton machen könnte, der holzartiger ist und der nicht brennbar ist. Vor allem für unsere Arbeiten in Arabien war wichtig, ob man aus Holz und Zement einen Beton herstellen könnte, der leicht ist und auch brandfest. So ein Produkt gab es ja auch, das Eternit, das früher mit Asbest, heute mit Holz hergestellt wird. Diesen Holzbeton, der je nach Dichte und Wasseranteil auch ganz flüssig gemischt werden kann, den kann man auch gießen.

R. Finsterwalder Dieser Holzbeton härtet wegen des enthaltenen Zements aus. | Du würdest daraus ein Halbzeug, zum Beispiel ein Plattenmaterial, herstellen, das die Eigenschaften von Holz und Beton miteinander verbindet?

F. Otto Man könnte bestimmt diesen Holzfaserbeton so flüssig halten, dass man ihn gießen könnte. Dieser Beton wäre dann haptisch angenehmer als Eternit, was eine sehr harte Faser hat. Zum Beispiel könnte man diesen Holzbeton mit hochwertigen Drähten versehen, also ihn zu einem Stahldrahtbeton bzw. Stahlbeton machen, und man kann ihn vorspannen. Das wäre am besten machbar, wenn man kleine Partikel als Zuschlagsstoff herstellen würde. Kleine vorgespannte Federn, die man einfriert, entweder mit einem sehr hochwertigen Lack oder hochwertigem Beton, als eine Art dünne Maccheroni, die hohl sind und die man dann zerschneidet und als Zuschlagstoffe für diesen Beton beimischt. Wenn so ein Beton Kräften ausgesetzt wird, werden die vorgespannten Federn, die eingefroren sind, das heißt, ihre Spannung ist eingefroren, aktiviert. Wenn also Mikrobrüche entstehen, bricht das System, aber die Feder ist nicht gebrochen, sondern zieht mit ihrer Spannung diesen Teil des Betons zusammen.

R. Finsterwalder Und schließt diese Brüche wieder.

F. Otto Wenn man einen Beton mit diesen vorgespannten Federn herstellt, dann bildet sich innerhalb des Betons ein Gerippe mit Vorspannung, also hat man einen vorgespannten Beton, einen vorgespannten Drahtbeton. Der müsste innerhalb eines Bauteils, gleichgültig, ob es ein dickes Volumen, zum Beispiel ein starker Träger, eine Platte oder ein Flächentragwerk ist, ein Gerippe bilden. Genau durch diese Kräfte, nämlich durch die Mikrofissuren, die Feinbrüche, die durch das System gehen, müsste sich ein inneres Gerippe, dem Kräfteverlauf entsprechend, bilden.

R. Finsterwalder Müssten die Federn dem Kräfteverlauf entsprechend ausgerichtet werden, oder können sie ungerichtet eingebracht werden?

F. Otto Man könnte den Beton so herstellen, dass die Grundstruktur leicht brüchig ist, deshalb habe ich von Mikrofissuren gesprochen. Durch die auftretenden Kräfte geben die Federn in dem Augenblick, in dem sie beansprucht werden, ihre Kraft an den Beton ab. Ich nehme eine Stahlfeder, hochwertigen Stahldraht, zum Beispiel Stahl 200, und so eine gespannte Stahlfeder bette ich in Beton, Glas oder hochwertigen Lack bzw. Kunststoff ein. Aber die klassische Variante wäre flüssiger Beton, ein Schnellbinder, der erhärtet, und ich schneide diese vorgefertigten Teile in kleine Scheiben. So bekomme ich kleine Elemente, aus jeweils einer vorgereckten Stahlfeder, ummantelt mit mineralischem Mikrobeton, der sehr fein ist. Diese Nüsse, ca. 50 mm lang, sind in diesem Gemisch von

1

R. Finsterwalder In an interview in AUT/Innsbruck, you told us about your vision for the future, liquid wood. What exactly do you mean by liquid wood?

F. Otto Well I've conducted several studies with materials that could potentially be used in architecture in the future. The question is whether it would be possible to make a fiber-reinforced concrete with wood-like characteristics that is not flammable. The ability to produce concrete made out of wood and cement that is both lightweight and fire resistant was especially important for our work in the Arabian Peninsula. And there was such a product, Eternit, which used to be made with asbestos but is now made with wood. You can also pour this wood concrete which can be mixed as more or less liquid depending on the concentration and water content.

R. Finsterwalder The cement contained within this wood concrete causes it to harden. | Would you then use it to produce a semi-finished product, sheet material for example, which combines the characteristics of wood and concrete?

F. Otto The wood fiber concrete could certainly be kept liquid enough to be able to be poured. This concrete would be more pleasant to the touch than Eternit, which has a very hard fiber. It could be made with high quality wires so it becomes a steel wire concrete or reinforced concrete, which could then be prestressed. This would work best with small particles produced as aggregates. Small, prestressed springs that are frozen, either with high quality lacquer or high quality concrete as a sort of thin macaroni that is hollow and can be cut up and mixed with the concrete as an aggregate. When this concrete is subjected to forces, the prestressed springs, which are frozen — meaning that their stress is frozen —, are activated. So if micro fractures occur, the system is broken but not the

Abb.1 Frei Otto und Rudolf Finsterwalder im Büro Otto

Fig.1 Frei Otto and Rudolf Finsterwalder in Ottos office

Holz und Beton enthalten. Wenn jetzt in einem gegossenen Volumen Kraftbeanspruchungen auftreten und die sogenannten Mikrofissuren eintreten, dann wird die Kraft dieser Feder frei. Dort, wo der Riss entlanggeht, zieht sie ihn zusammen, damit ist er haltbar und biegefest im Umfeld einer jeden Nuss. Als Namen für diese Nüsse hatte ich mir »Futurum« ausgedacht.

R. Finsterwalder Das heißt die Nüsse wären ungerichtet im Beton, man müsste nur auf eine gute Verteilung achten. Das Ganze wäre wie ein stahlfaserbewehrter Beton, nur eben vorgespannt.

F. Otto Richtig, mit diesen Federn wollte ich einen vorgespannten Beton herstellen, das heißt, ich friere die Vorspannung im Zuschlagsstoff ein. Die wird erst aktiv, wenn das fertige Produkt als tragender Beton aktiviert wird. Auf Grund der Beanspruchungen, insbesondere der Wechselbeanspruchungen, wie sie in der Natur, bzw. in Bauwerken vorkommen, zum Beispiel an Decken, treten dort Risse ein. Ich würde sogar ganz bewusst das Eintreten von Rissen provozieren.

R. Finsterwalder Du würdest diese Risse sogar provozieren?

F. Otto Ja, es wäre sogar fatal, wenn ein solches Bauwerk gar keine Beanspruchungen bekommen würde, dann kämen diese Elemente gar nicht zum Tragen. Aber ich glaube, das gibt es nicht, dass eine größere Struktur keine Lasten bekommt. | Das bedeutet, diese Mischung könnte man natürlich gießen, man müsste Zuschlagsstoffe herstellen, die eierartig sind.

R. Finsterwalder In etwa wie Kies.

F. Otto Ja, wie Kies, oder kiesähnlich, alle Techniken wie für normalen Schwerbeton wären einsetzbar, nur dass diese Teile im Inneren eine vorgespannte Feder hätten, die erst dann auslöst, wenn das System belastet wird. | Das habe ich alles bereits durchdacht und auch schon mit einigen Versuchen überprüft. Vor allem mit den Stählen habe ich mich beschäftigt. Ein hochwertiger Stahl wird, je dünner er ist, desto hochwertiger. Man kann zum Beispiel einen 5 mm starken Stahldraht durch mehrfaches Ziehen auf 2 mm, 1 mm, 0,5 mm usw. verjüngen, wobei bei jedem Mal Ziehen die Stahlgüte steigt, so dass man dann durch die Kaltverformung einen Stahl 200 bekommt, also 200 kp/mm^2 Festigkeit.

R. Finsterwalder Diese Mischung mit den Nüssen, dem Futurum, wäre für Normalbeton möglich, was wäre bei flüssigem Holz anders, es wäre in jedem Fall leichter?

F. Otto Ja, und beim flüssigen Holz kann man zusätzlich folgendes machen. Neben den Nüssen könntest du einen weiteren Zuschlagstoff hinzufügen, kleine Glaskugeln mit 3–10 mm Durchmesser, die es in der Massenproduktion gibt, die werden geblasen. Mich hat es gewundert, dass keine Stahlbetonfirma damit gearbeitet hat, um Beton leichter zu machen. Es gibt Schaumbeton, den man durch das Einblasen von Luftblasen erzeugt, aber es wäre bedeutend besser, wenn jede Luftblase in einer Glasblase sitzt, die eine gewisse Festigkeit hat. | Ich würde mein Futurum mit der Glasblasentechnik kombiniert sehen, dass also viele kleine Glasblasen mit ca. 3 mm Durchmesser beigemischt werden, die das Gewicht reduzieren.

R. Finsterwalder Wie viel Prozent Gewichtsreduzierung denkst du wären möglich?

F. Otto Eventuell bis zu 20 Prozent wären möglich.

R. Finsterwalder Mit dem Einsatz von Holzfaser wäre noch mehr möglich?

spring, which pulls this part of the concrete together with its stress.

R. Finsterwalder And seals the fractures back up.

F. Otto When you produce concrete with this prestressed spring, a prestressed framework is formed inside the concrete. So the result is prestressed concrete, prestressed wire concrete. It would form a framework within a structural element, regardless of whether it's a thick volume like a strong support, a plate or a surface structure. It is precisely these forces, specifically micro fissures — fine cracks that run through the system — that would form an inner framework according to the flow of forces.

R. Finsterwalder Do the springs have to be aligned according to the flow of forces or can they be inserted nondirectionally?

F. Otto The concrete could be produced in such a way that the basic structure is slightly brittle, which is where the micro fissures come in. When the springs are stressed, they transfer their force to the concrete through the forces that occur. I use a steel spring, high quality steel wire like steel 200, and I embed a tensioned steel spring into concrete, glass or high quality lacquer or a plastic. But the classic version would be liquid concrete, a quick binder that hardens and cuts the prefabricated parts into small strips. The result are small elements, each made from a pre-stretched steel spring covered with mineral micro-concrete that is very fine. These nuts, which are approximately 50mm long, are contained in this mix of wood and concrete. So now, when stressing occurs and these micro fissures result, this spring releases its force. The spring pulls cracks back together at the places where they occur so that they remain stable and resistant to bending in the vicinity of each nut. I thought of the name "Futurum" for these nuts.

R. Finsterwalder So that means that the nuts would be undirected in the concrete, but you would have to make sure they were evenly distributed. So the whole thing would be like a steel fiber reinforced concrete, but prestressed.

F. Otto Exactly. I wanted to use these springs to make prestressed concrete, meaning that the prestressing is frozen in the aggregate. It becomes active once the final product is activated as load-bearing concrete. Cracks occur due to stressing, particularly alternating stressing, as it occurs in nature and in construction, on ceilings for example. I would even try to provoke cracking on purpose.

R. Finsterwalder You would deliberately provoke cracking?

F. Otto Yes. It would actually be fatal if this kind of a building were not subject to stressing because these elements would not be put to use. But I don't believe that such a large structure would not be subject to stress. | This means of course that the mixture could be cast. You would have to produce aggregates that are like eggs.

R. Finsterwalder Like in gravel.

F. Otto Yes, like gravel, or something similar. Every technique could be used, like for normal heavyweight concrete, except that these parts would contain a prestressed spring that triggers once the system is strained. | I've already thought all of this through and performed some tests. I've been particularly concerned with steels. High-grade steel becomes higher grade the thinner it gets. So for example you can taper a 5mm thick steel wire down to 2 mm, 1mm and 0.5 mm by repeatedly pulling, while the steel grade increases every time you pull so that through cold-forming you get steel 200, so with 200 kp/mm^2 strength.

F. Otto Ja, und du könntest alles kombinieren, den vorgespannten Drahtfederbeton, den Glaskugelbeton und den Holzfaserbeton, in entsprechenden Kombinationen. Dann müsste man ein Produkt bekommen, das je nach Mischungsverhältnis sehr unterschiedlich fest ist, und auch sehr unterschiedlichen Wärmedurchgang hat. So ein Produkt könnte man auch mit einer Säge bearbeiten.

R. Finsterwalder Das wäre fantastisch, so ließen sich zum Beispiel Brandschutzeigenschaften von Beton mit der Leichtigkeit von Holz und Glaskugelbeton kombinieren, eventuell wäre so ein monolithisches Bauen denkbar, da alle benötigten Eigenschaften in einem Material vereint sind.

F. Otto Ich unterrichte ja nicht mehr, kann also auch keine Forschung mehr betreiben, außerdem wollte ich mich nicht nur mit einer einzigen Thematik beschäftigen. Vor allem haben mich aber die vorgespannten Drahtnetze interessiert. Man kann nicht leugnen, dass die Entwicklung des modernen flexiblen zeltartigen Daches aus gespanntem Stahl/Draht eine Revolution in der Architektur war. Wir haben Dächer mit sehr großen Spannweiten gebaut. Die großen Dächer, die heute gebaut werden, sind solche aus vorgespannten Netzen oder hängenden Netzen, sowohl schwergewichtig als auch vorgespannt. | Ein vorgespanntes Dach ist ja nichts anderes als die Idee mit dem Futurum, wenn du ein Hängedach nimmst, wie zum Beispiel die Schwarzwaldhalle, dann ist das das gleiche Prinzip. Wenn du ein vorgespanntes, betonummanteltes Hängedach aufschneidest, dann wäre ein kleiner Abschnitt genau diese Nuss, oder das Futurum.

2

Abb. 2 Modellstudien aus Ottos Archiv **Fig. 2** Model studies in Ottos archive

R. Finsterwalder This mixture with the nuts, Futurum, would be possible for normal concrete — what would be different with liquid wood? It would be lighter, at any rate?

F. Otto Yes and with liquid wood you can also do the following: in addition to the nuts, you could add an additional aggregate, small glass beads with a diameter of 3–10 mm, which are found in large-scale production and are blown. I was surprised that no reinforced steel company has used it to make their concrete lighter. There's foam concrete which is made by injecting air bubbles, but it would be a significant improvement if each air bubble were placed within a glass bubble with a certain strength. | I would like to see my Futurum combined with the glass bubble technique where several small, approx. 3 mm diameter glass bubbles are added which reduce weight.

R. Finsterwalder How much do you think the weight could be reduced by?

F. Otto Possibly by as much as 20 percent.

R. Finsterwalder And even more would be possible with wood fiber.

F. Otto Yes and you could combine everything, the prestressed wire spring concrete, the glass bubble concrete and the wood fiber concrete in appropriate combinations. The result should be a product with greatly varying degrees of consistency according to mixing ratios, as well as very different heat transition. Such a product could be worked with a saw.

R. Finsterwalder That would be fantastic, it would mean combining the fire resistant characteristics of concrete with the lightness of wood and glass bubble, and it would allow for a sort of monolithic style of building since all the necessary characteristics would be combined into one material.

F. Otto Well, as you know I don't teach anymore, so I also can't do any more research, and I got tired of being limited to one single subject. But the prestressed wire structure I found particularly interesting. You can't deny that the development of the modern, flexible, tent-like roof made from stressed steel/wire was a revolution in modern architecture. We built roofs with very large spans. The large roofs being built today are made out of prestressed nets or hanging nets, both heavyweight and prestressed. | After all, a prestressed roof is the same principle as the Futurum — if you take a suspended roof like the one in the Schwarzwaldhalle, it's the same principle. If you cut open a prestressed, concrete-encased suspended roof, a small section would be exactly this nut, or the Futurum.

R. Finsterwalder The difference being that the Futurum functions 3-dimensionally.

F. Otto Yes and you can specially produce the Futurum in small scale, but you can also produce it as a huge roof. It's exactly the same. Take our concrete pavillion roofs — it doesn't really make a big difference if you manufacture them from concrete or textiles. There is a continuous transition. So for example, you can replace the wire with real textiles, polymide fibers or carbon fibers, carbon fibers with a tension length of more than 500km still don't exist, but they are used with 200km, in the automotive industry for example. There is no difference between using this technology in the automotive industry or in construction; they are all fiber-reinforced lightweight structures.

R. Finsterwalder And that would all be possible with wood as well.

R. Finsterwalder Mit dem Unterschied, dass das Futurum dreidimensional funktioniert.

F. Otto Ja, und du kannst das Futurum extra im Kleinen herstellen, aber du kannst es auch als Riesendach herstellen. Es ist genau dasselbe. Nimm unsere Betonzeltdächer, ob du die aus Beton oder Textilien herstellst, ist kein so großer Unterschied. Es gibt einen kontinuierlichen Übergang. Du kannst zum Beispiel den Draht durch echte Textilien ersetzen, Polyamidfasern oder Kohlefasern, die Kohlefaser mit einer Reißlänge von über 500 km, die hat es bisher noch nicht gegeben, aber mit 200 km wird sie angewendet, zum Beispiel im Fahrzeugbau. Ob diese Technik für Fahrzeuge oder Bau verwendet wird, macht keinen Unterschied, es sind alles faserbewehrte leichte Flächentragwerke.

R. Finsterwalder All das wäre wiederum mit Holz vorstellbar.

F. Otto Sicher, das könnte man mit Holz als Masse realisieren, als Holzfaserbeton. Vor allem für die Brandsicherheit von Bauten und Fahrzeugen ist noch viel Potential da.

R. Finsterwalder Denkst du, dass man bei flüssigem Holz auf den Beton verzichten könnte, um Gewicht zu sparen?

F. Otto Das ist natürlich möglich, es gibt aber eine Grenze, wenn ich zum Beispiel einen fliegenden Bau realisieren möchte, wird man versuchen, dass er extrem leicht wird, aber wenn er auf der Erde steht, gibt es eine Grenze, wo er bei extremer Leichtigkeit

3

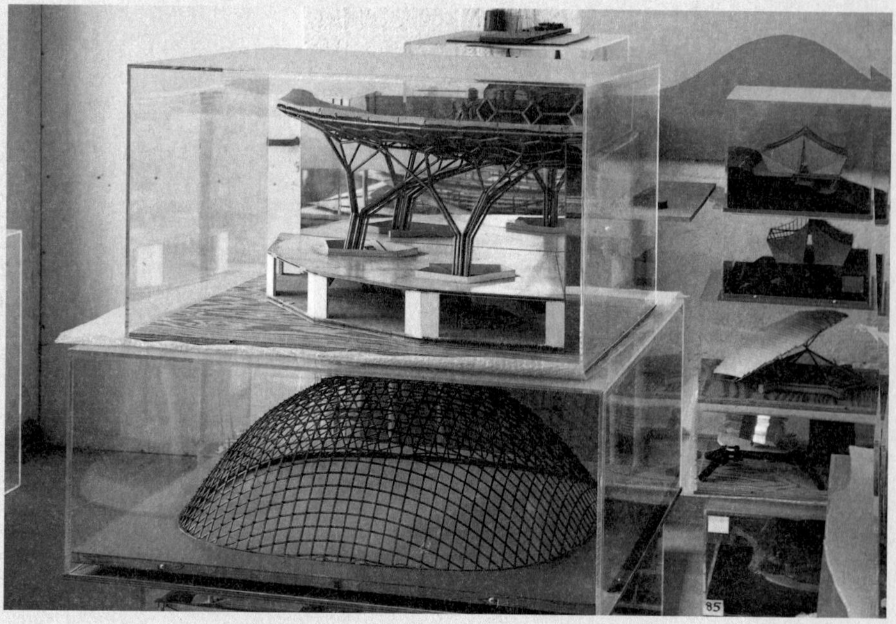

Abb. 3 Gitterschalenmodelle

Fig. 3 Gridshell models

F. Otto Definitely, it could be realized with wood as a mass, as wood fiber concrete. There's a lot of potential, especially when it comes to building and vehicle fire safety.

R. Finsterwalder Do you think that the concrete could be left out when using liquid wood to save weight?

F. Otto It's certainly possible, but there's no limit. If I want to build a temporary building, the idea would be to make it extremely light. But when it's on the ground there is a limit where it also becomes dangerous if it's too lightweight because the wind can easily knock over buildings. It's important that we have all the variations in order to selectively implement them. Wood fiber is very important because it's a renewable resource and because it's a building material which ideally combines with cement.

R. Finsterwalder Another application would be in furniture construction for example.

F. Otto Fire prevention would certainly be a concern with furniture building. I would have liked to build something with liquid wood but I never got the opportunity to do so.

R. Finsterwalder What did you have in mind for the liquid wood?

F. Otto A whole building structure cast in one piece would be interesting.

R. Finsterwalder How do you imagine the mold for that?

F. Otto By combining the tensile surface structures made from textiles with the mold, similar to how bells are cast. For example, you can form a textile air bubble — a so-called air hall with a doubled membrane — then you pump in the liquid wood and you have a nice bowl shape.

R. Finsterwalder So you would use a bivalve tire as a mold and then fill in the in-between space with liquid wood?

F. Otto I had that at an exhibition in Essen in 1962. We used concrete made from diatoms — glass in other words. It came about because we had a special research field on this subject.

R. Finsterwalder So to sum it up once more, it would be possible to make lightweight surface structures from Futurum-reinforced liquid wood that are then cast into bivalve tires. This mold would be expensive but could be recycled. An additional advantage would be that the super smooth surface would hold out well against the effects of weathering.

F. Otto Yes, you could say that these kinds of textile constructions actually represent high-tech construction, just as our textiles, the shirt for example is a high-tech model, indeed a rather complicated form.

R. Finsterwalder Another advantage of liquid wood would be good insulation.

F. Otto Absolutely, just the amount of air alone, you could adjust it as desired either through the glass bubbles or plastic parts, sometimes this reduces the strength but it's not always needed during construction.

R. Finsterwalder Your work was more concerned with basic research rather than result-oriented thinking.

F. Otto Correct. When conducting research it was important for me to set a yardstick in order to understand

auch gefährlich wird. Wenn zum Beispiel Wind Bauten einfach umpusten kann. Wichtig ist, dass wir alle Varianten haben, um sie gezielt einzusetzen. Holzfaser ist sehr wichtig, da es ein nachwachsender Rohstoff ist und weil es ein Baustoff ist, der eine ideale Kombination mit dem Zement bietet.

R. Finsterwalder Ein anderer Anwendungsbereich wäre zum Beispiel der Möbelbau.

F. Otto In jedem Fall, gerade bei Möbeln müsste man auf Brandschutz achten. Ich hätte gerne etwas mit flüssigem Holz gebaut, hatte aber bisher keine Gelegenheit dazu.

R. Finsterwalder Was hättest du dir vorgestellt, hast du eine Idee dazu?

F. Otto Eine ganze Bauwerksstruktur aus einem Guss würde mich interessieren.

R. Finsterwalder Wie stellst du dir eine Schalung dazu vor?

F. Otto Natürlich indem ich die zugbeanspruchten Flächentragwerke aus Textilien mit dem Guss kombiniere, ähnlich wie Glocken gegossen werden. Du kannst zum Beispiel eine textile Luftblase, also eine sogenannte Lufthalle mit einer doppelten Haut, ausbilden und da pumpst du flüssiges Holz hinein und bekommst so eine schöne Schale.

R. Finsterwalder Du würdest also einen zweischaligen Pneu als Schalung benutzen und den Zwischenraum mit flüssigem Holz ausgießen?

F. Otto Das habe ich zum Beispiel bei einer Ausstellung in Essen 1962 getan. Wir haben einen Beton aus Kieselalgen, also Glas, verwendet. Das kam daher, dass wir einen Sonderforschungsbereich zu diesem Thema hatten.

R. Finsterwalder Um es noch einmal zusammenzufassen, wären also leichte Flächentragwerke aus Futurum-bewehrtem flüssigem Holz denkbar, die in zweischaligen Pneus gegossen werden. Diese Schalung wäre zwar teuer, aber wiederverwertbar. Ein weiterer Vorteil wäre, dass die superglatte Oberfläche Witterungseinflüssen gut standhalten würde.

F. Otto Ja, man kann sagen, dass solche textilen Konstruktionen das Hightechbauen überhaupt wären, wie ja auch unsere Textilien, zum Beispiel das Hemd, ein Muster für Hightech sind, eigentlich eine superkomplizierte Form.

R. Finsterwalder Ein weiterer Vorteil des flüssigen Holzes wäre die gute Wärmedämmung.

F. Otto Aber sicher, alleine durch den Anteil von Luft, entweder über die Glaskugeln oder Kunststoffteile, kann man sie beliebig einstellen, manchmal wird dadurch die Festigkeit reduziert, diese wird aber auch nicht immer beim Bau benötigt.

R. Finsterwalder In deiner Arbeit ging es meist um Grundlagenforschung und weniger um ergebnisorientiertes Denken.

F. Otto Richtig, in der Forschung war es mir wichtig, einen Maßstab einzuführen, um zu wissen, was man mit diesen unendlich vielen Bauprodukten zur Verfügung hatte, und sie untereinander vergleichen zu können. Dazu habe ich den »bic« eingeführt, der die Frage nach der Masse in Bezug auf die Fähigkeit, Kräfte zu übertragen, also Gramm je Newtonmeter, also Masse bezogen auf die Fähigkeit, Newton über Strecken zu übertragen, klären sollte. Das hat

exactly what possibilities are available to me with these endless construction products and to be able to compare them with one another. So to do this I started using the "bic" in order to answer the question of mass relative to the ability of forces to transfer, in other words, gram per newton meter — mass relative to the ability of newtons to transfer across distances. This was very helpful to us as an internal basis for our research, as we were able to trace this value from the microscopic — a thousandth of a gram — up to the largest sizes and use this to make comparisons. Since we knew the bic values, we were able to determine what a specific design could accomplish for a specific construction, a bridge that would span 10 km for example. The bridge has to carry a force, traffic or its own weight. So you can define the stress in grams per newton meter meant to be supported by the bridge. If you know the bic, say for a bridge system, regardless of whether it's an arch bridge or a suspension bridge, you multiply the amount the bridge is meant to support by the bic and you get the minimum mass that the bridge has to have.

R. Finsterwalder So basically the ideal mass.

F. Otto Exactly, the ideal mass, below which you cannot fall. There is a defined mass which is mostly the span in the reciprocal form. The easiest conceptual model is the hanging thread, which is a reversal of the standing tower. The hanging thread has a maximum length that cannot be exceeded. The textile industry used to work with this value a great deal, the breaking length, and this breaking length is the same as the bic or the reciprocal value. You can convert the breaking length in grams per newton meter. I also published a thesis on all orders of magnitude from the smallest objects to the largest. This work has yet to be followed up on. But it has to happen; someone has to figure out how much material is needed for a given size, what is the largest order of magnitude. Can I build a 10 km tall tower, or not? Non-living nature has built on this scale in the Himalayas, for example, a mountain range 10 km above the Earth's surface. I believe that humans can build even higher than 10 km. If non-living nature is already capable of 10 km, better materials with a higher bic, where less material is required for the height or for the span, can reach even greater heights. Those were the questions that we pursued. I am very familiar with them so I could tell you on the spot which designs can be used to build the largest spans and where their limits are. There are limits everywhere and the idea is to expand these limits to make it possible for people to inhabit the Earth.

R. Finsterwalder Efficiency also plays an important role — only using the amount of material necessary. | That example shows how important it is to collaborate across disciplines.

F. Otto That's why we founded our collaborative research center.

R. Finsterwalder Unfortunately there aren't institutions like that any more. It's regrettable that there is an increasing specialization at universities and no more general studies.

F. Otto You need both, you need the specific as much as you need the general. I'm happy that we were able to discover as much as we did.

R. Finsterwalder Now it's up to our generation to continue that work.

uns viel bei unserer Forschung geholfen, als interne Grundlage der Forschung. Denn wir konnten diesen Wert vom mikroskopisch kleinen, also Tausendstelgramm, bis hin zu den grössten Grössen verfolgen und damit Vergleiche anstellen. Da wir die bic-Werte kannten, konnten wir für bestimmte Konstruktionen, zum Beispiel eine Brücke, die 10 km spannen sollte, sagen, was eine bestimmte Konstruktion leisten kann. Die Brücke soll eine Kraft, Verkehr oder das Eigengewicht einer stabilen Konstruktion tragen. Du kannst also die Aufgabe definieren in Gramm je Newtonmeter, was die Brücke leisten soll. Wenn du den bic kennst, zum Beispiel eines Brückensystems, egal ob als Bogenbrücke oder als Hängebrücke, dann multiplizierst du die Angabe, was eine Brücke tragen soll mit dem bic, dann bekommst du die Masse, welche die Brücke mindestens haben muss.

R. Finsterwalder Also sozusagen die Idealmasse.

F. Otto Richtig, die Idealmasse, die du nicht unterschreiten kannst. Es gibt eine definierte Masse, meist ist sie in der Reziprokform die maximale Spannweite. Das einfachste Gedankenmodell ist der hängende Faden, der in Umkehrung der stehende Turm ist. Der hängende Faden hat eine maximale Länge, die nicht überschreitbar ist. Das ist ein Wert, mit dem die Textilindustrie früher sehr viel gerechnet hat, die Reißlänge, und diese Reißlänge ist nichts anderes als der bic, bzw. als sein Reziprokwert. Du kannst die Reißlänge umwandeln in Gramm je Newtonmeter. Ich habe auch eine Abschlussarbeit veröffentlicht über alle Größenordnungen, von den kleinsten Objekten zu den größten. Diese Arbeit ist nicht weiter verfolgt worden. Das muss kommen, jemand muss fragen, wie viel Material brauche ich für eine gegebene Größe, wie ist die größte Größenordnung? Kann ich zum Beispiel einen Turm von 10 km Höhe bauen oder nicht? Die nicht lebende Natur baut ja mit dieser Grenzgrösse zum Beispiel im Himalaya, ein Gebirge von 10 km Höhe über der Erdnormaloberfläche der Kugel. Ich kann sagen, ein Mensch könnte höher bauen als die 10 km. Wenn die nicht lebende Natur bereits die 10 km schafft, können bessere Materialien, die einen höheren bic haben, also weniger Material für die Höhe verwenden oder für die Spannweite, noch größere Höhen erreichen. Das waren Problemstellungen, die wir verfolgten. Für mich sind sie sehr geläufig und ich kann dir aus dem Stegreif sagen, mit welcher Konstruktion man die größten Spannweiten bauen kann, wo ihre Grenzen sind. Grenzen gibt es überall und es geht darum, diese Grenzen auszuweiten, um den Menschen zu ermöglichen, die Erde zu bevölkern.

R. Finsterwalder Darin steckt ja auch der gewichtige Gedanke an Effizienz, nur so viel Material zu verwenden, als nötig ist. | Das Beispiel zeigt, wie wichtig es wäre, fachübergreifend zusammenzuarbeiten.

F. Otto Dafür wurde ja unser Sonderforschungsbereich gegründet.

R. Finsterwalder So eine Einrichtung gibt es leider nicht mehr. Bedauerlich ist auch, dass es an den Hochschulen eine immer deutlichere Spezialisierung gibt, kein *Studium generale* mehr.

F. Otto Es muss beides geben, das Spezielle wie das Allgemeine. Ich bin froh, dass wir so viel entdecken konnten.

R. Finsterwalder Jetzt ist es an unserer Generation, es weiterzuführen.

2

Chiesa in Olevano Romano

Italien, seit 2007 — ein Projekt mit Carsten Nicolai und Matthias Weischer, begleitet von Joachim Blüher

Chiesa in Olevano Romano

Italy, since 2007 — a project with Carsten Nicolai and Matthias Weischer, under the escort of Joachim Blüher

Rudolf Finsterwalder

Ein Kirchenbau erfordert, wie kein anderer Bau, eine intensive theoretische Auseinandersetzung mit dem Programm sowie im Falle einer Kirche auch mit der Liturgie. So war uns unter anderem das Betreten, das Hereinkommen in eine Kirche – der Übergang vom Profanen zum Heiligen – ein wesentliches Thema. Die Besucher sollen zur Ruhe kommen, indem sie einen, von uns bewusst lang angelegten Weg bis ins Innerste der Kirche zurücklegen müssen. Je weiter man in das Gebäude eindringt, desto ruhiger und dunkler wird es und die zunehmende Dunkelheit wird nur noch unterbrochen von den Lichteinfällen in die Seitenkapellen. | Der »Weg der Entschleunigung« mündet in den Altarraum, der eine tiefe Ruhe ausstrahlt; er erhält sein akzentuiertes Licht von oben, ist aber so dunkel, dass auch das Licht einer Kerze zur Geltung kommt. | Der Bau setzt sich aus zwei ineinander greifenden Schalen zusammen, die wie eine nach oben führende Schraube in sich gedreht sind und wie organisch gewachsene Gebilde anmuten. Sie weisen Analogien zur Anatomie eines menschlichen Herzens mit seinen Kammern oder zu einer gewendelten Muschel auf. Das Zentrum des auf diese Weise gebildeten Raums, der Punkt, um den sich alles dreht, bleibt gerade und ruhig, der Raum um ihn herum schließt sich. | Das akzentuiert eingesetzte Licht spielt eine zentrale Rolle in dieser Architektur. Je nach Stand der Sonne zeichnet das Licht unterschiedliche Bilder und Stimmungen. Die Lichtstrahlen in dem dunklen Kirchenraum werden fast materialisiert, wirken greifbar. Doch auch alle anderen Räume der Kirche werden vom gezielten Einsatz des Sonnenlichtes definiert, so gibt es bespielsweise eine Seitenkapelle mit einem Fenster zur Welt, das den Blick nach Westen auf einen gewachsenen Felsen freigibt, auf ein Stillleben der Natur. Ein anderer, sehr kleiner kreisrunder Raum, mit dem Durchmesser eines die Arme austreckenden Menschen, erhält sein Licht über eine Öffnung im Dach; von dort aus fällt es in dem gewundenen Raum nach unten. | Das Gebäude fügt sich in eine gewachsene räumliche Situation ein. Ein natürliches Plateau auf einem Bergsattel, flankiert von einem Felsen, bietet einen perfekten Bauplatz, der kaum verändert werden muss. Als Referenz für die Höhe und die Dimension des Gebäudes dienen auf dem Baugelände vorhandene Eichen sowie ein seitlich gelegener Granitfelsen. So wird die Kirche zu einem integralen Bestandteil des natürlich gewachsenen Ortes.

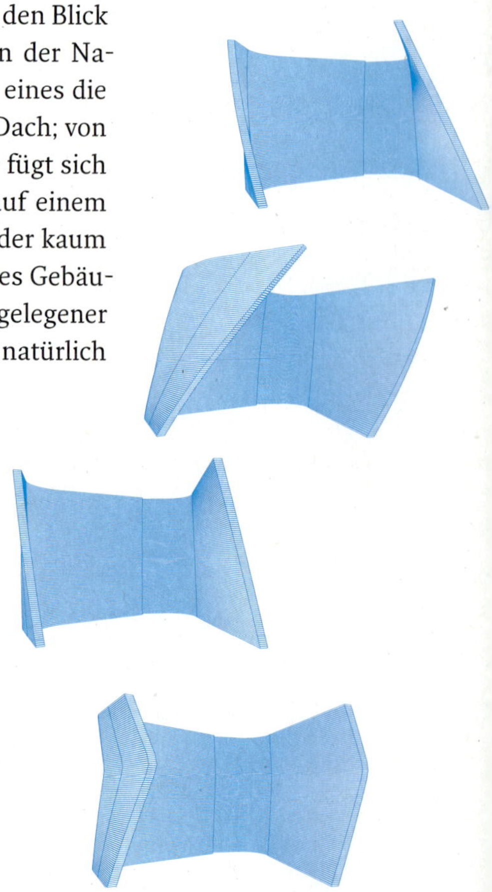

Abb.1 Rotationsstudie Nr.1 **Fig.1** Rotational studies No.1

Building a church needs, more than any other building, an intense theoretical examination of the program, as in the case of a church with the liturgy. So among other things, the way to enter a church, to come in — the step from profane to holy — was a main issue to us. The visitors should become more at peace with themselves, by being obliged to walk a long path to the inner most part of the church. The further one enters the building the quieter and darker it gets, with the increasing darkness only interrupted by the incoming light from the side chapels. | The "path of deceleration" ends in the sanctuary, offering an almost serene calmness with light entering from above, but is so dark that even the light of a candle can be emphasized. | The building is composed of two interlocking shells, which are rotated around a central point and look as though organically grown. They show some analogy to the anatomy of a human heart, with its chambers, or to a twisted seashell. The center of the space, the point around which everything is rotating, stays straight and quiet, the space is closing around it. | The accentuated light plays a central role in this architecture. Depending on the position of the sun the light is creating different moods and atmospheres. The rays of light in the dark church become almost materialized, almost palpable. All the other rooms of the church are defined by the proper use of sunlight, as in the side chapel with a window to the world that offers a view to the west of a large rock, a still life of nature. Another very small circular room, with the diameter of a man spreading his arms, gets its light from an opening in the roof, where it flows down into the meandering space below. | The building blends in with its surroundings. A natural plateau on a mountain ridge, accompanied by a rock offers a perfect site that will almost remain unchanged. Nearby oak trees and magnificent granite rock act as a reference for the height and the dimension of the building. In this way the church becomes an integral part of the naturally grown site.

Abb.2 Modell 1:33

Fig.2 Model 1:33

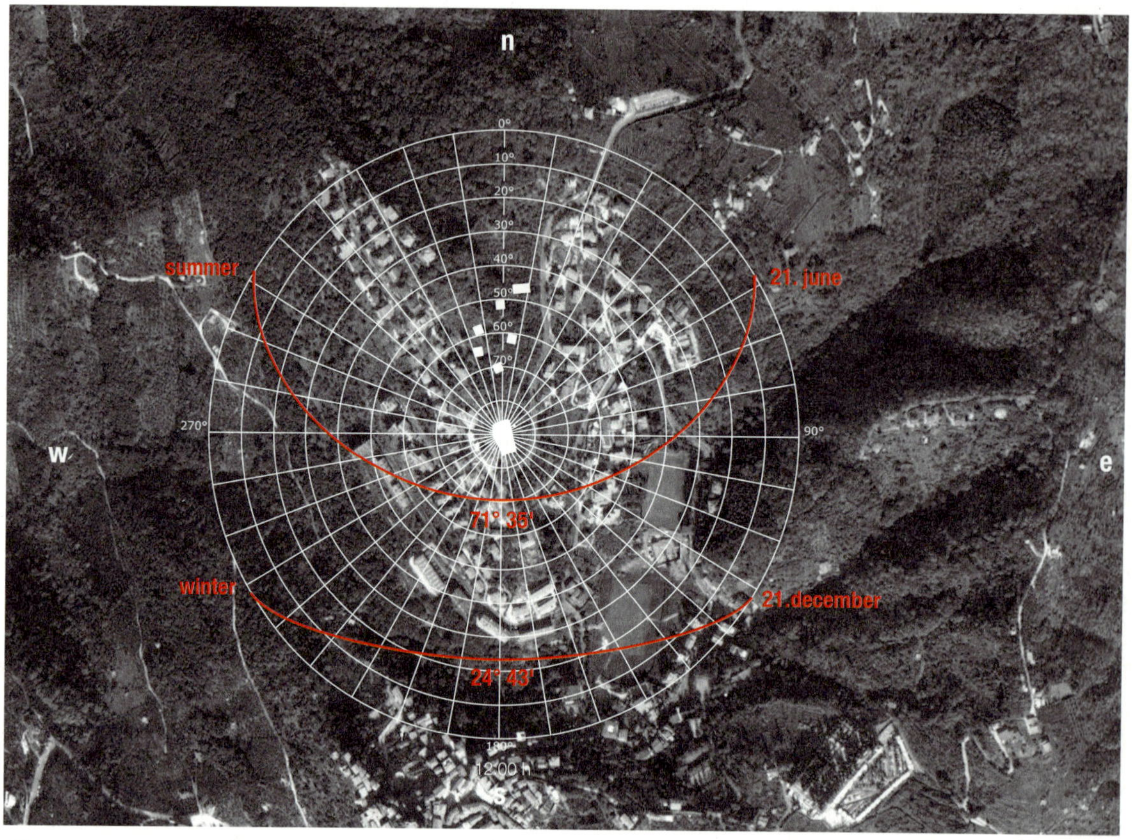

3

4

Abb.3 Analyse Sonnenstand
Fig.3 Sun position analysis

Abb.4 Außenansichten
Fig.4 Exterior views

5

Abb.5 Kirche auf dem Bauplatz in Olevano Romano

Fig.5 Church on the site in Olevano Romano

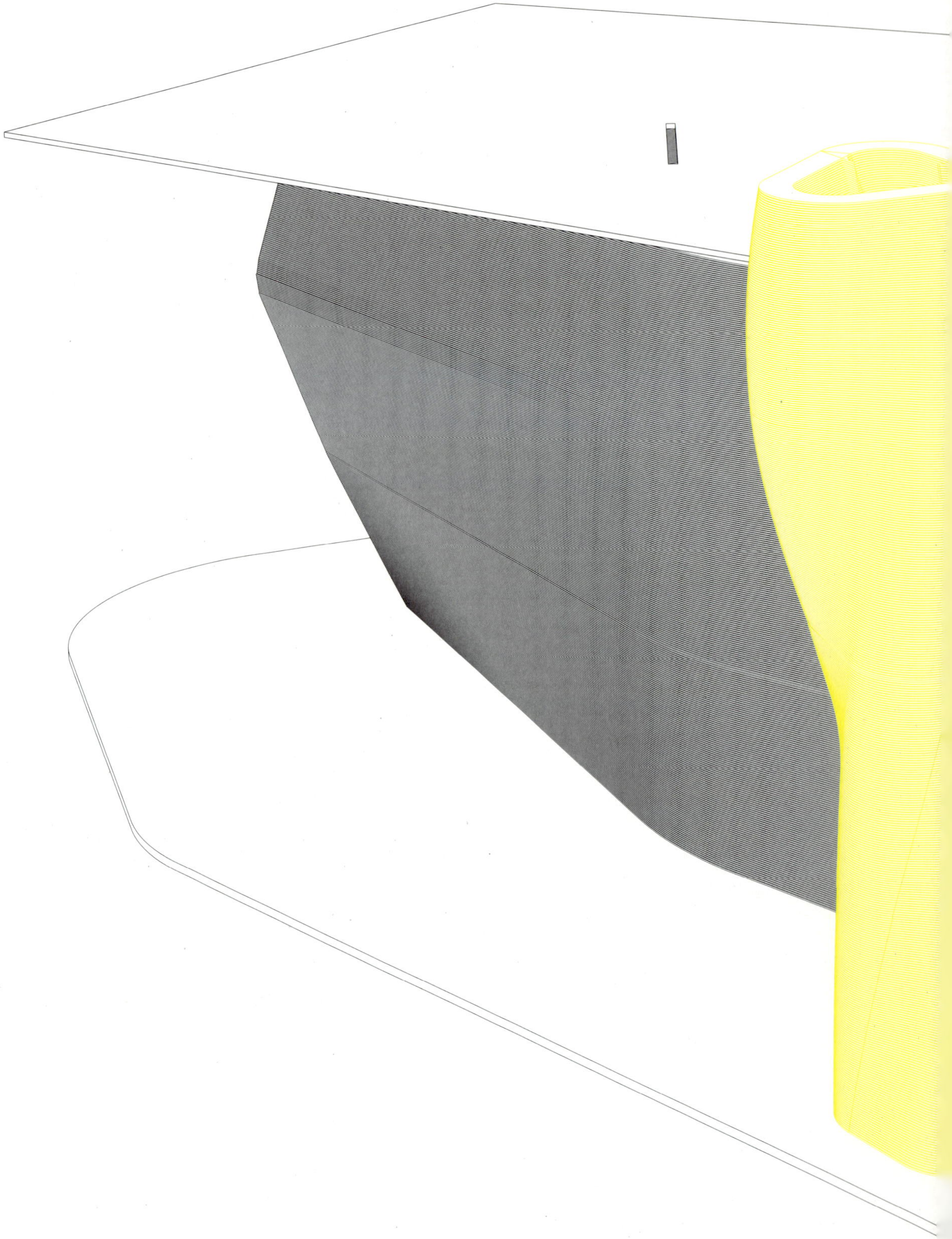

Abb.6 Dach mit Lichtöffnungen

Fig.6 Roof with light-openings

1 Eingang / Entrance
2 Lichtturm / Light tower
3 Innerer Altar / Interior altar
4 Altarraum / Sanctuary
5 Fenster zur Welt / Window to the world
6 Äußerer Altar / Exterior altar
7 Campanile / Bell tower
8 Weg der Entschleunigung / Way of deceleration
9 Baptisterium / Baptistery
10 Sakristei / Sacristy
11 Kleine Kapelle / Small chapel
12 Große Kapelle / Large chapel

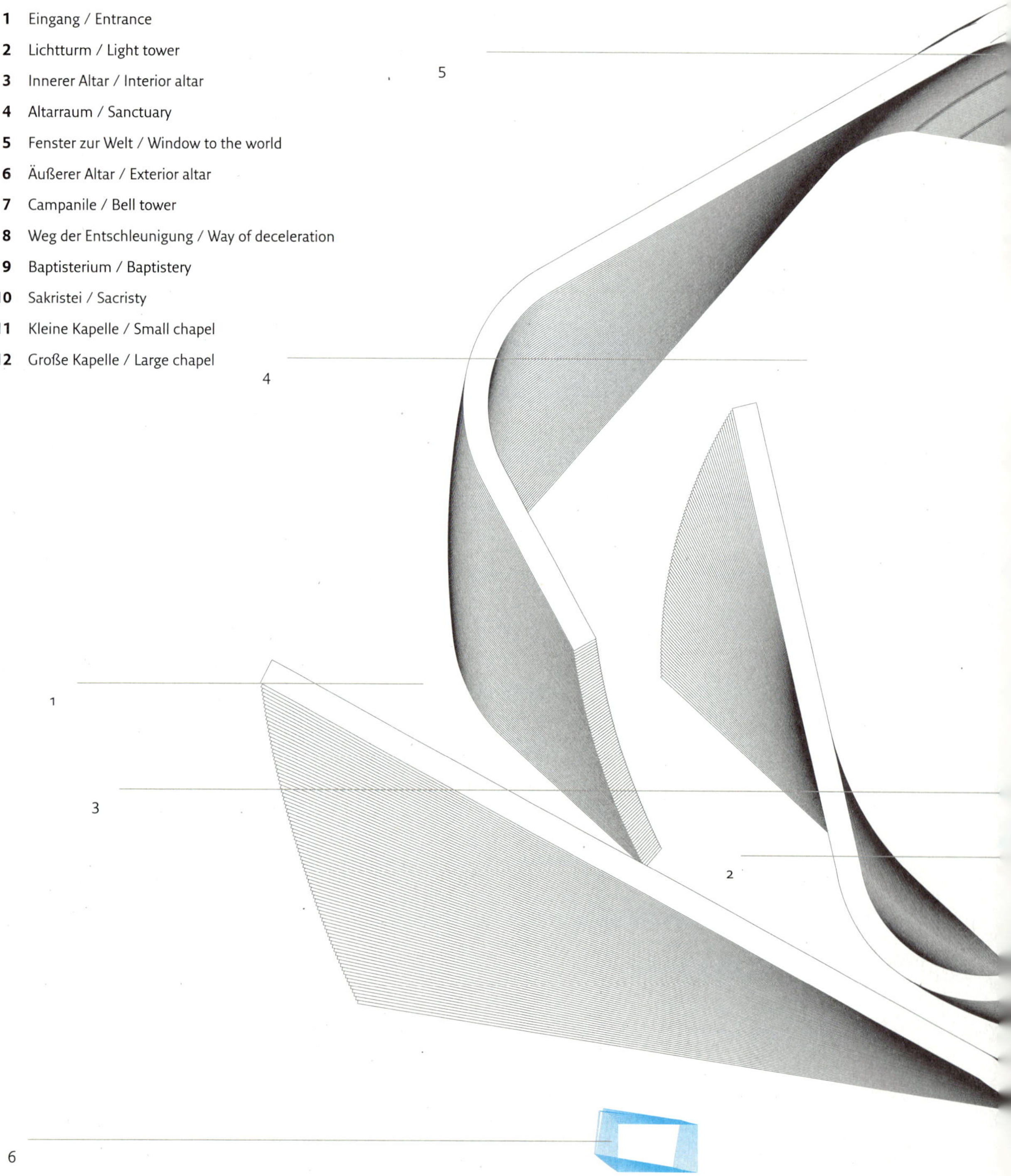

Abb.7 Grundriss

Fig.7 Ground floor

12
11
10
9
8
7

Abb.8 Rotation Nr.2

Fig.8 Rotation No.2

3

Formsachen.

Musik, Natur und Gedächtnis

Matters of form.

Music, nature and memory

Dieter Dolezel

Fallobst *Trois Morceaux en forme de poire*, drei Stücke in Birnenform. Erik Saties kompositorische Antwort auf den wohlgemeinten Ratschlag seines Freundes und Kollegen Claude Debussy, er solle in seinen Stücken mehr auf die Form achten. Ein augenzwinkernder Seitenhieb eines verschrobenen Exzentrikers, der jedoch die Frage nach der Definition von Musik in ihrem wohl ambivalentesten Punkt trifft: »Was ist musikalische Form?« | Wer sich mit dieser Fragestellung bereits eingehender beschäftigt hat, weiß um die Tatsache, dass es sich hierbei lediglich um eine abgeschwächte Variante einer bedeutend substantielleren Frage handelt: »Was ist Musik?« | Zieht man auf der Suche nach Antwort tradierte Quellen zu Rate, so stechen unmittelbar zweierlei Punkte ins Auge: Die meisten gängigen Definitionen betonen zum einen stark das struktural-gestalterische Element, also die formale Ebene von Musik, zum anderen bleiben sie aber in genau jenem Punkt eine präzisere Antwort schuldig. Es entsteht der Eindruck des merkwürdig Vagen, bisweilen sogar Hilflosen. Vom lapidaren »Musik sei gestaltete Zeit« über Hanslicks »tönend bewegte Form« hin zu Brockhaus-Riemanns »produktiver Gestaltung des Klingenden« oder auch Meyers »absichtsvoller Organisation von Schallereignissen«. So oder so ähnlich hätte man Musik aus einem ungefähren Gefühl heraus wohl auch selbst definiert. Im Zugzwang der Konkretisierung des Gesagten macht sich allerdings schnell Betretenheit breit. Produktiv, absichtsvoll, gestaltend, darüber ist man sich einig. Die genauere Bedeutung dieser Attribute bleibt allerdings unklar und als bloße Wertungen sind sie vom definitorischen Standpunkt aus auch prinzipiell schwer zu vertreten. Auch jede halbwegs gepflegte zwischenmenschliche Konversation ist nichts anderes als absichtsvolle Organisation von Schallereignissen. | Die Riemannsche Definition bemüht sich zwar um konkrete Erläuterung, driftet allerdings schnell ins nichtssagend Metaphysische ab:

> »Musik ist die produktive Gestaltung des Klingenden, das als Natur- und Emotionslaut die Welt und die Seele im Reich des Hörens in begriffloser Konkretheit bedeutet, und das als Kunst in solchem Bedeuten vergeistigt ›zur Sprache‹ gelangt kraft einer durch Wissenschaft (Theorie) reflektierten und geordneten und daher sinnvollen und sinnstiftenden Materialität.«

Bevor ein tiefergehender Versuch unternommen werden kann, der Definition von Musik und insbesondere ihrer formalen Gestaltung auf die Spur zu kommen, gilt es zuerst ein terminologisches Knäuel zu entwirren. Der Begriff ›Form‹ ist im musikalischen Sprachgebrauch keinesfalls eindeutig. Er kann sich sowohl auf die große, zusammenhängende Gesamterscheinung eines Musikstücks beziehen als auch ganz allgemein auf alles, das gestalthafte Züge in sich trägt. Das Erschei-

Windfall fruit *Trois Morceaux en forme de poire*, three pieces in the shape of a pear: Erik Satie's compositional response to his friend and fellow musician Claude Debussy's advice that he ought to pay more attention in his work to form. A dig with a twinkle in his eye by a quirky eccentric, who nonetheless hits the question as to the definition of music at its most ambivalent point: "What is musical form?" | Anyone who has already addressed this question in any detail will know that this is merely a diluted version of a significantly more substantial question: "What is music?" | If, in search of an answer, one draws on conventional sources, two points immediately come to light. Most current definitions strongly emphasize on the one hand the structural design element, i.e., the formal level of music, while on the other failing to provide a more precise answer to precisely that point. An impression emerges of remarkable vagueness, at times even helplessness. From the succinct "music is structured time" and Hanslick's "sonorously moving form" to Brockhaus-Riemann's "productive design of the resounding" or even Meyer's "intentional organization of sonic events". One way or another one would instinctively have defined music this way as well. However, under pressure to make concrete what has been said, embarrassment swiftly spreads. Productive, full of intention, structuring, there is agreement on that. The more precise significance of these attributes, however, remains unclear and as mere evaluations they are basically difficult to support from a definitional point of view. Even any halfway decent conversation between people is nothing other than the intentional organization of sonic events. | Though Riemann's definition endeavors to come up with a concrete explanation, it, however, quickly drifts into meaningless metaphysics:

> "Music is the productive design of what resounds, which as the sound of nature and emotion signifies the world and the soul in the realm of hearing in concreteness free of concept, and which sublimated as art in such implication becomes 'language' through a materiality that science (theory) has validated and lent order and that as such is meaningful and gives meaning."

Before there can be a more in-depth attempt to come up with a definition of music and in particular its formal structure, a terminology ball first of all needs to be disentangled. In the musical vernacular the term "form" is by no means unambiguous. It can refer both to the overall coherent appearance of a piece of music, as well as quite generally to everything that bears *gestalthaft* traits. As such the appearance of an eight-bar period is just a much covered by the term as an entire symphony. And if, as in precisely the latter, a certain "scheme" infuses the compositional common practice, it is joined by yet another level of interpretation. That of musical form (S). | So it is advisable to differentiate. To this end, Walter Gieseler

nungsbild einer achttaktigen Periode fällt also genauso in den Zuständigkeitsbereich des Begriffs, wie eine ganze Sinfonie. Und geht, wie bei eben jener, ein bestimmtes ›Schema‹ ins kompositorische Allgemeingut über, gesellt sich noch eine weitere Bedeutungsebene hinzu. Die der musikalischen Form(en). | Es ist also ratsam zu differenzieren. Walter Gieseler schlägt hierzu ein Modell vor, das aus drei aufeinander aufbauenden Schichten besteht. *Material-Struktur-Form*[1]. Jede dieser drei Schichten steht zwar phänomenologisch für sich selbst, ist aber in ein reziprokes Beziehungsnetz mit den beiden anderen verwoben. Struktur kann es also nur geben, wenn Material existiert, und Material wird erst dann zu solchem, wenn es strukturell in Beziehung gesetzt wird. Daraus lässt sich folgern, dass jedes akustische Phänomen prinzipiell als musikalisches Material dienen und durch strukturelle Mikro- wie Makrobeziehungen auf höchster Ebene in globaler Form aufgehen kann. Der Formbegriff wird also auf die sinnlich ästhetische Gesamterscheinung reduziert. Das Hauptaugenmerk liegt auf den strukturellen Beziehungen in den tieferen Schichten und von ihnen wird im Folgenden meist die Rede sein. || Um sich nun dem tatsächlichen Unterschied zwischen einem Musikstück und einer Unterhaltung zu nähern, sollte man sich der erstgenannten Definition nochmals ein wenig ausgiebiger widmen: »Musik ist gestaltete ZEIT«. | In einem Gespräch nehmen wir Zeit meist nur dann wahr, wenn es auf irgendeine Art und Weise nicht zu unserer Zufriedenheit verläuft. Das Resultat ist ein äußerst bewusstes und vor allem negatives Erleben von Zeit. Davon einmal abgesehen ist sie als strukturell bestimmender Faktor kaum von Bedeutung. Strukturbildend ist vor allem die grammatikalische und semantische Ebene von Sprache. | Die Beziehung von Musik und Zeit ist jedoch eine ganz andere. Denn mit dem Vergehen von Zeit beginnt musikalische Form überhaupt erst zu existieren. Musik *ist* nur in dem Moment, in dem sie auch erklingt. Eine Skulptur kann in ihrer Gestalt auf einen einzigen Blick erfasst und begriffen werden, Musik hingegen erfordert das Erleben ihrer zeitlichen Ausdehnung, egal ob aktiver oder passiver Natur, um sinnhaft zu werden. | Aus der Verbindung mit Gieselers Ansatz ergibt sich nun also folgende Konsequenz: musikalische Form als strukturelle Verknüpfung akustischer Phänomene, die in ihren Beziehungen zueinander nur durch das Erleben vergehender Zeit erfahrbar sind.

1 vgl. Gieseler, Komposition im 20. Jahrhundert, 1975 Moeck Verlag

Gedächtnisstrukturen Um vergehende Zeit überhaupt erleben zu können, bedarf es allerdings einer wichtigen evolutionären Errungenschaft. Man braucht dazu ein Gedächtnis. Erst wenn man in der Lage ist, ein musikalisches Ereignis sowohl im Kontext bereits erklungener als auch in Antizipation vermutlich noch erklingender Ereignisse wahrzunehmen, kann ein Netz aus strukturellen Ver-

proposes a model comprising three levels that are based one on top of the other: *Material - structure - form* [1]. Although phenomenologically speaking, each of these three levels stands for itself, each is nevertheless enmeshed in a reciprocal network of relationships with the two others. As such there can only be structure if material exists, and material only becomes such when it is related structurally. From this it can be concluded that any acoustic phenomenon can in principle serve as musical material, and through structural micro and macro relationships can merge in global form. The concept of form is thus reduced to the sensual aesthetic overall appearance. The main focus is on the structural relationships in the lower levels and will be the main point of discussion in the following. || To now get closer to the actual difference between a piece of music and a conversation, we ought to devote ourselves to the first definition in greater detail: "Music is structured TIME". | In a discussion we mostly only perceive time if somehow or other we are not satisfied with the way it is progressing. The result is an extremely conscious and primarily negative experience of time. Apart from that it is scarcely of significance as a structurally determining factor. It is the grammatical and semantic level of language that primarily gives structure. | The relationship between music and time, however, is totally different, as musical form only even begins to exist with the passing of time. Music "is" only ever at the moment at which it sounds. In its shape a sculpture can be registered and comprehended with a single glance. Music, on the other hand, needs to be experienced over a period of time, regardless of whether actively or passively, in order to become meaningful. | In other words, the following implication can be derived from the link to Gieseler's approach: Musical form is the structural connection of acoustic phenomena, which in their relationship to one another can only be experienced by sampling passing time.

1 see Gieseler, Komposition im 20. Jahrhundert, 1975 Moeck Verlag

Mnemonic structures Experiencing passing time at all requires an important evolutionary achievement. One needs a memory for it. Only if one is in a position to perceive a musical event both in the context of events that have already happened as well as in anticipation of events that supposedly are still to happen can a network of structural connections and as such musical form be experienced. | Present-day neuroscience distinguishes between three types of memory in total, which can all mainly be differentiated in terms of points of view relating to time, information theory, and content. The ultra short term memory, also known as the echoic memory, the short term or working memory and the long term memory. The latter can be sub-divided further still, depending on the type of information to be stored. To understand music it is primarily the semantic memory, in

knüpfungen und somit musikalische Form erfahrbar werden. | Die modernen Neurowissenschaften unterscheiden insgesamt drei Gedächtnistypen, die sich hauptsächlich nach zeitlichen, informationstheoretischen und inhaltlichen Gesichtspunkten differenzieren lassen. Das Ultrakurzzeitgedächtnis, auch Echogedächtnis genannt, das Kurzzeit- oder Arbeitsgedächtnis und das Langzeitgedächtnis. Letzteres lässt sich noch weiter unterteilen, je nach Art der zu speichernden Information. Für das Verstehen von Musik ist hier vor allem das semantische Gedächtnis, in dem allgemeines Wissen und Erfahrungen gespeichert sind, von Bedeutung. | Diese unterschiedlichen Gedächtnistypen stellen allerdings, nicht wie häufig fälschlicherweise angenommen, isolierte Speicherorte im Gehirn dar, sondern sie entsprechen den verschiedenen Informationsverarbeitungsstufen, die über ein Sinnesorgan eingehende Reize durchlaufen. Die Verarbeitungsstruktur ist hierbei jedoch keinesfalls streng hierarchisch, sondern bidirektional. Gespeicherte Inhalte des semantischen Gedächtnisses beeinflussen fortwährend die Arbeitsweise der früheren Verarbeitungsstufen. Im Echogedächtnis äußert sich das vor allem in einer Erhöhung der Verarbeitungsgeschwindigkeit, im Arbeitsgedächtnis wirkt es sich positiv auf die Speicherkapazität aus. || Unsere Sinne sind das Produkt von Evolutionsprozessen. Sukzessive Verbesserung, die nur einen Grund kennt: die beständige Assimilation an eine vorwiegend feindlich gesinnte Umwelt. Je besser wir uns in ihr zurechtfinden, desto größer sind unsere Chancen zu überleben. Je mehr uns also unsere Sinne über sie mitteilen, desto besser für uns. Es sind jedoch nicht unsere Sinnesorgane selbst, die im Laufe der Zeit immer besser geworden sind, sondern die Weiterverarbeitung des Wenigen, das sie in der Lage sind zu registrieren. Es sind unsere Gehirne, die dramatisch an Größe und Leistung zugenommen haben. | Wie bereits erwähnt, sind die Informationen, die unsere Sinnesorgane aufnehmen können, spärlich und vor allem äußerst ungenau. Das Einzige, was das Ohr überhaupt registrieren kann, sind Luftdruckschwankungen, die unablässig unser Trommelfell in Schwingung versetzen. Diese Schwingungen werden durch das Mittelohr verstärkt ans Innenohr weitergeleitet, wo sie nach Frequenzen sortiert in elektrische Impulse gewandelt und in Richtung Hörbahn geschickt werden. Das allerdings, was wir bewusst wahrnehmen, wenn wir hören, besteht zu größten Teilen aus geschickter Rekonstruktion, die unser Gehirn vornimmt. Es bedient sich dabei auf Erfahrungswerten beruhender Regelmäßigkeiten, welche sowohl rein physiologischer als auch ganz individuell subjektiver Natur sein können. Entscheidend dabei ist, dass gewisse Phänomene, die vom Ohr nicht wahrgenommen werden können, nachträglich ergänzt werden, da sie einer physikalischen Gegebenheit der natürlichen Umgebung entsprechen. Wir hören also genau das, was wir zur sicheren Orientierung in dieser Welt unbedingt brauchen. | Diese Prozesse spielen sich in

which general knowledge and experiences are stored that is of relevance. | However, these different types of memory do not represent isolated storage locations in the brain, as is frequently erroneously assumed, rather they correspond to the various levels of information processing, which incoming impulses run through via a sense organ. This processing structure is in no way strictly hierarchical but bi-directional. Contents stored in the semantic memory continually influence the manner in which the early processing levels work. In the echoic memory this is expressed primarily in an increase in the processing speed, in the working memory it has positive impact on the storage capacity. || Our senses are the product of evolutionary processes. Successive improvement, for which there is only one reason; the constant assimilation to an environment that is primarily hostile in nature. The better we cope with it the more chance we have of surviving. So the more our senses tell us about it the better it is for us. However it is not our sense organs themselves that have become better and better over the course of time, but the further processing of the little they are in a position to register. It is our brains that have increased dramatically in size and capacity. | As already mentioned, the information our sense organs can accommodate is meager, and primarily acutely imprecise. The only thing the ear can register at all is fluctuations in air pressure, which unremittingly cause our eardrums to oscillate. Heightened by the middle ear these oscillations are forwarded to the inner ear where, divided up by frequency, they are transformed into electric impulses and dispatched in the direction of the auditory pathway. However, what we consciously perceive when we hear something consists to a large extent of clever reconstruction, which our brain carries out. In doing so it uses instances of regularity based on experience, which can be purely physiological as well as wholly subjective in nature. What is decisive is that certain phenomena that the ear cannot perceive are subsequently complemented as they correspond to the physical circumstances of the natural surroundings. So we hear exactly what we absolutely need to find our way safely through this world. | These processes occur at different degrees of intensity in the individual processing levels of our memory. An important initial step occurs as soon as the echoic memory level. The continual acoustic flow is segmented by principles of linking and separating that are not yet fully understood and allocated to individual events or objects. When we hear someone speaking we claim to be able to perceive individual words quite clearly. That is, however, not the case. A sentence spoken without interruption is likewise an acoustic continuum. The perception of individual phonemes only occurs through segmentation. What with regard to a mother tongue appears totally automated becomes increasingly difficult when we are confronted with a language with which we are only vaguely or not at all familiar. Then it actually does seem to us to be a continuum. So the greater the amount

unterschiedlicher Ausprägung in den einzelnen Verabeitungsstufen unseres Gedächtnisses ab. Ein erster wichtiger Schritt geschieht bereits auf der Ebene des Echogedächtnisses. Der kontinuierliche akustische Strom wird nach noch nicht hundertprozentig verstandenen Prinzipien der Bindung und Trennung segmentiert und einzelnen Ereignissen oder Objekten zugeordnet. Wenn wir jemanden sprechen hören, so meinen wir ganz deutlich einzelne Wörter zu vernehmen. Dem ist aber nicht so. Ein ohne Unterbrechung gesprochener Satz ist ebenfalls ein akustisches Kontinuum. Die Wahrnehmung einzelner Phoneme entsteht erst durch Segmentierung. Was im Rahmen der Muttersprache vollends automatisiert erscheint, gestaltet sich in zunehmenden Maße schwieriger, wenn wir mit einer uns nur mäßig vertrauten oder gänzlich unbekannten Sprache konfrontiert werden. Sie erscheint uns dann tatsächlich als Kontinuum. Je größer also das im semantischen Gedächtnis gespeicherte Wissen ist, auf das wir zurückgreifen können, desto schneller und sicherer ist auch der Segmentierungsprozess. | Die für das Erfahren von musikalischen Strukturen wohl konstitutivsten Prozesse spielen sich auf der nächsthöheren Verarbeitungsstufe ab. Auf die sprachliche Ebene übertragen, entspricht dieser Schritt der Verbindung einzelner Phoneme zu Wörtern und Sätzen. Das Arbeitsgedächtnis besitzt also die Fähigkeit zur Gruppierung einzelner, zuvor segmentierter Ereignisse. Die musikalische Entsprechung wäre somit die strukturelle Verknüpfung akustischer Phänomene in zeitlicher Ausdehnung. Die Mechanismen, derer wir uns dabei bedienen, sind mittlerweile durch die Gestaltpsychologie ausreichend erschlossen. Zu nennen sind hier vor allem Prinzipien der Prägnanz, Ähnlichkeit, Nähe und Kontinuität. Erstaunlich ist dabei, dass sie für das Hören wie auch das Sehen gleichermaßen bedeutsam sind, wobei das Sehen diese Prozesse mehr oder weniger automatisch erledigt, da es sich hierbei um räumliche Prinzipien handelt, und diese können letztlich auf einen einzigen Blick erfasst werden. Musikalischer Raum entspricht jedoch nicht nur einem relativen Höher oder Tiefer von Frequenzen, sondern, viel wichtiger noch, vergehender Zeit. Aus räumlicher Nähe wird zeitliche. Das Hören wird also meist erst im Nachhinein entscheiden können, nach welchen Prinzipien das Gehörte gruppiert wird. Es sei denn es trifft auch hier auf bereits bestehendes Wissen, das wiederum im semantischen Gedächtnis gespeichert ist. | Musikalische Erfahrung ermöglicht aber auch einen wesentlich größer dimensionierten Umgang mit Gruppierungsprinzipien, da mehrere kleine Gruppen als Einzelbestandteile einer übergeordneten Gruppierung fungieren können. Dieses sog. Clustering erweitert die durchschnittlich nur auf etwa sieben Inhalte begrenzte Kapazität des Arbeitsgedächtnisses um ein Vielfaches und öffnet somit den Raum für sich weiter ausbreitende strukturelle Zusammenhänge musikalischen Materials. Diese elementaren kognitiven Fähigkeiten bilden die unabdingbare Voraussetzung, auf

of knowledge stored in the semantic memory that we can actually fall back on is the quicker and more certain the segmenting process is. | The most constitutive processes for experiencing musical structures occur at the next highest processing level. Transposed onto the level of language, this step corresponds to the combination of individual phonemes to form words and sentences. The working memory is therefore capable of grouping individual, previously segmented events. As such the musical equivalent would be the structural linking of acoustic phenomena over time. The mechanisms we use in the process have now been sufficiently investigated by gestalt psychology, primarily principles of conciseness, similarity, proximity, and continuity. What is astonishing is that they are equally significant for hearing and seeing, whereby seeing carries out these processes more or less automatically, as we are dealing with spatial principles and these can ultimately be registered with a single glance. Musical space, however, corresponds not only to a relative higher or lower of frequencies but rather, which is far more important, passing time. Spatial proximity becomes temporal. As such hearing mostly will only subsequently be able to decide on the principles by which what is heard is grouped. Unless, that is, it here too encounters existing knowledge, which in turn is stored in the semantic memory. | However, musical experience enables a far more extensive handling of grouping principles, as several small groups can act as individual parts of an overarching group arrangement. This 'clustering' increases several fold the working memory capacity, which on average is limited to approximately just seven items, thereby making space for structural contexts with musical material that spread out further. These elementary cognitive abilities form the indispensable prerequisite on which established formal compositional techniques can build on in the first place. Without the time-based information processing of acoustic events even the simplest structuring means as mere repetition or sequencing would not be perceivable as that and musical structure thus totally unimaginable. | The semantic memory is ultimately the culminating point at which all the threads come together. The general knowledge stored there that has been accumulated over a period of years is defined primarily through structures and schemes that have been learned. As mentioned above the schemes influence the processing mechanisms of the echoic and working memory to a large extent, depending on whether a new input correlates with a certain expectation or not. If the compliant part predominates, the result is not seldom predictability and thus boredom. If, on the other hand, non-compliant contents predominate, there is frequent excessive demand. Musical equivalents of these schemes are specific topoi, which can be found in all stages of development of western musical culture and in all other cultures as well, which can be determined by various factors. Frequently these are specific formal models, characteristic constellations of material,

der tradierte gestalterische Kompositionstechniken überhaupt erst aufbauen können. Ohne zeitbasierte Informationsverarbeitung akustischer Ereignisse wären selbst einfachste strukturbildende Mittel wie bloße Wiederholung oder Sequenzierung nicht als solche wahrnehmbar und musikalische Struktur somit gänzlich unvorstellbar. | Das semantische Gedächtnis ist schließlich der Kulminationspunkt, in dem alle Fäden zusammenlaufen. Das dort gespeicherte, über Jahre hinweg akkumulierte Allgemeinwissen definiert sich vor allem durch erlernte Strukturen und Schemata. Diese Schemata beeinflussen, wie oben bereits erwähnt, in hohem Maße die Verarbeitungsmechanismen des Echo- und Arbeitsgedächtnisses, je nachdem ob ein neuer Input mit einer bestimmten Erwartungshaltung korreliert oder nicht. Überwiegt dabei der konforme Anteil, ist das Resultat nicht selten Vorhersehbarkeit und somit Langeweile. Überwiegen hingegen nonkonforme Inhalte, so kommt es häufig zu Überforderung. Musikalische Entsprechungen dieser Schemata sind in allen Entwicklungsstadien der abendländischen Musikkultur, aber auch in allen anderen Kulturen zu findende Topoi, die sich an verschiedenen Gesichtspunkten festmachen lassen. Häufig sind es spezifische Formmodelle, charakteristische Materialkonstellationen, stereotype melodische Wendungen oder auch bestimmte Instrumentenkombinationen, die im Laufe der Zeit zu ästhetischen Habitus geworden sind. Allen gemein ist, dass sie im Rahmen ihrer oftmals auch stillschweigenden Allgemeingültigkeit in der Summe zutiefst stil- und vor allem auch formbildend sind. Sie stiften kulturelle Identität und zeichnen in der abendländischen Kunstmusik die Übergänge der verschiedenen Epochen nach. In der Popmusik sind es sogar oft nur marginale Unterschiede, die für stilistische Zuordnung sorgen. Kurzum, das semantische Gedächtnis repräsentiert für uns, wie Musik zu funktionieren und auch zu klingen hat, was wir glauben, von Musik zu verstehen.

Der wohl wichtigste, wenn auch in seiner struktur- und formbildenden Wichtigkeit häufig unterschätzte Topos der klassisch-romantischen, aber auch der zeitgenössischen populären Musik besteht in einem seit mittlerweile über dreihundert Jahren allgemeingültigen Beziehungssystem von Tonhöhen, das einem Sonderfall des musikalischen Materials in Einklang mit einer artifiziellen Stimmung geschuldet ist: der *Dur-Moll-Tonalität*. | Die diesem Tonsystem innewohnenden Gravitationskräfte ermöglichten formales Gestalten in zuvor nicht gekannten Ausmaßen. Zu den wesentlich allgemeineren Strukturprinzipien im hauptsächlich melodisch-rhythmisch Sukzessiven gesellte sich nun eine weitere formbildende Komponente, die nach Gieselers Modell in den Bereich des Materials fällt; die Ebene der Harmonik. | Bis ans Ende des 19. Jahrhunderts währte die Alleinherrschaft der Tonalität, ehe nach ihrer bereits länger anhaltenden

stereotype melodic phrases or certain combinations of instruments, which in the course of time have become aesthetic habitus. What they all have in common is that within the framework of their not seldomly implicit universal validity they are overall deeply influential and primarily determining in terms of style and form. They give cultural identity and in western art music trace the points of transition of the various epochs. In Pop music it is often only marginal differences that result in stylistic allocation. In short, the semantic memory represents for us how music should function and sound, what we believe we understand of music.

Presumably the most important topos in classical romantic as well as contemporary popular music, though one that is frequently underestimated in terms of its structural and form-determining importance, consists of a system of pitch-relationships, which has now been universally valid for over 300 years that is indebted to a special case of musical material in accordance with an artificial tuning-system: the *major-minor tonality*. | The gravitational forces inherent in this tone system enabled formal design to an extent formerly unknown. The considerably more general structural principles in the predominantly successive, melodic rhythmic domain, were now joined by a further form-determining component, which according to Gieseler's model belongs to the field of material; the plane of harmony. | The autocracy of tonality lasted until the late 19th century, before, following a lengthy crisis, its end was officially declared. The era of so-called free atonality began. However, just like most of the pieces written at that time, the epoch was not of long duration. With the abandonment of tonality one had not only banished a tone system that was regarded as outdated. With it the foundation of formal stability for structuring larger periods of time was also equally lost. Precisely the aphoristic shortness of many works from this era is eloquent proof of the increasing rampant formal uncertainty, which went with the supposed new freedom that had been acquired. | Of all things the return to structural principles, which had their heyday long before major-minor tonal music and had since almost sunk into oblivion, provided the desired way out. Via combinatorial procedures Schönberg's *Method of Composing with twelve Notes which are related only to one another* paved the way back to parametric *ars nova* thinking and heralded a century that in its abundant variety of forms eclipsed anything that had been previously known. Old contrapuntal techniques were once again put to copious use and the fascination of numbers was rediscovered. Not infrequently the vacant place of tonality was also replaced by elements borrowed from the visual arts which via montage, collage, minimalism and Fluxus even led to strict negation of any concept of form and at times even of music. With his oeuvre Karlheinz Stockhausen alone defined more new concepts of form than all of the history of music before him had done. | Pre-

Krise ihr offizielles Ende ausgerufen wurde. Es begann die Ära der sog. freien Atonalität. Diese war jedoch, genauso wie die meisten Stücke, die zu dieser Zeit geschrieben wurden, nur von kurzer Dauer. Mit der Preisgabe der Tonalität hatte man nicht nur ein als überholt geltendes Tonsystem verbannt. Man verlor mit ihr auch gleichermaßen die Grundlage formaler Stabilität zur Gestaltung größerer Zeitspannen. Eben jene aphoristische Kürze vieler Werke aus dieser Zeit ist beredtes Zeugnis der zunehmend grassierenden formalen Verunsicherung, die mit der vermeintlich gewonnenen neuen Freiheit einherging. | Ausgerechnet die Rückbesinnung auf strukturelle Prinzipien, die ihren Zenith lange vor der dur-moll-tonalen Musik hatten und seither beinahe in Vergessenheit geraten waren, brachte den ersehnten Ausweg. Schönbergs *Methode des Komponierens mit zwölf nur aufeinander bezogenen Tönen* ebnete über Verfahrensweisen der Kombinatorik den Weg zurück zum parametrischen Denken der *Ars nova* und läutete damit ein Jahrhundert ein, das in seinem überbordenden Formenreichtum alles bislang Gekannte in den Schatten stellte. Man bediente sich wieder ausgiebig alter kontrapunktischer Techniken und entdeckte die Faszination der Zahlen neu für sich. An die vakante Stelle der Tonalität rückten nicht selten auch gestalterische Mittel, die der bildenden Kunst entlehnt wurden und die über Montage, Collage, Minimalismus und Fluxus bis zur dezidierten Negation jedweden Form- und bisweilen sogar Musikbegriffs führten. Allein Karlheinz Stockhausen definierte mit seinem Schaffen wohl mehr neue Formbegriffe, als die gesamte Musikgeschichte es vor ihm getan hatte. | Gerade in den letzten Jahrzehnten kristallisierte sich eine formale Konstante heraus, die musikalische Struktur als graduellen, parametrisch gesteuerten Prozess von Zustand A nach B begreift. Permeabilität und Interpolation sind keine ausschließlich naturwissenschaftlichen Fachtermini mehr und das Denken in Algorithmen gehört längst schon zur guten Kinderstube des angehenden Komponisten. Seismografische Aufzeichnungen von Erdbeben werden genauso bereitwillig zur Strukturierung von Musik herangezogen wie die Verkehrssituation in New York oder die Maxwell-Boltzmann-Verteilung zur statistischen Bewegung von Gasmolekülen. Daraus lässt sich folgern, dass potentiell alles, was auf irgendeine Art und Weise formalisierbare Struktur besitzt, zu musikalischer Formgebung geeignet sein kann. Und nicht selten sind es Baupläne aus der Natur, die auch sinnlich ästhetisch überzeugende Ergebnisse liefern, ob es sich nun um die Proteinstruktur menschlicher DNS oder um virtuelle Simulationen pflanzlichen Wachstums handelt.

Kathedralen und Blumenkohl Abb. 1 zeigt den originalen Grundriss des Florentiner Doms *Santa Maria del Fiore*. Papst Eugen IV. beauftragte Guillaume

cisely over the past few decades a formal constant has crystallized that sees musical structure as a gradual, parametrically controlled process from condition A to B. Permeability and interpolation are not exclusively scientific terms anymore and thinking in algorithms has long since been part and parcel of any prospective composer's nursery. Seismographic recordings of earthquakes are used just as readily to structure music as the traffic situation in New York or the Maxwell-Boltzmann distribution law expressing the statistical motion of gas molecules. From this it can be concluded that potentially anything that in some way or other has a structure that can be formalized, can be suited to musical shaping. And not seldom it is construction plans from nature that also deliver convincing sensual, aesthetic results, be it a case of the protein structure of human DNA or virtual simulation of plant growth.

Cathedrals and cauliflower Fig. 1 reveals the original footprint of the cathedral *Santa Maria del Fiore* in Florence. Pope Eugene IV commissioned Guillaume Dufay, arguably the most influential composer of his day, to compose a festive motet for the cathedral's consecration. | *Nuper rosarum flores* was first performed on March 25, 1436 in the presence of the Pope and, given its stylistically far-reaching contrapuntal style, is regarded among experts as one of the most important church consecration motets of the late Middle Ages. Despite its visionary traits it is nonetheless absolutely a musical symbol of its time. | The view of music in the late 15th century, one that now seems slightly strange to us emphatically emphasized the antique Pythagorean idea of music as an integral part of mathematical disciplines. It maintained that music was based on proportional ratios of musical parameters such as pitches, duration or the entire length of a composition and as such was part of the likewise proportionally structured cosmic world order. | In Dufay's motet it is not difficult to recognize the ratios on which it is based, which structure the work on different levels. On the one hand the tempo relations of the four main sections behave like 6:4:2:3. This corresponds to the spatial relationship in the cathedral from nave to transept to apse to the height of the cupola. On the other the design of the individual voices in several parameters is based on the division ratio of the *Golden section* (ϕ = 1,618033989) respectively the closely related *Fibonacci sequence* (0 1 1 2 3 5 8 13 21 34 etc.), in which each new number is the total of the two previous numbers. For this reason it is related to the Golden section, as the ratio of two neighboring Fibonacci numbers with increasing size gets closer to it. | The Golden section and the Fibonacci sequence are proportional constants without which it is difficult to imagine the history of architecture and art, as since time immemorial they have been regarded as the epitome of har-

Dufay, den wohl tonangebenden Komponisten seiner Zeit, mit der Komposition einer Festmotette zur Einweihung des Doms. | *Nuper rosarum flores* wurde am 25. März 1436 in Anwesenheit des Papstes uraufgeführt und gilt unter Kennern aufgrund ihrer stilistisch weit vorausgreifenden Kontrapunktik als eine der wichtigsten Kirchweihmotetten des ausgehenden Spätmittelalters. Trotz ihrer visionären Züge ist sie dennoch ganz und gar musikalisches Sinnbild ihrer Zeit. | Das uns mittlerweile leicht befremdlich vorkommende Musikverständnis im 15. Jahrhundert bezog sich nachdrücklich auf die antik-pythagoreische Vorstellung von Musik als immanentem Bestandteil der mathematischen Disziplinen. Das Wesen von Musik gründete demnach auf proportionalen Zahlenverhältnissen musikalischer Parameter wie Tonhöhen, Tondauern oder der Gesamtdauer einer Komposition und war somit Bestandteil der ebenfalls proportional strukturierten kosmischen Weltordnung. | In Dufays Motette lassen sich unschwer die zugrundeliegenden Zahlenproportionen erkennen, die das Werk auf unterschiedlichen Ebenen gliedern. Zum einen verhalten sich die Temporelationen der vier Großabschnitte wie 6:4:2:3. Dies entspricht den räumlichen Verhältnissen im Dom von Langhaus zu Querschiff zu Apsis zu Kuppelhöhe. Zum anderen basiert die Gestaltung der Einzelstimmen in etlichen Parametern auf dem Teilungsverhältnis des *Goldenen Schnitts* (ϕ = 1,618033989) bzw. der ihm eng verwandten *Fibonacci-Reihe* (0 1 1 2 3 5 8 13 21 34 etc.), bei der jede neue Zahl die Summe der beiden vorhergehenden Zahlen ergibt. Verwandt ist sie deshalb mit dem Goldenen Schnitt, da sich ihm das Verhältnis zweier angrenzender *Fibonacci-Zahlen* mit zunehmender Größe immer mehr annähert. | Der goldene Schnitt und die *Fibonacci-Reihe* sind aus der Architektur- und Kunstgeschichte kaum wegzudenkende proportionale Konstanten, gelten sie doch seit jeher als Inbegriff harmonisch ästhetischer Disposition. So stehen erwartungsgemäß auch etliche Bauelemente des Florentiner Doms in Fibonacci-Proportionen zueinander, wie z. B. die komplette Kuppelarchitektur im Verhältnis 144:89:55. | Wie das Beispiel von Guillaume Dufay beweist, finden sich Fibonacci-Proportionen auch durchgehend im Fundus der musikalischen Weltliteratur. Allerdings ist in den meisten Fällen, insbesondere in der Zeitspanne vom 17. bis zum 19. Jahrhundert, wohl eher davon auszugehen, dass nicht etwa kompositorische Konstruktion wie bei Dufay, sondern vielmehr rein ästhetische Intuition dafür verantwortlich war. Gleiches gilt mit Sicherheit auch für eine Vielzahl von Kunstwerken. Das Wissen um die informationsverarbeitenden Mechanismen des menschlichen Gehirns und um die durch sie zu rekonstruierenden Strukturen der Umwelt schafft diesbezüglich einen wichtigen Anknüpfungspunkt. Durch das beinahe ubiquitäre Auftreten von *Fibonacci-Zahlen* in der Natur bekommt die unbewusste ästhetische Vorliebe für derartige Strukturen eine völlig andere Gewichtung. | Die nach dem italienischen Mathematiker

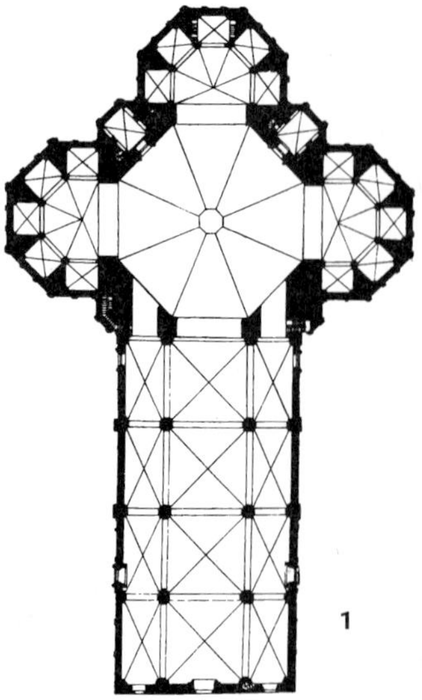

1

Abb. 1 Santa Maria del Fiori

Fig. 1 Santa Maria del Fiori

2

monious aesthetic arrangement. As can be expected, any number of construction elements in the cathedral in Florence reveal Fibonacci proportions to each other, for example the entire cupola architecture in a ratio of 144:89:55. | As the Guillaume Dufay example proves, Fibonacci proportions can also be found throughout musical world literature. In most cases, however, in particular between the 17th and 19th centuries, if anything one can assume that not compositional construction as with Dufay, but far more purely aesthetic intuition was responsible. The same surely applies to a multitude of art works. In this respect knowledge about the information-processing mechanisms of the human brain, as well as about the environmental structures to be reconstructed by them, creates an important link. Through the almost ubiquitous presence of Fibonacci numbers in nature the unconscious aesthetic preference for this type of structures is given a totally different weighting. | The Italian mathematician Leonardo Fibonacci, after whom the sequence is named, originally used it in 1202 to describe the growth of rabbit populations. It can also be used to illustrate the growth of a beehive population or the spread of algae. Likewise the gradual increase in speed of a falling table tennis ball. An estimated 95% of all plant inflorescence behaves accordingly. Lilies have three leaves, marigolds 13, daisies 34, 55 or, less frequently even 89 and the horsetail could pass for a miniature version of the cathedral in Florence as well. The two halves of a bee's body are divided up according to the Golden Ratio as

Abb.2 Romanesco

Fig.2 Romanesco

Leonardo Fibonacci benannte Reihe wurde von ihm selbst ursprünglich im Jahr 1202 dazu benutzt, um das Wachstum von Kaninchenpopulationen zu beschreiben. Auch das Wachstum eines Bienenstaats oder die Ausbreitung von Algen lässt sich damit darstellen. Ebenso das sukzessive Beschleunigen eines fallenden Tischtennisballs. Schätzungsweise 95% aller Blütenstände von Pflanzen verhalten sich danach. Lilien haben 3 Blütenblätter, Ringelblumen 13, Gänseblümchen 34, 55 oder seltener auch 89 und der Schachtelhalm könnte auch als Miniaturfassung des Florentiner Doms durchgehen. Die beiden Körperhälften einer Biene teilen sich ebenso im Goldenen Schnitt wie auch der Nabel den gesamten menschlichen Körper und die Längen der Fingerknochen des Mittelfingers verhalten sich wie 2:3:5:8. | Auf Abb. 2 lässt sich eine vollkommen andere Struktur erkennen, die dennoch bis ins kleinste Detail auf Fibonacci-Proportionen beruht und nicht minder häufig auftritt. | Der Romanesco, eine italienische Blumenkohl-Züchtung aus dem 16. Jahrhundert, bezieht seine Gestalt aus unzähligen sog. *Fibonacci-Spiralen*. Eine *Fibonacci-Spirale* lässt sich leicht durch ein *Goldenes Rechteck* approximieren, in dessen Quadrate jeweils ein Viertelkreis gesetzt wird (Abb. 3). || Eine einfache *Fibonacci-Spirale* zeigt große Ähnlichkeit mit dem Bau eines Schneckenhauses und eine geradezu frappierende Übereinstimmung mit Nautilus-Gehäusen. Ordnet man mehrere dieser Spiralen fächerförmig an, so ergibt sich eine strudelförmige Struktur, die dem Aufbau diverser Nadelbaum-Zapfen bereits recht nahe kommt (Abb. 4). | *Fibonacci-Spiralen* in freier Natur tauchen jedoch kaum in der obigen Form auf. Ausschließlich rechts- oder linksdrehende Spiralen sind eher eine Seltenheit. Der Normalfall ist eine Kombination aus beidem, wie auch im Falle des Romanesco Blumenkohls. Das wohl Bemerkenswerteste daran ist, dass sich die Anzahl von rechts- und linksdrehenden Spiralen ebenfalls zueinander im Verhältnis angrenzender Fibonacci-Zahlen befindet. Beim Romanesco für gewöhnlich 21:13 oder 13:8, beim Samenstand einer Sonnenblume sogar bis zu 144:89. Abb. 5 zeigt eine stark stilisierte Grafik einer solchen Struktur, die frei einer Sonnenblume nachempfunden wurde. || Der Romanesco hat noch eine weiter Eigenschaft, die ihn sogar zu einer absoluten Rarität der Pflanzenwelt macht. Er ist eines der wenigen Gewächse, deren Blütenstand *Fibonacci-Spiralen* mit fraktalen Strukturen vereint. Fraktale Strukturen sind für gewöhnlich nichts Besonderes, sie gehören zu den charakteristischen phänotypischen Merkmalen vieler Pflanzen. Maßgeblich geprägt wurde der Begriff durch den polnisch-französischen Mathematiker Benoît Mandelbrot. Er definiert eine natürliche oder artifizielle geometrische Struktur, deren Gestalt sich durch einen hohen Grad an Skaleninvarianz bzw. Selbstähnlichkeit auszeichnet. Als selbstähnlich werden Strukturen oder Gegenstände bezeichnet, die bei Skalierung ihre anfängliche Gestalt oder zumindest Teile davon reproduzieren. Dabei ist zwischen exakter und statistischer Selbstähn-

does the navel the entire human body and the lengths of the bones of the middle finger behave as 2:3:5:8. | In Fig. 2 a completely different structure is discernible that nonetheless, down to the very last detail, is based on Fibonacci proportions and occurs no less frequently. | Romanesco, an Italian cauliflower cultivation dating from the 16th century, acquires its shape from innumerable *Fibonacci spirals*. A *Fibonacci spiral* can easily be approximated by a *Golden Rectangle*, in each of whose squares a quadrant is placed (Fig. 3). || A straightforward Fibonacci spiral reveals great similarity with the structure of a snail shell and a downright striking congruence with nautilus shells. If one arranges several of these spirals in a fan-like shape a swirling structure emerges, already close to that of various conifer cones (Fig. 4). | *Fibonacci spirals*, however, hardly ever occur in the above shape in nature. If anything, solely clockwise or anti-clockwise spirals are a rarity. A combination of the two is normal, as is the case with the Romanesco cauliflower. The most remarkable thing about it is the fact, that the number of clockwise and anti-clockwise spirals is likewise in the ratio of neighboring Fibonacci numbers. In the case of the Romanesco, usually 21:13 or 13:8, for the seed level of a sunflower even up to 144:89. Fig. 5 shows a heavily stylized diagram of such a structure, based loosely on a sunflower. || The Romanesco has yet another characteristic that actually makes it an absolute rarity in the plant world. It is one of the few whose inflorescence combines *Fibonacci spirals* with fractal structures. Fractal structures are normally nothing special, they belong to the characteristic phenotypical features of various plants. The term was heavily influenced by the Franco-Polish mathematician Benoît Mandelbrot. It defines a natural or artificial geometric structure, the shape of which stands out for its high degree of scale invariance and self-similarity respectively. Structures and objects are classified as self-similar which, when scaled, reproduce their original shape or at least parts of it. Here a difference must be made between exact and statistical self-similarity as well as self-affinity, whereby in nature exact self-similarity categorically does not occur and self-affinity is probably encountered most frequently. It is characterized merely by reproductions of specific sections and a limited number of scaling factors. The Romanesco is an ideal study object for the exact observation of self-similar structures. Each individual elevation on a *Fibonacci spirals* in turn itself comprises clockwise and anti-clockwise *Fibonacci spirals*, whose elevations consist likewise of *Fibonacci spirals*. As there are logically limits to this process, one ought correctly to speak of self-affinity.

At first sight self-similarity and music would appear to be an unlikely couple. Listeners schooled in the traditional formal canon will presumably never associate fractal structures with a sonata. And in other respects too it would be relatively

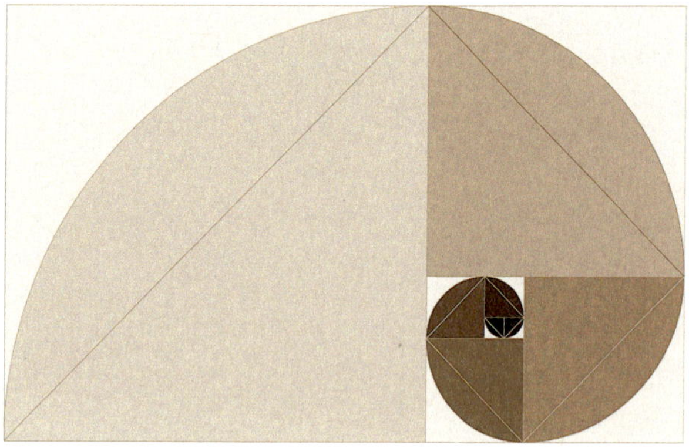

3

lichkeit sowie Selbstaffinität zu unterscheiden, wobei die exakte Selbstähnlichkeit in der Natur kategorisch nicht vorkommt und die Selbstaffinität wohl am häufigsten zu finden ist. Sie zeichnet sich durch lediglich ausschnitthafte Reproduktion und eine begrenzte Anzahl an Skalierungsfaktoren aus. Der Romanesco ist ein idealtypisches Studienobjekt, um selbstähnliche Strukturen genau zu beobachten. Jede einzelne Erhebung einer *Fibonacci-Spirale* besteht wiederum selbst aus rechts- wie linksdrehenden *Fibonacci-Spiralen*, deren Erhebungen ebenfalls wieder aus *Fibonacci-Spiralen* bestehen. Da diesem Vorgang logischerweise Grenzen gesetzt sind, müsste man auch hier korrekterweise von Selbstaffinität sprechen.

Selbstähnlichkeit und Musik scheinen auf den ersten Blick ein ungleiches Paar zu sein. Der am tradierten Formenkanon geschulte Hörer wird mit einer Sonate wohl niemals fraktale Strukturen assoziieren. Und auch sonst dürfte man sich verhältnismäßig schwer tun, selbstähnliche musikalische Strukturen zu finden, wenn man nicht ganz genau weiß, wo man zu suchen hat. | Doch es gibt sie. Manche davon, die seltenere Variante, geradezu ostentativ an der Oberfläche, die anderen, vergleichsweise häufig, tief vergraben im strukturellen Beziehungsnetz formaler Abhängigkeiten. Der bereits geschilderte Prozess des Clustering legt eine offensichtliche Affinität des menschlichen Gehirns zur sinnvollen Verarbeitung solcher Strukturen zumindest nahe. | Einfache, da nicht tiefer in die musikalische Struk-

Abb.3 Fibonacci-Spirale

Fig.3 Fibonacci spiral

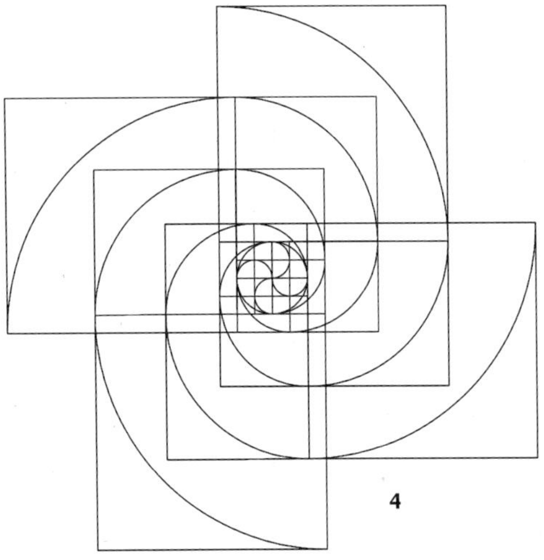

4

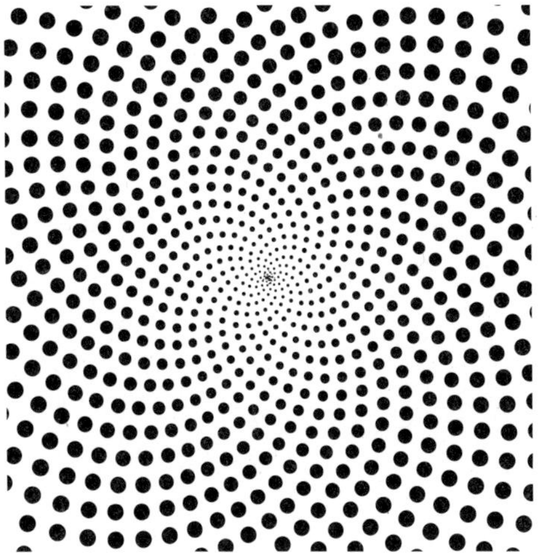

5

difficult to find self-similar musical structures if one is not entirely sure where to look. | Yet they exist. Many of them, the rarer variety, almost ostentatiously on the surface, the others, comparatively frequently, deeply buried in the structural relationship network of formal dependencies. The clustering process described earlier at least suggests a clear affinity between the human brain and the meaningful processing of structures such as these. | Self-similarity that is simple because it does not intrude on the musical structure to any great depth can occasionally be found in the work of Johann S. Bach. A "cantus firmus" accompanied by a faster version of itself. Reproduction of its own shape on another scale and therefore self-affine. | In order to track down the formal prototypes of open self-similarity one needs to immerse oneself in an epoch that has unfortunately almost sunk into oblivion. The *ars subtilior* marks the relatively short period of transition from the end of *ars nova* in the 1470s to the appearance of Guillaume Dufay and the Franco-Flemish school in the 1520s. What little has survived from this era bears witness to a highly complex musical language, access to which is not exactly made easy. Spearheading innovations forged a path for themselves, such as the inclusion of distant notes, which would still have been unimaginable in *ars nova*, and the rhythmical extension through new note values. In *ars subtilior*, precisely the rhythmic-metrical element acquired a previously unknown complexity through frequent mensuration changes and a preference for polymetric relations. | A form

tur eingreifende Selbstähnlichkeit findet sich gelegentlich bei *Johann S. Bach*. Ein Cantus Firmus, der von einer schnelleren Variante seiner selbst begleitet wird. Reproduktion der eigenen Gestalt in anderer Skalierung und somit selbstaffin. | Um den formalen Prototypen offener Selbstähnlichkeit aufzuspüren, ist es notwendig, sich in eine leider fast in Vergessenheit geratene Epoche zu begeben. Die *Ars subtilior* markiert den relativ kurzen Übergangszeitraum vom Ende der *Ars nova* in den 70er Jahren des 14. Jahrhunderts bis zum Auftauchen Guillaume Dufays und der franko-flämischen Schule in den 20er Jahren des 15. Jahrhunderts. Das Wenige, das aus dieser Zeit erhalten geblieben ist, zeugt von einer hochkomplexen Musiksprache, die einem den Zugang nicht gerade leicht macht. Wegweisende Neuerungen bahnten sich ihren Weg, wie die Einbeziehung entfernter Töne, die in der *Ars Nova* noch undenkbar gewesen wären und die Erweiterung des Rhythmischen durch neue Notenwerte. Gerade der rhythmisch-metrische Bereich erreichte in der *Ars subtilior* durch häufige Mensurwechsel und die Vorliebe für polymetrische Verhältnisse eine zuvor nicht gekannte Komplexität. | Ein gleichermaßen exzentrischer wie seltener Formtypus hatte vermutlich zu dieser Zeit seine Geburtsstunde: der sog. Mensuralkanon, gelegentlich auch Prolationskanon genannt. Alle beteiligten Stimmen haben dieselbe Stimme, nur jede einzelne davon steht in anderer Mensur, also in einem anderen Tempo. Das klangliche Resultat ist vordergründig ähnlich wie bei Bach, nur rhythmisch wesentlich komplexer, da sich alle Stimmen daran beteiligen und die Tempo-Relationen nicht ausschließlich binär sind. *Le ray au soleyl* einer der wenigen erhaltenen Mensuralkanons der Ars subtilior stammt von Johannes Ciconia und steht im polymetrischen Verhältnis von 1:3:4 (Abb. 6). | Mensuralkanons finden sich noch vereinzelt bis ins frühe 16. Jahrhundert, verschwanden dann aber komplett von der Bildfläche. Erst wieder im 20. Jahrhundert erlebten sie eine Renaissance in Conlon Nancarrows großartigen *Studies for player piano*. Studie 33 basiert beispielsweise auf einem Geschwindigkeitsverhältnis von √2:2, ein ›irrationales‹ Tempo... Ein weiteres Beispiel findet sich in Arvo Pärts weltbekanntem *Cantus in memoriam Benjamin Britten*.

Wenn man sich folgenden Bauplan eines einfachen Menuetts mit Trio ansieht (Abb. 7), so wird augenblicklich deutlich, dass es sich bei jeder dreiteilig zusammengesetzten Liedform prinzipiell um eine selbstaffine Struktur handeln muss. Selbstähnlichkeit, die im Verborgenen wirkt, da sie ausschließlich an der tiefergehenden Vernetzung musikalischer Strukturen beteiligt ist, aber selbst niemals an die Oberfläche tritt. | Derartige Vernetzungen finden sich zuhauf, bisweilen sogar da, wo man sie vielleicht am wenigsten erwartet. Die Struktur des ersten Viertakters aus ›*Hänschen klein*‹ projiziert sich beispielsweise auf den Verlauf der

that is eccentric as it is rare presumably saw the light of day around this time: the so-called "mensuration canon", sometimes referred to as prolation canon. All participating voices have the same line, though each one is written in different mensuration, in other words in a different tempo. Superficially the resulting sound is similar to Bach, only rhythmically far more complex, as all voices are involved and the relationships between the speed are not exclusively binary. *Le ray au soleyl*, one of the few surviving *ars subtilior* mensuration canons stems from Johannes Ciconia and has the polymetric ratio 1:3:4 (Fig. 6). | Mensuration canons could be found sporadically until the early 16th century before disappearing entirely. It was only in the 20th century that they underwent a renaissance in Conlon Nancarrow's superb *Studies for player piano*. Study 33, for example, is based on a tempo ratio of $\sqrt{2}:2$, an 'irrational' tempo... Another example is to be found in Arvo Pärt's world-famous *Cantus in memoriam Benjamin Britten*.

If you look at the following structure of a simple minuet with trio (Fig. 7), it immediately becomes clear that every compound ternary song-form has in principle to be a self-affine structure. Self-similarity, which impacts covertly, as it is part of the deeper networking of musical structures, but never appears on the surface. | This type of networking can be found in great numbers, on occasion even in places where it is perhaps least expected. The structure of the first four bars from '*Hänschen klein*'[2] is projected, for example, on the course of the song's entire 16 bars. And Eduardo Reck Miranda even distills a fractal from the old and dignified sonata form, which on top he even relates to Homer's Odyssey.[3] || Towards the end of the 20th century composers began systematically researching the potential of fractals in the genesis of structure. In this context György Ligeti in particular is worthy of mention, who with his 18 piano studies succeeded in producing a fascinating kaleidoscope in the field of tension between pygmy chant and chaos theory, Mandelbrot set and ars subtilior. | As a result of the swift advances in technology over the past few years, precisely with the support of computers there are now several new possibilities for integrating algorithm-based fractal structures in the compositional process. Particularly promising and furthermore easy to implement, the so-called *Lindenmayer systems* or *L-systems* for short have within a short space of time met with very positive response. They are now among the fractal algorithms most frequently used for compositional purposes.

2 A well known traditional German children's song, dating from the 19th century.

3 see Miranda, Composing Music with Computers, 2001 Focal Press

Virtual plants *L-systems* are a formal mathematical system for describing biological growth processes that emerged in 1968. The Hungarian biologist Aristid Lindenmayer developed it modeled on Noam Chomsky's *Formal Grammars* that

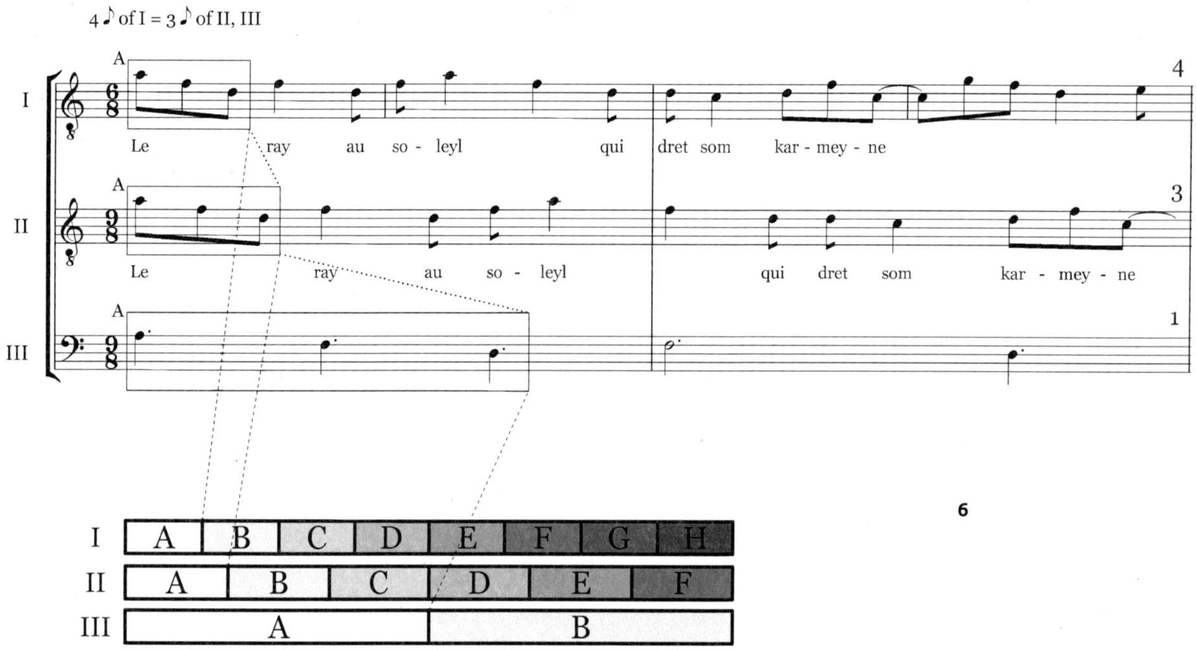

6

gesamten sechzehn Takte. Und Eduardo Reck Miranda destilliert sogar aus dem altgediegenen Sonatenhauptsatz ein Fraktal heraus, welches er obendrein auch noch zu Homers Odyssee in Bezug setzt.[2] || Gegen Ende des 20. Jahrhunderts begannen Komponisten das Potential von Fraktalen zur Strukturgenese systematisch zu erforschen. Insbesondere György Ligeti sei an dieser Stelle erwähnt, dem mit seinen 18 Klavieretüden ein faszinierendes Kaleidoskop im Spannungsfeld zwischen Pygmäen-Gesang und Chaostheorie, Mandelbrotschen Apfelmännchen und der Ars subtilior gelungen ist. | Bedingt durch den rasanten technologischen Fortschritt der letzten Jahre, bieten sich gerade durch die Unterstützung von Computern etliche neue Möglichkeiten, fraktale Strukturen auf algorithmischer Basis in den kompositorischen Prozess zu integrieren. Besonders vielversprechend und außerdem einfach zu implementieren, haben die sog. *Lindenmayer-Systeme*, kurz *L-Systeme*, in kurzer Zeit für große Resonanz gesorgt. Mittlerweile zählen sie mit zu den am häufigsten zu Kompositionszwecken verwendeten fraktalen Algorithmen.

2 vgl. Miranda, Composing Music with Computers, 2001 Focal Press

Virtuelle Pflanzen *L-Systeme* sind ein im Jahr 1968 entstandenes mathematisch formales System zur Beschreibung biologischer Wachstumsprozesse. Der ungarische Biologe Aristid Lindenmayer entwickelte es in Anlehnung an Noam

Abb.6 Le ray au soleyl — Johannes Ciconia **Fig.6** Le ray au soleyl — Johannes Ciconia

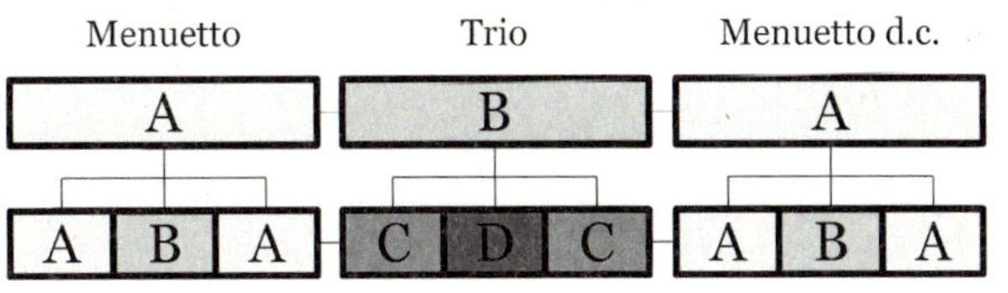

7

had emerged as early as the 1950s and, like the latter, is one of the so-called substitution systems. Systems such as these are based on the consistent derivation of a starting axiom, by means of one or several production rules. In other words, an, as a rule, simple symbol is iteratively replaced by others by means of the same procedure every time, whereas all the symbols stem from a common alphabet of arbitrary signs. A major difference from Chomsky's grammars is primarily that in them the production rules are exclusively sequential, whereas in the case of *L-systems* they are always used simultaneously.

A simple example:
Let the starting axiom $\omega = A$
and the production rules $P = \{(A \rightarrow AB),(B \rightarrow A)\}$

If the production rules are applied five times, they produce the following result (Fig. 8). | An *L-system* such as this is classified as context-free, as all the production rules are used without restriction. As such it represents the simplest case imaginable. This model mimics the growth of simple algae and apart from a small inaccuracy at the beginning creates a growth rate in Fibonacci proportions. One gets a complete Fibonacci sequence when, given constant production rules one begins with the axiom $\omega = B$. | Context-free *L-systems* can be expanded in vari-

Chomskys bereits in den 50er Jahren entstandene ›Formale Grammatiken‹ und es zählt ebenso wie diese zu den sog. Ersetzungssystemen. Solche Systeme basieren auf der konsequenten Ableitung eines Startaxioms, mittels einer oder mehrerer Produktionsregeln. D.h. ein in der Regel einfaches Symbol wird iterativ nach der immer gleichen Prozedur durch andere ersetzt, wobei sämtliche Symbole aus einem gemeinsamen Alphabet beliebiger Zeichen stammen. Ein großer Unterschied zu Chomskys Grammatiken besteht vor allem darin, dass dort die Produktionsregeln ausschließlich sequentiell, bei *L-Systemen* jedoch immer simultan zur Anwendung kommen.

Ein einfaches Beispiel:
gegeben ist das Startaxiom $\omega = A$
und die Produktionsregeln $P = \{(A \rightarrow AB),(B \rightarrow A)\}$

Lässt man die Produktionsregeln insgesamt fünf Mal anwenden, so fördern sie folgendes Ergebnis zutage (Abb. 8). | Ein solches L-System wird als kontextfrei bezeichnet, da immer alle Produktionsregeln uneingeschränkt zur Anwendung kommen. Es stellt somit den denkbar einfachsten Fall dar. Dieses Modell ist dem Wachstum einfacher Algen nachempfunden und erzeugt bis auf eine kleine Ungenauigkeit am Anfang eine Wachstumsrate in *Fibonacci-Proportionen*. Eine komplette Fibonacci-Reihe erhält man, wenn man bei gleichbleibenden Produktionsregeln mit Axiom $\omega = B$ beginnt. | Kontextfreie *L-Systeme* können auf verschiedene Arten erweitert werden. Beispielsweise durch die Integration stochastischer Elemente, um wachstumsbeeinflussende Umweltfaktoren zu simulieren. Eine andere Möglichkeiten stellen kontextsensitive *L-Systeme* dar, deren Produktionsregeln immer in Abhängigkeit der benachbarten Zeichen definiert werden und somit deutlich vielschichtiger eingesetzt werden können. | Eine grafische Umsetzung von *L-Systemen* war ursprünglich niemals geplant. Ihr ganzes Potential offenbarten sie allerdings erst im Laufe der 70er Jahre durch die Interpretation ihrer Zeichenketten mittels sog. Turtle-Grafik. Diese basiert auf der Vorstellung einer auf Papier sitzenden Schildkröte, die mit einfachen Befehlen gesteuert werden kann und mit jedem Schritt, den sie sich fortbewegt, einen Strich hinterlässt. Auf diese Weise lassen sich mit sehr einfachen Mitteln alle nur erdenklichen, selbstähnlichen Strukturen realisieren. Von einfachen Fraktalen bis zur Simulation komplex strukturierter Pflanzen. | Abb. 9 zeigt die Genese eines nicht allzu komplizierten Strauchs über vier Generationen hinweg. Eine einzige Produktionsregel sorgt dabei in jeder neuen Generation für die Projektion der Gesamtstruktur aus Generation 1 auf jeden einzelnen Strich der vorhergehenden Generation. Die Struktur des Strauchs verdichtet sich dadurch in rasend schnellem Tempo. Als

8

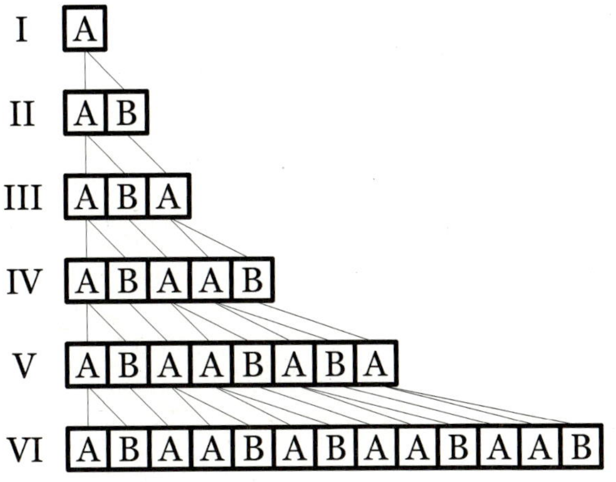

ous ways, for example by integrating stochastic elements in order to simulate growth-influencing environmental factors. Context-sensitive *L-systems* represent another possibility; their production rules are always defined depending on the neighboring symbols and as such can clearly be put to more diverse use. | There were originally never any plans for a graphical implementation of *L-systems*. It was, however, not until in the course of the 1970s that they revealed their full potential through the interpretation of their character-strings by means of the so-called Turtle graphics. These are based on the idea of a turtle sitting on a piece of paper that can be controlled by simple commands and leaves a trace with every step it moves forward. In this way with extremely simple means almost any self-similar structure imaginable can be realized. From simple fractals to the simulation of complex-structured plants. | Fig. 9 shows the genesis of a not too complicated shrub over four generations. In each new generation a single production rule projects the overall structure from the first generation to each individual line of the previous generation. This way the structure of the shrub becomes increasingly dense at an extremely fast rate. As an aesthetically convincing simulation, though, the whole thing is completely unusable so far. | A lack of authenticity in the appearance is, however, easy to remedy without a great deal of effort. Through the well-measured integration of a stochastic component, either at the production rule level or graphical interpretation itself, all those minor deviations necessary to

9

ästhetisch überzeugende Simulation ist das Ganze bislang aber völlig unbrauchbar. | Mangelnde Authentizität der Erscheinung lässt sich jedoch ohne großen Aufwand beheben. Durch die wohldosierte Integration einer stochastischen Komponente, entweder auf Ebene der Produktionsregel oder der grafischen Interpretation selbst, lassen sich all die kleinen Deviationen erzielen, die nötig sind, um dem Strauch ein wenig Leben einzuhauchen. | Abb. 10 zeigt denselben Strauch, diesmal allerdings weitere drei Generationen älter und mit einem Deviations-Spielraum von durchschnittlich 20% grafisch umgesetzt. || Der besondere Reiz, der von *L-Systemen* ausgeht, liegt in der Vielschichtigkeit ihrer Interpretationsmöglichkeiten. Als ursprünglich rein abstrakter Formalismus konzipiert, handelt es sich erstmal um nichts anderes als um strukturelle Verfahrensweisen mit aufeinander bezogenen Symbolen, die natürlichen Wachstums- und Vermehrungsprozessen nachempfunden sind. Das Resultat ist immer eine zu selbstähnlicher Wiederholung tendierende, mehr oder weniger komplexe Zeichenkette. Interpretiert man nun jedes einzelne Zeichen nicht mehr nur als einfachen Strich, so zeigt sich eindrucksvoll, weshalb Ersetzungssysteme mittlerweile auch Einzug ins Grafikdesign oder die Architektur gehalten haben (Abb. 11).

10

wildes FLEISCH Mit *L-Systemen* lässt sich prinzipiell alles Mögliche steuern. Die Leichtigkeit, mit der sich hochkomplexe Selbstähnlichkeit generieren lässt, fordert eine Auseinandersetzung mit der Steuerbarkeit musikalischer Prozesse förmlich heraus. Es verwundert also kaum, dass die letzten Jahre eine Reihe substantieller Werke hervorgebracht haben, die mit Hilfe von *L-Systemen* geschrieben wurden. Insbesondere das Werk von Hanspeter Kyburz verdient hier Erwähnung. Wie bei allen Algorithmen, die nicht primär dazu gedacht sind, zeitliche Prozesse zu steuern, stellt sich als Erstes die Frage, wie die zeitliche Komponente von Musik überhaupt sinnstiftend repräsentiert werden kann. Prinzipiell ist das sog. Mapping, also welcher Teil eines Algorithmus welchen musikalischen Parameter steuert, einer der wichtigsten konzeptionellen Schritte überhaupt für eine funktionierende Repräsentation. Er entscheidet maßgeblich über Gelingen oder Misslingen, denn durch ihn werden die Weichen gestellt, wie sich ein Algorithmus in Musik darstellt bzw. ob er in seiner strukturellen Wirkung überhaupt wahrnehmbar ist. | Zieht man hierbei die Übereinstimmung von Gruppierungsprinzipien optischer und akustischer Informationen in Betracht, dann lässt sich daraus eine interessante Analogie gerade für das Mapping von Zeitparametern ableiten. Würde ein Algorithmus auf visueller Ebene die Skalierung des Raums verändern, so sollte er es in einer musikalischen Repräsentation gleichsam mit der Zeit tun. Verkleinerter Raum gleich verkürzte Zeit und umgekehrt. Somit würden, egal ob auf visueller oder akustischer Ebene, die selben Verarbeitungs- und Gruppierungsmechanismen angesprochen werden. Die algorithmische Struktur wäre somit unabhängig vom Ausbreitungsmedium ideal repräsentiert. | *Wildes FLEISCH* ist ein größer dimensionierter Zyklus von Solowerken mit Elektronik, der sich ausschließlich der Erforschung des musikalischen Potentials von L-Systemen widmet. Er besteht zum jetzigen Zeitpunkt aus drei Stücken, von denen jedes für sich versucht einen anderen Blickwinkel zur Thematik einzunehmen. Allen drei Stücken gemein ist allerdings der Versuch, musikalische Verdichtungs- bzw. Wucherungsprozesse zu inszenieren. Der Titel ist also Programm. Wildes Fleisch ist die umgangssprachliche Bezeichnung für ein sog. Keloid, eine nicht selten angeborene, benigne Hautwucherung. | Im Rahmen dieser Betrachtung bietet sich vor allem der zweite Teil als Studienobjekt an, da er den Versuch unternimmt, mittels eines *L-Systems* einen auf Fibonacci-Proportionen aufbauenden Mensuralkanon zu konstruieren, der sich sukzessiv verdichtet, bis er schließlich zu wuchern beginnt. Die Repräsentation des Algorithmus wird hierbei ausschließlich der Elektronik anvertraut, das Solo-Instrument, in diesem Fall eine Posaune, kann somit vernachlässigt werden. | Ausgangsbasis für alle formalen Überlegungen war das *L-System* auf Abb. 8. Um eine allmähliche strukturelle Verdichtung und nicht nur eine einfache Beschleunigung wahrzunehmen, ist das beständige Fortklingen

breathe a touch of life into the shrub can be achieved. | Fig. 10 shows the same shrub, this time, indeed, three generations older and depicted with deviation leeway of an average 20%. || The special appeal *L-systems* exude lies in the diversity of their interpretation possibilities. Originally conceived as purely abstract formalism they are initially nothing other than structural procedures with symbols relating to each other, which mimic natural growth and propagation processes. The result is always a more or less complex string with a tendency to self-similar repetition. If one no longer interprets each individual symbol just as a simple line it is impressively revealed why substitution systems have now also found their way into graphic design and architecture (Fig. 11).

wildes FLEISCH [wild MEAT] In principle *L-systems* enable all sort of things to be controlled. The ease with which highly-complex self-similarity can be generated, verily calls for an investigation of the controllability of musical processes. As such it is scarcely surprising that the last few years have produced a series of substantial works written with the help of *L-systems*. The work of Hanspeter Kyburz in particular is worthy of mention here. | As with all algorithms not primarily meant to control temporal processes the first question to be posed is how the temporal component in music can be represented at all in a meaningful way. In principle so-called mapping, in other words which part of an algorithms controls which musical parameter, is one of the most important conceptual steps at all for a functioning representation. It plays a decisive role in success or failure, as through it the foundations are laid for how an algorithm can be depicted in music or whether it is at all perceivable in a structural impact. | If in this one considers the congruence of grouping principles for visual and acoustic information, one can derive an interesting analogy precisely for mapping time parameters. Were an algorithm to change the scale of space at the visual level, in musical representation it ought to do so with time, as it were. Downsized space equals shortened time, and vice versa. This way, regardless of whether at a visual or acoustic level, the same processing and grouping mechanisms would be addressed. As such the algorithmic structure would be ideally represented, independent of the distribution medium. | *Wildes FLEISCH* is a larger cycle of solo works with electronics devoted entirely to investigating the musical potential of *L-systems*. At the present time it comprises three pieces, each of which attempts to address a different perspective on the topic. However, what all three pieces have in common is the attempt to orchestrate processes of musical densification and proliferation. As such the title says it all. Wild meat is the vernacular for a so-called keloid, a benign excessive growth of skin that is not seldom hereditary. | Within the scope of this study, the second part in

jeder einzelnen Generation jedoch von größter Bedeutung. Ein Mensuralkanon würde also einer idealtypischen Repräsentation eines derartigen Prozesses sehr nahe kommen. Demnach wurde es notwendig, den ursprünglichen Algorithmus geringfügig zu modifizieren. Aus dem vormals kontextfreien wurde ein kontextsensitives *L-System* gewonnen, das ausgiebig Gebrauch von Verzweigungsstrukturen macht und erst damit den strukturellen Anforderungen eines Mensuralkanons genügt (Abb. 12). | Nur auf diese Weise lässt sich ein ständiges Nachwachsen der unteren Schichten und somit ein sukzessiver Aufbau des Kanons gewährleisten. Auf visueller Ebene entspräche dies ebenfalls, ganz ähnlich einem Romanesco Blumenkohl, permanent nachwachsenden *Fibonacci-Spiralen*. | Ein weiterer konstitutiver Gestaltungsfaktor findet sich bereits weiter oben erklärt. Die Korrelation zwischen räumlicher und zeitlicher Verdichtung. Konkret auf dieses Stück angewandt bedeutet dies, dass jeder neuen Generation nur dieselbe Zeitspanne zur Verfügung steht wie Generation 1. Daraus resultiert eine nach und nach aufbauende Tempo-Proportion von 1:2:3:5:8:13 etc. | Nüchtern betrachtet hat das gesamte Stück also nur zwei unterschiedliche Formteile, die allerdings in ständig wechselnden Konstellationen und Tempi zueinander stehen. | Ein Blick auf die Mikrostruktur, also die inhaltliche Gestaltung der beiden Formteile, offenbart eine gänzlich andere, parametrische Herangehensweise. Die zeitliche Struktur der Einzelereignisse wurde von der Aminosäuren-Sequenz menschlicher Keloid-DNS abgeleitet. Jeder der vier Aminosäuren wurde mehr oder weniger willkürlich eine bestimmte Zeitdauer zugeordnet. Die unterschiedlichen Dauern stehen untereinander aber selbstverständlich ebenfalls im Verhältnis von 1:2:3:5. Strukturiert durch einen winzig kleinen Abschnitt DNS, wurden somit apriori sämtliche Einsatzabstände der musikalischen Ereignisse definiert. | Das musikalische Material besteht ausschließlich aus gesampelten und beinahe bis zur Unkenntlichkeit verfremdeten Posaunenklängen, die für jede Temposchicht des Kanons auf einen bestimmten Frequenzbereich beschränkt bleiben. Je schneller die Schicht, desto höher auch der Frequenzbereich. Mit zunehmend struktureller Verdichtung weitet sich also auch der gesamte Frequenzraum des Stücks. | Abb. 13 zeigt einen Gesamtüberblick der ersten sechs Kanon-Generationen und einen Ausschnitt aller Stimmen zu Beginn von Generation 5. Deutlich lässt sich erkennen, wie sich die Makrostruktur eins zu eins auf die Mikrostruktur projiziert. Die einzelnen Ereignisse verhalten sich also genauso zueinander, wie es auch die Formteile tun, in denen sie sich jeweils befinden. | Abb. 14 ist ein Versuch, über grafische Momentaufnahmen der Generationen 2–6, diesen allmählichen Verdichtungsprozesssichtbar zu machen. Die x-Achse repräsentiert dabei Zeit, die y-Achse Frequenz. Jeder einzelne Kreis symbolisiert ein klangliches Ereigniss, wobei der Kreisdurchmesser für die jeweilige Lautstärke steht. Die Zeitspanne ist dabei in

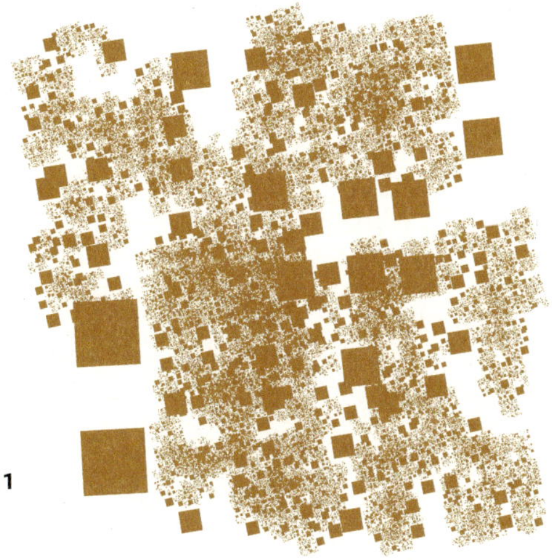

11

particular is suitable as a study object, as it attempts by means of an *L-system* to construct a mensuration canon based on Fibonacci proportions, which gradually becomes concentrated until it ultimately begins to proliferate. Here, representation of the algorithm is entrusted exclusively to the electronics, and thus the solo instrument, in this case a trombone, can be disregarded. | The starting point for all formal considerations was the *L-system* in Fig. 8. In order to perceive gradual structural densification and not just simple acceleration, the constant existence of each individual generation is, however, of immense importance. As such a mensuration canon would come very close to an ideal type of representation of a process of this nature. Accordingly it was necessary to modify the original algorithm slightly. From the previously context-free *L-system* a context-sensitive one was obtained that makes extensive use of branching structures and only then meets the structural requirements of a mensuration canon (Fig. 12). | Only in this way can the re-growth of the lower layers and thus the gradual buildup of the canon be guaranteed. On a visual level this would likewise correspond, just like a Romanesco cauliflower, to permanently re-growing Fibonacci spirals. | Another constitutive formal factor is explained further up. The correlation between spatial and temporal densification. Applied in concrete terms to this piece this means that each new generation only has the same time span available as the first generation. This results in a gradually evolving tempo proportion of 1:2:3:5:8:13 etc. | Ob-

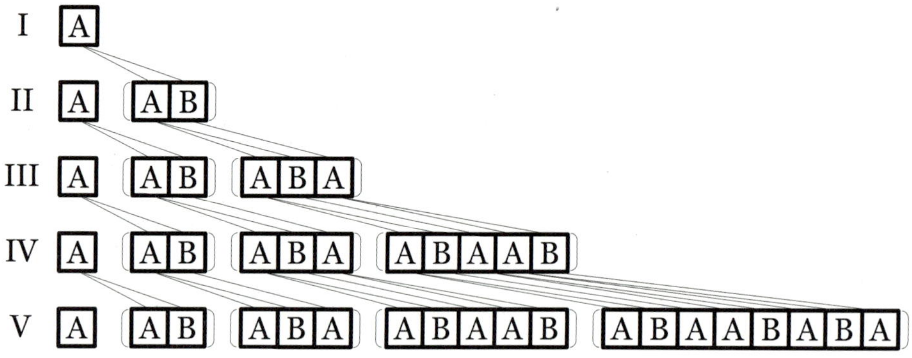

12

allen Grafiken identisch. | Die musikalische Ereignisdichte ist nach der vierten Generation bereits derart hoch, dass die einzelnen Ereignisse nicht mehr isoliert wahrgenommen werden können. Der Eindruck unkontrollierten Wucherns beginnt zu entstehen. Es bilden sich neue, übergeordnete Strukturen heraus, denn das Ohr hat keine andere Wahl mehr, als andere, gröbere Gruppierungsmechanismen heranzuziehen. Ab diesem Zeitpunkt beginnt unterstützend auch eine weitere, stochastische Komponente des Algorithmus zu greifen, die mit 25-prozentiger Wahrscheinlichkeit in den höheren Ebenen akustische Ereignisse durch Pausen ersetzt, um einer allzu statischen Ausbreitung entgegenzuwirken. | Das zeitliche Auflösungsvermögen des Gehörs ist zwar dem aller anderen Sinne deutlich überlegen, dennoch sind seine Kapazitäten ab einem bestimmten Punkt erschöpft. Ab einer Ereignisdichte von ca. 16 - 20 Ereignissen pro Sekunde beginnt jegliche Differenzierungsmöglichkeit zu versagen und das gehörte Resultat verwandelt sich in einen statischen Einzelton mit der Frequenz der Ereignisdichte. Der zeitlichen Ausdehnung des gesamten Prozesses sind somit physiologische Grenzen gesetzt und das Stück bricht nach der siebten Generation einfach ab. Es kommt gleichsam zu einer zeitlichen Implosion. || Würde man nun tatsächlich davon ausgehen, ein Musikstück könnte das Wachstum von Organismen hörbar machen, so stellt sich auch automatisch die Frage nach deren Aussehen. Abschließend, und selbstverständlich ohne Gewähr, drei Versuche einer grafischen Deutung (Abb. 15).

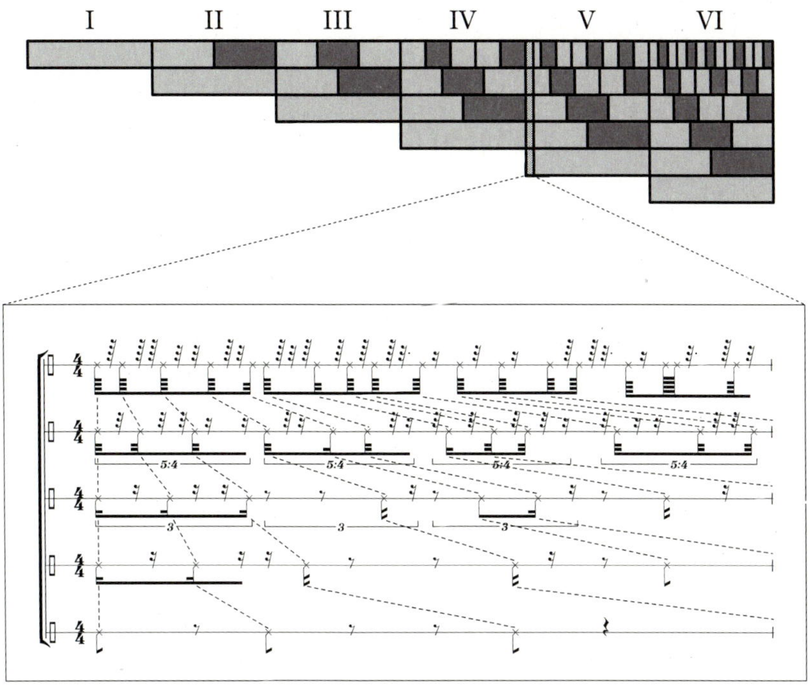

13

served clinically, the entire piece only has two different sections which, however, are in constantly changing constellations and times to each other. | A glance at the microstructure, in other words the shaping of the musical contents of both sections reveals a totally different, parametric approach. The temporal structure of the individual events was derived from the amino acid sequence of human keloid DNA. Each of the four amino acids was more or less arbitrarily allocated a particular time interval. These different durations, however, are of course likewise in a ratio of 1 : 2 : 3 : 5 to each other. As such, structured by a minute section of DNA, the entry delay [4] of each musical event was defined a priori. | The musical material consists exclusively of sampled trombone sounds electronically manipulated almost to the point of being unrecognizable, which for each tempo layer of the canon remain restricted to a certain frequency-area. The faster the layer, the higher the frequency as well. With increasing structural densification the entire frequency range of the piece thus is extended. | Fig. 13 shows an overall view of the first six canon generations and an excerpt of all voices at the beginning of the fifth generation. It is clearly recognizable how the macrostructure is projected one-to-one on the microstructure. The individual events behave exactly the same way to each other as do the particular sections in which they are to be found. | Fig. 14 is an attempt to make this gradual densification process visible by means of graphical snapshots of generations 2 - 6. The x-axis represents time, the y-axis frequency.

4 The onset time of an event, calculated as the time passed since the previous event's onset.

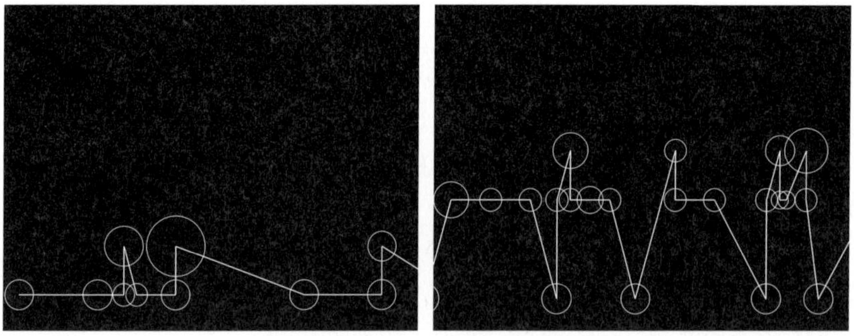

14

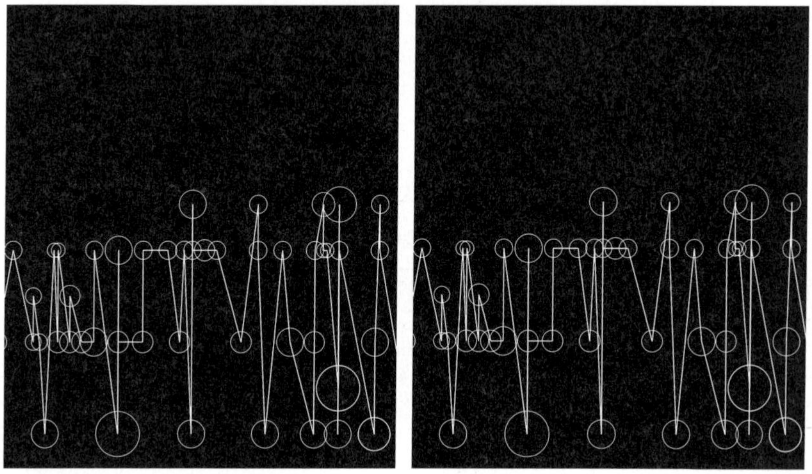

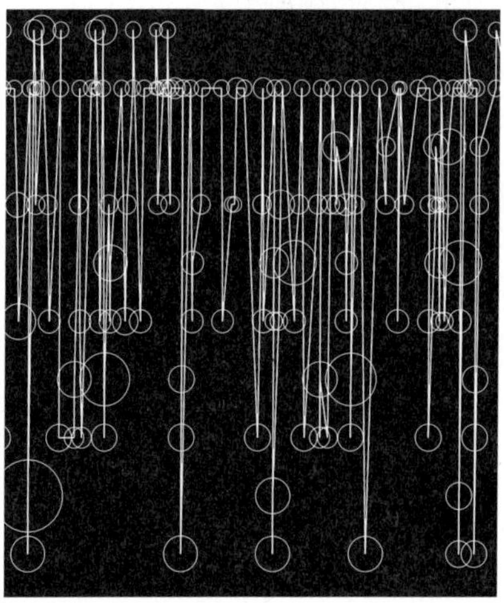

Each individual circle symbolizes a sound event, with the diameter of the circle standing for the particular loudness. The time span in each case is identical. | After the fourth generation the concentration of musical events is already so intense that the individual events can no longer be perceived in isolation. An impression of uncontrolled proliferation begins to emerge. New, overriding structures form, as the ear no longer has any choice other than to draw on different, coarser grouping mechanisms. As of this point in time another stochastic component of the algorithm begins to impact in a supporting function, which with 25 percent probability replaces acoustic events in the higher layers with pauses in order to counter an all too static spread. | Though the hearing's temporal resolution is considerably superior to that of all the other senses, as of a certain point its capacities are nonetheless exhausted. As of a concentration of approx. 16 - 20 events per second any possibility of differentiation begins to fail and the result heard morphs into a single static pitch with the frequency of the concentration of events. As such there are physiological boundaries to the temporal expanse of the entire process and after the seventh generation the piece simply comes to a halt. There is, as it were, a temporal Implosion. || If one were now to actually assume that a piece of music could make the growth of organisms audible, the question automatically arises as to their appearance. In conclusion and, it goes without saying, with no guarantee, three attempts at a graphic interpretation (Fig. 15).

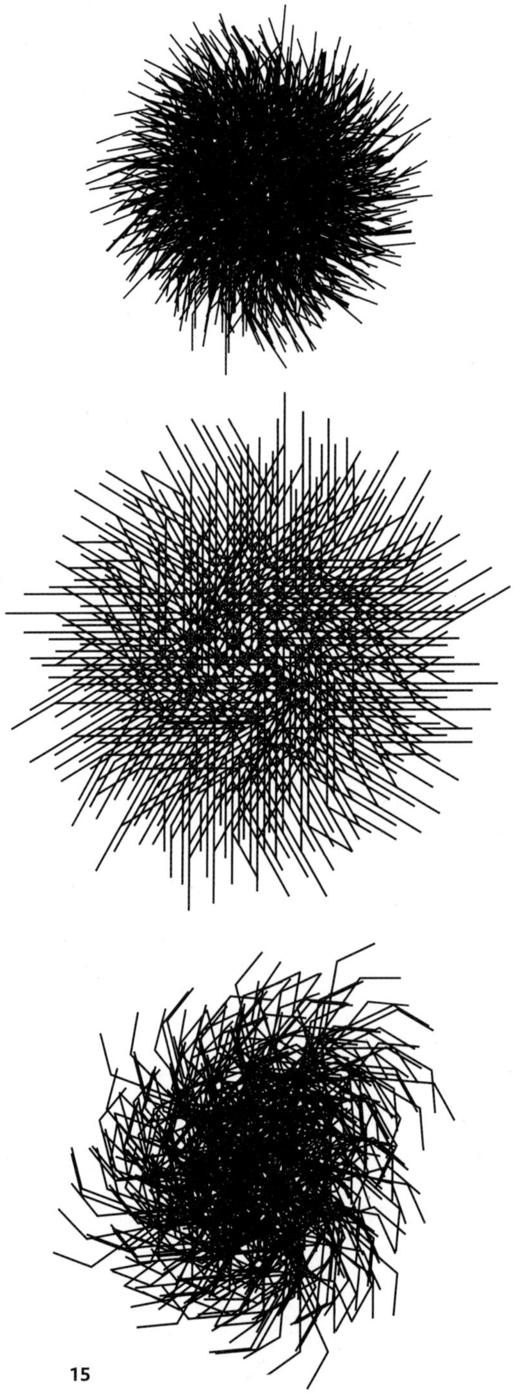

15

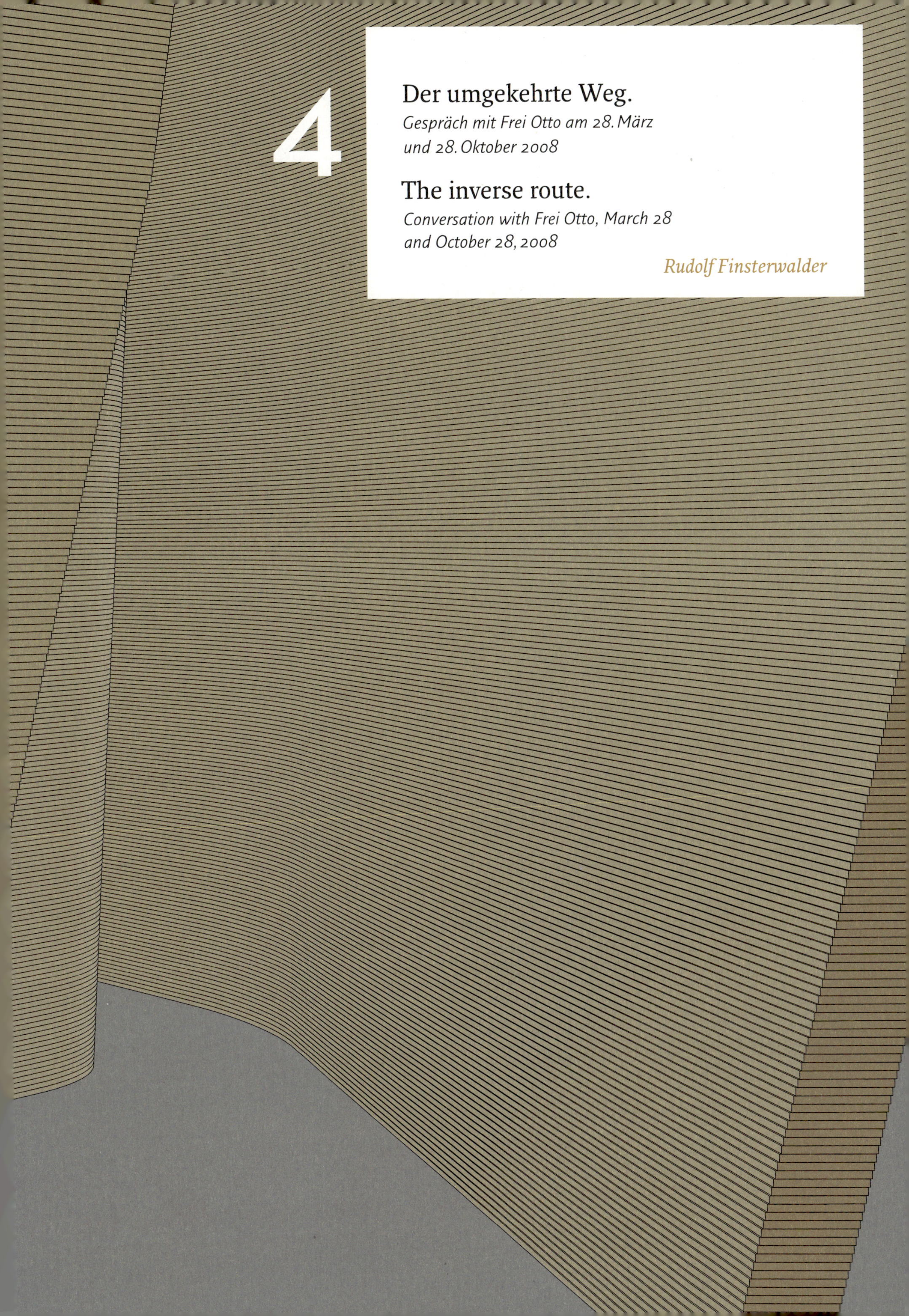

4

Der umgekehrte Weg.

Gespräch mit Frei Otto am 28. März und 28. Oktober 2008

The inverse route.

Conversation with Frei Otto, March 28 and October 28, 2008

Rudolf Finsterwalder

Ausgangspunkt für das Gespräch war ein Fragenkatalog aus 14 Fragen:

1. Wie würdest du den Einfluss von Natur auf deine Arbeit beschreiben und wie ist dein Naturbegriff?
2. Eines deiner Bücher trägt den Titel »Der umgekehrte Weg«, – kannst du mir erklären, was du darunter verstehst?
3. Wie würdest du das Verhältnis eines schaffenden Menschen zur Natur beschreiben?
4. Inwieweit siehst du in der Architektur eine Möglichkeit zum ganzheitlichen Denken und Planen, so wie die Natur ganzheitliche Systeme und nicht isolierte Einzellösungen hervorbringt?
5. Inwieweit hast du dich in deiner Arbeit von der Natur als Lehrmeisterin emanzipiert?
6. Was bedeutet der Selbstfindungsprozess für die Form in deinen Arbeiten?
7. Hast du Formen bewusst gesucht oder nur »zufällig« gefunden?
8. Hat ein Wille zur Form in deinen Augen eine Bedeutung in der Architektur?
9. Was bedeuten die Begriffe Schönheit und Ästhetik in deinen Augen für die Architektur?
10. Hegel sah in der Symmetrie und Harmonie der Natur das Maximum an Schönheit, Adorno im Gegensatz dazu in einer »zufälligen« Störung der Perfektion.
11. Wo wird an dein Lebenswerk angeknüpft?
12. Siehst du dich mehr als wissenschaftlichen Grundlagenforscher oder als planenden Architekten?
13. Wo siehst du die Aufgabe der Architektur heute?
14. Warum, glaubst du, wird deine Arbeit nur von wenigen Kollegen weitergeführt?

The starting point for the conversation was a set of 14 questions:

1. How would you describe the influence of nature on your work and what concept do you have of nature?
2. One of your books bears the title "Der umgekehrte Weg" (the inverse route) – can you please explain what this means to you?
3. How would you describe the relationship of a creative person to nature?
4. To what extent do you see a possibility for holistic thinking and planning in architecture in the way that nature produces holistic systems and not isolated individual solutions?
5. To what extent have you liberated yourself in your work from nature as master?
6. What does the self discovery process mean for the form of your works?
7. Did you consciously seek out shapes or just find them by chance?
8. Do you consider the will to create form is important in architecture?
9. What, in your opinion, do the terms beauty and aesthetics mean in architecture?
10. Hegel saw ultimate beauty in the symmetry and harmony of nature, Adorno on the other hand in a "chance" disruption of perfection.
11. What continues your life work?
12. Do you consider yourself more a basic scientific researcher or a planning architect?
13. What do you consider architecture's role today?
14. Why so you think your work is only carried on by a few colleagues?

R. Finsterwalder In der Villa Massimo in Rom habe ich mit meinen Studenten einen Pavillon selbst gebaut. Mit einfachen Mitteln einen Pavillon zu bauen, der Platz für 60 Personen bietet, war das Ziel des Projektes. An einer Plattform aus Holzbalken werden Flacheisen 30 x 10 mm befestigt. Jeweils zwei gegenüberliegende 6m-Stangen werden nach innen gebogen und mit Schellen verbunden, wodurch die Konstruktion ihre Stabilität erhält. Das sich so ergebende Skelett wird mit einer dehnbaren Kunststofffolie, wie sie zur Verpackung in der Industrie verwendet wird, umwickelt. (Abb. 1)

F. Otto Das sieht beinahe aus wie ein aufliegender Tropfen. Die Form kommt heraus, wenn man eine Latexfolie mit Wasser füllt. Wenn man aber diese Form als luftgefüllte Hülle baut, kann man nicht verhindern, dass hier sowohl Zugkräfte als auch Biege- und Knickkräfte auftreten, die berücksichtigt werden müssen und die scheinbare Eleganz des flüssigen Modells stören. (Abb. 2)

R. Finsterwalder Welche Erkenntnisse gab es aus der Beschäftigung mit der lebenden und der nicht lebenden Natur?

F. Otto Uns interessieren Prozesse, die mit der Gestaltentstehung in der nicht lebenden und in der lebenden Natur zu tun haben, die von Schwerkraft, Festigkeit und tragfähigen Konstruktionen abhängen und zusammenhängen, wobei unser Augenmerk gleichen und ähnlichen Entwicklungsprozessen galt, die in der

1

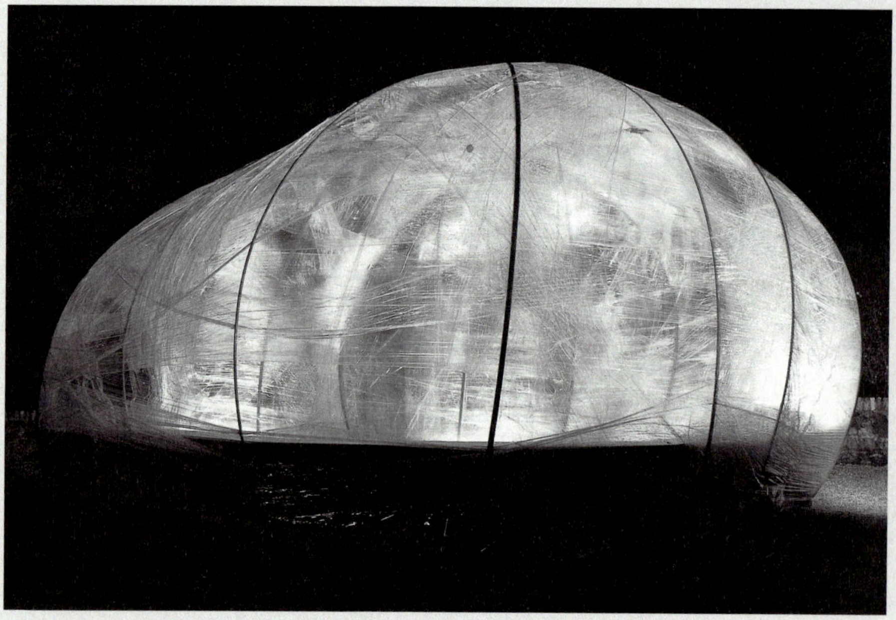

Abb. 1 Polyethylene, Pavillon Finsterwalder

Fig. 1 Polyethylene, Pavilion Finsterwalder

R. Finsterwalder In Villa Massimo in Rome I built a pavilion together with my students. The aim of the project was to use simple means to build a pavilion with enough space for 60 people. Panels of flat steel, 30 x 10 mm, were mounted on a platform made of wooden beams. Two opposing 6 m rods were bent inwards and joined with brackets to lend the construction stability. The skeleton formed in this way was swathed in a stretchable plastic foil such as is used for industrial packaging. (Fig. 1)

F. Otto It almost looks as if it is covered with an overlying drop. The shape emerges when a latex foil is filled with water. But when you construct this form as an air-filled sheath you cannot prevent tensile forces as well as bending and buckling forces arising: these have to be taken into account and disturb the apparent elegance of the fluid model. (Fig. 2)

R. Finsterwalder What findings did you arrive at after investigating living and non-living nature?

F. Otto We are interested in the processes related to the creation of shape in non-living and living nature which are dependent on and linked to gravity, solidity, and supporting constructions, whereby we focused on identical and similar development processes observable in living and non-living nature. We found a range of comparable processes which could by all means serve as thought models for the most important forms and structures of living and non-living objects.

2

Abb. 2 Aufliegender Tropfen

Fig. 2 Laying drop

nicht lebenden und in der lebenden Natur beobachtbar sind. Und wir fanden eine Reihe von vergleichbaren Prozessen, die durchaus als Gedankenmodelle für die bekanntesten Gestalten und Strukturen nicht lebender und lebender Objekte dienen können.

R. Finsterwalder Walter Bauersfeld hat zusammen mit Ulrich Finsterwalder für das Zeiss-Planetarium in Jena eine Konstruktion, eine Gitterschale wie die geodätische Kuppel Buckminster Fullers erdacht, nur 30 Jahre früher. (Abb. 3)

F. Otto Bei meinem Studium an der TU Berlin 1948–52 war die Halle in den Baukonstruktionsvorlesungen schon ein Thema, das mich besonders interessierte. Später habe ich die Jenakuppel, die auch Fuller und Candela kannten, im IL 10 veröffentlicht.

R. Finsterwalder Die Kugel war wohl für die Projektion gewollt, außen war sie keine Notwendigkeit.

F. Otto Die Kuppel wurde Mitte der 20er Jahre gebaut. Der Auftrag war, die steife Innenfläche einer perfekten Kugelschale zu bauen. Eine Kugelschale ist nicht die perfekteste Konstruktion. Die annähernd perfekte Kugelform entsteht in der nicht lebenden Natur nur im Bereich von Flüssigkeit begrenzenden Membranen, wie z. B. beim frei schwebenden Wassertropfen oder dem Luftbläschen im Wasser. | Die Jenaer Gitterschale ist ein hoch komplexes menschengemachtes Konstrukt, das aber Verwandtschaft zu den Radiolarien in der Natur hat. | Die Jena Kuppel wie auch die Kuppel von Bucky Fuller symbolisierten ein technisches Produkt mit klarer Aufgabe, während bei Produkten lebender Natur selten Parallelen zur Erfüllung technischer Aufgaben gefunden werden können. | Ein Lebewesen kann nur dann weiter- oder überleben, wenn es mit der ihm gegebenen Gestalt, die meist nur wenig veränderbar ist, zurechtkommt.

3a

3b

3c

3d

Abb.3a–d Zeiss-Planetarium, Jena

Fig.3a–d Zeiss-Planetarium, Jena

Kugelförmige Radiolarien sind nicht als Vorbild für Planetarien entstanden, auch wenn sie von Mathematikern dafür benutzt werden.

R. Finsterwalder Inwiefern hast du dich in deiner Arbeit von der Natur als Lehrmeisterin emanzipiert, wie z. B. Frank Lloyd Wright? Für ihn war »form follows nature«, er nannte das »organic architecture«, ein wichtiger Leitsatz. Er wollte die Ursprünge verstehen und nicht nur nachahmen.

F. Otto Die »organic architecture« war für Wright vor allem die Idee des ganzheitlichen Denkens, das sich auf alle Bereiche des Lebens bezog.

R. Finsterwalder Was ist dein Naturverständnis, dein Verhältnis zur Natur?

F. Otto Ich war immer auf der beobachtenden Seite und habe dabei technische Erfindungen in der Natur wiedergefunden. Natur ist ja äußerst komplex. Die Entwicklung der lebenden Natur erfährt seit 4,5 Milliarden Jahren bis heute tägliche Veränderungen, die nicht immer als Optimierungen angesehen werden können. Ich glaube nicht, dass die lebenden Objekte in sich optimal sind. Manche glauben, dass sie sich ganz bestimmten Aufgaben manchmal angenähert haben. Doch wenn man sie einfach nachmacht, steht die Lösung oft nicht in Beziehung zur ursprünglichen Aufgabe. | Wir haben einen Sonderforschungsbereich gegründet, um solchen Phänomenen nachzugehen.

R. Finsterwalder Ist der Mensch gezwungen zu vereinfachen und eigene Wege zu finden, weil er diese über Jahrmillionen gewachsene Komplexität nicht erreichen kann?

F. Otto Nein, die menschliche Technik fängt jetzt erst richtig an. Für mich ist die Geschichte des Konstruierens wichtiger als die Architekturgeschichte. Architektur ist vom Menschen durch Konstruktion in Form gebrachtes Material. (Abb. 4)

R. Finsterwalder Die Ursprünge der Architektur und des Konstruierens waren nicht getrennt.

F. Otto Wir glauben, dass es so war. Sehr früh war der Wille des Menschen zur Form da. Bei der Realisierung entstand dann der Konflikt mit der Herstellung und den Materialien. Später kam das Wissen über das Material, das sowohl aus der nicht lebenden als auch aus der lebenden Natur genommen wurde. | Es gibt bei uns eine ganz präzise Unterscheidung zwischen nicht lebend und tot. Tot ist das, was mal gelebt hat, und behält die Form, in seinem mehr oder minder langen Leben, auch wenn es nur eine Sekunde gelebt hat. Diese sauberen Unterscheidungen werden von den wenigsten Kollegen gemacht. Viele sprechen von toter Natur, das ist falsch, manche sprechen von belebter Natur, mir ist auch das zu ungenau. | Klarer waren für mich die Begriffe lebende und nicht lebende Natur. Zwischen ihnen es gibt nur wenige Übergänge, wenn überhaupt. | Beim Beginn unserer Arbeiten waren die Bildungsprozesse bei der Entstehung von Formen und Konstruktionen im Bereich der nicht lebenden Natur viel weniger erforscht als die in der lebenden Natur. Deshalb haben wir unsere Forschungen zunächst hauptsächlich auf Objekte der nicht lebenden Natur konzentriert. Zum Beispiel hat ein Wassertropfen – ob aufliegend oder hängend – eine bestimmte Form. Wenn er größer wird, tropft er ab und ist dann nur in dieser Kugelform existent. (Abb. 5) | Die Formen der nicht lebenden Natur findet man auch in der lebenden Natur, die alle Entstehungsprozesse der Formen der nicht lebenden Natur nutzt. Keine Maus gräbt eine

R. Finsterwalder Walter Bauersfeld together with Ulrich Finsterwalder conceived a construction, a latticed shell like Buckminster Fuller's geodetic dome, a mere 30 years earlier. (Fig. 3)

F. Otto When I was studying at the Technical University in Berlin from 1948 to 1952, the hall was one of the subjects in our structural design lectures which really interested me. Later, I published the Jena Dome, which Fuller and Candela also knew, in IL 10.

R. Finsterwalder The sphere was desired for projection but was not necessary on the outside.

F. Otto The dome was built in the mid 1920s. The brief was to build the rigid inner surface of a perfect spherical shell. A spherical shell is not the most perfect construction. The almost perfect spherical form exists in non-living nature only in membranes confining fluids, for example in floating drops of water or air bubbles in water. The latticed shell in Jena is a highly complex, man-made construct related to radiolarians in nature. The Jena dome, like Bucky Fuller's, symbolize a tcchnical product with a clear purpose, while in living nature parallels for fulfilling technical purposes are seldom found. | A creature can only continue to live or survive if it can manage with its given form, which is usually not very changeable. Spherical radiolarians were not created as models for planetariums, even if mathematicians use them as such.

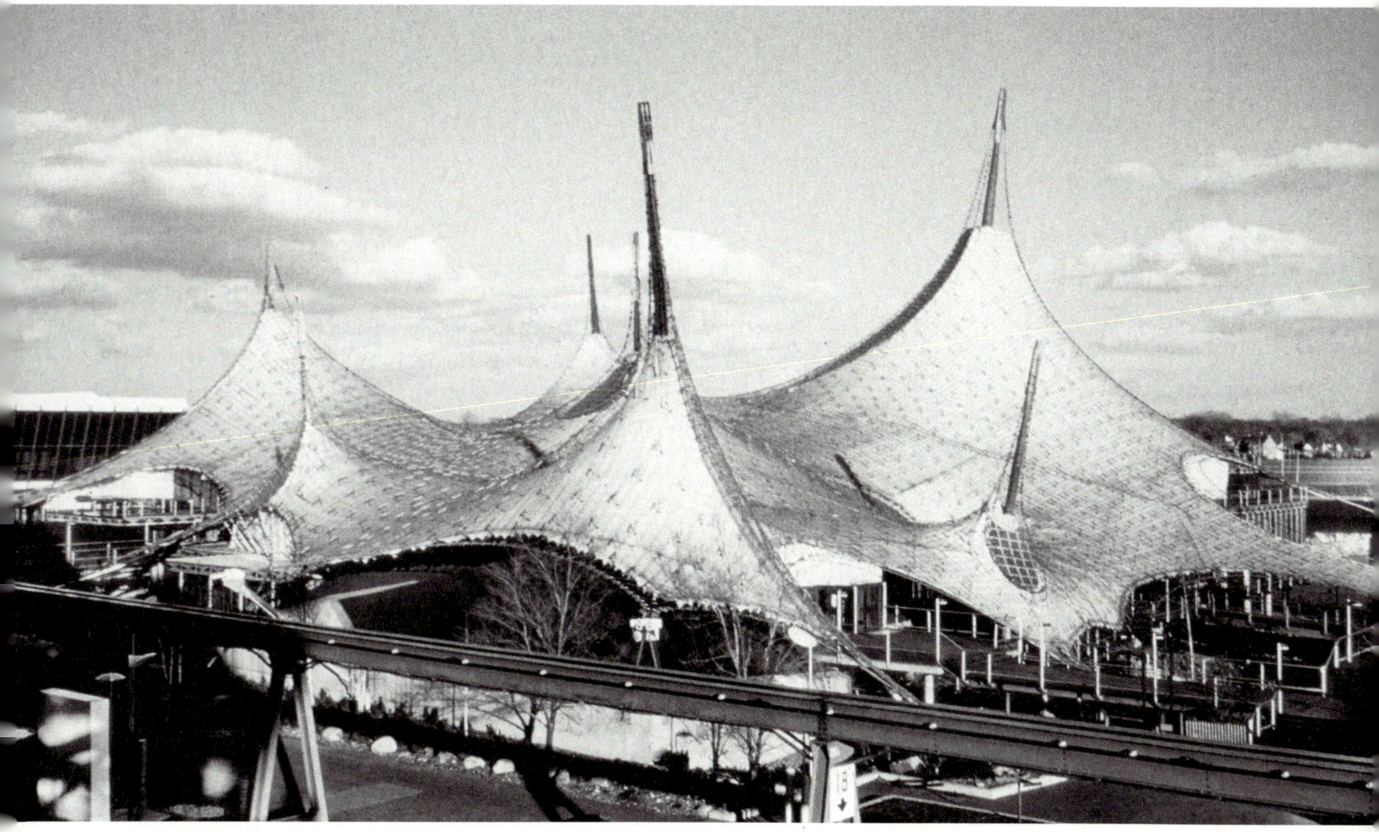

4

Abb.4 Expo Pavillon Montreal

Fig.4 Expo Pavilion Montreal

Höhle, die dauernd einfällt. Sie nutzt die Bildungsprozesse der nicht lebenden Natur. | Ein Selbstbildungsprozess, wie er in der nicht lebenden Natur vorkommt, ist z. B. eine Blutansammlung im Binde- oder Knochengewebe von Lebewesen. Die Blutteile organisieren sich unter Innendruck. Wenn jetzt zusätzliche Kräfte wirken, können die Einflüsse pulsierender oder statischer Kräfte eine neue, möglicherweise tragfähigere Form entstehen lassen. Diese Kräfte fördern Knochenentstehung zunächst als schaumartige Strukturen (Spongiosa). | In einem Schaumpaket gleicht keine Blase 100% exakt der anderen, das ist nicht möglich, weil die Schaumblasen nur unter gleichem Winkel zusammenkommen können, das sind 120°. | Wenn die Blasen nur unter diesem Winkel von ca.108° zusammenkommen können, der Winkel ist stabil, dann kommen sie nicht formgleich zusammen, sondern jede Form ist ein bisschen anders, sie zwängen sich hin. Deshalb kann man auch z. B. bei Rasierschaum dauernde Bewegungen feststellen, es ist eine Art mobiles Gleichgewicht. Sie arrangieren sich, das ist dieses Selbstarrangement. (Abb. 6)

R. Finsterwalder Du hast so ephemere Dinge wie Wassertropfen untersucht und daraus Erkenntnisse für Bauwerke und Konstruktionen gewonnen.

F. Otto Ich habe mir Gedanken darüber gemacht, dass nicht nur die Oberflächenspannung eine Form bestimmt, sondern dass es vor allem echte Netze sind, die die Gestalt in der lebenden Natur bewirken.

5

Abb.5 Wassertropfen

Fig.5 Water drop

R. Finsterwalder To what extent have you liberated yourself in your work from nature as master, like for example, Frank Lloyd Wright for whom "form follows nature" — he called it "organic architecture" — was an important guideline. He wanted to understand origins and not just copy.

F. Otto For Wright "organic architecture" was predominantly the idea of holistic thinking which was related to all areas of life.

R. Finsterwalder What is your idea of and relationship to nature?

F. Otto I have always been someone who likes to observe, and in so doing have come across technical discoveries in nature. Nature is extremely complex. For 4.5 billion years living nature has been experiencing daily changes which cannot always be seen as improvements. I do not believe that living objects are perfect in themselves. Some people believe that they have at times gradually adapted to very specific tasks. If you simply copy them, however, the solution is often not related to the original task. | We have established a special research department to study such phenomena.

R. Finsterwalder Is mankind compelled to simplify and find his own way because he cannot achieve this complexity, which has developed over millions of years?

F. Otto No, human technology is really only starting now. For me the history of construction is more important than architectural history. Architecture is material formed by man through construction. (Fig. 4)

R. Finsterwalder The origins of architecture and construction were not separate.

F. Otto We believe that is how it was. Mankind had the will to form very early on. Along with the realization came conflict with regard to production and materials. Knowledge about the materials came later on, derived from both living and non-living nature. | We make a very fine distinction between not living and dead, dead refers to something that has lived and maintains its form in its life, long or short, even if it only lived just for a second. These clear distinctions are made by only very few fellow colleagues. Many speak of dead nature, this is wrong, some speak of animated nature but for me this is also too imprecise. | I consider the terms living and non-living nature to be clearer. Between them there are only few transitions, if any at all. | When we started our works the formative processes in the creation of forms and constructions in the realm of non-living nature were much less researched than those in living nature. As such we initially focused predominantly on objects from non-living nature. For example, a water drop — whether lying on something or hanging — has a certain form. When it becomes larger it drops off and then exists only in a spherical form. (Fig. 5) | The forms in non-living nature can also be found in living nature which uses all the genesis processes of forms in non-living nature. No mouse would build a hole which always collapses. It uses the forming processes of non-living nature. An example of a self-formation process such as occurs in non-living nature is the accumulation of blood in the connective or bone tissue of a creature. The blood components organize themselves under internal pressure. Now if additional forces take effect, the influences of pulsating or static forces could give rise to a new, possibly more stable form. These forces promote bone formation, initially as foam-like structures (spongiosa). | In a foam package no two bubbles are exactly the same, this is not possible as the foam bubbles can only meet at the same angle, that is, 120°. When the bubbles meet

Eine der vielen Fragen, die sich heute stellen, ist die, ob die DNA dieses Netz direkt ist oder nur ein Element dieses Netzes. Diese Gedanken habe ich in einem kleinen Buch skizziert (»Das Netz der lebenden Gestalt«)[1]. | Erst wenn man sich mit Konstruktionen, also zusammengesetzten Strukturen, intensiv beschäftigt hat, kann man Entstehungsprozesse in nicht lebender und auch in lebender Natur erkennen, untersuchen und vergleichen. Ohne die Selbstbildungsprozesse der nicht lebenden Natur ist auch die lebende Natur nicht denkbar. | Die wissenschaftliche Erkenntnis, dass es möglich sein kann, dass Leben ohne Reproduktion aus Leben auch heute noch entstehen kann, ist nicht von der Hand zu weisen. Allerdings hätte es wenig Überlebenschancen. Diese Meinung vertritt z. B. der Paläontologe Seilacher.

R. Finsterwalder Wenn die Theorie stimmt, hat dieser Prozess irgendwann eingesetzt. War er irgendwann zu Ende oder besteht er weiter?

F. Otto Der Prozess hat irgendwann eingesetzt und die irdische Umwelt ist dazu geeignet, dass er bis heute fortbesteht. Manchmal entsteht eine neue Lebensvariante durch eine Veränderung oder einen Fehler im Netz.

1 »Das Netz der lebenden Gestalt« von Frei Otto erschien 2006 im Verlag Ulrich Keicher.

R. Finsterwalder Der, der zuerst da war, ist nicht unbedingt der Optimalere.

6

Abb.6 Knochenquerschnitt

Fig.6 Bone cross section

below an angle of approx. 108°, The angle is stable and they do not meet with the same form, rather each form is slightly different, they squeeze together. This is also the reason why you can detect constant motion e.g. in shaving foam, it is a sort of mobile equilibrium. The bubbles arrange themselves, this is self-arrangement. (Fig. 6)

R. Finsterwalder You have examined such ephemeral things as water drops and in so doing acquired knowledge which can be used in buildings and construction.

F. Otto I have thought about the fact that it is not only the surface tension which determines a form; it is primarily real meshes that bring about shapes in living nature. One of the many questions which arise today is whether DNA directly constitutes this network or is just a component of this mesh. I have outlined these thoughts in a small book "Das Netz der lebenden Gestalt" (The mesh of the living form)[1]. | You can only recognize, examine and compare origination processes in non-living and living nature once you have investigated constructions or compound structures in depth. Living nature is not conceivable without the self formation processes in non-living nature. The scientific finding that it is possible even today for life to arise from life without reproduction cannot be dismissed. It would, however, have little chance of

1 "Das Netz der lebenden Gestalt" from Frei Otto, published in 2006 by Publisher Ulrich Keicher.

7

Abb.7 Schneckenschale

Fig.7 Seashell

F. Otto Bestehende Lebewesen haben fast alle ein intaktes, funktionierendes Immunsystem. Lebewesen im sichtbaren Bereich sind einfacher zu erforschen. Das Leben im Bereich der Nanobiologie ist noch weitestgehend unerforscht. Um herauszufinden, wie sich Lebewesen den Umweltbedingungen anpassen, braucht man das Wissen über die Umbildungs- und Selbstbildungsprozesse sowohl der lebenden als auch der nicht lebenden Natur. | Mit den Studien über die Entstehung eines Wassertropfens, eines möglichen Vorläufers des Lebens, haben wir die Suche nach Erklärungsmodellen zur biologischen Gestaltentstehung begonnen. Komplexe Dinge wie eine Muschelschale sind viel schwieriger zu untersuchen. | Adolf Seilacher legte mir zwei Schneckenschalen auf den Tisch und wir überlegten gemeinsam, ob man durch mechanische Prüfungen herausfinden könne, welche der beiden lebensfähiger sei, wobei er vermutete, dass eine der beiden Arten in der Kreidezeit nur deshalb ausgestorben sei, weil die Schale nicht so gut ist. (Abb. 7)

R. Finsterwalder Wie ist so etwas entstanden?

F. Otto Wir fanden keine lebensnahe Prüfmethode zur Beantwortung dieser Frage. Es gab keine Hinweise auf die Entstehungsgeschichte und auf lebensbedrohende Gefahren. | Eine Antwort haben wir gefunden. Die Erhärtung dieser Objekte ist irgendwann erfolgt, als sie wahrscheinlich schon hoch entwickelt waren, nämlich als Weichtiere. Sie zeigen an ihrer Form ihre Entwicklung und ihre Herkunft. | Ein Luftballon könnte die Vorentwicklung eines starren Luftschiffes gewesen sein, das nur eine Zeit lang technisch effektiv war. (Abb. 8) | Die lebende Natur entstand im gewichtslosen Bereich, im Wasser. Es gab eine Entwicklung in der Entstehung der weichen Lebewesen, bei der auch harte Teile erzeugt wurden. Die Entstehung der weichen Lebewesen ist älter als die der harten und ohne die Entstehungsgeschichte der weichen Lebewesen sind solche mit teilweisen Erhärtungen nicht denkbar. | Es gibt kein Lebewesen, das total hart ist, nicht einmal eine Auster, die durch viel Materialansammlung statisch am Meeresboden liegt. Der lebende Teil aber ist weich. Das gesamte Leben ist eigentlich vom Ursprung weich und mobil. Aber Mobilität ist nicht unbedingt mit Weichheit gekoppelt, wie man an der Auster sieht. | Mit den einfachsten nicht lebenden Elementen kann man einen Zugang zu Entstehungsprozessen in der lebenden Natur finden, mit dem Gedanken an das Netz der lebenden Gestalt, wenn es auch bis heute nur eine Vermutung ist. | Mit beliebig gearteten Netzen voller Flüssigkeit kann man unendlich viele Formen gestalten. Dies sagt noch nichts über deren Fähigkeit lebendig zu sein aus. Dazu müssen sie z. B. ein Immunsystem haben oder die Fähigkeit Feinde abzuwehren. Ein Lebewesen existiert nur so lange, wie sich sein Organismus gegen Angriffe wehren kann.

R. Finsterwalder Erfolgt der Übergang, die Herleitung der harten Erscheinungen aus einem weichen Organismus sprunghaft?

F. Otto Da ist mir keine schlüssige Antwort bekannt. In der Entwicklung kann es weiche Lebewesen geben, die nicht erhalten sind. Das war eine Erkenntnis von Adolf Seilacher, der herausfand, dass man in manchen Phasen der Entwicklungsgeschichte keine harten Schalen, keine Vorläufer findet. | Weiche Organismen, die nicht erhärten, sind nicht mehr auffindbar. Seilacher fand Spuren im Sediment, aus denen man Schlüsse auf die Form der Objekte ziehen kann, die diese Spuren verursacht haben, seien es lebende Würmer oder Weichtiere oder nicht lebende Objekte.

survival. This opinion is shared, for example, by the paleontologist Seilacher.

R. Finsterwalder If the theory is correct, this process started at some time or other. Did it come to an end at some point or does it still exist?

F. Otto The process started one day and the terrestrial environment has been suitable for it to continue to the present. Sometimes a new variation arises due to a change or a defect in the mesh.

R. Finsterwalder The one which is there first is not necessarily the optimal.

F. Otto Almost all existing creatures have an intact, functioning immune system. Creatures in the visible realm are easier to research. Life in the field of nano-biology is still largely unresearched. To find out how creatures adapt to the environmental conditions you need knowledge about the adjustment and self formation processes of both living and non-living nature. | In studying the origination of a water drop, a possible precursor to life, we have begun the search for explanatory models for biological forms of origin. Complex objects such as a seashell are much more difficult to examine. | Adolf Seilacher put two snail shells on the table in front of me and together we considered whether we could apply mechanical tests to find out which of the two was more likely to survive although he already suspected that one of the two only died out in the Cretaceous Period because the shell was not so good. (Fig. 7)

8

Abb. 8 Lufthalle

Fig. 8 Pneumatic construction

R. Finsterwalder Wird so etwas weiter passieren, wo stehen wir heute?

F. Otto Alle gehen leichtfertig davon aus, dass die Natur Vorbild sei, und man fragt sich: Wofür, für Totschlag oder Ehrenwertes? In der Bioethik sucht man Antworten. Und bei ethischen Fragen kann man auch die Religionen nicht umgehen. | Zur Bioethik gehört auch das Verhalten der Lebewesen untereinander. Die Formulierung des Begriffs der Ökologie, der heute oft nur noch als Synonym zu Energieeinsparung begriffen wird, bezeichnet eigentlich das Verhalten von Lebewesen untereinander zum Teil zur gegenseitigen Förderung des Lebens, zum Teil auch als Abwehr gegen Feinde. | Leider gibt es nur wenige Allgemeinbiologen wie Helmcke, Kull oder Seilacher, die eine Entwicklung in einer Gesamtsituation betrachten. Die Entwicklung der lebenden Natur geht ständig weiter. Die Objekte der nicht lebenden Natur wären auch ohne Lebewesen existent, würden sich vielleicht verändern, aber weiterhin existieren. | Heute ist die Denkweise in der Gesellschaft vorherrschend: Was in der Natur ist, ist gut. Ist es wirklich gut? Die Frage nach Gut und Böse hat mit den Formen, Gestalten und den Fähigkeiten der Objekte der lebenden Natur wie auch der nicht lebenden Natur nichts zu tun.

R. Finsterwalder Das Darwinjahr wird als Anlass genommen für Diskussionen. Es gibt die Theorie der bedingungslosen Optimierung in der Natur.

F. Otto Es gibt eine Optimierung, klar, aber nicht bedingungslos. Wenn man versucht den heutigen Darwinismus wissenschaftlich zu betrachten, so entwickelt sich ein Lebewesen aus dem Vorhergehenden und gibt einer Art die Chance zum Überleben. Als man festgestellt hat, dass es nicht allein das Überleben des Stärkeren gibt, hat man den Begriff »fit« eingeführt, der die geistigen Fähigkeiten zum Teil mit einschließt. Der größte Generalangriff auf die Menschheit sind die Mikroorganismen, die große Prozentsätze der Menschheit ausgelöscht haben, wie z. B. die Pest und andere.

R. Finsterwalder Ein altes Thema ist die Abgrenzung zwischen Natur und Mensch. Wie stehst du dazu?

F. Otto Es gibt keine klar definierbare Abgrenzung zwischen Natur und Mensch. Bei der Aufgabe, eine riesige unterirdische Halle zu bauen, begann die Entwicklung mit einer flüssig-weichen Struktur, die dann als Gewölbe erhärten sollte. | Wenn man eine Form gefunden hat, die durch eine Linie gleicher Krümmung in einer druckbeanspruchten Fläche mit einem Druckstab beschrieben wird, genannt Gewölbe, ist es möglich, die gleiche Form in bereits auf der Erde existierenden natürlichen Steinformationen wie z. B. am Paulssee in den USA oder in Utah wiederzuerkennen. (Abb. 9)

R. Finsterwalder Ist nicht jede Höhle oder Kaverne so eine Schale?

F. Otto Die Innenfläche einer Höhle kann als Schalentragwerk angesprochen werden. Nicht alle natürlichen Höhlen sind als stabile Tragwerke zu bezeichnen. Viele neigen dazu z. B. bei Erdbeben einzufallen.

R. Finsterwalder Es wäre auch vorstellbar, dass ihr etwas entwickelt habt, was es nicht gibt.

F. Otto Ja, wichtig ist, dass eine Form entsteht, die allen Angriffen der nicht lebenden Natur begegnen kann.

R. Finsterwalder Gibt es noch andere Beispiele in deiner Arbeit?

R. Finsterwalder How did something like that develop?

F. Otto We could not find a naturalistic testing method to answer that question. There was no information on the origination history and life-threatening dangers. | We did find one answer. The hardening of the object probably took place at a time when it was already highly developed, namely as a mollusk. They show their development and their origins in their form. | A balloon could be the predevelopment of a rigid airship which was only technically effective for a period of time. (Fig. 8) | Living nature arose in the weightless realm, in water. There was a development in the emergence of soft creatures in which hard parts were also developed. The appearance of soft creatures is older than that of hard creatures and without the evolutionary history of soft creatures those with hard parts would not be conceivable. | There is no creature which is totally hard, not even an oyster, which lies static on the seabed by accumulating lots of material. The living part is soft. Life overall is originally soft and mobile. But mobility is not necessarily coupled with softness, as you can see from the oyster. You can find an access point into evolutionary processes with the most simple non-living elements through the idea of the mesh of the living form, even if it is still regarded as an assumption today. | With various sorts of mesh full of fluid you can create an infinite number of forms. But this still does not say anything about their ability to live. For this they need an immune system or the ability to repel enemies. A creature exists only as long as it can defend its organism from attacks.

R. Finsterwalder The transition, the derivation of hard appearances from soft organisms — does this happen gradually?

F. Otto I do not know the conclusive answer. There can be soft creatures in the course of development which have not survived. This was the conclusion Adolf Seilacher came to. He discovered that in some phases of the history of development you could not find a hard shell, a predecessor. | Soft organisms which do not harden can no longer be traced. Seilacher found traces in the sediment from which conclusions can be drawn as to the form of the object, what caused these traces, be they living worms or soft creatures or non-living objects.

R. Finsterwalder Will these things continue to happen, where do we stand today?

F. Otto Everyone frivolously believes that nature is a model, and you have to ask what for? For homicide or honorableness? We look for answers in bioethics. And in the case of ethical questions you cannot circumvent religions. | Bioethics also includes the behavior of creatures among one another. | The formulation of the term ecology, which today is often understood as a synonym for energy-saving, actually describes the behavior of creatures among one another, in part mutual promotion of life, in part also defense against enemies. | Unfortunately there are only a few general biologists such as Helmcke, Kull and Seilacher, who observe development in the overall situation. Living nature continues to develop steadily. Objects from non-living nature would also exist without living creatures, they might change but would continue to exist. | The overruling mindset in society today is: What is in nature is good. But is it really good? The issue of good and bad has nothing to do with the forms, shapes and capabilities of objects of living nature or of non-living nature.

R. Finsterwalder The Year of Darwin gives us an opportunity to discuss these things. There is the theory of unconditional optimization in nature.

F. Otto Der Bahnhof Stuttgart ist wohl das extremste Beispiel, da die meisten anderen Formen in der nicht lebenden Natur bekannt sind, als Schüttkegel, als Düne oder Deich. | Es ist möglich, dass diese Form auch in der lebenden Natur gefunden werden kann, z. B. bei der Spongiosa. | Ich habe Hemmungen, wenn alle sagen: Die Millionen Jahre Entwicklungszeit der lebenden Natur brauchen wir nur zu nutzen und schon haben wir eine bessere Umwelt. | Unsere Zeit ist gefärbt durch die Worte Ökologie und Nachhaltigkeit. Der früheste Ursprung des Wortes Nachhaltigkeit stammt von einem Oberforstmeister, der bei August dem Starken arbeitete. Er hatte Holz an die Bergwerke zu liefern und hatte beim König durchgesetzt, dass nur so viel Holz geschlagen wird, wie nachwächst, d. h. wie nachgehalten werden kann. | Richard Burton und ich haben darüber nachgedacht, als wir einen Preis für Nachhaltigkeit im Städtebau erhielten, weil wir frisches Stangenholz für den Bau einer Holzbauschule verwendeten, das wieder nachwächst.

R. Finsterwalder Die Begriffe müssen klar definiert sein. Natur ist ein schwieriger Begriff.

F. Otto Was ist denn Natur? Ist vielleicht auch Kunst Natur, denn sie ist vom Menschen gemacht.

R. Finsterwalder Ist der Mensch Teil der Natur oder steht er außerhalb?

F. Otto Der Mensch als Teil der Natur schafft auch Kunst. | Ist Kunst Natur? Auf diese Frage mag es keine vollständige Beantwortung geben.

R. Finsterwalder Das ist auch in der Philosophie nicht einheitlich.

F. Otto Von der Antike an bis zu Kant und bis heute wird das diskutiert. Man sollte nicht zu oberflächlich damit umgehen. Auch Religionen spielen eine Rolle.

9

Abb. 9 The Delicate Arch, Utah

Fig. 9 The Delicate Arch, Utah

F. Otto There is optimization, of course, but not unconditional optimization. If you attempt to observe Darwinism today, scientifically, an organism develops from its predecessor and gives a species the chance to survive. When it was ascertained that there is not just a survival of the strongest, the term "fit" was introduced which in part also includes mental capabilities. The greatest general attack on humanity comes from microorganisms which have eradicated large percentages of humanity, e.g., the plague.

R. Finsterwalder An old theme is the boundary between man and nature. What is your opinion on this?

F. Otto There is no clearly definable boundary between man and nature. The assignment of building a giant subterranean hall began with developing a soft-fluid structure which was supposed to harden as a dome. Once you have found a form which is described by a line with a regular curvature in a compressed surface with a compression strut, a vault in other words, it is possible to recognize the form in stone formations which exist naturally on Earth, e.g. at Lake Paul in the United States, or in Utah. (Fig. 9)

R. Finsterwalder Are not all caves and caverns such a shell?

F. Otto The inner surface of a cave can be referred to as a shell structure. Not all natural caves can be described as stable structures. Many tend to collapse, for example in earthquakes.

R. Finsterwalder It is also possible that you have developed something which does not otherwise exist.

F. Otto Yes, the important thing is that you make a form which can stand up to all onslaughts from non-living nature.

R. Finsterwalder Are there other examples in your work?

F. Otto The Stuttgart train station is probably the most extreme example as most other forms in non-living nature are known — as an angle of repose, as a dune or as a dike. | It is possible that these forms can also be found in living nature, e.g. in the cancellous bone. | I have my reservations when everyone says: we only have to make use of the million years in which living nature has developed and hey presto we already have a better environment. Our era is colored by the words ecology and sustainability. The earliest origins of the word sustainability hail from a chief forest superintendent who worked for Augustus the Strong. He had to deliver wood to the mines and convinced the king that only as much wood should be cut as could grow back, i.e. as was sustainable. | Richard Burton and I had thought about this when we received a prize for sustainable urban development because we used fresh pole wood which grows back again, to build a wooden school.

R. Finsterwalder Concepts have to be clearly defined. Nature is a difficult concept.

F. Otto What is nature? Is art perhaps also nature because it is made by people?

R. Finsterwalder Is man part of nature or does he stand outside it?

F. Otto Man as part of nature also creates art. There may be no conclusive answer to this question.

R. Finsterwalder It is also not consistent in philosophy.

F. Otto This has been discussed from antiquity to Kant and through to the present day. One should not deal with it too superficially. Religions also play a role.

R. Finsterwalder Der Begriff ist schwer zu definieren. Wichtig ist mir, diese Vielfalt zu diskutieren. | Die Frage ist, wie weit kann man sich einander annähern. Ich plane gerade eine Kirche in Italien zusammen mit drei Künstlern. Das kann nur klappen, wenn sich beide Seiten, Kunst und Religion, einander annähern.

F. Otto Das Hinterfragen von Natur und Darwinismus ist wichtig, denn er hat die Generation unserer Eltern geprägt und ist auch wichtig für die Beurteilung der Standpunkte in der heutigen Generation. Ungewiss ist, wie in der Zukunft darüber gedacht wird.

R. Finsterwalder Die Inhalte müssen auch transportiert werden. Mir ist es wichtig, an ein, zwei Stellen in die Tiefe zu gehen, den Dingen auf den Grund zu kommen, nicht in die Breite.

F. Otto Ich werde in aller Welt in allen Literaturen zur Bionik als Kronzeuge angeführt, weil man sich einfach nicht vorstellen kann, dass sich zwar bei manchen meiner leichten Bauten zu Objekten der lebenden Natur eine formale Analogie aufdrängt, dass sie aber ohne Vorbild entwickelt wurden und Ähnlichkeiten in der Natur nachträglich festgestellt wurden. | Es ist eine Chance, Begriffe wie Ökologie oder Nachhaltigkeit und auch den Modebegriff Bionik genauer zu hinterfragen, wie du es tust, und auch die Meinung, dass man lebende Objekte als Vorbilder für menschliche Konstruktionen nutzbringend verwenden kann, was manchmal möglich ist. (Abb. 10)

R. Finsterwalder Vielleicht wäre es gut, ein Interview mit Nachtigall zu unserem Thema zu führen.

F. Otto Ich habe mit ihm manche interessante Diskussion geführt. Wir schätzen uns auf unseren Gebieten.

R. Finsterwalder Das Problem ist, dass zu wenig in Grundlagenforschung investiert wird, sondern nur in Anwendung.

F. Otto Das ist gerade auch in der Biologie der Fall. Die Grundlagenbiologie wird vernachlässigt genau wie in der Architektur. Alle machen nur noch Details. Wer kann schon das menschliche Tun in der Gesellschaft als Ganzes begreifen? Für mich ist auch Kunst als menschliches Tun ein Teil der Natur. | Es gibt eine Meinung in der Akademie der Künste Berlin, dass Kunst politisch ist. Aber man kann die freie Kunst auch so sehen, dass sie das Einzige im menschlichen Leben ist, was nicht politisch ist. Kunst ist ein Gegner von Politik. Wenn man den Faden auch zur Natur zieht, muss man eigentlich feststellen, dass Kunst vom Menschen veränderte Natur ist. | Baukunst im öffentlichen Raum jedoch ist immer politisch. Als Teil des menschlichen Tuns hat Baukunst aber immer auch eine Beziehung zur Natur. Aber der Einfluss der Politik kann so stark werden, dass er die Kunst beengt. Ein typisches Beispiel ist der Abriss der Berliner Schlosses, der Bau des Palastes der Republik und wiederum dessen Abriss.

R. Finsterwalder Das Zurück zum Wilhelminismus war gewollt wie der Abriss des Palastes der Republik.

R. Finsterwalder Warum, glaubst du wird deine Arbeit so wenig fortgeführt, in meinen Augen viel zu wenig?

F. Otto Die meisten Architekten wollen heute bauen, Ideen verwenden und sie nicht wissenschaftlich weiterführen. Sie scheuen den mühseligen Weg der Forschung.

R. Finsterwalder Wenn es Neuland für die Beteiligten ist, wird es schwierig. Unglaublich wie du München oder Montreal realisieren konntest, das war ja wirklich ein revolutionärer Ansatz damals.

R. Finsterwalder The term is difficult to define. It is important for me to discuss this diversity. | The question is how close can we come together. At the moment I am planning a church in Italy together with three artists. This can only work if both sides, art and religion, converge.

F. Otto Scrutinizing nature and Darwinism is important because this characterizes our parents' generation and is also important for evaluating the standpoint of today's generation. What is uncertain is how people will think in the future.

R. Finsterwalder The content also has to be transported. For me it is important to address one or two things in depth, to get to the bottom of things and not to be superficial.

F. Otto All over the world, in any literature to do with bionics, I am described as a chief witness simply because people cannot imagine that although some of my lightweight constructions were imposed with a formal analogy to objects of living nature, they were developed without a model and similarities with nature were subsequently determined. It is a chance to question concepts such as ecology and sustainability more precisely, and also the fashionable concept of bionics, as you do, not to mention the opinion that living objects can be usefully employed as models for human constructions, which is sometimes possible. (Fig.10)

R. Finsterwalder Perhaps it would be a good idea to conduct an interview with Nachtigall about our topic.

F. Otto I have had many and interesting discussions with him. We hold each other in high esteem in our fields.

R. Finsterwalder The problem is that there is too little investment in basic research, only in application.

F. Otto Precisely in biology too that is the case. In biology the basics are being ignored, just like in architecture. All people are doing now is details. Who is capable of comprehending human activity in society as a whole? For me art, as a human activity, is also part of nature. | There is a mindset in the *Akademie der Künste* in Berlin that art is political. One can, however, also see free art as the only thing in human life that is not political. Art is an opponent of politics. If you also follow the thread to nature you actually come to realize that art is nature that has been altered by man. | Architecture in public spaces is, however, always political. As part of human activity, though, architecture always has a relationship with nature. The influence of politics can, however, become so strong that it constricts art. A typical example is the demolition of the *Stadtschloss* in Berlin, the construction of the Palace of the Republic, and in turn its demolition.

R. Finsterwalder There was a wish for a return to the Wilhelmine style, as there was for the Palace of the Republic to be demolished.

R. Finsterwalder Why do you think your work is continued so little, to my mind far too little?

F. Otto Nowadays most architects want to build, use ideas and not pursue them academically. They shy away from the arduous research path.

R. Finsterwalder If it is new territory for those involved it gets difficult. Incredible how you were able to build Munich and Montreal, that really was a revolutionary approach at the time.

F. Otto Wir stellten der Fachwelt zum ersten Mal weit spannende wetterfeste textile Konstruktionen vor als Fortführung des uralten Zeltbaus. Mit einem guten Team von Architekten, Ingenieuren und Ausführenden. Zeltbauten sind in ihrer Tragwirkung auch als Schalen anzusprechen, die jedoch durch ihre Neigung zum Ausknicken und Beulen größere Schwierigkeiten beim Bau machen. Lange Zeit meinte man, dass das hyperbolische Paraboloid die Lösung sei, bis ich entdeckte, dass entlang der so genannten geraden Erzeugenden die Schwachstelle liegt, die wir nun mit dem so genannten Kontinuum vermieden haben, d. h. mit einem Objekt, das an jeder Stelle gekrümmt ist, obwohl eine Sattelfläche an zwei Linien gerade Schnitte hat. | Meine erste Realisation einer weit spannenden echten Schale war die Multihalle Mannheim 1971–75. Die Tendenz zur Selbsterzeugung der Form ist schwer zu verstehen. Es gehört beinahe ein Lebenswerk dazu, um die Natürlichkeit der Schale zu begreifen, die das Ergebnis eines Prozesses ist. (Abb. 11)

R. Finsterwalder Die Konstruktion ist so weit entwickelt, dass man erwarten muss, dass sie noch in der nicht lebenden Natur zu finden ist.

F. Otto Ja, besonders da sie eine sehr wirtschaftliche Lösung ist, kann man erwarten, dass sie in der lebenden Natur noch gefunden wird. Mit dieser Vermutung stehen wir am Anfang einer neuen Erkenntnisgeschichte, die auch die Beziehung zur Ästhetik mit einbezieht. Man kann nicht einfach sagen: Folgt den Formen der Natur, dann macht ihr schöne Bauten.

10

Abb. 10 Olympiastadion München

Fig. 10 Olympic stadium Munich

F. Otto For the first time ever we presented the industry wide-span weatherproof fabric constructions as a continuation of ancient tent building. With an excellent team of architects, engineers, and construction workers. In their load-bearing properties tent structures can also be referred to as shells which, however, given their tendency to buckle and bulge create greater difficulties during construction. For a long time it was thought the hyperbolic paraboloid was the solution, until I discovered that the weak point was along the so-called straight generatrix, which we managed to avoid with the so-called continuum, in other words with an object that is curved everywhere, although a saddle surface has straight cuts on two lines. | The first time I built a genuine wide-span shell was for the *Multihalle* in Mannheim (1971–75). The tendency for the shape to generate itself is difficult to understand. You almost need a life's work to comprehend the natural properties of the shell, which is the result of a process. (Fig.11)

R. Finsterwalder The construction is so advanced you have to expect it to be still found in living nature.

F. Otto Indeed, particularly as it is a very economical solution one can expect it to be found in living nature. This supposition puts us at the beginning of a new set of findings, which also includes the relationship with aesthetics. You cannot just say: follow nature's shapes and you will make beautiful buildings.

11

Abb.11 Multihalle, Mannheim

Fig.11 Multihalle, Mannheim

R. Finsterwalder Es ist unglaublich, was für Anstrengungen unternommen werden, damit Gebäude natürlich aussehen. Aber eigentlich passiert das Gegenteil, sie sind zutiefst unnatürlich. | Das ist Design und nicht Architektur, wenn es nicht den Anspruch hat, ganzheitlich zu denken, Konstruktion, Architektur und Funktion zusammenzubringen. Wenn der Formwille dominiert, dann ist es keine Architektur, dann ist es Design.

F. Otto Der Unterschied zwischen Design und Architektur ist ein Thema seit den 1920er Jahren bis heute. Am Ende einer ehrlichen, funktionalen Arbeit kann durchaus ein Objekt stehen, das als schön empfunden wird. Das ästhetische Element kann man nicht direkt planen. (Abb. 12) | Eine ästhetische Form steht am Ende eines Prozesses. Allein mit dem Willen zur Schönheit wird man sie nicht erreichen. Wenn wir ehrlich gearbeitet haben, bekommen wir sie manchmal geschenkt. Das war auch Leitlinie des Deutschen Werkbundes, der hundert Jahre alt ist. | Diese Leitlinie ist ein bedeutender Teil unserer heutigen Kultur, nicht nur in der Architektur, auch z. B. in der Musik.

R. Finsterwalder Ehrlichkeit muss man weglassen, da es in die Irre führt. Mies hat keine konstruktive Ehrlichkeit propagiert und sich dennoch intensiv damit auseinander gesetzt. | Die Abgrenzung von Architektur ist durchaus sinnvoll. Design ist keine Architektur. An den Unis werden 3D Bilder nach dem Prinzip »try and error« erzeugt und so werden Projekte entwickelt.

F. Otto Beim Thema der Architekturbetrachtung interessiert mich vor allem die Frage, was gut ist und ob die Begriffe schön und gut voneinander abhängen, weniger der alte Leitsatz: »form follows function«. An den amerikanischen Unis endet eine Diskussion oft mit dem nicht zu widersprechenden »but I like it«. Es hat viele Lehrer gegeben, die dieses Mögen auf ihre Fahne schrieben, so auch Yamasaki, der wie ich 1958 an der Washington University in St. Louis lehrte. | Ich habe mich damals mit Yamasaki über technische und auch ästhetische Schwächen der Entwurfsmodelle für die Towers des World Trade Center in New York gestritten, woran ich nach dem Einsturz der Towers sehr denken musste. Mit dem Einsturz war die Architekturbetrachtungsweise »I like it« beendet.

R. Finsterwalder An so einem Beispiel sieht man, dass es um mehr geht in der Architektur als Fun und »I like it«. | Die Ursprünge und Grundlagen werden nicht mehr vermittelt. Mit den Möglichkeiten der Computerplanung wird ein »try and error« gefördert, das »I like it, I like it not«.

F. Otto Man guckt auf den Monitor. Was einem nicht gefällt, wird weggelöscht, übrig bleibt, was gerade schön zu sein scheint.

R. Finsterwalder Meist ist so ein Objekt nur aus einer Perspektive schön, es ist nicht um die Ecke gedacht. Das sind nicht die Probleme, um die es in der Architektur geht. Vielleicht ist das ein Grund für den Erfolg der Bionik, die Heilsuche in der Natur, die Suche nach Werten.

F. Otto Das ist so logisch. Man erfährt bereits in der Schulbiologie von der Entwicklung der Arten und der des Menschen. Dann erscheint der Rückschluss, dass heute lebende Objekte optimal seien und Vorbild sein können für Tätigkeiten des Menschen, oberflächlich gesehen logisch.

R. Finsterwalder Wenn man es als Abziehbild versteht und nicht versteht, was dahinter liegt, funktioniert das nicht.

R. Finsterwalder It is incredible what effort goes into making buildings look natural. But actually quite the opposite happens, they are utterly unnatural. | It is design and not architecture, if it does not think holistically, bringing together construction, architecture, and function. If the will to form dominates then it is not architecture but design.

F. Otto The difference between design and architecture has been a topic ever since the 1920s through to the present day. At the end of the day honest, functional work can indeed produce a building that is considered beautiful. The aesthetic element cannot be planned directly. (Fig.12) | An aesthetic form emerges at the end of a process. It cannot be achieved through the will to beauty alone. If we have worked honestly it sometimes comes as an extra present. This was also the guideline used by the Deutscher Werkbund, which is one hundred years old. | This guideline is an important part of our current-day culture, not just in architecture but for example in music as well.

R. Finsterwalder You have to forget honesty, as it is misleading. Mies did not propagate constructive honesty yet still addressed it intensively. | The demarcation of architecture is certainly meaningful. Design is not architecture. At many universities 3D images are created by trial and error and that is how projects are developed.

F. Otto With regard to the topic of considering architecture, what I am most interested in is the question of what is good and whether the terms beautiful and good are dependent on one another, and less the old guiding principle "form follows function". At US universities discussions often end with "but I like it", which there is no contradicting. There have been several teachers who have championed this liking, Yamasaki for example, who like myself in 1958 was teaching at Washington University in St. Louis. | Back then I argued with him about technical as well as aesthetic weaknesses of the draft models for the towers of the World Trade Center in New York, to which I was forced to give much thought following their collapse. This put an end once and for all to the "I like it" school of architecture.

R. Finsterwalder This is a good example of there being more to architecture more than fun and "I like it". | The origins and principles are no longer taught. The opportunities computer planning provides champion the try and error, I like it, I don't like it approach.

F. Otto You look at the screen and just delete whatever you don't like so that you are just left with what looks nice there and then.

R. Finsterwalder For the most part buildings like that that are attractive just from one perspective, they have not been thought through. These are not the problems that architecture deals with. Perhaps that is a reason for the success of bionics, the search for healing powers in nature, the search for values.

F. Otto That is so logical. Even in school biology you learn about the development of the species and that of man. On a superficial level the conclusion that objects living today are perfect and can serve as a role model for man's activities can seem logical.

R. Finsterwalder That does not work if you see it as a peel-away image and fail to understand what it involves.

F. Otto Nowadays ecology and bionics are regarded as a guarantee for good results. However, you repeatedly have to prompt critical questioning. I am not interested in falsification à la Popper. Even if you have

F. Otto Die Themen Ökologie oder Bionik werden heute als Garant für gute Arbeitsergebnisse angesehen. Aber man muss immer wieder zum kritischen Hinterfragen anregen. Die Falsifikation nach Popper interessiert mich nicht. Auch wenn man etwas Falsches entdeckt hat, muss man Neues suchen. Das ist entscheidend.

R. Finsterwalder Wie ist dein Verhältnis zu Shigeru Ban? Als wir in deinem Atelier die Knoten gesehen haben für die Expo-Halle, die ihr entwickelt habt, habe ich mir gedacht, das ist die eigentliche Leistung.

F. Otto Shigeru Ban ist ein Erfinder, mit dem man gut Gedanken austauschen kann. Zum Beispiel hat er sich, wie ich auch, um katastrophengerechtes Bauen bemüht.

R. Finsterwalder Das hat Frank Lloyd Wright schon gemacht mit dem Imperialhotel in Tokyo. Er hat ja ähnlich gearbeitet wie du. Erstaunlich, sich so etwas grundlegend Neues zuzutrauen.

F. Otto Ja, das Imperialhotel habe ich auch mal besucht. Wrights erste Arbeiten sind von heutiger Warte sehr ähnlich wie meine Arbeiten. Dabei habe ich sie nie nachgemacht. Ich fand eher seine Auskunft über Broadacre City durchaus bedenklich, jedoch nachdenkenswert. Sie hat eine Komponente zur Thematik Raumortlabor in Hombroich[2], die von Karl-Heinrich Müller ins Leben gerufen wurde. (Abb. 13)

2 »Raumortlabor« ist der Titel eines städtebaulichen Projektes, das von der Stiftung Insel Hombroich initiiert wurde. unter dem Thema »Architektur parallel zur Natur« haben 14 Architekten und Künstler Vorschläge zur Bebauung des Geländes um die Insel und die Raketenstation erarbeitet. Unter den Planern sind auch Frei Otto und Rudolf Finsterwalder.

12

Abb. 12 Institut für leichte Flächentragwerke, Stuttgart

Fig. 12 Institute for Lightweight Structures, Stuttgart

discovered something wrong you have to look for something new. That is decisive.

R. Finsterwalder What is your relationship with Shigeru Ban like? When we saw the nodes you developed for the Expo hall in you studio I thought that is the real achievement.

F. Otto Shigeru Ban is an inventor with whom you really can exchange ideas. By way of example he, like me, has addressed buildings that can withstand disasters.

R. Finsterwalder Frank Lloyd Wright also did that with the Imperial Hotel in Tokyo. He worked in a similar manner to you. Having the self-confidence to do something so fundamentally new is astonishing.

F. Otto Yes, I have also been to the Imperial Hotel. From today's point of view Wright's first works are very similar to my works, though I never copied them. If anything I found his information about Broadacre City questionable, though certainly worth thinking about. It has a component on the topic of the Raumortlabor (spaceplacelab) in Hombroich[2], which Karl-Heinrich Müller initiated. (Fig.13)

R. Finsterwalder What are your misgivings about Broadacre City?

F. Otto The project was perhaps too abstract. How much ground does a family need to be self-sufficient, in both architectural and agricultural terms. The assumed need fluctuates between 4,000 and 10,000 m^2. In Europe that is much more difficult to realize than in the USA.

R. Finsterwalder I see even more parallels with your work: the growth he imagined, traditional garden cities were limited in size. Frank Lloyd Wright thought of his structure as infinite, an organically growing structure, organic growth structures that remind me of your minimum distance systems.

F. Otto Everything is inter-linked, for example even what Scharoun and Martin Wagner, the Director of Urban Planning in Berlin, were drawing in the 1920s.

F. Otto Architecture can also be seen as a person's third skin. The second is clothing. If the third skin is perfect, there is no need for the second. We are getting closer to the original image of the human environment.

R. Finsterwalder Hegel saw maximum beauty in the symmetry and harmony in nature, Adorno in contrast in an incidental disruption of perfection.

F. Otto There is a lot to be said about symmetry. There was a major symposium about it in Darmstadt. The art historian Gombrich assumed the main role for design and architecture. His main work is called "The Sense of Order". In his opinion symmetry in architecture makes sense because it enables one to remember the shape better. | If anything I see a meaning for symmetry in function and reproduction biology. If, for example, from the point of view of another animal the originally asymmetrical appearance of a mammal suddenly becomes symmetrical, it triggers the greatest of attention, as in the case of a male animal it has perhaps adopted an attacking position or in that of a female animal is in heat. And vice versa the naked

2 The "spaceplacelab" is the title of an urban planning project initiated by the Insel Hombroich oundation in 2004. Taking as their theme "architecture parallel with nature" 14 architects and artists came up with proposals for developing the area around hombroich island and the missile base. The planners include Frei Otto and Rudolf Finsterwalder.

R. Finsterwalder Was sind deine Bedenken zu Broadacre City?

F. Otto Vielleicht war dieses Projekt zu abstrakt gedacht. Wie viel Erdoberfläche braucht eine Familie zur Selbstversorgung sowohl im architektonischen als auch im landwirtschaftlichen Sinn! Der angenommene Flächenbedarf schwankt zwischen 4.000 und 10.000 m². Das ist in Europa viel schwieriger zu verwirklichen als in den USA.

R. Finsterwalder Ich sehe noch mehr Parallelen zu deiner Arbeit: das Wachstum, das er sich vorstellte, Gartenstädte waren in der Größe begrenzt. Frank Lloyd Wright hat seine Struktur unendlich gedacht, eine organisch wachsende Struktur, organische Wachstumsstrukturen, die mich an deine Minimalwegesysteme erinnern.

F. Otto Es hat alles miteinander zu tun, z. B. auch das, was Scharoun und Martin Wagner, Stadtbaudirektor in Berlin, in den 1920er Jahren gezeichnet haben.

F. Otto Architektur kann auch als die dritte Haut des Menschen angesehen werden. Die zweite ist die Kleidung. Wenn die dritte Haut perfekt ist, braucht man die zweite Haut nicht. Man nähert sich dem Urbild der menschlichen Umwelt an.

R. Finsterwalder Hegel sah in der Symmetrie und Harmonie der Natur das Maximum an Schönheit, Adorno im Gegensatz dazu in einer zufälligen Störung der Perfektion.

F. Otto Zur Symmetrie ist schon einiges zu sagen. Es hat darüber ein großes Symposium in Darmstadt gegeben. Für Design und Architektur hatte der Kunsthistoriker Gombrich den Hauptpart übernommen. Sein Hauptwerk heißt: »the sense of order«. Er vertrat die Meinung, dass die Symmetrie in der Architektur deshalb einen Sinn hat, weil man sich die Form besser merken kann. | Ich sehe einen Sinn der Symmetrie eher in der Funktions- und Reproduktionsbiologie. Wenn z. B. das ursprünglich asymmetrische Erscheinungsbild eines Säugetieres aus der Sicht eines anderen Tieres plötzlich symmetrisch wird, löst es höchste Aufmerksamkeit aus, weil es als männliches Tier möglicherweise in Angriffsposition gegangen ist oder als ein weibliches Tier gerade empfangsbereit ist. Und umgekehrt bewirken die nackten asymmetrischen Gestalten im Tympanon des völlig symmetrischen griechischen Tempels höchste Aufmerksamkeit. | Zellen können sich nicht streng symmetrisch in der Bedeutung »Gleichmaß« organisieren. Sie sind nicht stapelbar. Ich weiß nicht, ob das ein Vor- oder

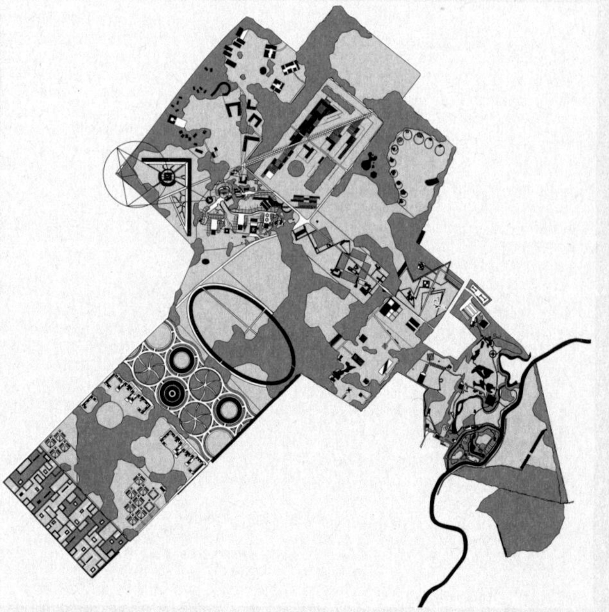

13

Abb.13 Lageplan Raumortlabor Planstand 2004 mit den Planungen u.a. von Frei Otto, Alvaro Siza, Tadao Ando und Finsterwalder Architekten

Fig.13 Siteplan Spaceplacelab, planning of 2004 with the projects of Frei Otto, Alvaro Siza, Tadao Ando and Finsterwalder Architekten

asymmetrical figures in the tympanum in completely symmetrical Greek temples drew attention. | Cells are unable to form strictly symmetrical structures. They cannot be stacked. I don't know whether that is an advantage or a disadvantage in terms of the ability to survive. So there are no symmetrical multi-cell beings; single-cell beings, perhaps, as for example spherical blue algae. | Symmetry in the meaning of equal proportions appears in innumerable varieties in architecture. Exact symmetry makes it boring. Gombrich was aiming to show that in architecture there are repeated attempts to score success with symmetry, but that the art lies in introducing asymmetry now and again, as in the tympanum.

R. Finsterwalder Johannes Kepler, for example, investigated the most compact of packages and listed various types of symmetry. Ian Stewart, a Scottish physicist, investigated symmetry in nature, the refraction of symmetry down to fractals. Johannes Kepler once wrote an essay about the symmetry of the snow flake.

F. Otto But there are hundreds of types of snowflakes. The basic crystal is indeed symmetrical.

R. Finsterwalder That is not necessarily an area of application for architecture, but it is interesting for constructions.

F. Otto In the case of asymmetrical membrane buildings and meshes inaccuracies are less important than in symmetrical ones. With a shape like the *Tanzbrunnen* fountain, for example, which has six axes of reflection and thus twelve sections, a small error in the layout can add up to a fatal error overall. (Fig.14) For the *Tanzbrunnen* I had been expected to come up with an asymmetrical design, which would have followed the fashion at the time for organic architecture. We nevertheless decided in favor of a symmetrical design, which needs less planning than an asymmetrical. | Nowadays I am of the opinion that in my most important structures of this era, the entrance arch and the *Tanzbrunnen*, the symmetry was right. There are reasons for the symmetry that should not be ignored. The production method played an important role in the decision for or against symmetry. Human functions rarely enforce symmetry, human needs do not require perfect symmetry in buildings, though at times a certain amount of equal proportions. The game between asymmetry and symmetry or equal proportions is always interesting and is the basis for designing architecture.

R. Finsterwalder Equal proportions is better. What we perceive to be beautiful in nature is this sense of equal proportions, the repetitive, regular inherent in an object.

F. Otto The symmetry in the unequal proportions of shells and membranes is difficult to understand. The equally supported surface in the design for the new station in Stuttgart is an example both of unequal and equal proportions, as it can be assembled from equal parts and thus produce a continuum, which for example in the case of formwork with the construction of the grid shell sets itself and overall adopts the shape of a so-called minimum surface, which at no point has any direct curvature but which nevertheless is curved overall. Nowadays the successful use of minimum surfaces is regarded as the fine art of shell and vault construction in architecture. | My first use of a continuum was the *Multihalle* in Mannheim. A wooden grid with movable nodes produces a continuum.

R. Finsterwalder What about the *Multihalle*, does it have symmetry?

Nachteil ist für die Lebensfähigkeit. Es gibt also keine symmetrischen Vielzeller; Einzeller vielleicht schon wie z. B. eine kugelige Blaualge. | Symmetrie in der Bedeutung von Gleichmaß tritt in der Architektur in unzähligen Varianten auf. Exaktes Gleichmaß macht sie langweilig. Gombrich wollte darauf hinaus, dass in der Architektur immer wieder versucht wird, mit Symmetrie Erfolg zu haben, dass es aber die alte Kunst ist, immer wieder Asymmetrie hineinzubringen wie z. B. beim Tympanon.

R. Finsterwalder Johannes Kepler hat z. B. die dichtestmöglichen Packungen untersucht und er hat verschiedene Arten von Symmetrie aufgeführt. Ian Stewart, ein schottischer Physiker, hat die Symmetrie in der Natur, die Symmetriebrechung bis zu Fraktalen untersucht. Von Johannes Kepler gibt es einen Aufsatz über die Symmetrie der Schneeflocke.

F. Otto Es gibt über hundert Typen von Schneeflocken. Das Grundkristall ist tatsächlich symmetrisch.

R. Finsterwalder Das ist kein zwingender Anwendungsbereich für Architektur, aber es ist interessant für Konstruktionen.

F. Otto Bei asymmetrischen Membranbauten oder -netzen spielen Ungenauigkeiten eine geringere Rolle als bei symmetrischen. Bei einer Form wie dem Tanzbrunnen z. B., der sechs Spiegelachsen hat und damit zwölf Teile, kann sich ein kleiner Fehler im Zuschnitt zu einem fatalen Fehler im Gesamten addieren. (Abb. 14) | Von mir hatte man für den Tanzbrunnen einen asymmetrischen Entwurf erwartet, der der damaligen Mode der organischen Architektur gefolgt wäre. Dennoch haben wir uns für einen symmetrischen Entwurf entschieden, der weniger Planungsaufwand verlangt als ein asymmetrischer. | Heute bin ich der Meinung, dass die Symmetrie bei meinen wichtigsten Bauten dieser Zeit, dem Eingangsbogen und dem Tanzbrunnen, richtig war. Es gibt Gründe für die Symmetrie, die man nicht vernachlässigen sollte.

14

Abb. 14 Tanzbrunnen, Köln

Fig. 14 Tanzbrunnen, Köln

F. Otto It is symmetrical inasmuch as the grid has equal distances of 50 cm. It is largely a continuum and a minimum surface. In the case of grids such as this all instances of curvatures are elastic. Without the outer roof skin the shell can be folded flat and transported with little volume, as we did for Montreal.

R. Finsterwalder For a small roof Shigeru Ban joined up roofing battens to form a grid, raised the whole structure and then fixed the nodes.

F. Otto That is what we did for the Japanese Pavilion in Hanover. Symmetry in the form of equal proportion is indispensable. If you look at a bubble as a model of the biological cell as well as a construction element in living nature, it provides the symmetry. (Fig. 15)

R. Finsterwalder Does bionics in the case of Nachtigall include the not living nature?

F. Otto I don't know. He would not deny that it is the starting point for living nature. However, just what impact the environment has is unclear. Previously opinion was that the environment had no direct influence on the genetic code. Now it is thought that it can possibly be manipulated by man. | The question of whether nowadays new forms of life can be created is still relevant. Is creation still occurring? According to Adolf Seilacher that is certainly possible. Perhaps nowadays just as many species are being created as are dying out, nobody knows. It is probably the case that any new species is immediately got rid of. Nonetheless it is possible for a new species with long-term stability to emerge.

15

Abb. 15 Japanischer Pavillon, Expo

Fig. 15 Japanese Pavilion, Expo

Bei der Entscheidung für oder gegen Symmetrie spielt die Herstellungsmethode eine wichtige Rolle. Menschliche Funktionen zwingen selten zur Symmetrie, menschliche Bedürfnisse verlangen keine perfekte Symmetrie an Bauten, wohl aber zuweilen eine gewisse Dosierung von Gleichmaß. Das Spiel zwischen Asymmetrie und Symmetrie oder Gleichmaß bleibt interessant und ist Basis des Entwerfens von Architektur.

R. Finsterwalder Gleichmaß ist besser. Das, was man in der Natur als schön empfindet, ist dieses Gleichmaß, das sich Wiederholende, Regelhafte, was einem Objekt innewohnt.

F. Otto Schwer zu verstehen ist Gleichmaß im Ungleichmaß von Schalen und Membranen. Die gleichmäßig gestützte Fläche des Entwurfes für den neuen Stuttgarter Bahnhof ist sowohl ein Beispiel für Ungleichmaß als auch für Gleichmaß, denn man kann sie aus gleichen Teilen zusammensetzen und daraus ein Kontinuum herstellen, das sich beispielsweise bei einer Schalung mit der Konstruktion der Gitterschale von selbst einstellt und insgesamt die Form einer sog. Minimalfläche bildet, die an jedem Punkt keinerlei direkte Krümmung hat und dennoch insgesamt gekrümmt ist. Gelungene Anwendungen von Minimalflächen gelten heute als hohe Kunst des Schalen- und Gewölbebaus in der Architektur. | Meine erste Anwendung eines Kontinuums ist die Multihalle in Mannheim. Ein Holzgitterrost mit drehbaren Knoten ergibt ein Kontinuum.

R. Finsterwalder Wie ist es bei der Multihalle, hat sie eine Symmetrie?

F. Otto Sie ist insofern symmetrisch, als der Gitterrost ein Gleichmaß von 50 cm hat. Sie ist weitgehend ein Kontinuum und eine Minimalfläche. Bei solchen Gitterschalen sind alle Krümmungen elastisch. Ohne die äußere Dachhaut lässt sich die Schale flach zusammenfalten und mit geringem Volumen transportieren, wie wir das für Montreal gemacht haben.

R. Finsterwalder Shigeru Ban hat bei einem kleinen Dach Dachlatten zum Gitterrost verbunden, das Ganze angehoben und dann die Knoten fixiert.

F. Otto So haben wir es für den Japanischen Pavillon in Hannover erarbeitet. Symmetrie als Gleichmaß ist unverzichtbar. Wenn man eine Blase als Modell der biologischen Zelle auch als Bauelement der lebenden Natur ansieht, dann liefert es das Gleichmaß. (Abb.15)

R. Finsterwalder Schließt die Bionik bei Nachtigall die nicht lebende Natur mit ein?

F. Otto Ich weiß es nicht. Er würde wohl dem nicht widersprechen, dass sie Ausgangspunkt für die lebende Natur ist. Wie aber die Einflüsse der Umwelt einwirken, das ist offen. Bisher war man der Meinung, dass die Umwelt keinen direkten Einfluss auf den genetischen Code hat. Inzwischen gilt der genetische Code als möglicherweise vom Menschen manipulierbar. | Die Frage bleibt aktuell, ob heute noch neue Lebewesen entstehen können. Ist die Schöpfung heute noch präsent? Nach Adolf Seilacher ist das durchaus möglich. Vielleicht entstehen heute noch ebenso viele Arten wie vergehen, keiner weiß es. Wahrscheinlich ist, dass jede neue Art sofort beseitigt wird. Dennoch ist es möglich, dass eine neue Art entsteht, die eine Stabilität von großer Dauerhaftigkeit besitzt.

R. Finsterwalder Es gibt auch Mutationen.

F. Otto Mutationen sind bleibende Veränderungen der genetischen Blaupause bestehender Arten, um mit Nachtigall zu sprechen. Es waren eigentlich die Bio-

R. Finsterwalder There are also mutations.

F. Otto Mutations are permanent changes in the genetic blue print of existing species, to quote Nachtigall. It was actually biologists who noticed that as far as the conditions for the development of membranes and load-bearing surfaces were concerned we were dealing with the same media as in the emergence of things in living nature. Results such as these saw the light of day through curiosity about tires, tents, light-weight constructions, though without clear direction.

R. Finsterwalder That was what was unusual, working together in various disciplines. Nowadays there is far too little inter-disciplinary work and too much specialization.

F. Otto Results can only be achieved if interested parties from various disciplines become involved in the work of the others.

R. Finsterwalder When I was at the German Academy in Rome there were people from four different fields. For me the exchange with the other disciplines was very fruitful, because you notice that the others are working on similar things to yourself.

F. Otto You often cannot plan things like that. Good research results frequently depend on chance and are achieved purely through friendly discussions or competing views.

R. Finsterwalder In Germany such inter-disciplinary examples should be promoted, especially at universities. Nowadays are there similar facilities to the Institute for lightweight Structures?

F. Otto Not really. | You also have to have the courage to make mistakes. If something is more than half right it is worth pursuing.

R. Finsterwalder There is a theory of failure, to preferably fail spectacularly now and again rather than never have attempted anything. Without this sort of risk lots of innovations such as your pioneering achievements would never have come about.

F. Otto There was, however, a long line of development prior to our research. The origins are thousands of years old.

R. Finsterwalder What were your clients like? Hardly anything is possible if they are hesitant and lacking in courage.

F. Otto It goes without saying that special projects are only possible with clients that are open to new ideas. For Montreal, for example, we had Karl Merz as the government representative and a courageous building contractor Peter Stromeyer, the tent specialist.

R. Finsterwalder How old were you when you built the pavilion in Montreal?

F. Otto A good 40.

R. Finsterwalder The technical opportunities available to architects nowadays are fantastic. Advances in the development of building material and technology provide a huge opportunity for developing sustainable buildings that lead the way forward. The fact that developments in various natural science and technical disciplines are available through worldwide publication immediately favor swift development. The objective should be intelligent buildings, which meet present-day requirements. We must once again conduct more fundamental research in order to develop innovative products and not get bogged down in out of date technology. Architects must free themselves and once assume a spearheading role. The time for this has long since come.

logen, die gemerkt haben, dass wir mit den Bedingungen für die Entwicklung von Membranen und Flächentragwerken die gleichen Medien hatten wie bei der Entstehung von Objekten in der lebenden Natur. Durch die Neugier für Pneus, Zelte, Leichbauten kamen solche Ergebnisse zutage, jedoch ungezielt.

R. Finsterwalder Das war das Außergewöhnliche, dass man zusammengearbeitet hat unter den verschiedenen Disziplinen. Inzwischen wird zu wenig fachübergreifend gearbeitet, es gibt zu viel Spezialisierung.

F. Otto Es kann nur Ergebnisse geben, wenn sich Interessierte aus verschiedenen Disziplinen auf die Arbeiten der anderen einlassen.

R. Finsterwalder Als ich in Rom an der deutschen Akademie war, waren dort Leute aus vier verschiedenen Bereichen. Für mich war der Austausch mit den anderen Disziplinen sehr fruchtbar, weil man merkt, dass die anderen an ähnlichen Dingen arbeiten wie man selbst.

F. Otto Solche Dinge kann man oft nicht planen. Gute Forschungsergebnisse sind oft vom Zufall abhängig, ergeben sich manchmal nur aus freundschaftlichen Gesprächen oder auch aus konkurrierenden Ansichten.

R. Finsterwalder Man müsste in Deutschland solche fachübergreifenden Beispiele fördern, vor allem an den Unis. Gibt es heute ähnliche Einrichtungen wie das Institut für leichte Flächentragwerke?

F. Otto Eigentlich nicht. Man muss auch den Mut zum Fehler haben. Wenn etwas mehr als zur Hälfte richtig ist, dann lohnt es sich weiterzuarbeiten.

R. Finsterwalder Es gibt eine Theorie des Scheiterns. Lieber ab und zu grandios scheitern als nie etwas versucht zu haben. Ohne dieses Risiko hätte es viele Innovationen nicht gegeben wie eure Pionierleistungen.

F. Otto Es gab aber schon einen langen Entwicklungsgang vor unseren Forschungen. Ursprünge sind tausende Jahre alt.

R. Finsterwalder Wie waren deine Bauherren? Mit zögerlichen und mutlosen Bauherren ist so etwas kaum möglich.

F. Otto Natürlich sind besondere Projekte nur mit neuerungsfreundlichen Auftraggebern zu machen.
Für Montreal hatten wir beispielsweise Karl Merz als Regierungsvertreter und einen mutigen Bauunternehmer Peter Stromeyer, den Zeltbauer.

R. Finsterwalder Wie alt warst du damals beim Bau des Pavillons in Montreal?

F. Otto Gute vierzig.

R. Finsterwalder Die technischen Möglichkeiten, die sich heutzutage Architekten bieten, sind phantastisch. Die Fortschritte in der Entwicklung von Baustoffen und Techniken bieten eine große Chance zur Entwicklung zukunftsweisender und nachhaltiger Bauten.
Die Gleichzeitigkeit von Entwicklungen in verschiedenen naturwissenschaftlichen und technischen Disziplinen und deren weltweit verfügbare Publikation begünstigen eine rasche Entwicklung. Intelligente Bauten, die den Anforderungen unserer Zeit gerecht werden, sollten das Ziel sein. Wir müssen wieder vermehrt Grundlagenforschung betreiben, um Neues zu entwickeln und nicht in längst überholten Techniken stecken zu bleiben. Die Architekten müssen sich emanzipieren und wieder eine Vordenkerrolle übernehmen. Die Zeit ist mehr als reif dafür.

5

Testlabor Natur

Entwickeln und Konstruieren nach biologischen Vorbildern

Nature — a test lab

Developing and constructing according to biological archetypes

Werner Nachtigall

Wenn ein Ingenieur die Natur kopieren will, erleidet er Schiffbruch. Die Natur liefert nun mal keine Blaupausen für die Technik. Doch hält sie ein unerschöpfliches Reservoir an Konstruktionen, Verfahrensweisen und Evolutionsprinzipien bereit, das der Mensch, wenn er's richtig macht, mit großem Vorteil für seine Technik nutzen kann. Wie aber macht er's richtig? Ganz einfach. Er versucht nicht, sozusagen sklavisch jedes Detail eines Naturvorbilds nachzubauen, sondern er bemüht sich, die zu Grunde liegenden Prinzipien herauszufinden. Diese kann der Ingenieur dann in sein eigenes technisches Gestalten einbauen – als Anregung vielleicht nur oder doch als unersetzlichen Wegweiser. || Die japanische Firma »Sony« hat einen hübschen kleinen Kunsthund auf den Markt gebracht. Er legt sich hin, wenn man ihn streichelt, er rollt die Augen und gibt Laut, wenn er Hunger hat, und demnächst wird er wohl auch einen Haufen machen können. Er sieht aus wie ein kleiner Hund, ein technischer eben. Und ist doch nicht eigentlich eine Konstruktion der Bionik. Es handelt sich eben um eine möglichst genaue Kopie eines Hundes »mit technischen Mitteln«, eine Kopie seiner Form und seiner Verhaltensweise. | Im Stuttgarter Flughafen tragen grazile »Baumstützen« das Dach. Ausgegangen ist diese Art von Stützenbau von Frei Ottos legendärem Stuttgarter »Institut für leichte Flächentragwerke«. Diese verzweigten Tragekonstruktionen heißen »Baumstützen«, sehen aber überhaupt nicht aus wie Bäume. Und sie sind dennoch streng bionische Konstruktionen. Wie das? || Es gibt in der Zwischenzeit vielerlei solcher Stützenkonstruktionen. Bei diesen ist eben kein Baum kopiert worden, sondern es sind einige Prinzipien der Tragestruktur von Bäumen zunächst einmal erforscht und dann technisch umgesetzt worden. Es ist dies unter anderem die Art und Weise, wie vom Stamm aus immer feiner verzweigte Äste ausgehen, wie ihre Länge mit feinerer Verzweigung abnimmt, wie sich dabei auch ihr Längen-Durchmesser-Verhältnis ändert, wie sich die Verzweigungswinkel bei jeder weiteren Zweig-Generation darstellen und anderes mehr. Das sind die Randbedingungen. Sie bestimmen die eigentliche Zielfunktion. Und dies heißt: Für die gegebene Maximallast, die diese Stützen abzufangen haben, sollen sie die geringstmögliche Eigenmasse aufweisen. Die genannten Parameter werden nun im Computer so lange verändert und hin und her geschoben, bis die Zielfunktion das erstrebte Minimum aufweist. Leichter kann man solche »Baumstützen« für ihre gegebenen Anforderungen dann nicht mehr bauen: Folglich hat man auf die bestmögliche Weise Material gespart und damit auch Geld. | Außerdem wirken diese »Baumstützen« viel graziler und schöner als langweilige Rundsäulen vom Boden bis zum Dach. Für den Bauingenieur ist das ein Nebeneffekt, für den Architekten kann das aber der gestalterische Haupteffekt sein – wobei wir nun schon fast beim Design wären. || Doch lässt sich Bionik-Design besser verstehen und einordnen, wenn man zunächst selbst ein wenig über bionische Vorgehensweise nachdenkt. || Oben

Pflanzen – und hier besonders die Bäume – sind die unbestrittenen Meister bei der Entwicklung dauerhafter und stabiler Trage- und Stützenkonstruktionen. So haben Bäume (wie der Baumfarn) optimierte Lösungen für das Problem gefunden, eine möglichst große Last mittels einer passenden Anzahl abzweigender Äste von geringer Eigenmasse abzustützen. Für eine intelligente technische Lösung des Problems sich verdrillender Kabelstränge steht die Ranke der Zaunrübe.

Wer die Natur einfach nur kopieren will, erleidet Schiffbruch.

Technische Biologie und Bionik verhalten sich wie Bild und Spiegelbild.

When an engineer wishes to copy nature, he is bound to fail. Nature does not provide any blueprints for technology. But it provides an inexhaustible reservoir of structures, procedures and evolutionary principles which man, when he gets it right, can use to great advantage for technological advances. But how does he get it right? Quite simple. He should not slavishly attempt to rebuild every detail of a natural model but instead to discover the principles it is based on. The engineer can then incorporate these into his own technical design – perhaps just as a stimulus or maybe as an irreplaceable indicator. || Japanese company Sony launched a sweet little artificial dog on the market. He reclines when you pat him, rolls his eyes and barks when he is hungry and soon he will probably even be able to defecate. He looks like a small dog, but is simply a technical one. And he is actually not a bionic construction – rather, an exact-as-possible copy of a dog made "by technical means", a copy of its form and behavior patterns. | In Stuttgart Airport, graceful tree supports hold up the roof. This sort of support structure started with Frei Otto's legendary Stuttgart "Institut für leichte Flächentragwerke". These branched support structures are referred to as "tree supports" but do not look at all like trees. And they are nevertheless strictly bionic constructions. How is this so? || There are now many such support constructions. These do not copy trees, rather research is first conducted on the principles of support construction in trees and then this knowledge is put into practice on a technical level. | These principles include the way in which finely ramified branches spread out from the trunk, how the length decreases the finer the branches, how the length: diameter ratio changes, how the angle of branching changes with every new branch generation etc. These are the basic conditions. They determine the actual target function. And this means that the supports themselves should weigh as little as possible for the maximum given load they have to bear. The above-mentioned parameters are altered back and forth in the computer until the target function offers the desired minimum. Such tree supports can not be built lighter than this under the given requirements: Material is consequently spared in the best possible way, and money is thus also saved. | These "tree supports" are also a lot more graceful and beautiful than boring, floor-to-ceiling round columns. For the engineer this may be a side-effect, but for an architect it is probably the main creative effect – which more or less takes us on to design. || But bionic design can be better understood and classified if one thinks a little about the bionic procedures up front. || As stated above, you first have to study the trees, before you can technically translate their constructive principles. This sounds pretty trivial. But tell this to an economist or even a politician. Everyone likes business ideas with which you can proliferate sales and make money. But research costs money and before long, lots of money. You first have to invest before you can get anything back – something job-

Plants – and particularly trees – are the undisputed masters when it comes to developing lasting and stable, load-bearing and support structures. Trees (such as the tree fern) have discovered optimal solutions for the problem of supporting as large a load as possible by means of the appropriate number of branches which themselves do not weigh very much. The tendrils of the byrony, for example, represent an intelligent technical solution to the problem of twisting strands of cable.

Those who want to simply copy nature are bound to fail.

Technical biology and bionics behave like image and mirror image

wurde gesagt, man müsse die Bäume erst mal erforschen, bevor man ihre Bauprinzipien technisch umsetzen kann. Das klingt reichlich trivial. Aber erzählen Sie das mal einem Wirtschaftler oder gar einem Politiker. Ideen, mit denen man in der Wirtschaft Umsätze und Geld machen kann, die hätte man schon gerne. Aber Forschung kostet Geld, rasch auch viel Geld. Das muss man erst mal reinstecken, bevor man etwas rausholen kann. Und das sieht vor allem der auf Arbeitsplätze fixierte Politiker nicht so gern. Für den Wissenschaftler freilich ist das selbstverständlich. Max Planck hat einmal sinngemäß gesagt, vor dem Umsetzen müsse die Erforschung stehen. Für unser Thema heißt das: »Vor der bionischen Übertragung in die Technik muss die technisch-biologische Erforschung der Grundlagen stehen.« | Technische Biologie und Bionik verhalten sich also wie Bild und Spiegelbild, wie Kopf und Zahl einer Münze. Das eine kann man nicht ohne das andere haben. || Technische Biologie erforscht die belebte Welt, indem sie Vorgehensweisen und Kenngrößen der Technik und Ingenieurswissenschaften einbringt, um ihre Untersuchungsobjekte angemessen beschreiben und verstehen zu können. Wenn ich einen Wasserkäfer untersuche, der so schön strömungsschnittig aussieht – ein gutes »biologisches Design«, eine gute »Funktionsmorphologie« –, benutze ich den technischen Begriff des Widerstandsbeiwerts (cW-Wert), um ihn strömungsmechanisch zu kennzeichnen. Natürlich denke ich dabei auch schon an die Möglichkeit einer technischen Umsetzung, die ja ohne eine solche Kenngröße gar nicht auskommt. Wasserkäfer-Rümpfe eignen sich nach neuen Messungen hervorragend als Vorbilder für Riesen-Unterseeboote für den interkontinentalen Transport von Handelsgütern, fernab der Wasseroberfläche mit ihren Stürmen und anderen Störungen. | Bionik untersucht nun all die Kenngrößen, die den Wasserkäfer »technisch« beschreiben, spiegelt sie an den Anforderungen einer Großausführung und macht dem Schiffsbauer Designvorschläge. Mit anderen Worten: Technische Biologie betreibt diejenige Grundlagenforschung, die Bionik für ihre angewandte Weiterführung und für die technische Umsetzung benötigt. | Wenn die Bionik aber Designvorschläge macht, muss man wohl davon ausgehen, dass es auch so etwas wie ein »Biologisches Design« gibt, das die belebte Welt prägt und von den Grundprinzipien in die Welt des Menschen übertragbar ist. Ist dem so? Macht man solche Grenzvergleiche, so landet man rasch im Sumpf des Unverbindlichen, wenn man sich nicht an klaren Definitionen ausrichtet. Wir sollten also erst einmal fragen, wie »Design« überhaupt definiert ist und ob diese Definition auch für die Konstruktionen der belebten Welt einen Erklärungscharakter aufweist. | Das lateinische Wort *designare* bedeutet »bezeichnen, kennzeichnen«. Im Lexikon finden wir für »Design« auch die Begriffe »Entwurf, Planung, Muster, Modell«. Das Kunstlexikon spricht von der »Mitarbeit des Künstlers bei der Gestaltung einer Form«. Auf einen ebenso kurzen wie praktikablen Nenner hat es die

Die Wasseramsel zeigt in Perfektion eine Fähigkeit, die unsere Ingenieure noch heute vor größte Probleme stellt: ein Körper, dessen Stromlinienform sowohl für höchste Beweglichkeit in der Luft sorgt als auch ein sicheres Tauchen und wendiges Schwimmen erlaubt. Die technische Umsetzung dieses »bionischen Designs« in fliegende Unterseeboote dürfte noch in weiter Ferne sein.

focused politicians do not particularly like to hear. Of course for scientists it goes without saying. Max Planck once said on the subject that research should take priority over sales figures. In our discipline this means: "The technical-biological research must have been thoroughly conducted before it can be transferred bionically to technology." || Technical biology is an exploration of the vital world, into which it brings technical procedures and parameters and engineering sciences in order to be able to adequately describe and understand the object of study. When I examine a water beetle which looks so streamlined — a good "biological design", a good "functional morphology" — I use the technical term drag coefficient (cd) in order to identify it in terms of fluid dynamics. Naturally I also consider whether it is possible to translate this technically — which would not be possible without such a parameter. According to recent tests, the torsos of water beetles appear very suitable as models for submarines for the intercontinental transport of trade goods, far from the surface of the water with its storms and other disturbances. | Bionics investigates all the parameters which "technically" describe the water beetle, mirrows these in the requirements of a large-scale execution and offers the shipbuilder design suggestions. In other words, technical biology carries out the basic research which bionics requires for its applied extrapolation and for the technical implementation. | But if bionics can make design suggestions, one should also probably assume that something like a "biological design" exists, which shapes the vital world and is transferrable from the basic principles to the world of humans. Is that not so? In making such comparisons you can quickly land in the mire of the non-committal, if you do not stick to clear definitions. So first of all we should ask whether "design" can be defined in the first place and whether this definition is also able to explain structures in the animated world. | The Latin word *designare* means to "describe" or "identify". In the lexicon we also find words such as "draft, planning, sample and model" for "design". The art lexicon speaks about the "participation of the artist in creating a form". The former Ulm School of Design also summed it up for all practical purposes as "product design within the framework of a practical aesthetic." There are also products in nature but no aesthetic. This arises in the human brain as a reflection on what is perceived with the senses. | And thus the Ulm definition is suitable for the world of humans but cannot be transposed back on to the world of nature — if we want to keep these things at all separate. Here, we need another definition, namely a phrase which incorporates the concept of design in biology. About ten years ago I said the following on the subject which I can now confirm: "Biological design means creating organismic shapes in terrain defined by various functional requirements." This sounds pretty abstract, as do most natural scientific definitions. But in principle it is as clear as it is simple. It is about how we shape the living world

Technical biology and bionics behave like image and mirror image, like the head and tail sides of a coin. You cannot have one without the other.

The water ouzel perfectly displays a skill which still baffles our engineers today: a body whose streamlined form allows both the highest level of mobility in the air as well as safe diving and agile swimming. It will probably still take some time before this "bionic design" is translated into flying submarines.

ehemalige Ulmer Hochschule für Gestaltung gebracht. Sie spricht von der »Produktgestaltung im Rahmen einer praktischen Ästhetik«. Produkte gibt es auch in der Natur, aber keine Ästhetik. Diese entsteht im Gehirn des Menschen als Reflexion auf das mit Sinnesorganen Wahrgenommene. | Damit ist die Ulmer Definition für die Welt des Menschen geeignet, lässt sich aber nicht auf die »Welt der Natur« – wenn wir einmal so trennen wollen – zurück übertragen. Hierfür brauchen wir eine andere Definition, nämlich einen Satz, der den Designbegriff in der Biologie mit einbezieht. Vor etwa zehn Jahren habe ich dazu Folgendes gesagt, das seine Gültigkeit bestätigt hat: »Biologisches Design bedeutet organismische Formgestaltung im Kräftefeld unterschiedlicher funktioneller Anforderungen.« Das klingt nun reichlich abstrakt, wie fast jede naturwissenschaftliche Definition. Aber im Grunde ist es ebenso klar wie simpel. Es geht um die Gestaltung in der belebten Welt (daher das Wort »organismisch«), also nicht um eine Produktgestaltung durch den Menschen. Und es geht nicht um ein herausgegriffenes Einzelproblem. Die Natur befasst sich immer mit der Ausgestaltung ganzer Problemkomplexe. Und die sind einfach deshalb so »komplex«, weil ganz unterschiedliche Anforderungen auf den betrachteten Problemkreis einwirken. | Der letzte Punkt ist besonders wichtig. Es gibt keine »freie Gestaltung« in der belebten Welt, sondern stets funktionelle Gestaltungen als Antwort der Evolution auf funktionelle Anforderungen. »Funktionelles Design« im technischen Bereich entspricht aber den Begriffen »Funktionsanatomie« und »Konstruktionsmorphologie« in der Biologie. Somit lässt sich technisches und biologisches »Design« durchaus vergleichen. Der Begriff der Funktionalität hält beide zusammen. Und es ist unerheblich, ob eine überlegt entwerfende Technik oder eine »blind ablaufende« Evolution zu funktionellen Konstruktionen führt. | Was sind nun die ganz typischen und sozusagen einmaligen Aspekte eines »Biologischen Designs«, von dem der Mensch lernen kann? Zum einen ist es die Komplexheit im Detail, die sich immer dann ergibt, wenn man näher in die Symmetrie eindringt (obgleich es oft im Großem und Ganzen »recht einfach« aussieht). Zum anderen ist es die Vielfalt der Lösungsmöglichkeiten, welche die Natur bei ein und derselben Fragestellung anbietet. Beide Aspekte finden sich häufig kombiniert. Eine biologische Formkonstruktion muss stets Mehrfachansprüche abdecken, ein Gesichtspunkt, auf den sich Designer im Zuge immer stärker ökologisch ausgerichteter Designansprüche erst einstellen müssen. Zwei Beispiele: || Die Eischale der Schmeißfliege besteht »nur« aus Chitin, allerdings in sehr unterschiedlicher Ausformung. Damit werden die folgenden vier Ansprüche abgedeckt: Leichtbau, gute Elastizität, Wasserabweisung und Gasdurchtritt. | Der Stachel des Stachelschweins besteht »nur« aus Hornsubstanz, die allerdings zwei ineinander laufende Lamellensysteme bildet, welche durch eine Art Ausschäumung auf Abstand gehalten werden. Damit werden die folgenden vier

Eine biologische Form stellt stets eine optimierte Kompromissantwort auf unterschiedliche Funktionsanforderungen dar.

(thus the word "organismic"), in other words not about products designed by man. And it is not about an individual problem singled out from all the rest. Nature always focuses on addressing whole complexes of problems. And the reason they are so "complex" is because quite different demands have an influence on the observed cycle of problems. | The latter point is particularly important. There is no "free design" in the living world but always functional designs as evolution's response to functional demands. "Functional design" in the technical field corresponds to the terms "functional anatomy" and "structural morphology" in biology. In this way technical and biological "design" can certainly be compared. The term functionality brings both together. And it is irrelevant whether a considered design technology or a "blindly proceeding" evolution leads to functional constructions. | So what are the typical and, as it were, unique aspects of a "biological design" from which man can learn? For one, the complexity of detail which always arises when one gets closer to the symmetry (although by and large these often look really simple). On the other hand it is the diversity of solution possibilities which nature offers for one and the same question. Both aspects are often found in combination. A biological form construction always has to cover multiple demands, a criterium which designers have to take into account in the course of design aspirations which are becoming increasingly ecological. Two examples: || The shell of bluebottle eggs consists "solely" of chitin, although in a very different shape that meets the following four criteria: light structure, prime elasticity, water resistance and the gas permeability. | The spines of the porcupine consist "solely" of horn which, however, forms two intertwined lamella systems kept apart with a sort of foam. It covers the following four demands: light construction, stable penetrative form, high buckling strength and a high degree of elasticity. | A biological form always represents an (optimized) compromise for the demands of different functions. For example with the manifold suction apparatuses in the animal realm, as a biologist you could continue to represent the tremendous diversity of possible functional solutions which — independent of the various external demands — have developed in the most different of ways. Similarly, the no less impressive complexity of detail could be described by taking the example of sea urchin teeth. The function was always in the forefront. | Ideally, technical design should link the components of a technical system in a functionally optimal way and also conceive the external formal design (sheath) in such a way as to, on the one hand, enclose the inner part in a functionally optimal way and, on the other, satisfactorily consider the criteria of appearance, that is, aesthetic aspects. Many everyday products which you can buy in the supermarket are beautified in design to have an effect, a pure formal pseudo-aesthetic effect, lacking the function of good design. Good design is, without exception, characterized

A biological form always represents an optimized compromise for various functional demands.

Ansprüche abgedeckt: Leichtbau, stabile Eindringform, hohe Knickfestigkeit und große Elastizität. | Eine biologische Form stellt stets einen (optimierten) Kompromiss dar im Anforderungsfeld unterschiedlicher Funktionen. Am Beispiel der vielfältigen Saugapparate im Tierreich könnte man als Biologe weiter die ungeheuere Vielfalt der funktionellen Lösungsmöglichkeiten darstellen, die sich – abhängig von den unterschiedlichen äußeren Anforderungen – in verschiedenartigster Weise herausgebildet haben. In gleicher Weise ließe sich am Beispiel der Seeigelzähne die nicht minder beeindruckende Komplexheit im Detail herausarbeiten. Immer stünde dabei die Funktion im Vordergrund. | Im Idealfall sollte technisches Design die Komponenten eines technischen Systems funktionell-optimal verknüpfen und auch die äußere Formgestaltung (»Hülle«) so konzipieren, dass sie einerseits das Innere funktionell-optimal umschließt, andererseits aber auch Gesichtspunkte des Erscheinungsbilds, also ästhetische Aspekte, befriedigend mit bedenkt. Viele Alltagsprodukte, die man im Supermarkt kaufen kann, sind designerisch »auf Wirkung« geschönt, auf reine Formeffekte einer Pseudoästhetik, der die dienende Funktion eines guten Designs fehlt. Gutes Design ist ausnahmslos dadurch gekennzeichnet, dass diese beiden Grenzaspekte unaufdringlich ineinander laufen, gut verzahnt sind. Gutes Design wirkt eben deshalb stets auch »einfach« und »schön«. Mit Ausnahme des Ästhetik-Gesichtspunkts gilt dies ebenso für das »Biologische Design«. | Die Vergleichbarkeit von Technik und Biologie ergibt sich einfach auch daraus, dass die »schöpferische Natur« vor genau denselben Problemen war, ist und sein wird, vor denen im zivilisatorischen Bereich der »schöpferische Designer« steht: Es gilt, das (für die Besetzung einer ökologischen Nische, für die Ausfüllung einer Marktnische etc.) am besten Geeignete zu finden. Dieses ist stets das System, bei dem die Optimierung des Ganzen Vorrang hat vor einer Maximierung einzelner Elemente. Es geht also um einen möglichst günstigen Kompromiss. Ein solcher »Optimalkompromiss« war auch die eingangs geschilderte »bionische Baumstütze«. Sie hat zum Beispiel nicht mit möglichst wenigen oder möglichst dünnen Einzelstäben gearbeitet (das wäre eine »Maximierung des Einzelelements«), sondern mit einer optimalen Verzahnung aller Randbedingungen. Das Optimum war hier ein Minimum, nämlich die Stütze möglichst geringer Gesamtmasse. Es gibt bereits viele Designbüros, die nach diesen Überlegungen »natürliche Prinzipien« in ihr funktionell-gestalterisches Wirken mit einbeziehen. Als Beispiel genannt sei Carmelo di Bartolo, vormals *Istituto Europeo di Design* in Mailand, der auch zum methodischen Vorgehen im Bereich des Bionik-Designs Stellung genommen hat. || Die Strukturen, Funktionen und Strategien der Natur bilden ein riesiges Reservoir für technologische Zukunftsaspekte. Das »Anzapfen« dieses Reservoirs kann aber, wie eingangs geschildert, nicht kritiklos geschehen, nicht durch schlichtes Abkupfern (»Naturkopie«) – das wäre reine Scharlatanerie.

by the fact that these two delineating aspects discreetly merge and cog together well. This is why good design looks "simple" and "beautiful". This also holds true for biological design with the exception of the criterion of aesthetic appeal. | The comparability of technology and biology also simply arises from the fact that "creative nature" is, was and will continue to be faced with the same problems as the "creative designer" is faced with in the field of civilization: It is essential (for the occupation of an ecological niche, or the filling of a market niche etc.) to find what is most suitable. This is always the system — in which optimization of the whole holds precedence over maximizing individual elements. In other words it is about finding the most favorable compromise. The "bionic tree support" which we mentioned at the beginning is also such an "optimal compromise". It does not work with as few as possible or a thin as possible individual rods (that would be a "maximization of individual elements") but by optimally linking all basic conditions. In this case the optimum was a minimum, namely that the supports feature as low a total volume as possible. There are already many design offices which incorporate "natural principles" in their functionally creative effects, in accordance with these ideas. One example is Carmelo di Bartolo, formerly of the "Istituto Europeo di Design" in Milan, who has also commented on methodological procedures in the field of bionic design. || The structures, functions and strategies of nature form an enormous reservoir of future technological aspects. Tapping into these reservoirs, however, as was outlined in the beginning, cannot be uncritical, nor by merely copying nature — that would be pure charlatanism. Rather one should critically comb through the tremendous potential of "nature's discoveries", research them and implement them technically, thus making them useful for humanity. The design of our future technical structures — and here I refer to an inner and outer structural and functional context — could be increasingly geared in a bionic direction. This should happen everywhere it makes sense. And there is a tremendous number of technical facets for which a natural orientation would most certainly make sense. | We can formulate abstract issues and statements drawing on practical experience with such a creative effect and the inclusion of natural models. The issue of multi-disciplinarity: Often it is the teamwork between the disciplines which first leads to success. Issue: progress and recourse: A discovery sometimes has an innovative impact in unexpected areas. Issue: innovation: Conceptual integration of already known effects can also be innovative. || There are also other possible discernible links between nature and design: Nature offers the material possibilities for a near-natural design. Nature is the source of all man's inspirations and expertise. Nature is also the source of inspiration for creative design. | A lot of thought has gone into this and the interconnections between ecology and design. For example, designer Günter Horntrich from

Simply "copying" from nature is nothing but pure charlatanism.

Man sollte vielmehr das ungeheure Potential der »Erfindungen der Natur« kritisch durchforsten, erforschen sowie technisch umsetzen und damit der Menschheit nutzbar machen. Das Design unserer zukünftigen technischen Gebilde – und damit meine ich einen inneren und äußeren strukturellen und funktionellen Zusammenhang – könnte dann mehr und mehr bionisch ausgerichtet werden. Das sollte überall dort geschehen, wo es sinnvoll ist. Und es gibt eine ungeheuere Zahl technischer Facetten, für die eine Naturausrichtung durchaus sinnvoll wäre. | Aus der Praxis eines solchen gestalterischen Wirkens und des Einbindens natürlicher Vorbilder kann man Problemkreise und Aussagen abstrahieren. Problemkreis Interdisziplinarität: Oft führt erst die Zusammenarbeit der Disziplinen zum Erfolg. Problemkreis Fortschritt und Rückgriff: Eine Entdeckung wirkt sich bisweilen an unerwarteten Enden innovativ aus. Problemkreis Innovation: Auch konzeptionelles Einbinden bereits bekannter Effekte kann innovativ sein. || Darüber hinaus lassen sich mögliche Querverbindungen zwischen Natur und Design erkennen: Die Natur bietet die materialtechnischen Möglichkeiten für ein naturnahes Design. Die Natur ist die Quelle aller Inspirationen und Erkenntnisse des Menschen. Die Natur ist somit Anregungsquelle auch für ein kreatives Design. | Auch darüber und über Querverbindungen zwischen Ökologie und Design wurde vielfach nachgedacht. Beispielsweise sagt der Designer G. Horntrich vom Fachbereich Design der FH Köln Folgendes: »Das Design ist in ein Spannungsfeld geraten, das es selbst mitgeschaffen hat. An der Überproduktion und dem ständigen Wachstum unserer Gesellschaft hat auch das Design mitgewirkt. Entwicklungen wie diese erfordern heute Konsequenzen auch für den Designer. Ganzheitliches Denken ist gefragt. Die Herausforderung des Gestalters heißt ›Systemdesign‹. Für die Ausbildung des Designers bedeutet dies eine Verschiebung der Schwerpunkte. Über die klassische Produktentwicklung hinaus gilt es heute, ganzheitliche Konzepte und Kreisläufe zu gestalten.« || G. Horntrich führt aus: »Eine ökologische ganzheitliche Betrachtungsweise beinhaltet neben der aktiven auch die passive Ökologie des Produkts. Aktive Einflussfaktoren sind der Einsatz umweltgerechter Werkstoffe, eine umweltgerechte Verarbeitung und die Berücksichtigung von Umweltbelastungen und Energieverbrauch. Unter der passiven Ökologie eines Produkts ist seine Umwelt zu verstehen. Sinnvolle Konzepte zur Ökologie von Produkten befassen sich auch mit Aspekten wie Mehrfachnutzung oder Wiederverwendung. Bisher bestand die Aufgabe des Designers darin, neue Produkte zu entwerfen. Veränderte Bedürfnisse erfordern jedoch die Konzentration der Kreativität auf grundsätzliche Aspekte.« | Die Sichtweise des hier zitierten Designers und die Sichtweise des Technischen Biologen und Bionikers verzahnen sich gut. Wenn man natürliche Konstruktionen als reines Formvorbild nimmt, so ist das selbstredend zulässig, aber reines Formdesign führt häufig nicht sehr weit. Ein Beispiel ist die Entwicklung

Schlichtes »Abkupfern« aus der Natur ist nichts als Scharlantanerie.

Biologische Systeme bilden ein riesiges Reservoir für technologische Zukunftsaspekte.

the Design Department at the Cologne University of the Applied Sciences, comments: "Design has become a field of polar tension, one it helped get itself into." Design has also played a part in the excess production and the constant growth of our society. Developments such as these have their consequences today, and also for designers. What we need is holistic thinking. The designer's challenge is one of "system design". This means there need to be a shift in the focus of design education. It is essential to design holistic concepts and cycles." || Günter Horntrich remarks: "An ecological, holistic perspective includes the active as well as the passive ecology of a product. Active influential factors are: the use of environmentally sound materials, environmentally friendly processing and taking into account the environmental impact and energy consumption. The passive ecology of a product can be understood as its environment. Meaningful concepts on the ecology of products also take into account aspects such as multiple usage or recycling. To date, the task of designers has consisted of designing new products. Changed needs, however, require that creativity be focussed on fundamental issues." | The perspective of the designers quoted herein and that of technical biologists and bionic scientists dovetail together well. To take natural constructions as a pure form model is of course permissible, but pure formal design often does not lead very far. One example is the design of the "Amarys" telephone shape, by France Telecom – created by graphically distorting the shape of a nautilus shell. The product looks more futuristic than a normal telephone, but functions just the same. | In general we can speak of a functional design and thus a bionic design if the functional aspect is also considered. Rudolf Bannasch's concept for a "self-aligning grip" is a good example: the automatic aligning function derives from the tail fins of fish. | And only if the object to be designed is not considered on its own, but in conjunction with its passive and active ecological impacts, can you call it an ECOLOGICAL DESIGN. An example of this would be U. Küpper's concept of waste-free packaging which systematically considers all the links in the life cycle of an optimized packaging – from the low-energy material supply and energy-optimized transport all the way to the complete composting if possible. | Paying as much consideration as possible to whole system chains in an ecological regard, including bionic stimuli, is the challenge to the design of the future.

Biological systems make up a giant reservoir of future technological aspects.

der Telefonform »Amarys« von France Telecom aus der graphischen Verformung einer Nautilus-Schale. Das Produkt sieht futuristischer aus als ein normales Telefon, funktioniert aber genauso. | Erst wenn der Funktionsaspekt mitbetrachtet wird, kann im Allgemeinen von einem Funktionsdesign und damit von einem bionischen Design gesprochen werden. Als Beispiel kann das Konzept von R. Bannasch eines »sich selbst anschmiegenden« Greifers dienen, dessen automatische Anschmiege-Funktion der Schwanzflosse von Fischen abgeschaut worden ist. | Und erst, wenn der zu gestaltende Gegenstand nicht für sich betrachtet wird, sondern in seinen passiven und aktiven ökologischen Auswirkungen systemisch hinterfragt wird, kann man von einem ÖKOLOGISCHEN DESIGN sprechen. Ein Beispiel dafür wäre U. Küppers Konzept einer abfallfreien Verpackungswirtschaft, das die Gesamtketten im Leben einer optimierten Verpackung – von der energiearmen Materialbereitstellung über den energetisch optimierten Versand bis hin zu einer möglicherweise vollständigen Kompostierung – systemisch mitbetrachtet. | Das Durchdenken ganzer Systemketten in ökologischer Hinsicht unter weitgehender Einbeziehung bionischer Anregungen ist die Herausforderung an ein Design der Zukunft.

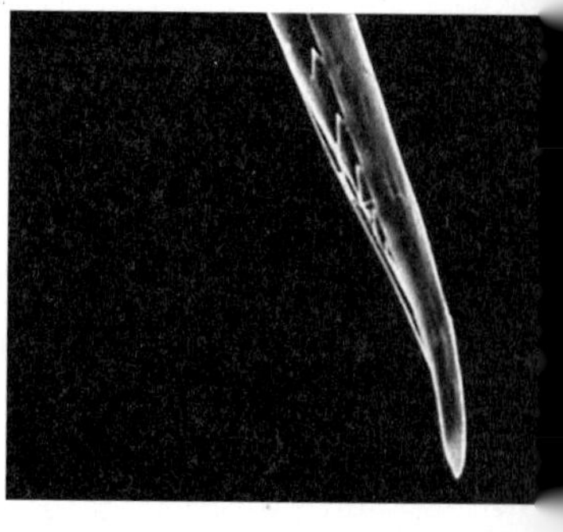

Wie Natur die Technik inspiriert…
How nature inspires technology…

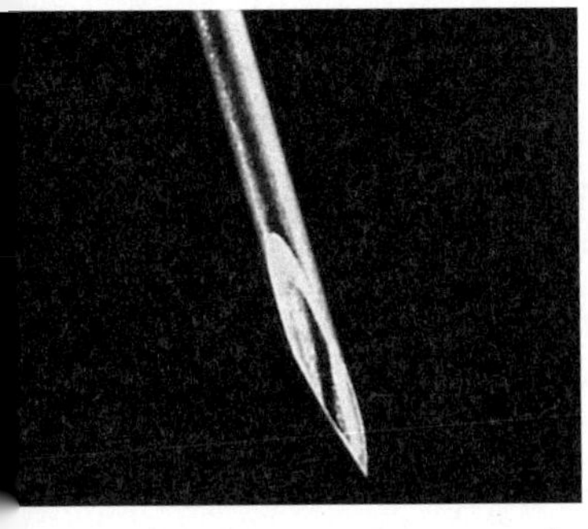

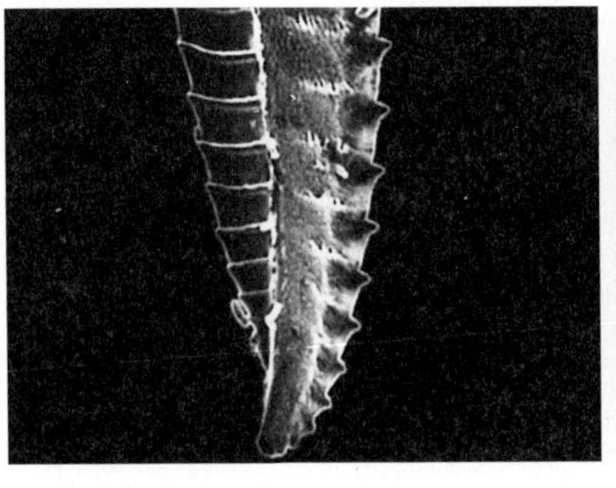

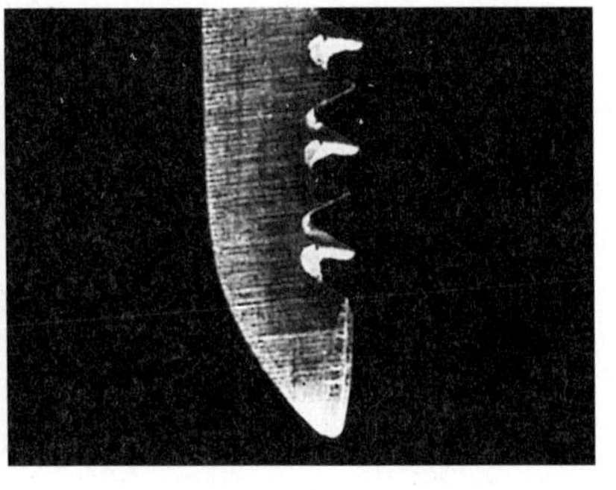

1 2

Abb.1 Stichsäge bei Blattwespen und technische Taschenmesser-Säge
Die Hinterleibsanhänge der Blattwespen zeichnen sich durch sägeblattartige Auswölbungen aus, mit denen sie sich beim Gegeneinanderarbeiten der einzelnen Teile rasch in Pflanzenmaterial einsägen. Es gibt auch Mehrfachzahnungen, ganz analog den Klappsägen bei modernen Vielzweck-Taschenmessern.

Abb.2 Giftstachel der Honigbiene und Injektionskanüle
Der Giftstachel, besteht aus dunkelgefärbtem, sehr hartem Chitin, ist außerordentlich dünn und langgestreckt, schmalkonisch. Am Ende trägt er eine seitliche Mündung. Technische Injektionskanülen erreichen auch in feinster Ausführung nicht die Zartheit und Stabilität des biologischen Gegenstücks.

Fig.1 Saber saw on saw flies and technical penknife saw
The saw flies' rear body appendages are characterized by saber saw-like cambers, with which, with the individual parts working against each other, they can they swiftly saw into plant material. There are also several degrees of saw tooth, just like the folding saws on modern penknives.

Fig.2 Honey bee poison sting and Injection cannula
The poison sting consists of dark-colored, very hard Chitin, is extraordinarily thin, long, slender, and conical. At its end there is a side outlet. Even highly developed technical injection cannulas do not have the gentleness and stability of their biological counterpart.

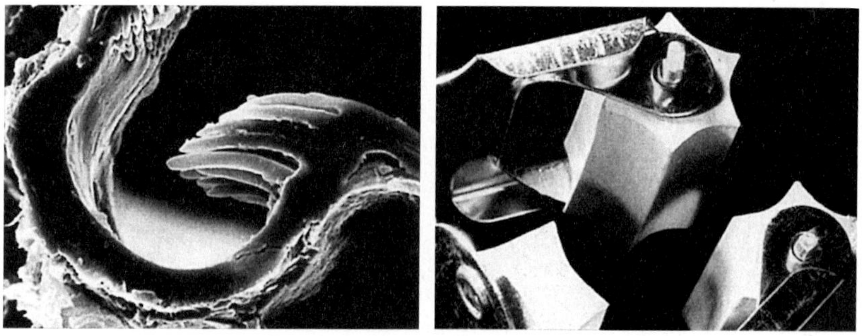

3

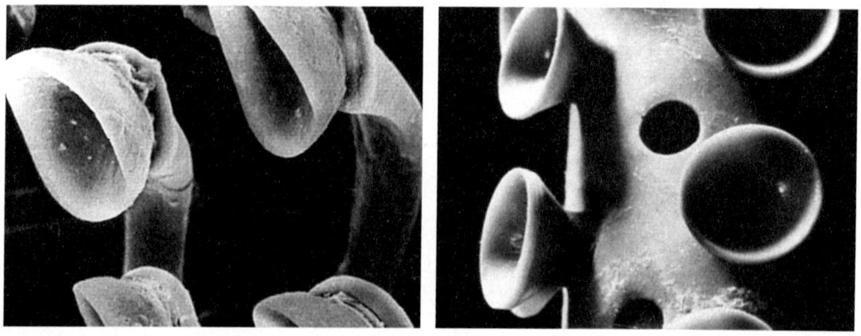

4

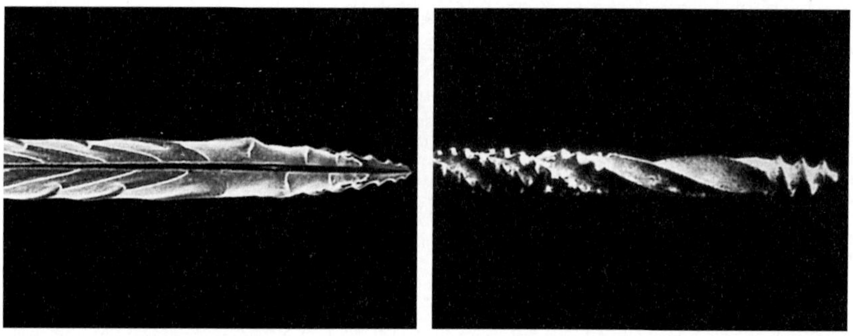

5

Abb.3 Flügelkoppelung bei Landwanzen und Besenklemme

Bei zahlreichen fliegenden Insekten werden die Vorder- und Hinterflügel durch Koppelungsmechanismen zusammengehalten. Die Besenklemme umfasst in ähnlicher Weise den Rundstiel des Besens.

Abb.4 Saugnäpfe des Gelbbrandkäfers und Saugnapf-Seifenhalter

Das Männchen des Gelbbrandkäfers besitzt an seinen Vorderbeinen Saugnäpfe. Entsprechende technische Saugnäpfe finden sich beispielsweise an Badezimmermatten und Seifenhaltern.

Abb.5 Holzbohrer der Riesenholzwespe und Bohrraspel

Die Riesenholzwespe bohrt sich zur Eiablage mit einem stecknadeldünnen, dreiteiligen Legebohrer 1 bis 1,5 cm tief ins Holz ein. Analoge technische Bohrwerkzeuge für den Heimwerker sind als Bohrraspeln bekannt.

Fig.3 Wing coupling in bugs and brush clip

Numerous kinds of flying insects have their front and rear wings held together by coupling mechanisms. In a similar way the brush clip surrounds the circular stem of the brush.

Fig.4 Suckers of the great diving beetle and suction soap holder

The male species of the great diving beetle has suckers on its front leg. Corresponding technical suckers can be found for example on bathroom mats and soap holders.

Fig.5 Wood drills of the giant ichneumon wasp and drill rasps

To deposit its eggs the giant ichneumon wasp uses its slender triple-section ovipositor to drill 1 to 1.5 cm deep into the wood. Similar technical drilling devices for DIY are known as drill rasps.

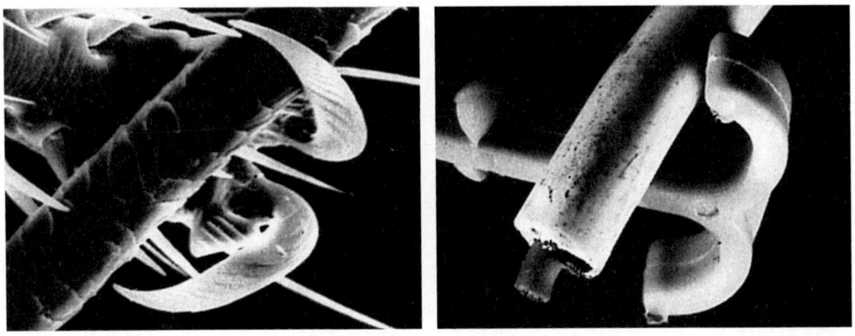

6

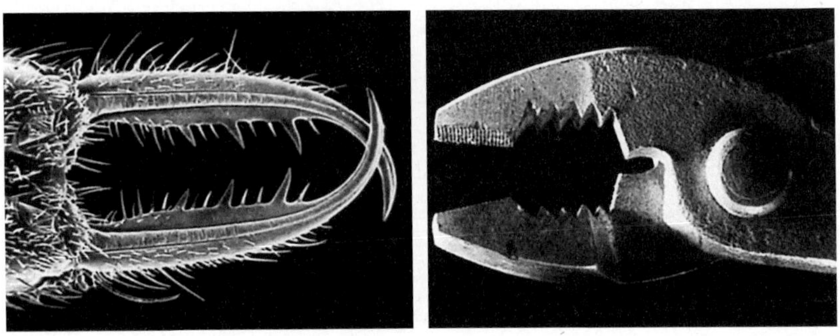

7

Abb.6 Klammerklaue des Hundeflohs und Spezialdübel zur Kabelverankerung

Beim Hundefloh enden die Beine in rückgebogenen Hafthaken, deren Wölbung sich dem Durchmesser eines Hundehaares anpasst. Ähnlich angepasst sind bestimmte Untergreifdübel für Elektroleitungen.

Abb.7 Oberkiefer des Ameisenlöwen und Kombizange

Die Oberkiefer des Ameisenlöwen sind Vielzweckgeräte, die als Sandwurfschaufel, Festklemmeinrichtung, Anstechapparat und Saugvorrichtung dienen. In ähnlicher Weise ist eine Kombizange für mehrere Funktionen geeignet.

Fig.6 Dog flea's hooked claw and special dowel for securing cables

A dog flea's legs end in a backward-curved claw, the curvature of which is adapted to the diameter of a dog's hair. Certain bottom-clasp dowels have been adapted to electric cables.

Fig.7 Upper jaws of the ant lion and combination pliers

The upper jaws of the ant lion are multi-purpose devices, which serve as a sand shovel, clamping gadget, stinger and suction facility. Combination pliers are also suitable for several uses.

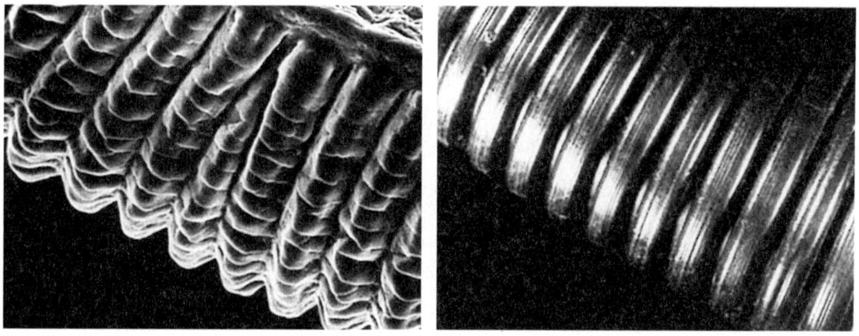

8

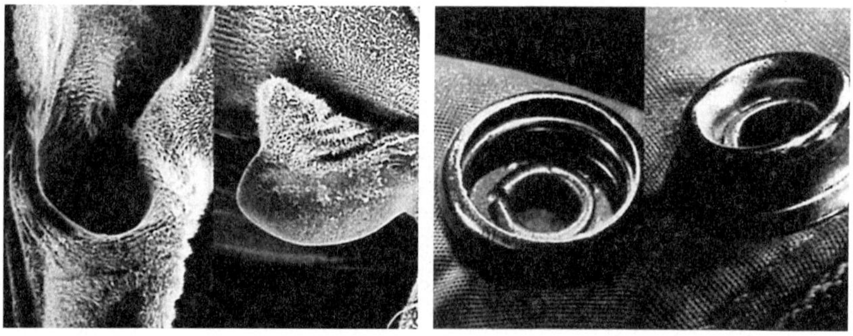

9

Abb.8 Biegeader am Sandlaufkäfer-Flügel und Biegesäule einer Taschenlampe
Der Sandlaufkäfer besitzt im Hinterflügel Biegeadern. Sie können spalt-, knick- und beulungsfrei verbogen werden und entsprechen so Metall-Duschschläuchen oder Trägern für Taschenlampen.

Abb.9 Druckknopfsystem beim Rückenschwimmer und technischer Druckknopf
Der Rückenschwimmer besitzt eine wohlausgebildete Druckknopfeinrichtung zum Ankoppeln der harten Vorderflügel an den Rumpf. Funktionell entsprechen sie den Druckknöpfen der Bekleidungsindustrie.

Fig.8 Bending vein on tiger beetle wing and bending column on a torch
The tiger beetle has bending veins in its rear wing. These can be bent without splitting, buckling and denting and as such are similar to metal shower hoses and torch holders.

Fig.9 Push-button system for back swimmers and technical press-stud
The back swimmer has a highly developed push-button facility for coupling its hard front wings to its trunk. In terms of function they correspond to press-studs in the clothing industry.

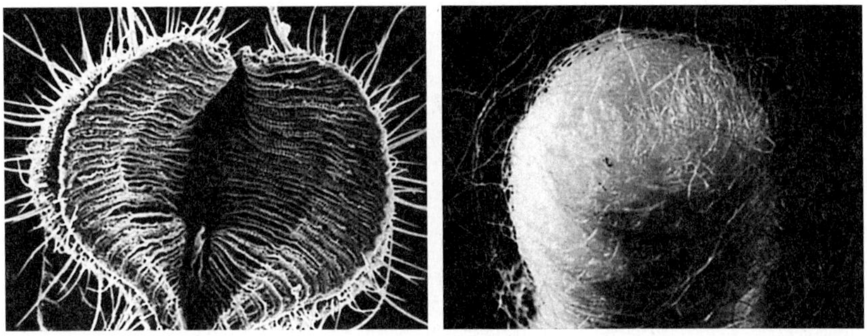

10

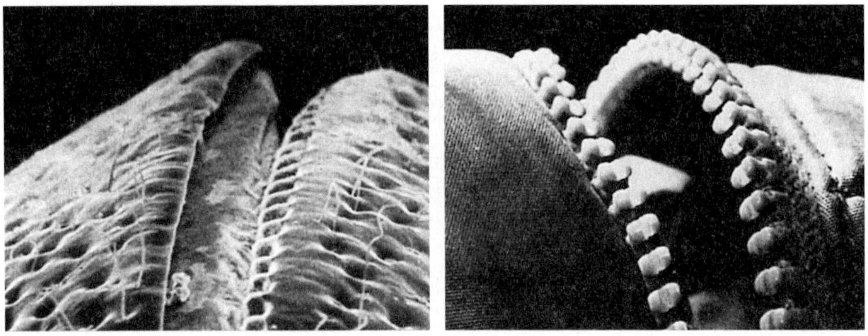

11

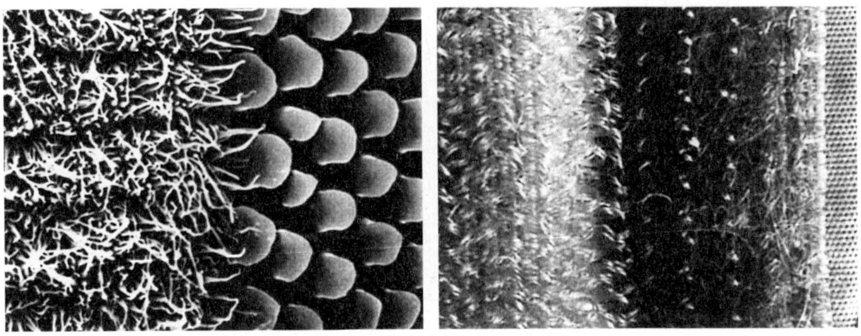

12

Abb.10 Saugrüssel bei der Schmeißfliege und »Wattestäbchen«
Die Schmeißfliege besitzt einen Tupfrüssel. Das Prinzip der Oberllächenverbreiterung ist im technischen Bereich bei Wattestäbchen verwirklicht.

Abb.11 Reißverschlusssystem beim Zwergrückenschwimmer und technischer Reißverschluss
Mit Hilfe einer Art Reißverschlusssystem, nämlich nacheinander einrastender Nuten und Noppen, werden beim Zwergrückenschwimmer die Flügeldecken gegeneinander fixiert. Die chitinösen Strukturen werden beim Einrasten eines einzelnen Zackens ein wenig gedehnt und schließen sich dann wieder.

Abb.12 Statische Verhakung am Koppelungssystem des Rückenschwimmers und technischer Klettverschluss
In eine feine, chitinöse »Wolle« des einen Teils greifen feinste Widerhakenschichten des anderen Teils ein. Dies reicht entsprechend den technischen Klettverschlüssen für eine solide Verbindung.

Fig.10 Trunk on blow-fly and cotton swab
The blow fly has a trunk. The principle of surface expansion can be found in a technical sense in cotton swabs.

Fig.11 Zip system on dwarf back swimmer and technical zip
The wing covers on the dwarf back swimmer are fixed against each other using a type of zip system involving notches and knobs that inter-link one after the other. When just a single prong clicks into place the chitinous structures are stretched a little before closing again.

Fig.12 Static inter-hooking on the back-swimmer's coupling system and technical Velcro
Ultra-fine barbed hook layers on one section insert themselves in a fine, chitinous "wool" on the other. Like Velcro, this produces a sturdy connection.

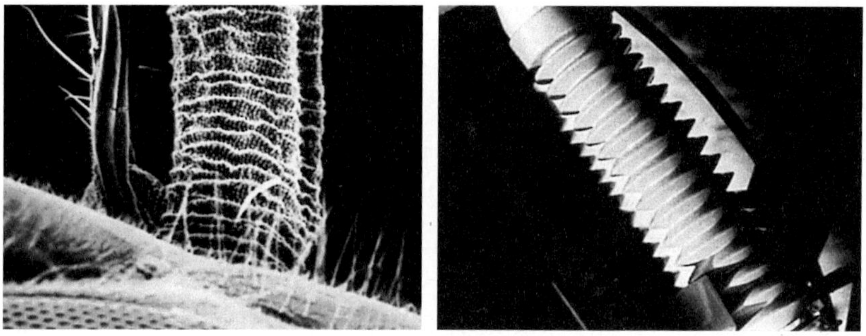

13

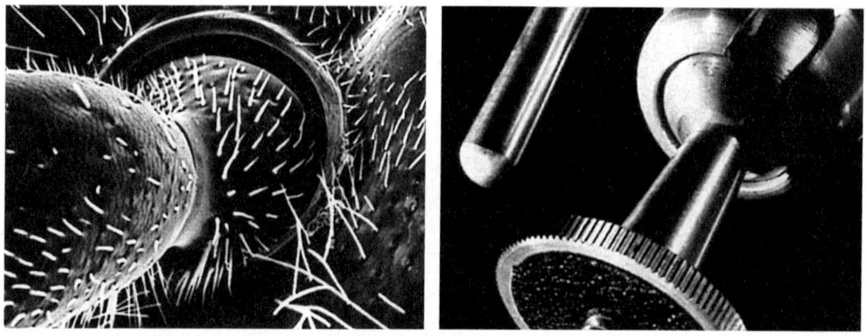

14

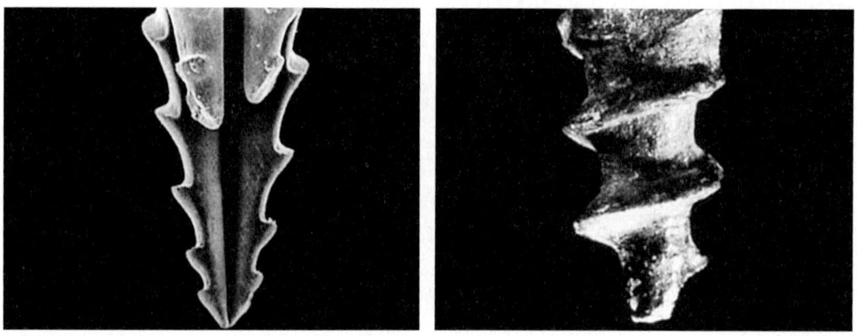

15

Abb.13 Stoßdämpfer am Fliegenrüssel und an einer Motorradgabel
Die Schnabelschwebfliege besitzt an der Rüsselbasis ein stoßdämpferartiges Gebilde, das möglicherweise die Erschütterungen beim kräftigen Aufsetzen abdämpft. Technische Stoßdämpfer finden sich beispielsweise als Federbeine bei Motorrädern.

Abb.14 Kugelgelenkiger Fühleransatz beim Bienenwolf und Foto-Kugelgelenk
Insektenfühler müssen in alle Richtungen bewegt werden können. Diesem Zweck dient eine kugelige Einlenkung des basalen Fühlerglieds. Gleiches gilt für Foto-Kugelgelenke.

Abb.15 Legebohrer und Bohrraspel
Auf gleiche Vergrößerung gebracht, überzeugt das Naturgebilde durch seine hochfeine Ausführung. Das technische Bildungsprodukt begnügt sich mit einer Grobspirale.

Fig.13 Shock absorber on fly trunk and a motorbike fork
The hover fly has a shock-absorber-like structure on the base of its trunk that possibly cushions the force of heavy landing. Technical shock absorbers can be found as shock absorbing struts on motorbikes.

Fig.14 Ball-joint feelers on bee wasp and photo ball joint
Insects' feelers need to be able to be moved in all directions. Nodular-based motion of the basal feeler serves this purpose. The same applies for photo ball joints.

Fig.15 Laying drill and drill rasps again
Enlarged to the same size, the natural structure is noticeable for its refined design. The technical product makes do with a basic spiral.

6

Form Follows Nature

Rudolf Finsterwalder

Die Natur zu verstehen und von ihr zu lernen ist ein alter Menschheitstraum. Die Mischung von Regelmäßigkeit und unendlicher Vielfalt fasziniert uns und stellt ein schier unerschöpfliches Thema dar, die in der Natur vorkommenden Muster scheinen unendlich. Dabei fällt auf, dass bestimmte Muster häufiger anzutreffen sind als andere. Manchmal besitzen zufällig anmutende Erscheinungen eine versteckte Ordnung, die zu entdecken ist. Die Mathematik ist das Werkzeug, das uns hilft, diese Dinge zu erklären und zu verstehen. | Viele der komplexen Erscheinungsformen der Natur sind auf die gleichen oder ähnliche mathematische Regeln zurückzuführen. Auch scheinbar chaotische und unregelmäßige Erscheinungen beruhen oftmals auf klaren Regeln und Strukturen. Die Suche nach einer Matrix des Lebens, nach einer Formel, die alles erklärt, ist eine Illusion, aber es ist möglich, bestimmte Regeln und Gesetze zu entdecken, die immer wieder auftauchen. || Vor 2000 Jahren postulierten griechische Philosophen einen Dualismus von Ordnung und Chaos. Die moderne Mathematik und Physik hat uns gezeigt, wie das Chaos zu begreifen ist und welches kreative Potential und welche Regelhaftigkeit in ihm steckt, dass auch das Chaos bestimmten Regeln gehorcht, dass auch in ihm Ordnungen stecken, die verstanden werden können. || Das Chaos erzeugt komplizierte Strukturen, die aus einfachen Regeln entstehen. Eine ungeheure Vielfalt an Formen, auch sehr komplexe und komplizierte Strukturen können aus einfachen mathematischen Formeln entstehen. | Das Leben organisiert sich selbst und entwickelt sich mit Hilfe bestimmter Regeln. Dieses Prinzip wird in die Entwicklung von Formen und Skulpturen umgesetzt, eine mathematische Regel oder ein Prinzip sind die Grundlage, auf der mit Hilfe von scheinbar willkürlichen Prozessen eine Form entsteht. Dieser Prozess soll jedoch nicht völlig zufällig ablaufen, sondern bewusst gesteuert. In den reinen mathematischen Prozess wird eingegriffen, der Ablauf wird verändert oder gelenkt und so Neues erschaffen, das nicht die perfekte Regelmäßigkeit der Mathematik besitzt, sondern Brüche und Spannungen in diese allzu perfekte Matrix einbringt. Es ist der Versuch, eine Matrix der Natur nutzbar zu machen für einen Gestaltungs- und Formungsprozess, ein abstrahiertes Abbild der Natur oder von Prinzipien natürlichen Wachstums zu zeigen. || Eine Serie von Skulpturen, alle mit diesem Entwurfsansatz während eines einjährigen Aufenthaltes an der Villa Massimo in Rom entstanden, zeigt die Anwendung dieses Prinzips auf die Architektur und Skulptur.

To understand nature and to learn from nature is an age-old dream of mankind. The mixture of regularity and endless variety is fascinating and offers an almost endless theme; that the patterns you can find in nature seem infinite. It is noticeable that certain patterns occur more often than others. Sometimes seemingly random phenomena have a hidden order behind it, waiting to be discovered. | Mathematics is the key that can help to explain and understand this phenomenon. Many of these complex entities of nature follow the same or similar mathematic rules. Even apparently chaotic and irregular phenomena are often based on clear rules and structures. The search for a matrix of life, for a formula that explains everything is an illusion, but it is possible to discover certain rules and laws that appear again and again. || 2000 years ago, Greek philosophers postulated a dualism of order and chaos. Modern mathematics and physics have shown us how to understand chaos, the creative potential it generates and what kind of regularity governs it. This seemingly chaotic structure follows certain rules, a certain underlining order, that can be understood. || Chaos creates complicated structures that grow from simple rules. An enormous variety of forms, even very complex and complicated structures, occur from simple mathematical formulas. | Life organizes itself and develops with the help of certain norms. This principle is implemented in the creation of forms, using a mathematical rule or order, resulting in structures assuming an apparently casual process. This process should not be completely casual, but deliberately controlled. Adjustments in the pure mathematical formula can alter the sequence or rules and create something new that does not have the perfect regularity of science, but brings about tension and disruption in this probably too-perfect matrix.It is an attempt to apply a matrix of nature to a design process, to show an abstract image of nature or the principles of natural growth. || A series of sculptures, all created with this concept during a one-year scholarship at the Villa Massimo in Rome, show the application of this principle on architecture and sculpture.

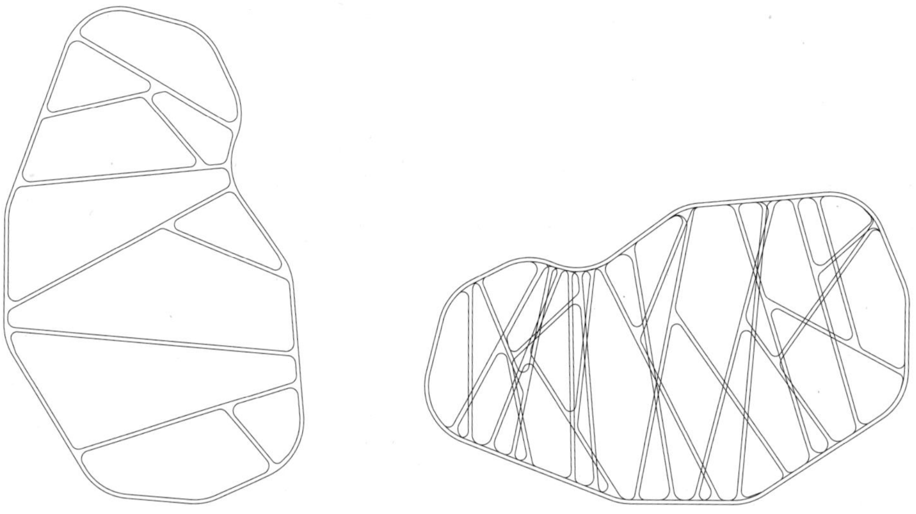

Dragonfly Vorbild für die Skulptur sind organische Strukturen, wie z. B. ein Libellenflügel oder die Rippen eines Blattes. Eine optimierte Struktur, die auf nichts mehr verzichten kann und durch ihren perfekten Minimalismus besticht.

Dragonfly An organic structure is a paradigm for the sculpture, such as a dragonfly's wing or the ridge of a leaf. An optimized structure, which is reduced to the essentials, creates a perfect minimalistic form.

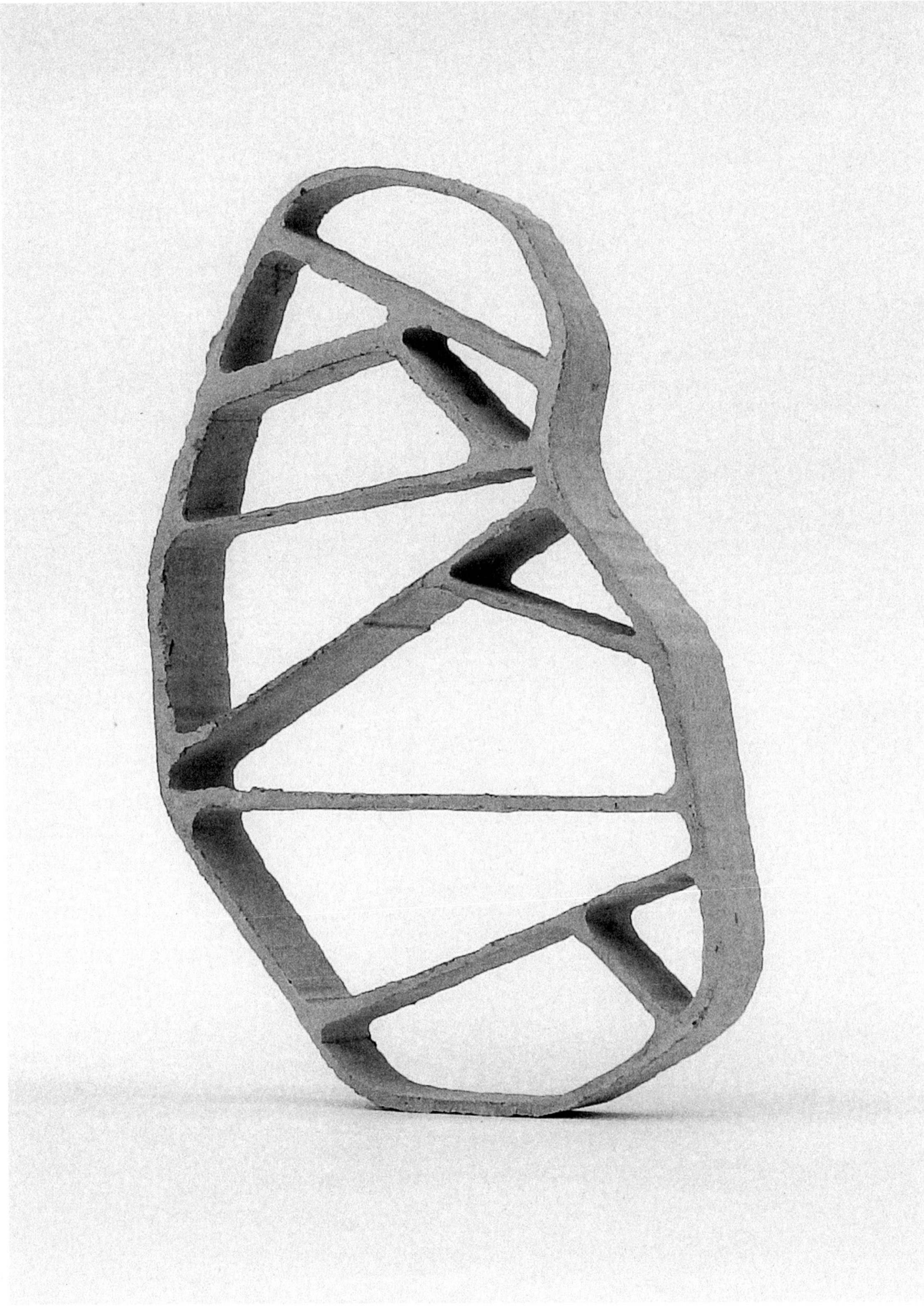

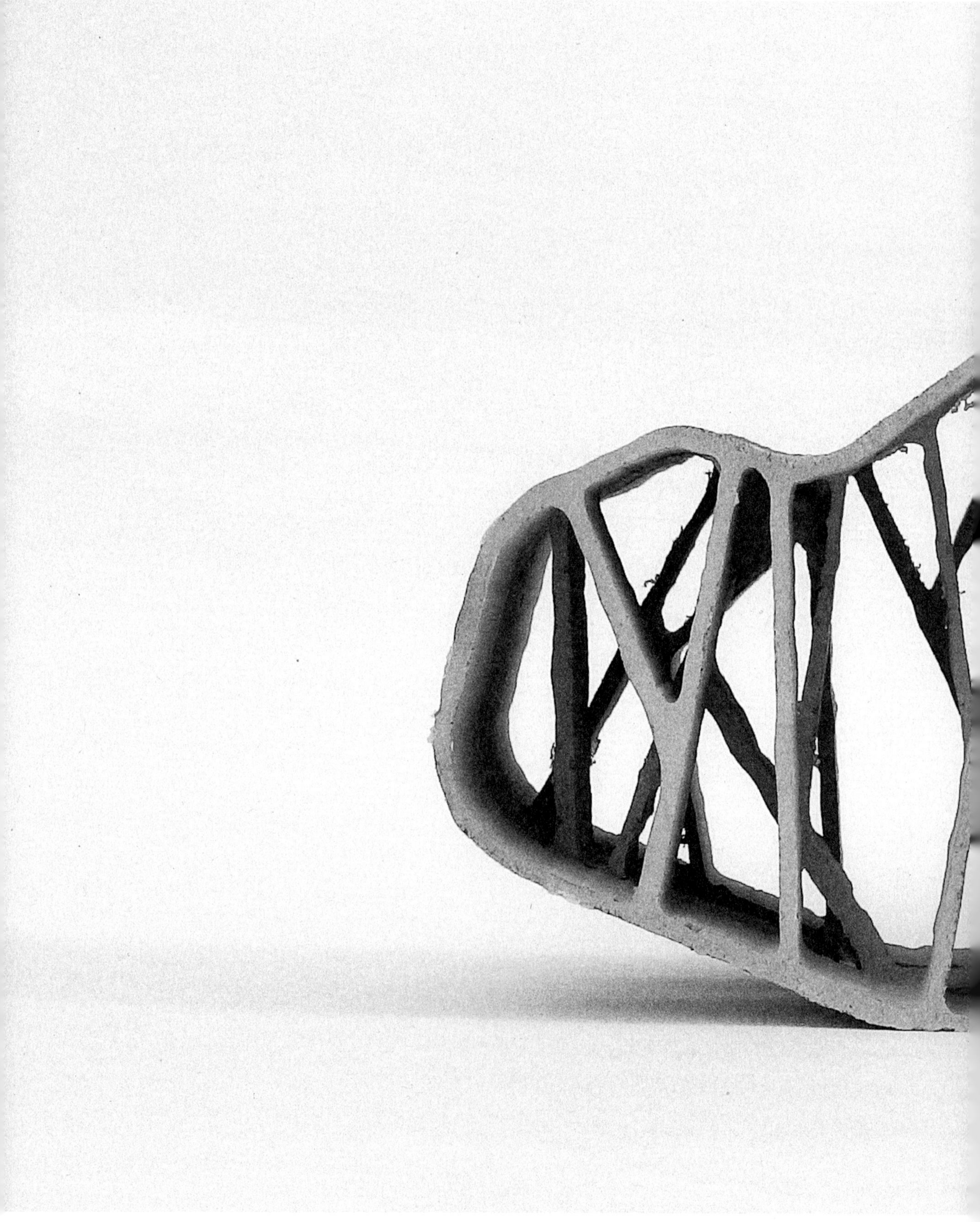

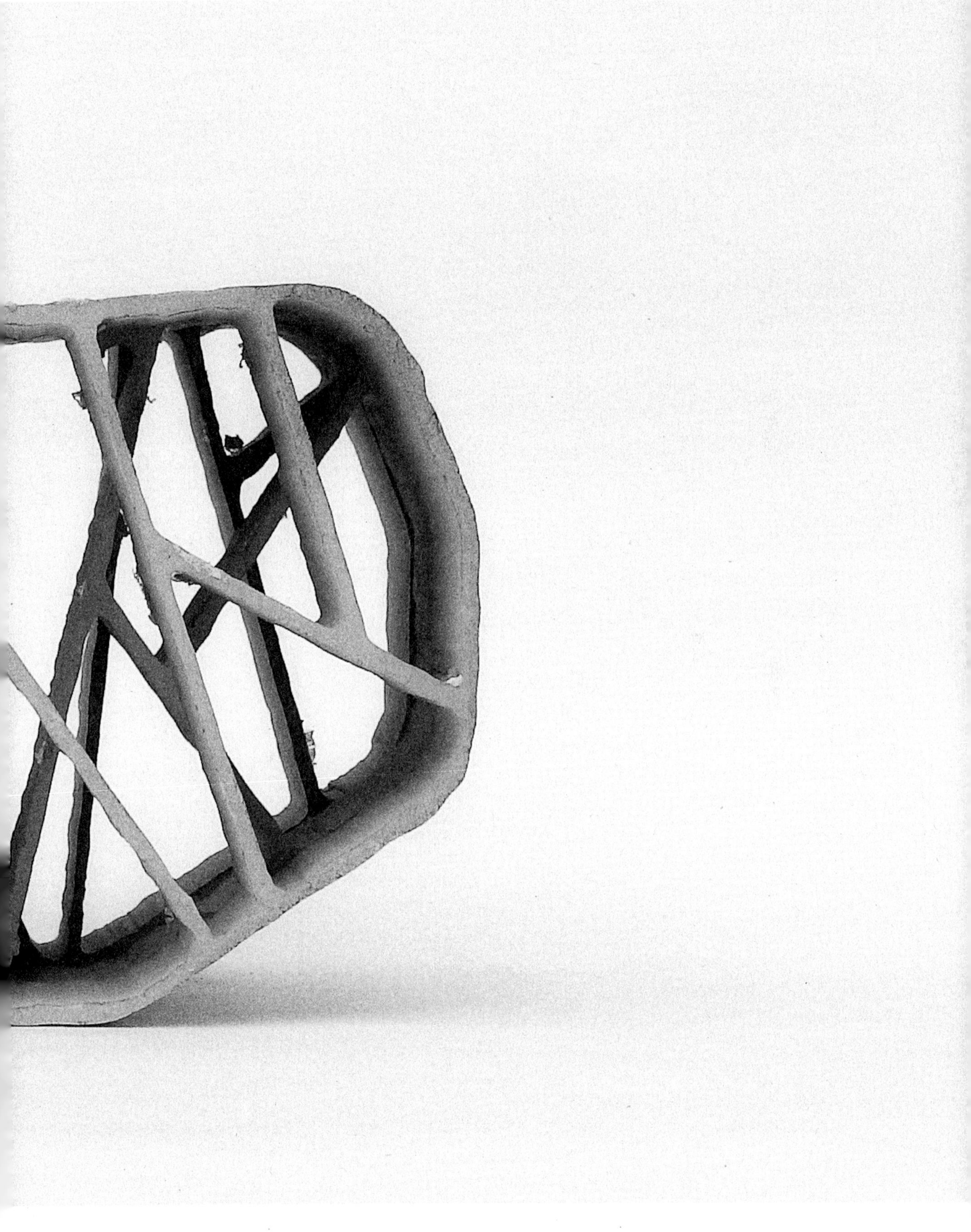

Dragonfly

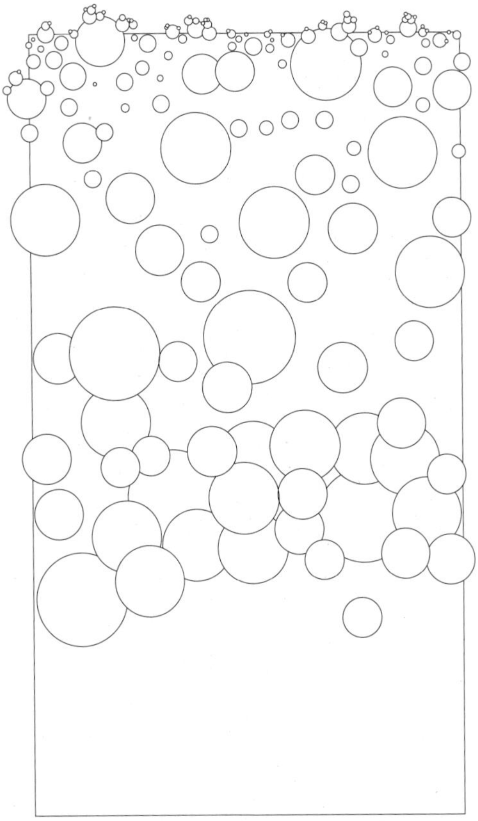

Foam Ein Objekt, das einen hohen Grad der Selbstähnlichkeit aufweist, das aus mehreren verkleinerten Kopien seiner selbst besteht, ein Objekt, das einen Wachstumsprozess abbildet.

Foam An object with a high degree of self-reference. Comprised of several diminished copies of itself, it is an object that shows the process of growing.

Strings Geknickte Barren, die scheinbar chaotisch angeordnet sind, doch alle nach dem gleichen Prinzip strukturiert sind und einer unsichtbaren Ordnung gehorchen.

Strings Folded beams, apparently placed chaotically, but all structured with the same principle, following an invisible rule.

Honeycomb Aus einer Matrix von Hexagonen ist eine Form herausgearbeitet. Ein Prinzip, wie man es in der Natur häufig findet. Grundlage ist eine perfekte, mathematische Struktur, die durch einen chaotischen Prozess verändert wird, ein Abbild eines Wachstumsprozesses, wie er in der Natur vorkommt.

Honeycomb A form is shaped out of a matrix of hexagons, a principle that is often be found in nature. The basis is a perfect mathematic structure that is transformed by a chaotic process. A copy of a growth process like tat found in nature.

Strings II

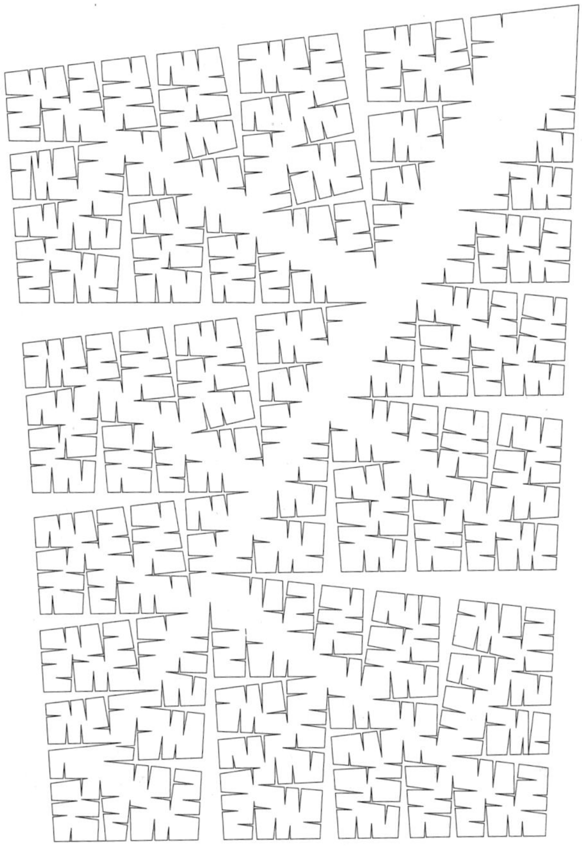

Koch-Kurve Ein Fraktal, das ein Derivat der Koch-Kurve ist, erzeugt das Muster. Ein Objekt, das einen hohen Grad der Selbstähnlichkeit aufweist, das aus mehreren verkleinerten Kopien seiner selbst besteht.

Koch-curve A fractal, that is a derivative of the Koch-curve. An object that shows a high degree of self-reference and is composed of several diminished copies of itself.

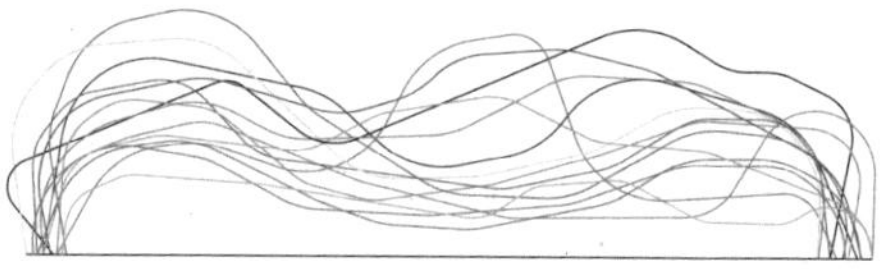

Frequenz Wie phasenverschobene Wellen von Tönen oszilliert das Objekt, scheinbar in der Bewegung erstarrt.

Frequency Like dephased waves of tones the object is oscillating, apparently frozen in motion.

Cell Eine Zellstruktur mit Zellen unterschiedlicher Größe, die in dichtester Packung ein schaumartiges Gebilde ergeben, ist das Vorbild der Form. Sie erinnert an Zellstrukturen unter dem Mikroskop oder an Schnitte durch Knochen.

Cell A foam-like entity, formed with the densest possible packing, made up of a cellular structure with cells of different sizes. It evokes images of cellular structures under the microscope, or of sections through bones.

Rock

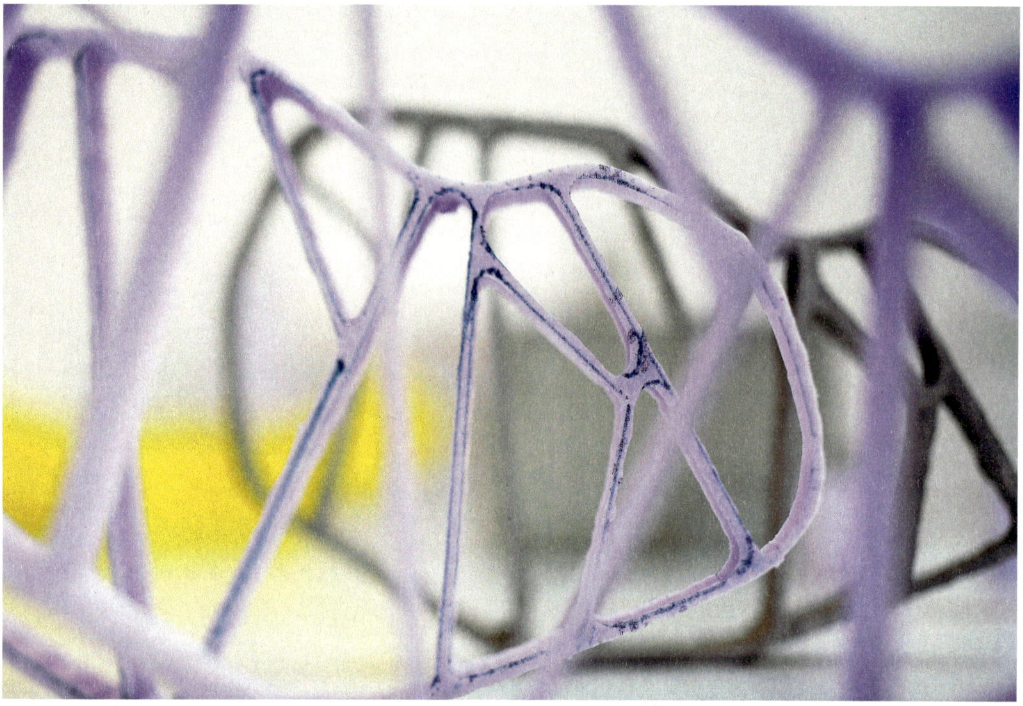

7

Organische Architektur

7 Projekte von Finsterwalder Architekten

Organic Architecture

7 Projects by Finsterwalder Architekten

Rudolf Finsterwalder

Organische Architektur hat einen ganzheitlichen Anspruch, sie will das Zusammenspiel der einzelnen Teile und die Architektur mit der Umwelt verbinden zu einem organischen Ganzen. Nicht aus einem Stilwillen heraus wird geplant, sondern es werden jeweils unterschiedliche Bauten aus den relevanten Parametern entwickelt. || Vergleichbar zu den Entwicklungen der Natur werden Lösungen gesucht, die alle relevanten Themen, die Umgebung, die Funktion und die Menschen sowie andere relevante Dinge miteinbezieht. | Entwicklung in der Natur folgt den Naturgesetzen, ist also nicht willkürlich, sondern gezielt und daher weitestgehend optimiert. So wächst eine spezifische Architektur, die sich aus der Funktion, den biologischen, sozialen und psychologischen Anforderungen sowie der Konstruktion und dem Material entwickelt. | Eine organische Konstruktion versucht z.B. die entstehenden Lasten fließend abzutragen, wie es Konstruktionen in der Natur auch tun. So kann nicht nur mit einem Minimum an Material gebaut werden, sondern es entstehen auch fließende Formen, die ihre innere Logik zeigen und damit auch »schön« sind. || Die gezeigten sieben Projekte versuchen jeweils, eine optimale Lösung für Funktion, Konstruktion und Material zu entwickeln. Die entstehenden Architekturen haben auch einen hohen Anspruch an die Schönheit der Form. Der steht jedoch nicht am Anfang, sondern ist das Ergebnis einer Entwicklung. Das Ziel ist eine Integration der Teile, ein organisches Zusammenwachsen von Architektur und Natur, von Mensch und Technik. || Frei Otto formuliert es so: »Das heutige Ziel ist, Häuser und Städte zu bauen, die natürlich sind. Um natürlich zu sein, muss ein menschliches Produkt nicht aussehen wie eine Pflanze oder ein Baum. Das große Ziel (...) ist, dass Häuser und Städte zusammen mit Pflanzen und Tieren ein natürliches Biotop darstellen, dass also das Haus nicht gegen die Natur gerichtet ist, sondern dass der Mensch und seine Technik ein untrennbarer Teil der Natur sind.«[1]

1 Frei Otto: *Natürliche Konstruktionen. Formen und Konstruktionen in Natur und Technik und Prozesse ihrer Entstehung.* 1982. S.7-9.

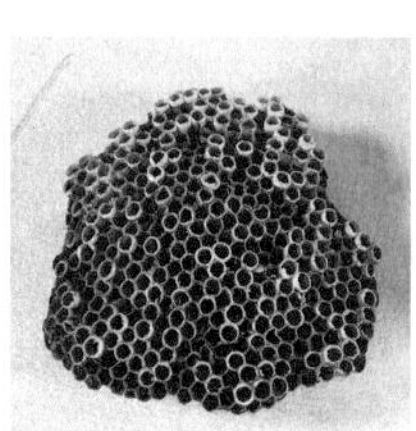

Organic architecture has a holistic thrust, which seeks to be the interaction of individual parts connecting architecture and the environment. It does not wish to create structures in keeping with a particular style, but to develop a specific architecture from the respective parameters given. || As with developments in nature, we search for solutions that incorporate all aspects of architecture, the surroundings, the function and the users, as well as other relevant matters. | Development in nature follows the laws of nature, and is thus not arbitrary, but geared to a specific goal and therefore largely optimized. In this way, a specific architecture evolves from the functional, biological, social and psychological requirements, as well as the structure and materials. | An organic structure attempts, for example, to find a fluid way of bearing the loads that arise in a manner similar to that used in nature. This enables not only the input of materials to be minimized, but also reveals their intrinsic logic, such that they are also "beautiful". || The seven projects showcased here each attempt to develop an optimal solution for functionality, structure, and materials, with the resulting architecture aspiring to constitute beautiful shapes. That is not the starting point, however, but the result of a development. The goal is an integration of parts, an organic fusion of architecture and nature, between humans and technology. || As Frei Otto put it: "The goal today is to build houses and cities that are natural. To be natural, a human product need not look like a plant or a tree. The major goal is (...) for houses and cities to grow together with plants and animals to constitute a natural biotope, i. e., for the house not to be aimed against nature, but for man and technology to form an inseparable part of nature." [1]

1 Frei Otto: *Natürliche Konstruktionen. Formen und Konstruktionen in Natur und Technik und Prozesse ihrer Entstehung.* 1982. p.7–9.

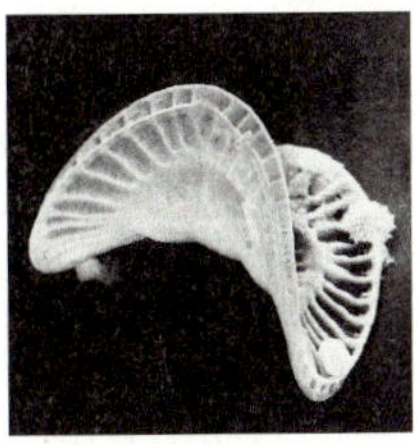

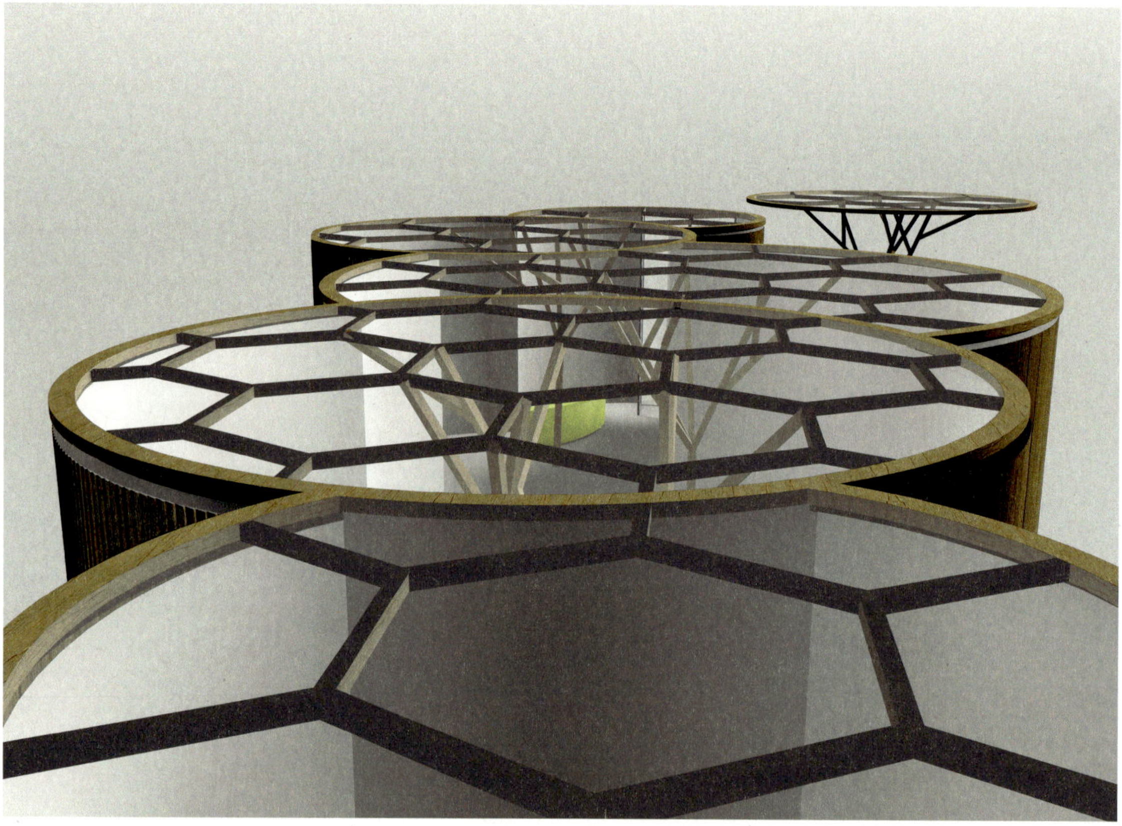

1

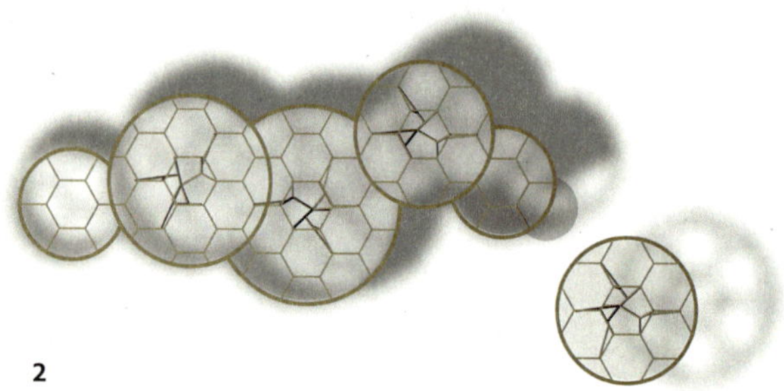

2

Honeycomb (Rosenheim, 2007) Holzstützen, die große Pinienbäume zum Vorbild haben, tragen das Dach des Ausstellungsbaus. das Tragwerk ist optimiert und kommt mit schlanken Querschnitten aus. Ähnlich wie ein natürlicher Baum verzweigt sich das Tragwerk bis in den Wipfel, leitet die Lasten von dort fließend bis nach unten in die Fundamente. Das Dach selbst wird gebildet aus einer Struktur von hexagonalen Holzrahmen, die ein Flächentragwerk bilden. Auch das Hexagon kommt in der Natur vor und ist eine optimierte Form, wie sie z. B. Bienen zum Bau ihrer Waben verwenden. || In den Knoten der Hexagone greifen die Zweige der Baumstützen an und tragen das Dach. | Die auf der Tragstruktur liegende transluzente Folie lässt Licht durch und dient als Witterungsschutz. Das Licht- und Schattenspiel, das sich im Inneren zeigt, schafft eine ähnliche Atmosphäre wie in einem natürlichen Wald und relativiert so den Unterschied zwischen Außen- und Innenraum.

Honeycomb (Rosenheim, 2007) Wooden columns, similar to large pine trees, carry the roof of the exhibition pavilion. The bearing structure is optimized and operating with slim dimensions. Much like a natural tree the structure derives from the top, carrying the weight down to the foundations in a fluent way. The roof itself is formed from a structure of hexagons that form an ideal surface structure. Even the hexagon can be found in nature and is another optimized form, for example those used by bees for the construction of their honeycombs. || The columns connect the nodes of the hexagons and carry the roof. | The translucent membrane lying on the hexagonal structure lets the light through in order to illuminate the interior. Light and shadow in the interior create an atmosphere similar to that of a natural forest, in this way establishing a relationship between interior and exterior space.

3

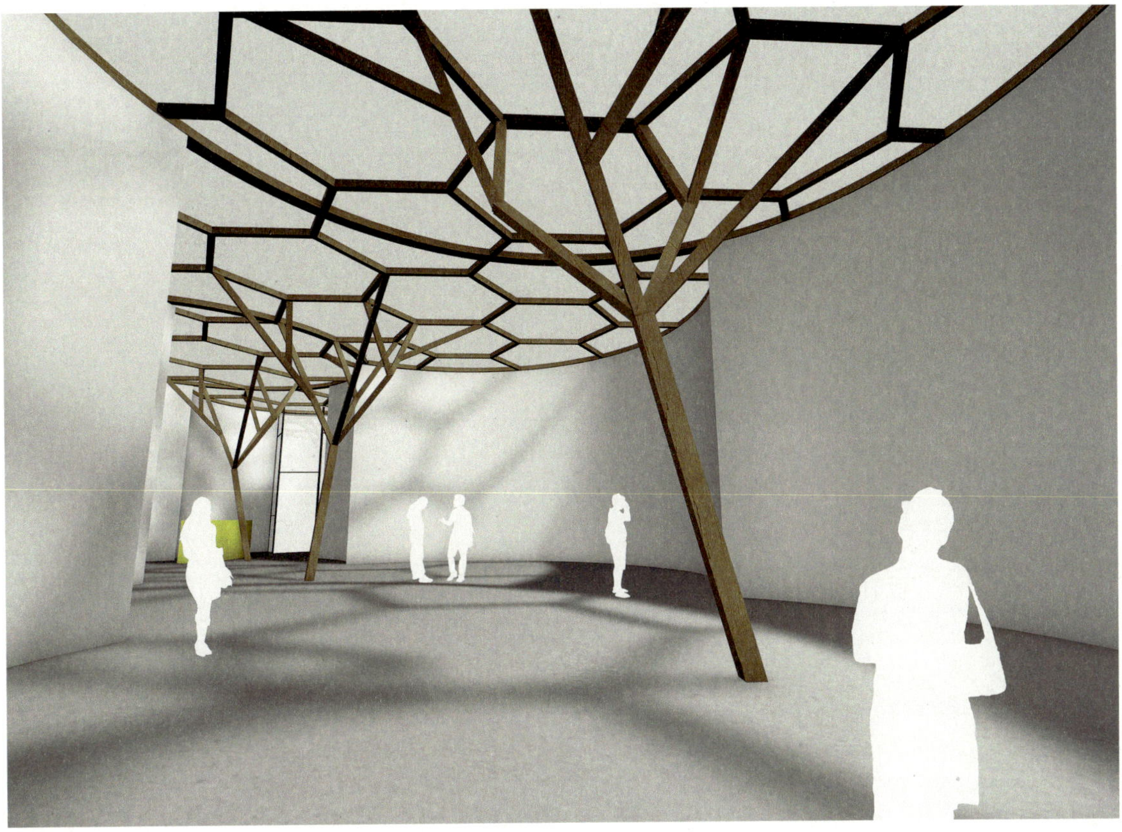

4

Abb.1 Dach des Pavillons
Abb.2 Aufsicht
Abb.3 Längs- und Querschnitt
Abb.4 Innenraum

Fig.1 Roof of the pavilion
Fig.2 Birds eye view
Fig.3 Section
Fig.4 Interior space

Abb.5 Modell
Abb.6 Holzschalung
Abb.7 Formstudie, Peter Pohl
Abb.8 Grundriss

Fig.5 Model
Fig.6 Wooden boarding
Fig.7 Design study by Peter Pohl
Fig.8 Floor plan

6

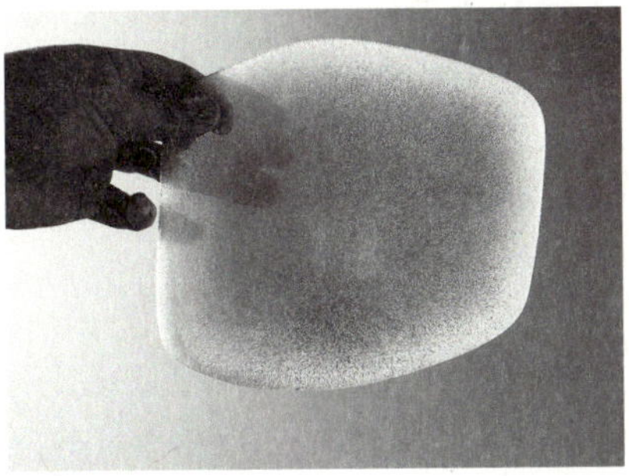

7

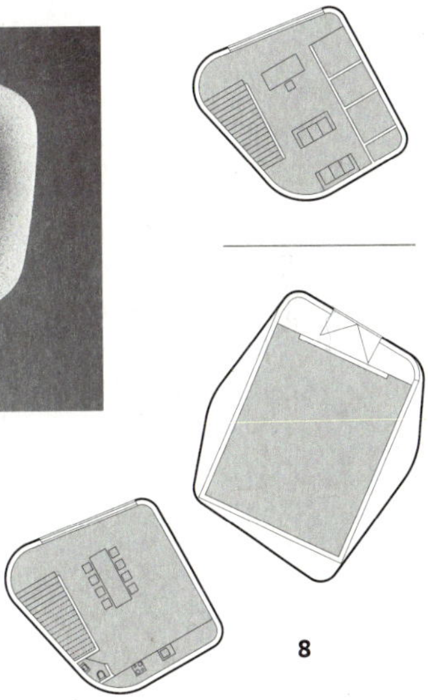

8

Atelier Pohl (Beuerberg, 2007) Das Projekt ist in einem intensiven Dialog mit dem Maler und Bildhauer Peter Pohl entstanden, der sich in seiner Kunst mit der Morphologie von Insekten auseinandersetzt. | Aus dieser Formensprache ist ein Entwurf für ein Atelier mit Wohnhaus entstanden. Die organischen Formen der Insekten werden in eine moderne Bauweise, ein verschalte Holzelementbauweise übertragen. Die hochkomplexen Formen der Insekten können nicht 1:1 nachgebaut werden, sondern werden übersetzt in eine Form, die konstruktiv überzeugend baubar ist. Da die Flächen der Baukörper jeweils lediglich in einer Richtung gekrümmt sind, ist eine Beplankung mit einer vertikalen Stulpschalung nicht nur möglich, sondern auch einfach herstellbar. Eine Form ensteht, in der das Vorbild Natur spürbar ist, jedoch nicht nachgeahmt wird. Die biomorphe Form ist ein kompakter Baukörper, der auch energetisch optimiert ist. | Eine umlaufende Stulpschalung erzeugt optische Effekte, die auch manche Insekten aufweisen. Die in einer Richtung laufende Schalung bricht die Symmetrie der Körper und schafft ein asymmetrisches Erscheinungsbild mit wechselnden Ansichten.

Studio Pohl (Beuerberg, 2007) The project was developed in an intense dialog with the painter and sculptor Peter Pohl, who is working with the morphology of insects. | The concept has been developed with this vocabulary. The organic forms of the insects are transformed into a modern wooden construction. The highly complex forms of the insects cannot be copied 1:1, but are translated in a form that can be built. Since the curved surface is always in only one direction, the cladding with wooden boards is not only possible, but also easy to manufacture. It creates a form in which one can feel the closeness of nature, but not being mimicked directly. A biomorphic form generates a compact volume that is also optimized in an energetic way. | The vertical boarding creates visual effects, which can also be seen at some insects. The cladding, which is fitted in the same direction throughout the building, breaks the symmetry of the buildings and creates an asymmetric image with changing perspectives.

9

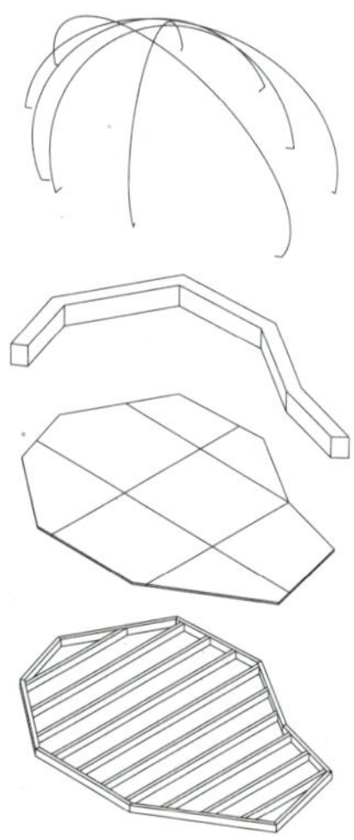

10

Polyethylene (Rom. Villa Massimo - Berlin. Gropius-Bau, 2007) Mit einfachen Mitteln einen Pavillon zu bauen, der Platz für ca. 60 Personen bietet, war das Ziel des Projektes. Entstanden ist der Bau als begehbare Skulptur anlässlich der Präsentation »finale« an der Villa Massimo in Rom. Der Pavillon, komplett aus preiswerten Materialien gebaut, kann innerhalb weniger Stunden auf- und abgebaut werden. An einer Plattform aus Holzbalken werden Flacheisen 30×10 mm befestigt. Jeweils 2 gegenüberliegende 6 m Stangen werden nach innen gebogen und mit Schellen verbunden. Durch die flexible Verbindung kann die Form des Pavillons bestimmt werden. Die Form erinnert an einen aufliegenden Tropfen oder eine Blase. Die 5 Bögen werden an ihren Kreuzungspunkten miteinander verbunden, wodurch die Konstruktion ihre Stabilität erhält. Das sich so ergebende Skelett wird mit einer dehnbaren Kunststoffolie, wie sie zur Verpackung in der Industrie verwendet wird, umwickelt. Diese Hülle ist Witterungsschutz und zugleich eine weitere Versteifung der Konstruktion. Die vielen Schichten der Folie ergeben ein lebendiges Bild der Hülle, die je nach Licht und Blickwinkel sehr verschieden wirkt.

Polyethylene (Rome. Villa Massimo - Berlin. Gropius-Bau, 2007) The aim of this project was to build a pavilion with clear techniques that offers a place for 60 people. The building was created as a sculpture to walk through, for the presentation "finale" at the Villa Massimo in Rome. The pavilion, which can be built entirely with inexpensive materials, can be mounted and dismounted in a few hours. Flatiron profiles 30×10 mm are fixed on a platform built with wooden beams. At one time, 2 opposing 6 m bars are bent inwards and joined with clamps. Because of the flexible joints, the shape of the pavilion can be defined. The five arcs are joined at their crossing points, making the construction more stable. The structural skeleton is wrapped with a stretchable film, which is used for packaging in various uses. This skin is used as protection against the weather, while also acts as further reinforcement of the construction. The many layers create a vivid, living image of the skin that changes its appearance depending on light and point of view.

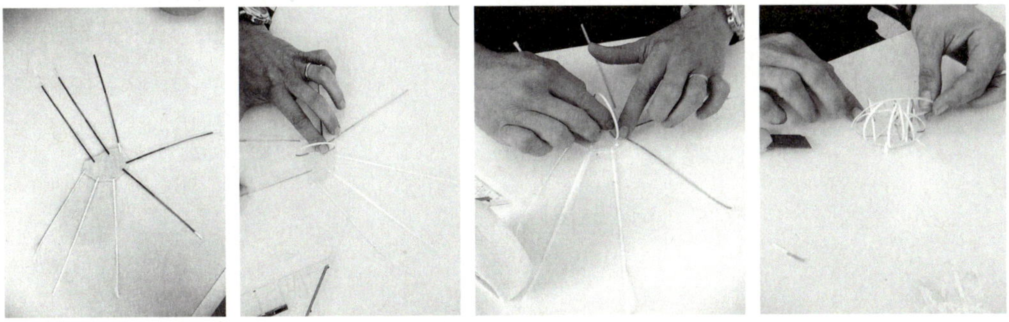

11

12

13

Abb.9 Polyethylene in Rom
Abb.10 Struktur des Pavillons
Abb.11 Formfindung
Abb.12/13 Gropius-Bau, Berlin

Fig.9 Polyethylene in Rome
Fig.10 Structure of the pavilion
Fig.11 Form finding
Fig.12/13 Gropius-Bau, Berlin

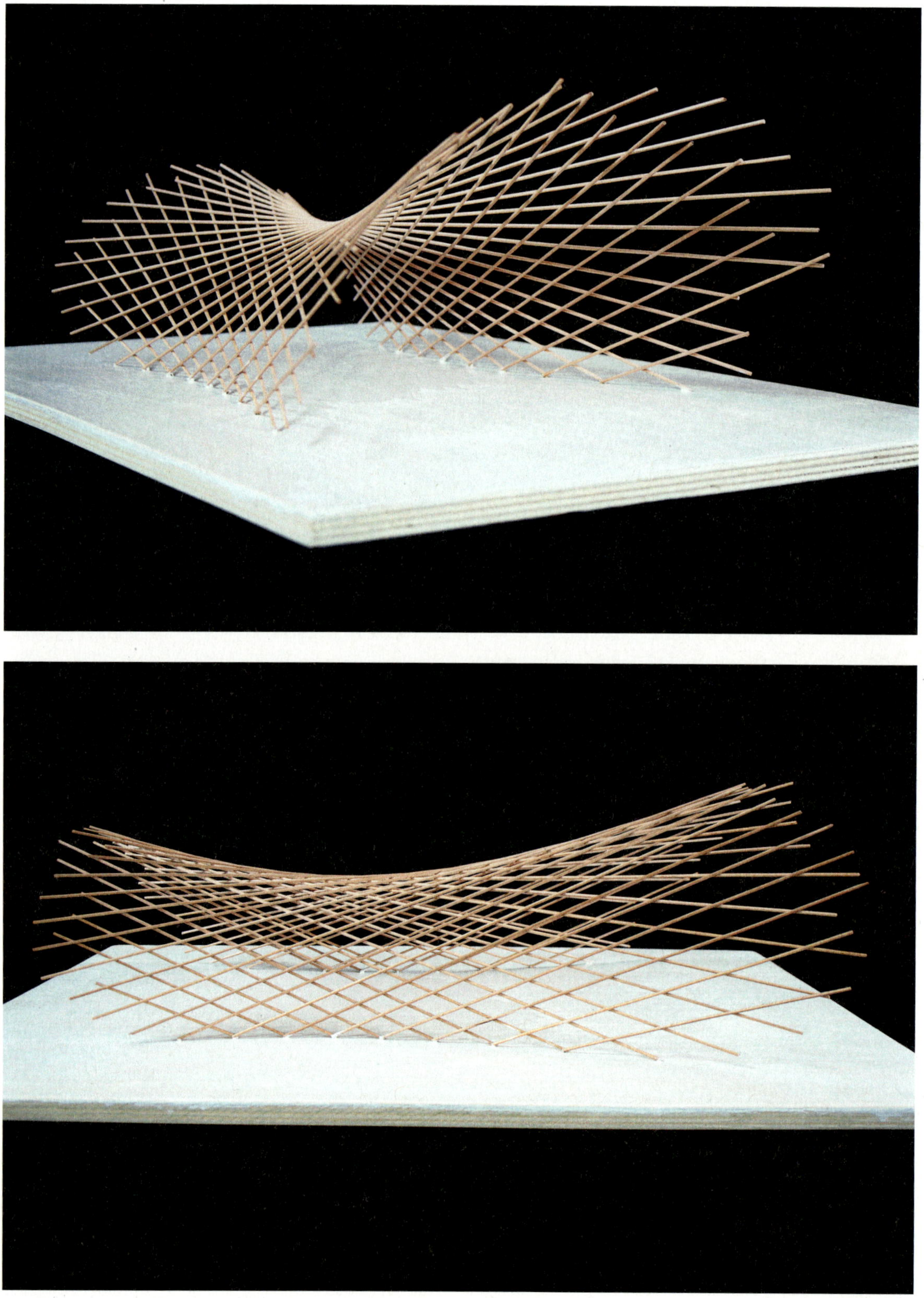

14 / 15

Abb.14 / 15 Modell der Gitterschale
Abb.16 Formstudie
Abb.17 Grundriss mit Schlachtfesthaus
Abb.18 Querschnitt

Fig.14 / 15 Model of the gridshell
Fig.16 Study of form
Fig.17 Floor plan with butchery
Fig.18 Cross section

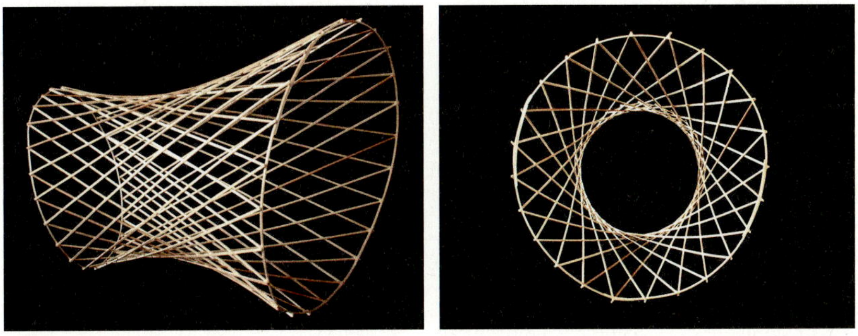

16

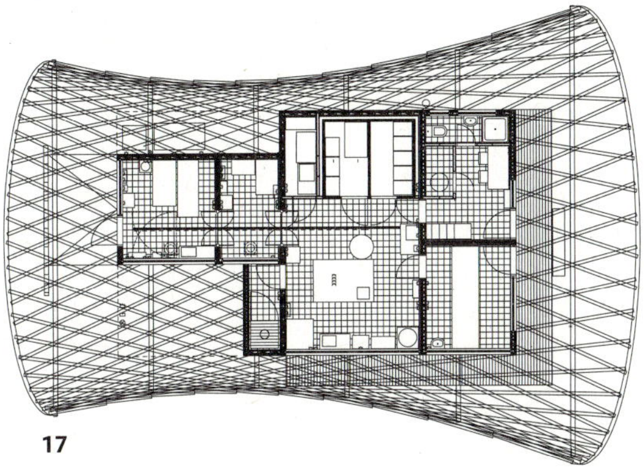

17

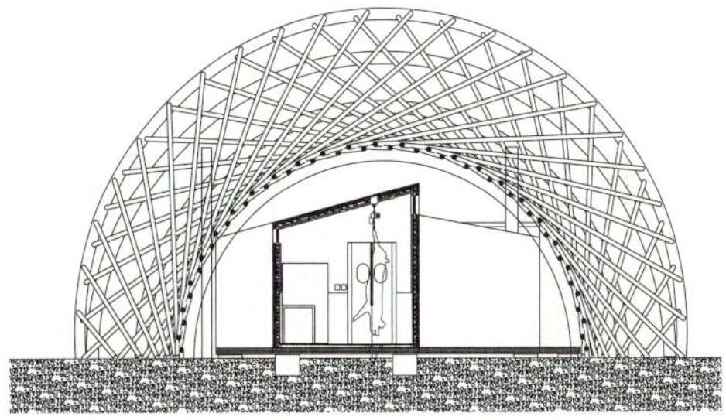

18

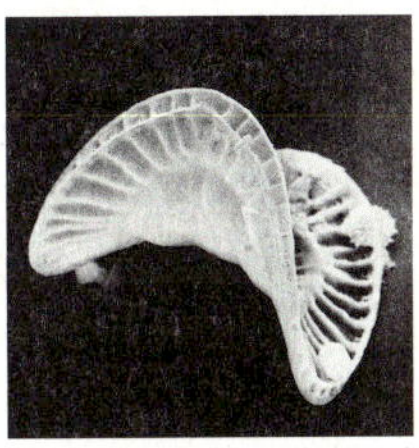

Hull / Holzgitterschale (Herrmannsdorfer Landwerkstätten, 2010) Für einen landwirtschaftlichen Nutzbau, eine Feldscheune, sollte eine möglichst nachhaltige und naturnahe Konstruktion entwickelt werden. Der Bauherr, Karl-Ludwig Schweisfurth, der Betreiber einer ökologischen Landwirtschaft, wollte eine vollständig recyclebare Konstruktion, bei der kein Beton verwendet werden und der ökologische Fußabdruck möglichst klein sein sollte.
Der Bau ist eine Holz-Gitterschale aus geraden Erzeugenden, die auf ihre Gesamtlänge um einen Viertelkreis verdreht an vertikalen Rundbögen angeschlagen werden, ein hyperboles Paraboloid. Trotz der geraden Sparren entsteht eine runde Form, eine Schale, wie sie auch in der Natur oft zu finden ist. Überspannt wird die Konstruktion mit einer leichten Membrane, in diesem Fall mit einem *Gore tenara*.
In Planung ist ein weiterer Bau mit Standort in Ägypten, der mit einer Membran mit Fotozellen zur Energiegewinnung überspannt werden soll. Die Fundamente bilden mit Steinen gefüllte Drahtkörbe, so genannte *Gabionen*, die auch wiederverwertet werden können. Die gesamte Konstruktion ist äußerst material- und kostensparend und findet eine Formensprache, die mit der Nutzung des Baus korreliert.

Hull / Wooden Gridshell (Herrmannsdorfer Landwerkstätten, 2010) A sustainable and ecological construction should be developed for an agricultural building, in particular a barn. The client, Karl-Ludwig Schweisfurth, the owner of an ecological farm wanted a completely recyclable construction, with no concrete and a very small carbon footprint. The building is a wooden grid shell with straight beams, which are twisted a quarter turn over their entire length and fixed to vertical hyperbolic paraboloid arcs. Although the beams are straight it creates a curved form, a shell, which can be found in nature. The construction is covered with a light membrane; in this case a *gore tenara*. A second building is planned for Egypt and shall be covered with a membrane integrating photocells to produce energy. The foundations are made of wire-cages filled with stones, called *gabions*, which are also recyclable. The whole construction is highly efficient in the sense of material and costs, and finds a form that correlates with its purpose.

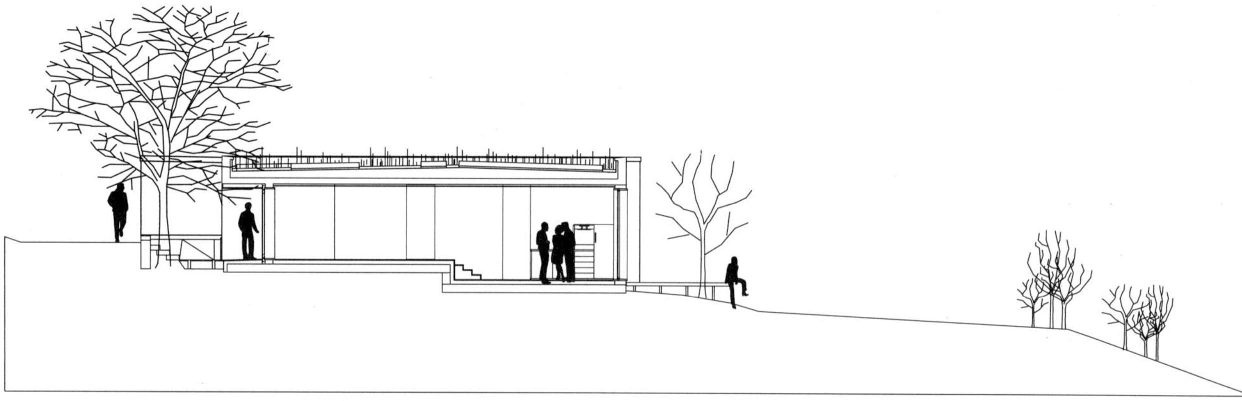

19

Mutterhaus (Bad Endorf, 2011) Das eingeschossige, nicht unterkellerte Haus liegt auf einem Hanggrundstück in Bad Endorf, Bayern. Seine organische, amorphe Form reagiert auf die Bedingungen des Außen-, wie auch des Innenraumes, das Haus verschränkt die Landschaft mit dem Inneren des Hauses. Es formt z. B. einen Patio um einen bestehenden riesigen Ahornbaum, der nachts beleuchtet wird und als ein Stillleben der Natur vom Wohnzimmer aus betrachtet werden kann. | Die Innenwände bilden das orthogonale Rückgrat des Hauses, die Wände berühren kaum die fließende Außenwand. Die inneren Wände tragen das Dach und erlauben der nichttragenden Fassade ihr organisches Eigenleben. | Die Rundungen des Baukörpers werden außen mit einer vertikalen Stulpschalung aus unbehandeltem Lärchenholz verkleidet, die für dieses Projekt entwickelt wurde. | Aus Dickholz sind die Decke und die Innenwände gebaut, einzelne Bretter werden aneinander gelegt und zu einer Platte verbunden, vertikale Nuten bilden die Oberfläche der Platten. Die so gebauten Innenwände werden in drei verschiedenen Farben lasiert, diese farbigen Holzwände werden beleuchtet. | Der Boden ist aus Eschenholz, die Sockelleisten aus einem spiegelnden, gebogenen Edelstahlband. | Das Haus ist sehr effizient wärmegedämmt und allein mit dem Kachelofen gut zu beheizen. Die Dachterrasse ist als begrünter Dachgarten angelegt und bietet einen phantastischen Ausblick auf die Alpen und den nahegelegenen Simssee.

Mutterhaus (Bad Endorf, 2011) The one-storey house, without basement, is situated on a hillside in Bad Endorf, Bavaria. Its organic, amorphous form responds to the conditions of the site, the house interconnects the landscape with the interior spaces. For example, it forms a patio around an existing huge maple tree that is illuminated at night and can be viewed as a still life from the living room. | The interior walls build the orthogonal backbone of the house; the walls are barely touching the flowing exterior wall. The interior walls are carrying the roof and allow the non-structural façade an organic life of its own. | The curvature of the volume is covered with a vertical cladding, made of untreated larch boards, which was developed specifically for this project. Ceiling and interior walls are built with solid wood. This solid wood consists of single boards joined together with vertical channels milled out to form the surface. The single interior walls are finished in 3 different colours; these coloured wooden blocks are then illuminated. The floor is finished in ash wood, while the skirting boards are made of curved stainless steel bands. | The house is highly insulated and can be heated just using the tile stove. The roof terrace is designed as a landscaped roof garden and offers a fantastic view of the Alps and the nearby lake Simssee.

1 Garderobe / Wardrobe
2 Technik / Technical room
3 Arbeitsplatz / Office
4 Gäste / Guest room
5 WC
6 Bad / Bathroom
7 Schlafen / Bedroom
8 Wohnküche / Kitchen and Dining room
9 Wohnen / Living room
10 Lager / Storage
11 Bad und WC / Bathroom and WC
12 Speisekammer / Pantry
13 Eingang / Entrance
14 Terrasse / Terrace
15 Werkstatt / Workshop
16 Stellplatz / Parking

20

21

Abb.19 Schnitt
Abb.20 Baustelle aus der Vogelperspektive
Abb.21 Grundriss

Fig.19 Section
Fig.20 Bird's eye view of construction site
Fig.21 Floorplan

22 / 21

Abb.22 Fassade Südost
Abb.23 Ansicht Süd

Fig.22 Southeast facade
Fig.23 South elevation

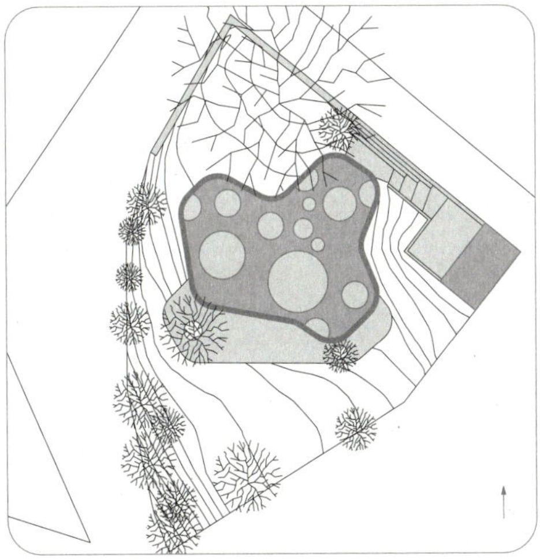

25

24

26

Abb.24 Dachterrasse
Abb.25 Lageplan
Abb.26 Begehbares Dach extensiv begrünt

Fig.24 Roof terrace
Fig.25 Siteplan
Fig.26 Accesible roof with extensive green

27

28

Abb.27 Wohnküche mit zentralem Küchenblock
Abb.28 Das Innere ist orthogonal

Fig.27 Living room with kitchen table
Fig.28 The interior is all rectangular

29

30

Abb.29 Weg entlang der Aussenwand
Abb.30 Übergang Essbereich zu Bibliothek

Fig.29 Corrido along the outer wall
Fig.30 passage from dining room to library

31

Smarthaus (Stephanskirchen, 2014) Das Smarthaus ist ein Vorzeigemodell in punkto Nachhaltigkeit, Energieeffizienz und ökologischem Bauen sowie ein Beispiel für modernes Wohnen. || Ein low-tech Haus, das seinem Benutzer das Wohnen leicht und preiswert macht. Eine intelligente Planung senkt die Baukosten auf ein Maß, das es zu einem erschwinglichen Objekt macht. || Das Holzhaus wird hauptsächlich aus Profil-Holz-Elementen, einem Massivholz ohne Leim, errichtet. Dieses tragende Material mit seiner wunderschönen genuteten Struktur ist sichtbar verwendet und prägt die Erscheinung im Innenraum, Außenwände, Innenwände, Decke und das Dach sind daraus gefertigt. Außen ist der gesamte Baukörper, inklusive Dach mit einer Stulpschalung aus Lärchenholz verkleidet. | Die komplette Südfassade ist verglast und macht das Haus zu einem lichtdurchfluteten Wohnerlebnis. Geplant ist das Smarthaus mit bisher 5 verschiedenen Grundrissvarianten bei immer gleichen Außenmaßen, z.B. eine effiziente Variante für Familien, eine luxuriöse Loftvariante sowie eine Version für Arbeiten und Wohnen. || Energetisch ist das Haus so effizient geplant, dass es in der Standardversion ausschließlich mit einem Kachelofen beheizt wird. || Das Wesentliche für Interessenten ist die Kostensicherheit. Da ein Musterhaus bereits gebaut wurde, sind die Kosten bekannt und das gleiche Haus ist zu den gleichen Kosten wiederholbar.

Smart house (Stephanskirchen, 2014) The smart house is exemplary when it comes to sustainability, energy efficiency and green building, as well as being a model for modern living. || A low-tech house that facilitates easy living at low cost. The intelligent planning keeps construction expenses to a level that makes the building affordable. || The timber construction is primarily composed of wood sections of unglued solid timber. This supporting material with its beautiful exposed grooved structure shapes the appearance of the interior, exterior walls, inner walls, ceiling and the roof. The entire building exterior including the roof features larch lap siding. | The south facade is completely glazed and creates a light-filled living ambiance. The smart house has been planned with five different layouts of equal exterior dimensions, e.g. an efficient design for families, a luxurious loft layout as well as a mixed living-working arrangement. || In terms of its energy use, the house has been planned so efficiently that the standard model is heated with only a tiled stove. || One of the decisive criteria for interested parties is the predictability of expenses. A show house has already been completed, and the same building can be replicated at equal costs.

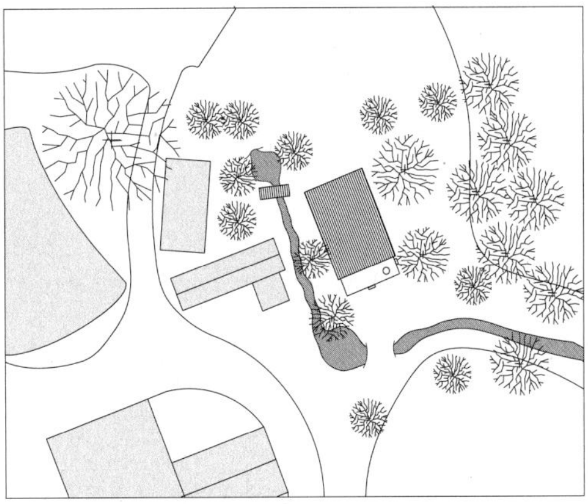

33

32

34

Abb.31 Vollverglaste Südseite
Abb.32 Zugang mit Brücke
Abb.33 Lageplan
Abb.34 Blick in den Garten

Fig.31 full glass facade to the south
Fig.32 siteplan
Fig.33 Section
Fig.34 view to the garden

35 /36

Abb.35 Eingang mit weitem Ausblick
Abb.36 Der Kachelofen ist das Zentrum des Hauses

Fig.35 Entrance with wide view
Fig.36 The large wood stove is the center of the house

37

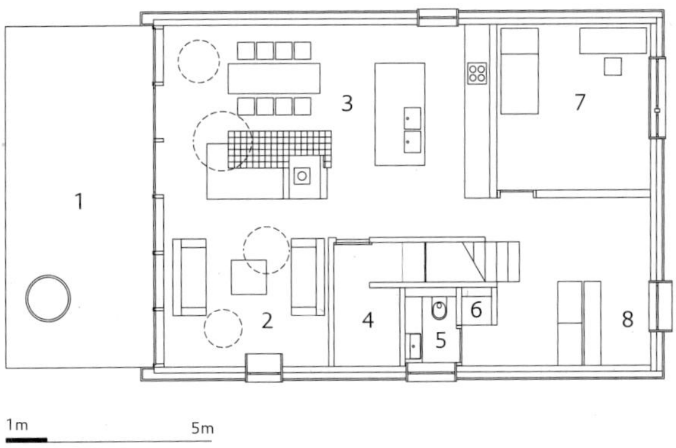

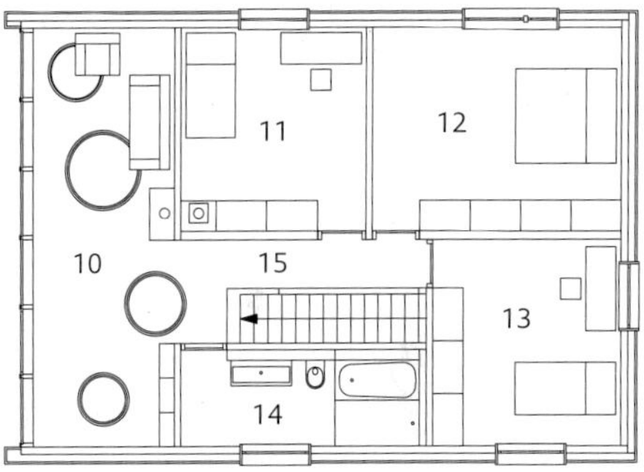

38 /39

1 Terrasse / Terrace

2 Wohnen / Living room

3 Wohnküche / Kitchen and dinig room

4 Speis / Larder

5 WC

6 Waschmaschine / Washing machine

7 Arbeitsplatz / Office

8 Diele / Hall

9 Eingang / Entrance

10 Lounge

11 Zimmer I / Bedroom I

12 Zimmer II / Bedroom II

13 Zimmer III / Bedroom III

14 Bad / Bathroom

15 Flur / Corridor

Abb.37 Bibliothek im Obergeschoss
Abb.38 Grundriss EG
Abb.39 Grundriss OG

Fig.37 Library in the upper level
Fig.38 Floorplan groundlevel
Fig.39 Floorplan upper level

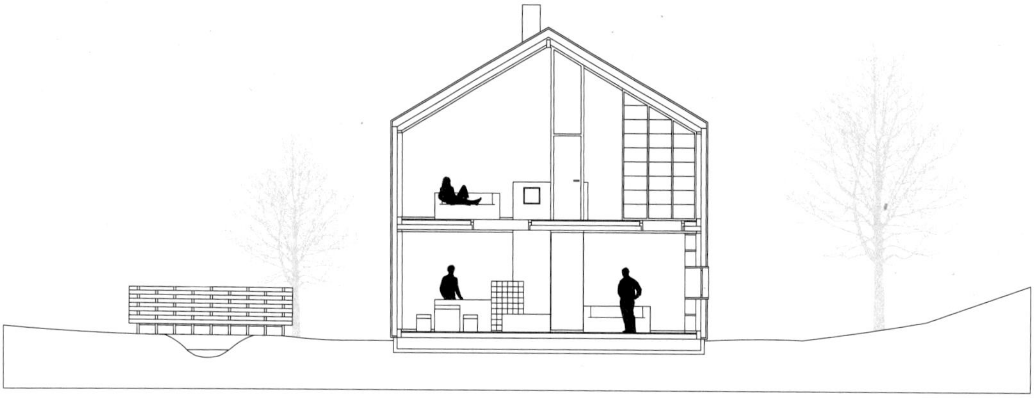

40

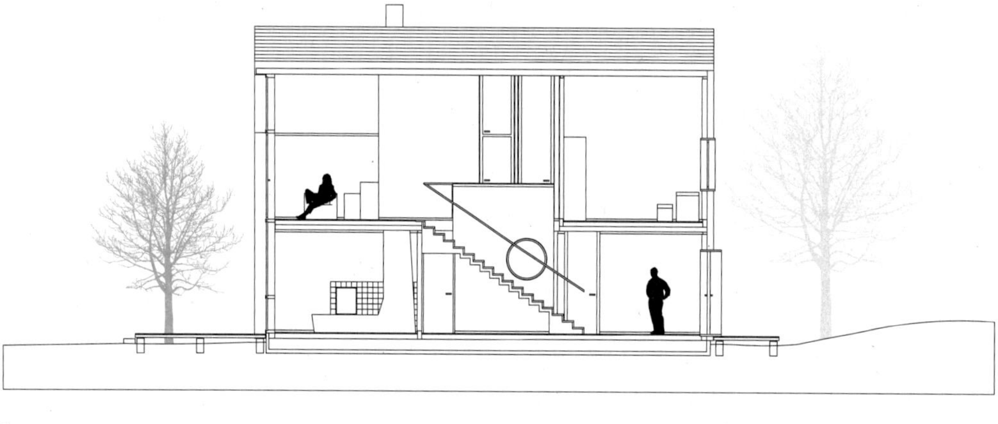

41

Abb.40 Querschnitt
Abb.41 Längsschnitt

Fig.40 Cross section
Fig.41 Longitudinal section

42 /43

Abb.42 Die vollverglaste Südseite
Abb.43 Wohnküche mit Essbereich

Fig.42 Full glass southfacade
Fig.43 Living and dining room

Abb.44 Ginger und Fred
Abb.45 Modellstudie

Fig.44 Ginger and Fred
Fig.45 Model study

45

Ginger und Fred (Stephanskirchen, 2015) Ginger reagiert mit ihrer Form auf die vor ihr stehende Birke. Um die Wurzeln nicht zu gefährden und dennoch zu ermöglichen, dass die Terrasse im 2. OG und die gesamte Südfassade dem Baum nahekommen, kippt die Fassade in den Baum. Die Bewohner leben mit dem Baum und in dem Baum. || Fred reagiert mit seiner Südseite auf Ginger. Beide gegenüberliegenden Seiten klappen voneinander weg und erlauben den Bewohnern des dahinterliegenden Hauses einen gerahmten Ausblick und eine Besonnung von Südwesten. Beide Gebäude verzahnen sich mit ihren organischen Körpern mit der Natur, der Erde, dem Baum und dem Himmel. Sie werden Teil ihrer Umgebung, obwohl sie sehr selbstbewusst ihre eigenwillige Form haben. Es ist nicht eine typische architektonische Form, sondern eigenwillig wie ein Stück Natur. Zwei skulpturale Baukörper, die wie Monolithen auf der Erde sitzen und zu versinken scheinen, wie große Steine oder Bäume. Die Holzschalung unterstreicht diesen Charakter, die Fassade mit ihren bewusst unregelmäßigen Stößen hat ein Erscheinungsbild, das nicht glatt und langweilig ist, sondern dem Auge Halt bietet und lebendig wirkt. || Beide Häuser weisen eine Analogie zu Baumhäusern auf, sie sind auf einer sehr kleinen Grundfläche von 6 m x 9 m errichtet, kragen nach oben hin aus, wie ein Baum mit seiner Krone und sind mit 3 Etagen relativ hoch. || Beide Häuser sind ein Vorzeigemodell in punkto Nachhaltigkeit, Energieeffizienz und ökologischem Bauen sowie ein Beispiel für modernes Wohnen. || Im Inneren der beiden Häuser wird ebenso konsequent mit den ungewöhnlichen Winkeln gearbeitet. Die genutete Oberfläche des Holzes läuft nicht senkrecht, sondern parallel zur gekippten Außenwand. Das durchgehende Treppenhaus, wie auch die Lufträume, lassen die Bewohner die vertikale Ausrichtung des Gebäudes auch im Innenraum erleben. Die Fenster sind bewusst nicht in einem Raster gesetzt, sondern bieten gezielte Ausblicke auf die Umgebung, vor allem die Bäume, aber auch die Wiese oder den Himmel. Die Fenster in den schrägen Wänden schaffen ungewohnte Perspektiven, Blicke schräg nach oben oder unten.

Ginger and Fred (Stephanskirchen, 2015) With her form Ginger responds to the birch tree in front. To avoid damaging the roots and allow the terrace on the second floor as well as the entire south facade to lean into the tree, the facade is tilted into it. The building occupants live with and in the tree. || With his south side, Fred responds to Ginger. Both opposite sides fold away from each other and afford the inhabitants of the house behind it a framed view and light entering from the south-west. Both buildings with their organic structures are intermeshed with nature, the earth, the tree and the sky. They become part of their surroundings, while confidently showcasing their unusual forms. It is not a typical architectural form but rather unconventional, much like a part of nature itself. Two sculptural structures that appear to sink in like large rocks or trees rest like monoliths on the ground. The wooden sheathing enhances this feature, while the appearance of the facade with its irregular protrusions is not smooth and boring but catches the eye and exudes liveliness. || The two buildings evoke the style of tree houses; their floor space is very small, measuring only 6 m x 9 m, and they project out like a tree with its crown, reaching relatively high at three stories. || Both houses are exemplary when it comes to sustainability, energy efficiency and green building, and serve as a model for modern living. || The unusual angles are continued in the building interior as well. The grooved surface of the timber is not arranged in a vertical pattern, but rather parallel to the slanted exterior wall. Thanks to the full-length staircase and various openings, the building occupants are able to experience the vertical orientation of the building even in the interior. The windows are deliberately not placed in a grid formation, but allow views of the surrounding landscape, especially the trees, and also the meadow and the sky. Windows in inclined walls offer unusual perspectives, sloping views of above and below.

46

48

47

Abb.46 Balkon Ginger
Abb.47 Balkon Fred
Abb.48 Zugang Ginger

Fig.46 Balcony Ginger
Fig.47 Balcony Fred
Fig.48 Approach Ginger

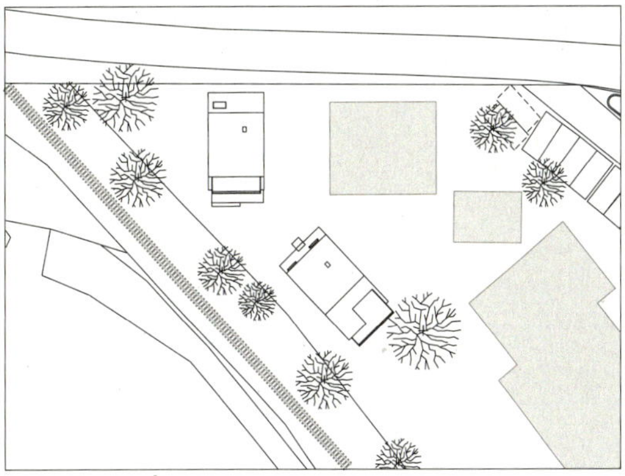

50

49

51

Abb.49	Eingang Fred	**Fig.49**	Entrance Fred
Abb.50	Lageplan	**Fig.50**	Siteplan
Abb.51	Blick von Süden	**Fig.51**	View from south

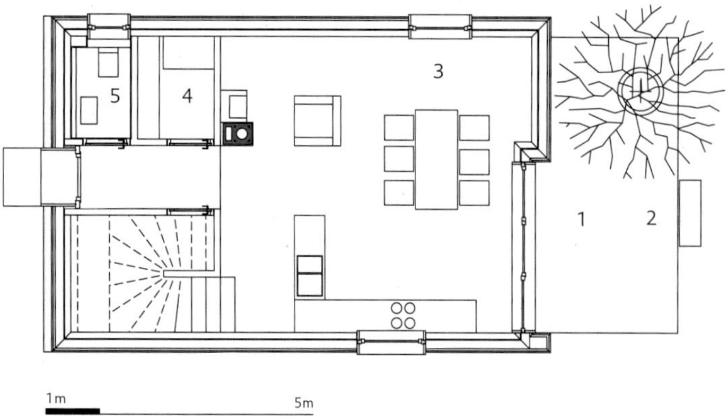

52

1 Eingang / Entrance
2 Terrasse / Terrace
3 Wohnküche / Kitchen and dinig room
4 Speis / Larder
5 WC

6 Zimmer I / Bedroom I
7 Bad / Bathroom

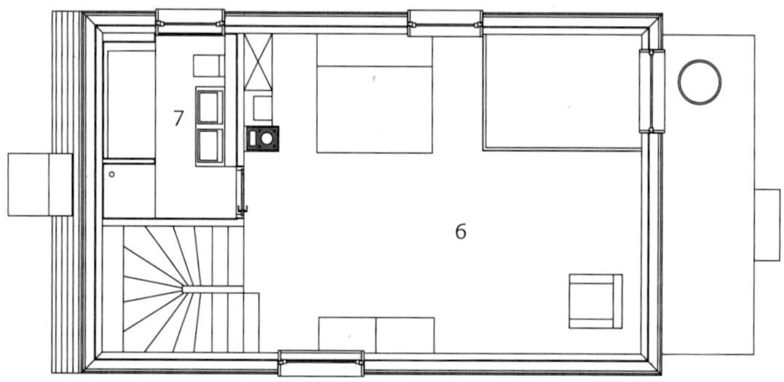

53

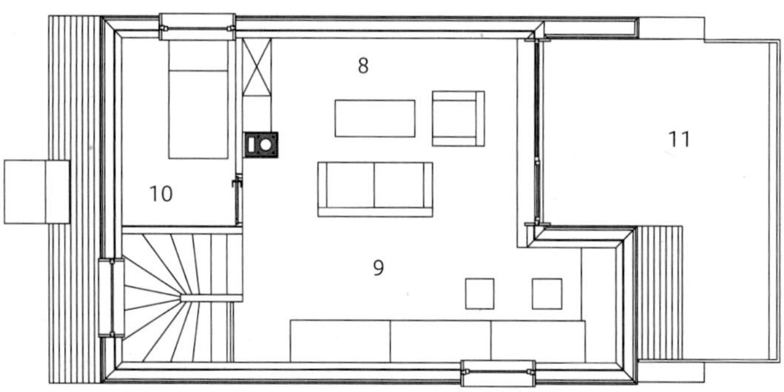

54

8 Wohnen / Living room
9 Arbeitsplatz / Office
10 Zimmer II / Bedroom II
11 Terrasse / Terrace

55

Abb.52 Grundriss Ginger Erdgeschoss
Abb.53 Grundriss Ginger 1. Obergeschoss
Abb.54 Grundriss Ginger 2. Obergeschoss
Abb.55 Schnitt Ginger

Fig.52 Groundlevel Ginger
Fig.53 First floor Ginger
Fig.54 Second floor Ginger
Fig.55 Section Ginger

56 / 57

Abb.56 Dachgeschoss Ginger
Abb.57 Obergeschoss Ginger

Fig.56 Top floor Ginger
Fig.57 Upper level Ginger

58 / 59

Abb.58 Ausblick von Fred
Abb.59 Dachgeschoss Fred

Fig.58 Outlook Fred
Fig.59 Top level Fred

1 Eingang / Entrance

2 Terrasse / Terrace

3 Wohnküche / Kitchen and dinig room

4 Speis / Larder

5 WC

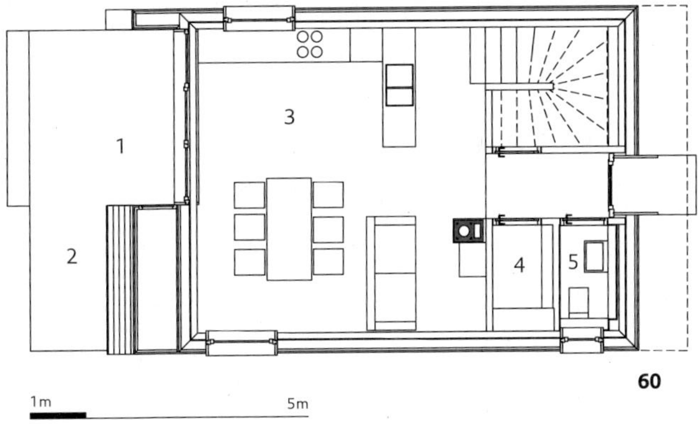

60

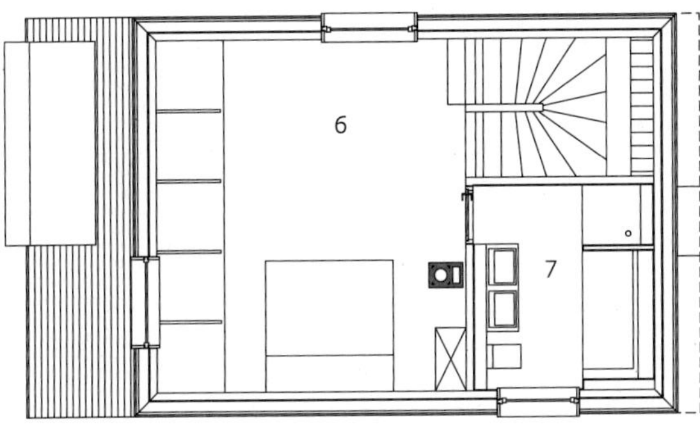

61

6 Zimmer I / Bedroom I

7 Bad / Bathroom

8 Wohnen / Living room

9 Arbeitsplatz / Office

10 Terrasse / Terrace

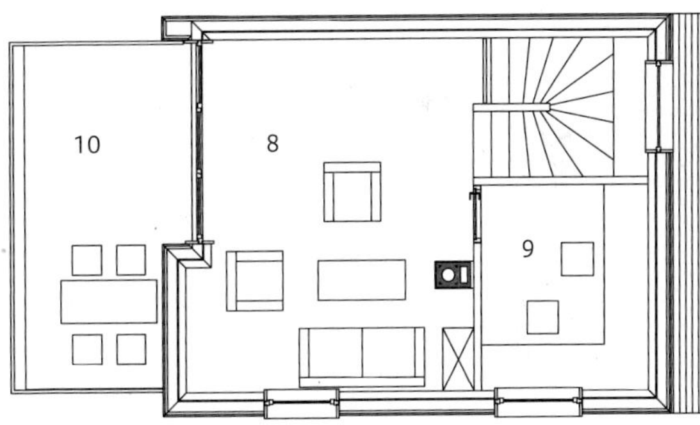

62

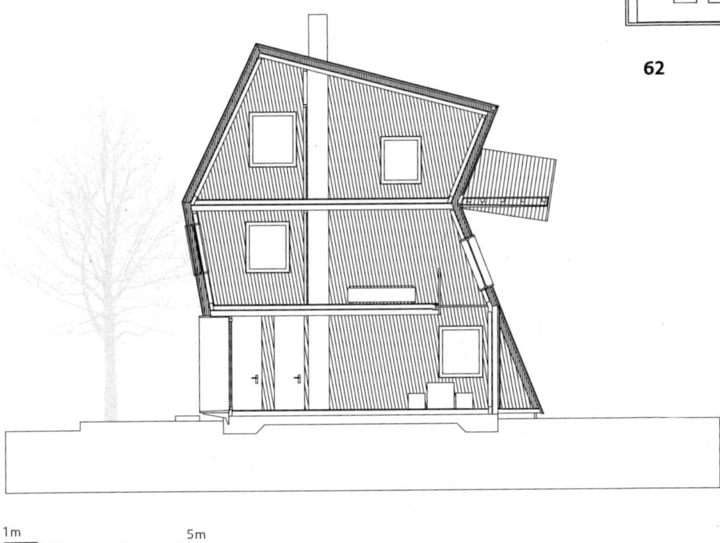

63

Abb.60 Grundriss Fred Erdgeschoss
Abb.61 Grundriss Fred 1. Obergeschoss
Abb.62 Grundriss Fred 2. Obergeschoss
Abb.63 Schnitt Fred

Fig.60 Groundlevel Fred
Fig.61 First floor Fred
Fig.62 Second floor Fred
Fig.63 Section Fred

64

65

Abb.64 Brüstung Fred
Abb.65 Brüstung als negativ alter Geländer

Fig.64 Railing Fred
Fig.65 Rail as a negative of ancient rails

8

milch, 2000

Carsten Nicolai

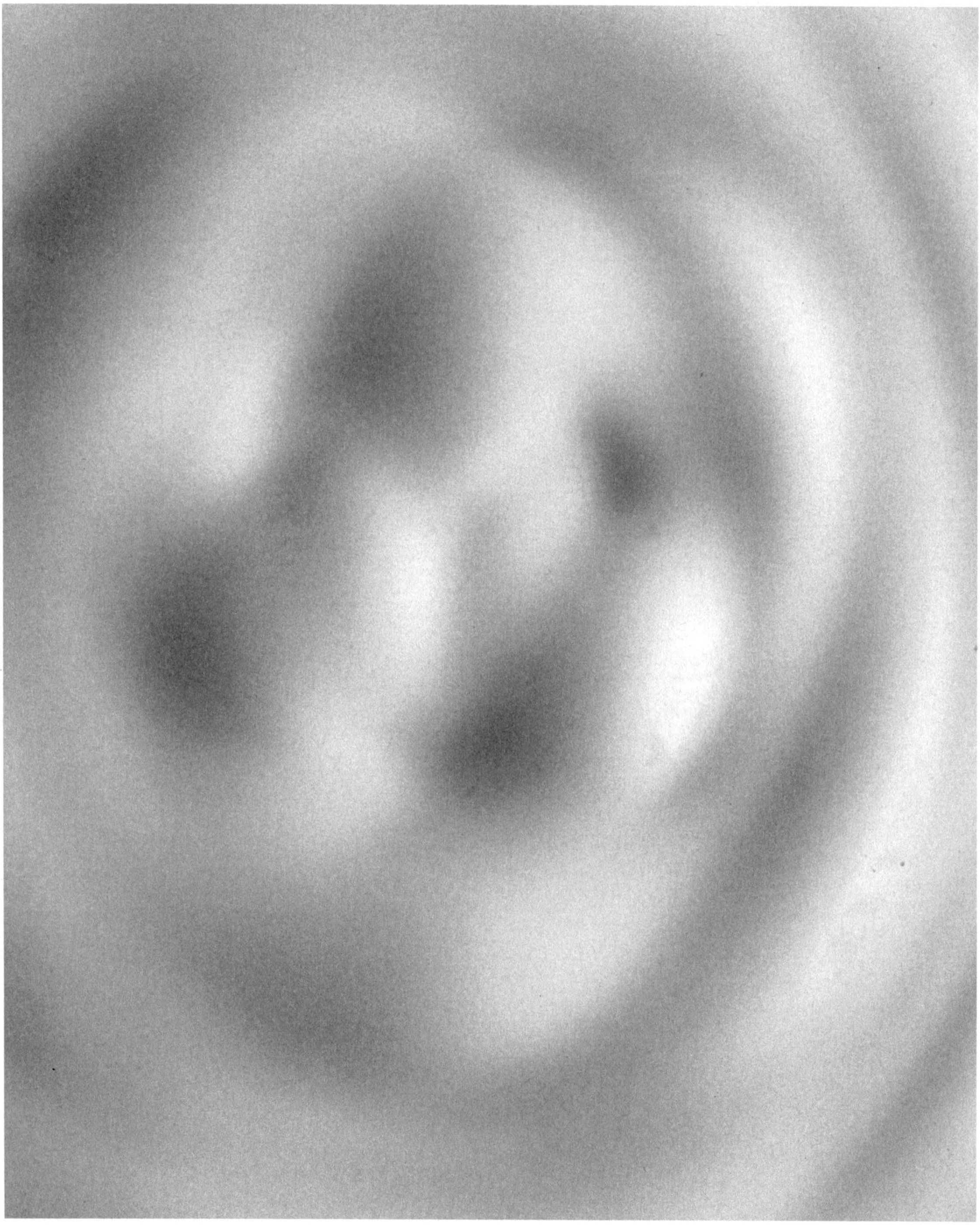

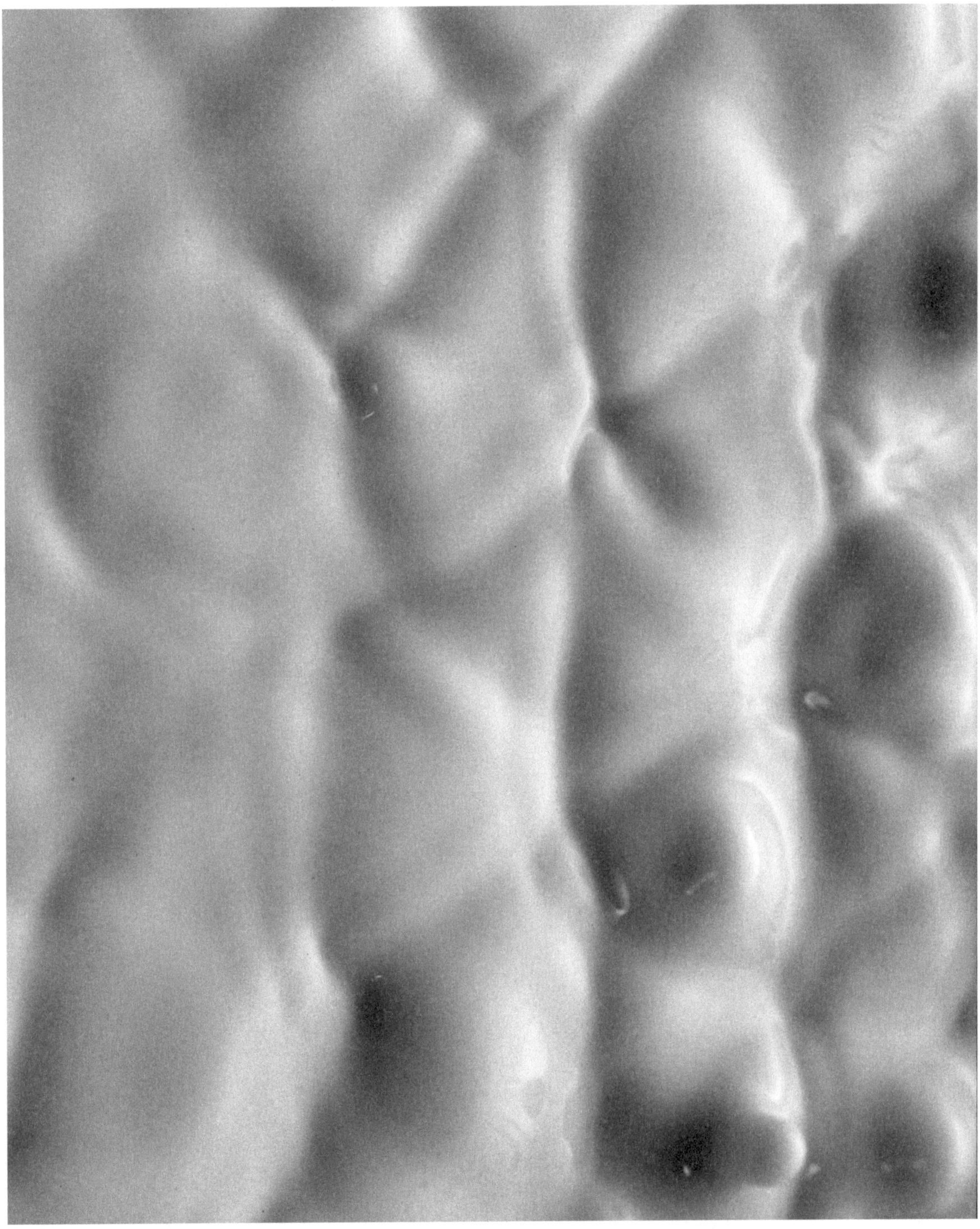

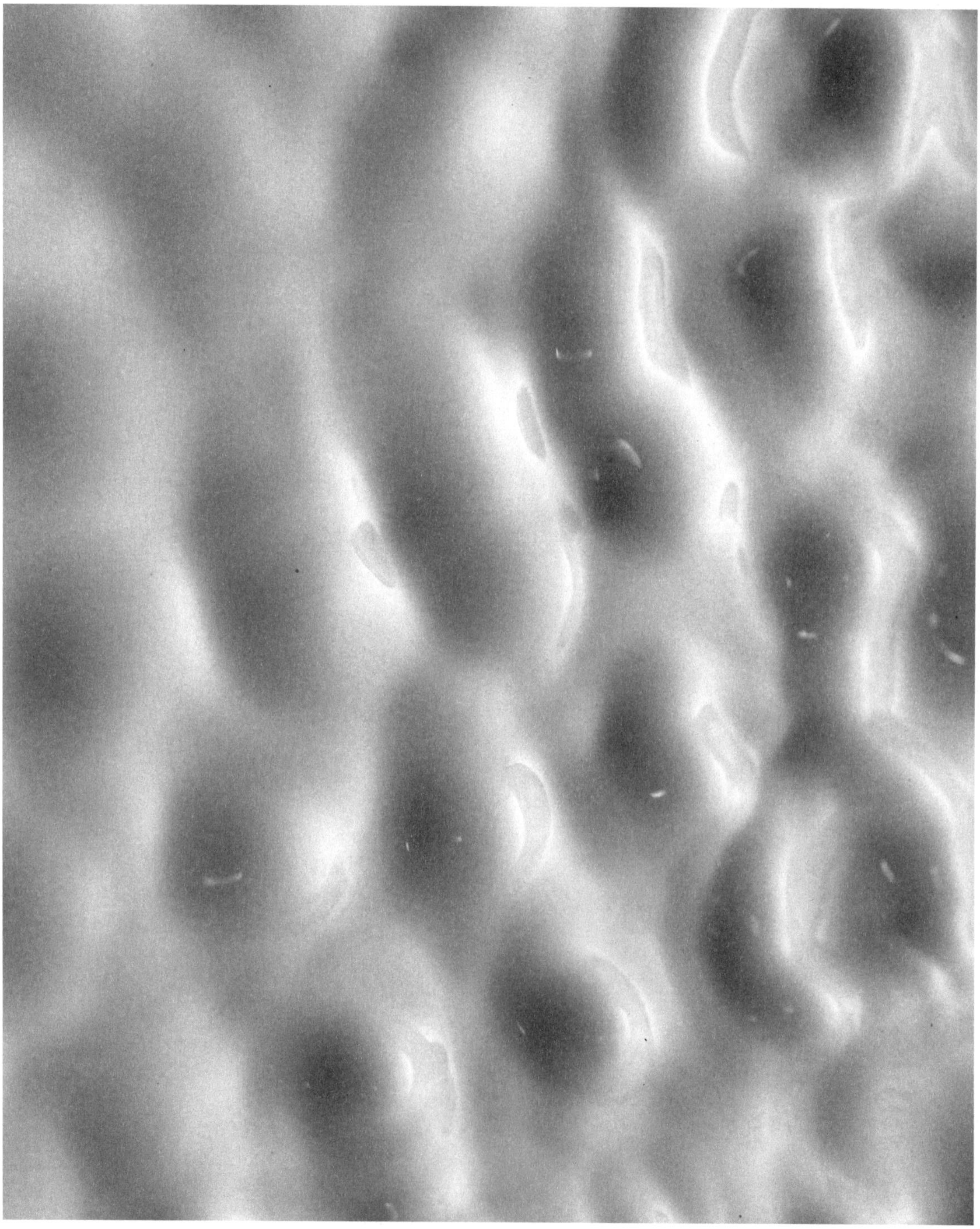

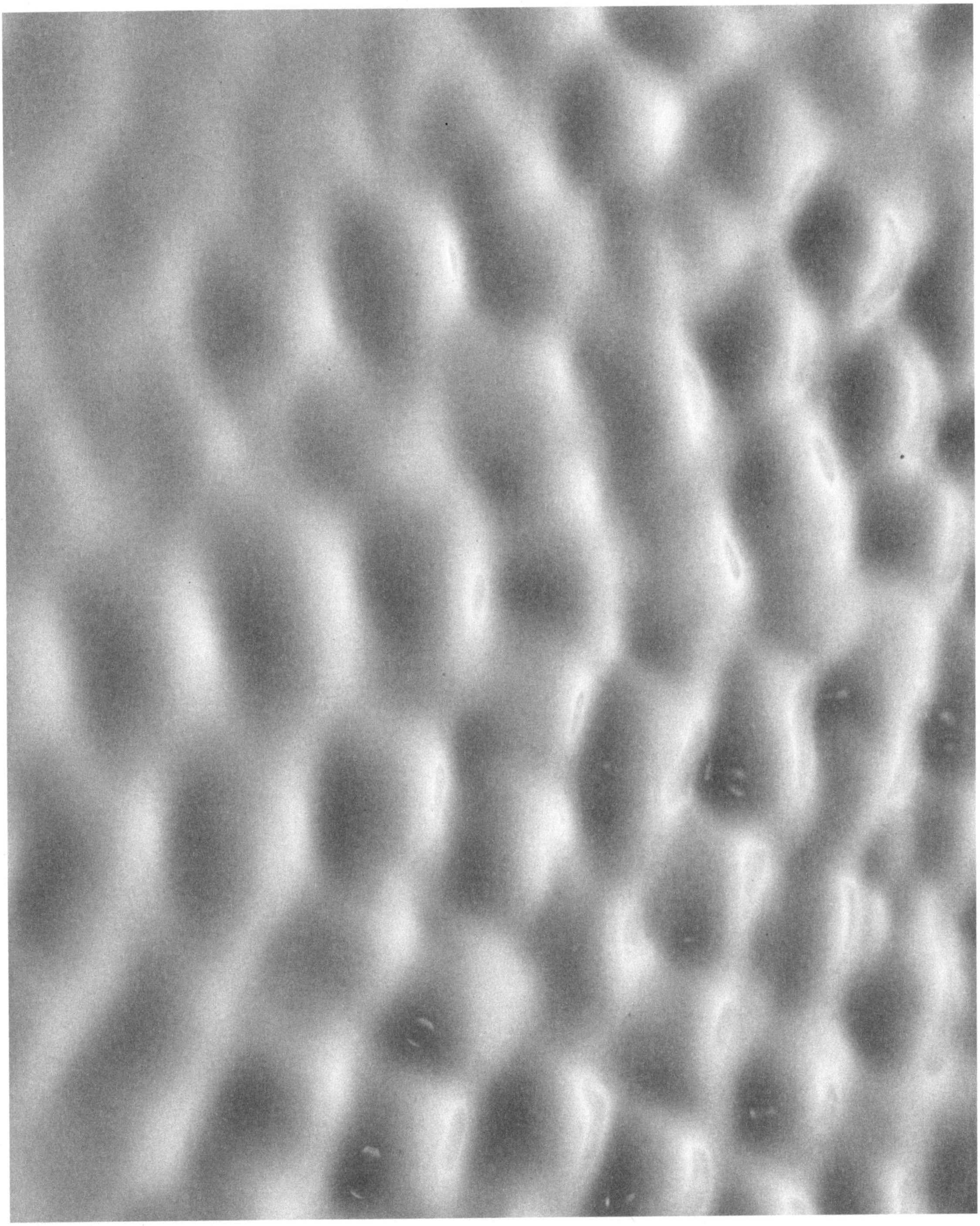

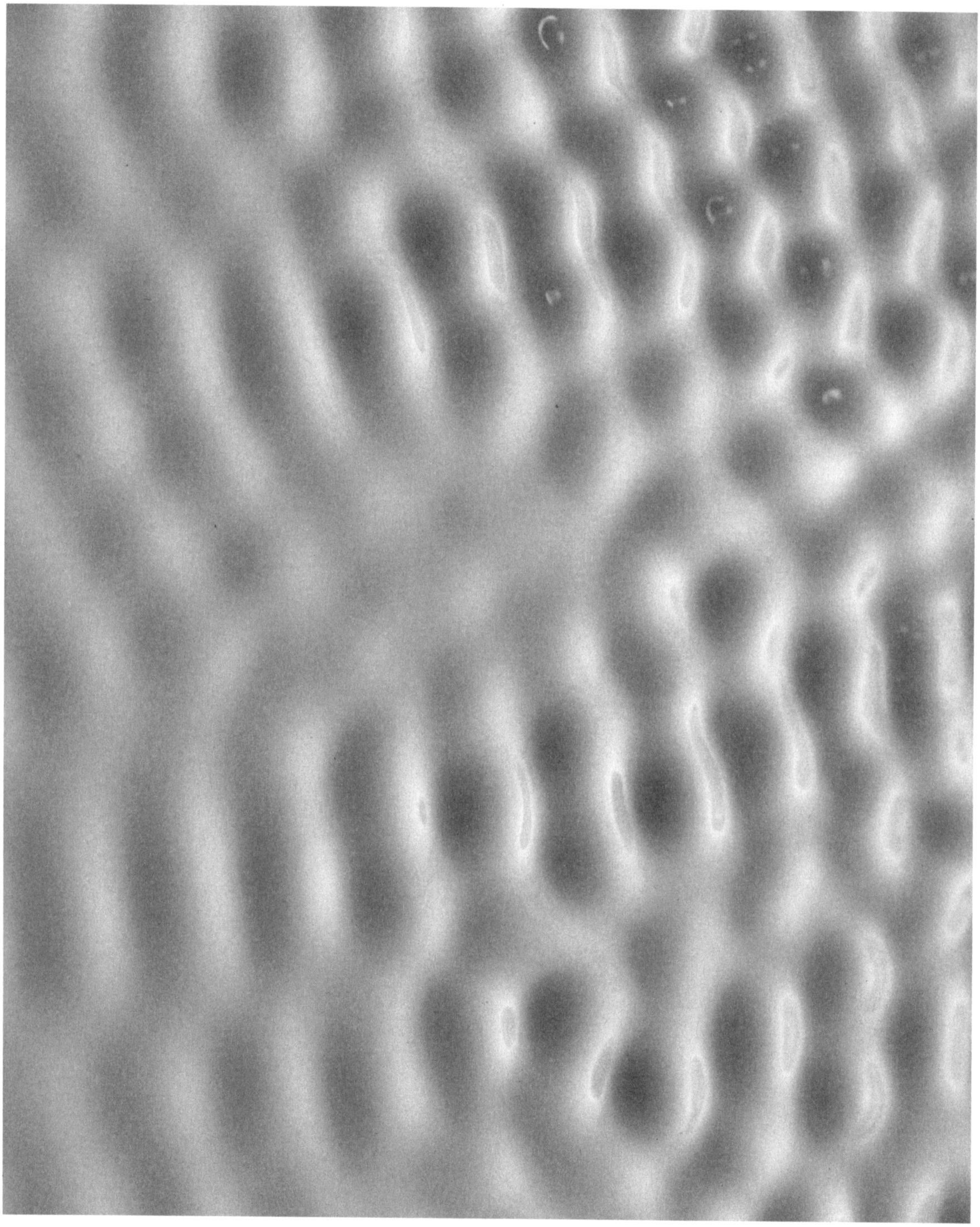

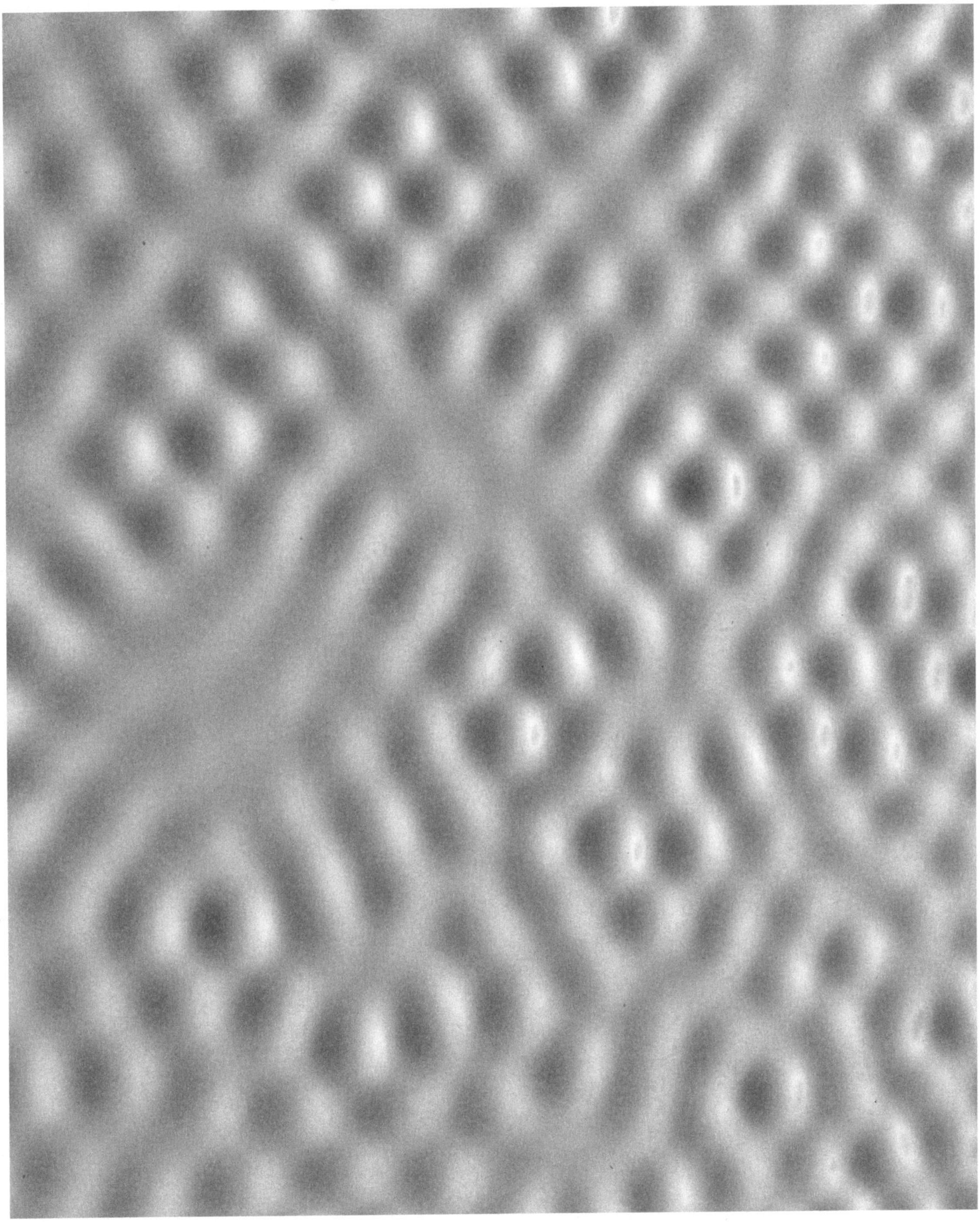

milch 75 Hz | 2000

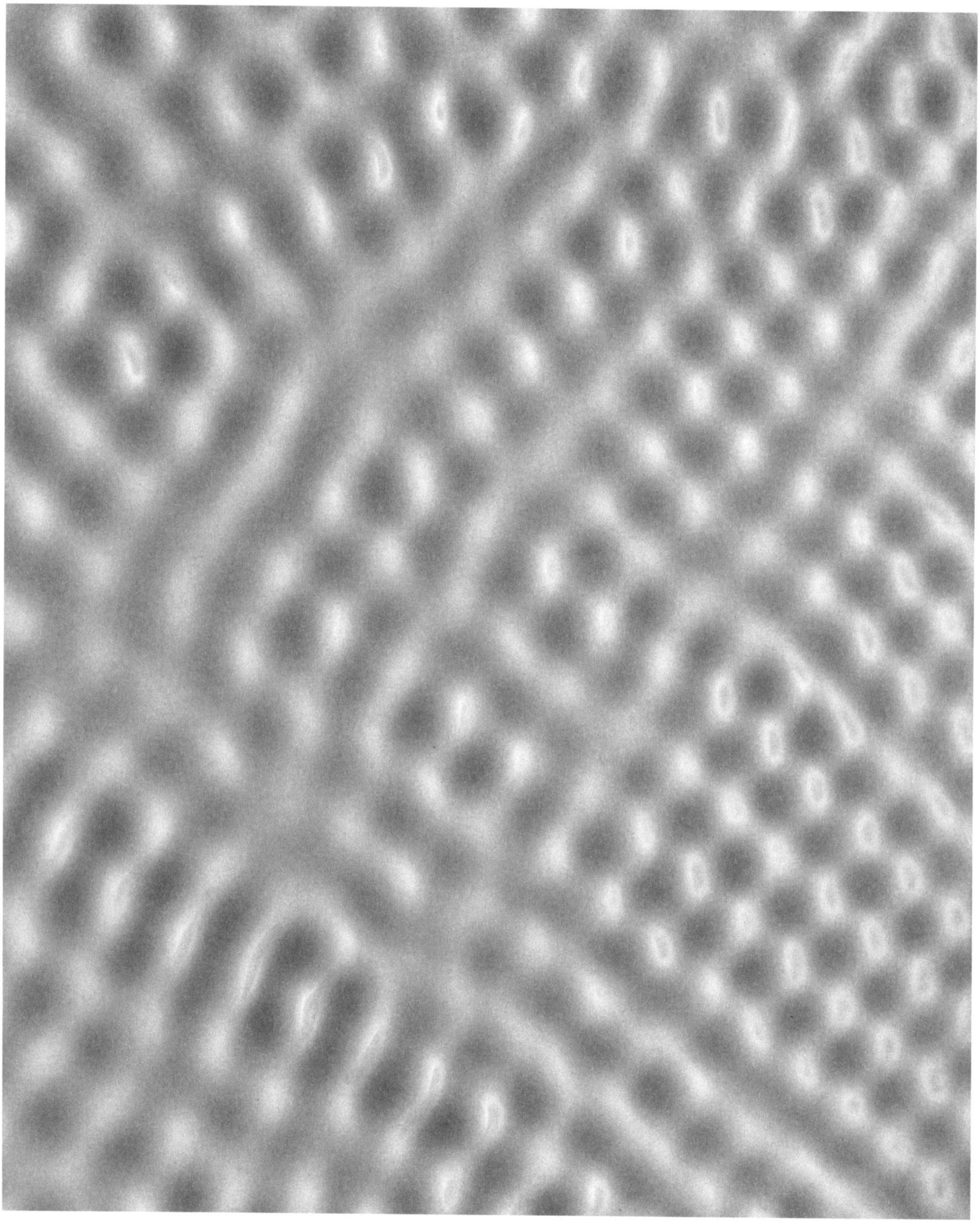

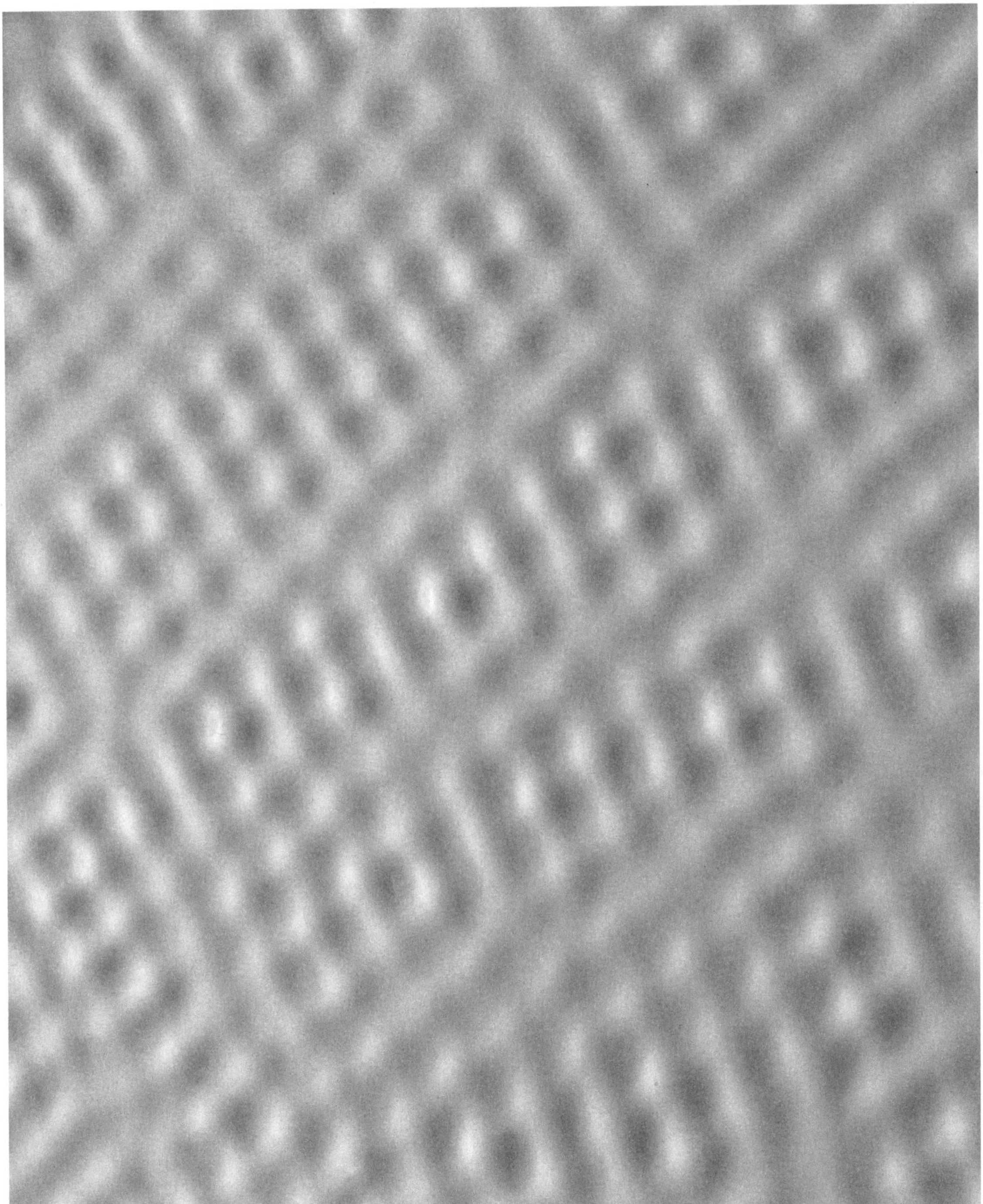

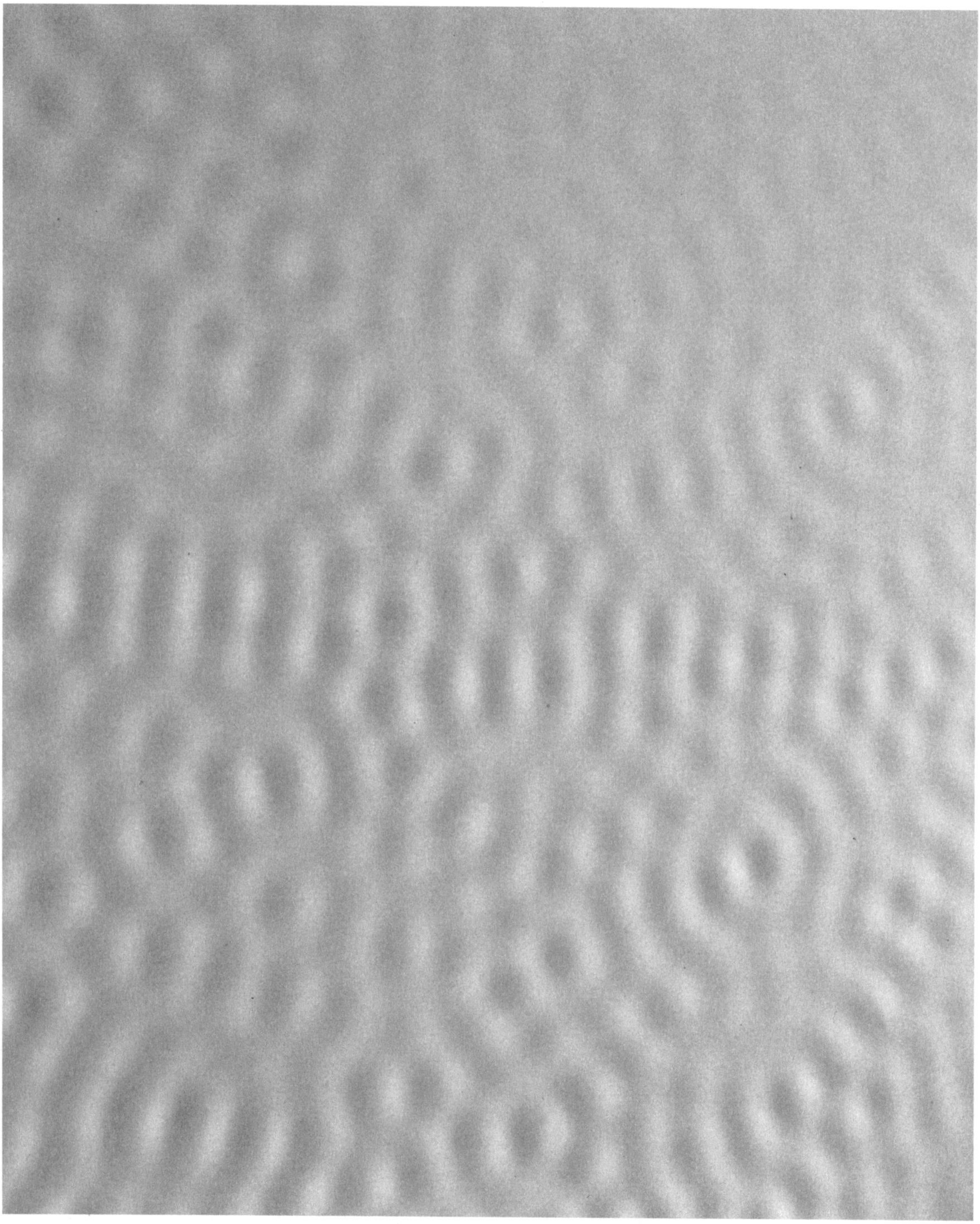

milch 140 Hz | 2000

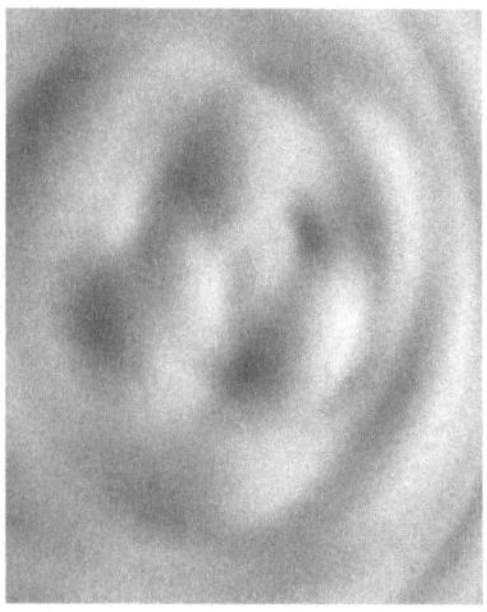

milch 10 Hz

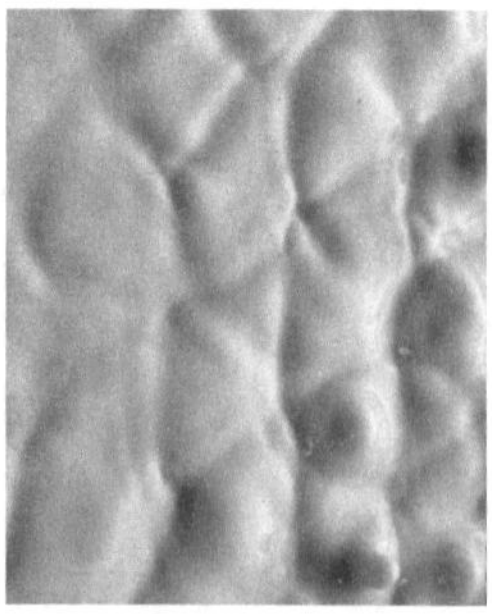

milch 20 Hz

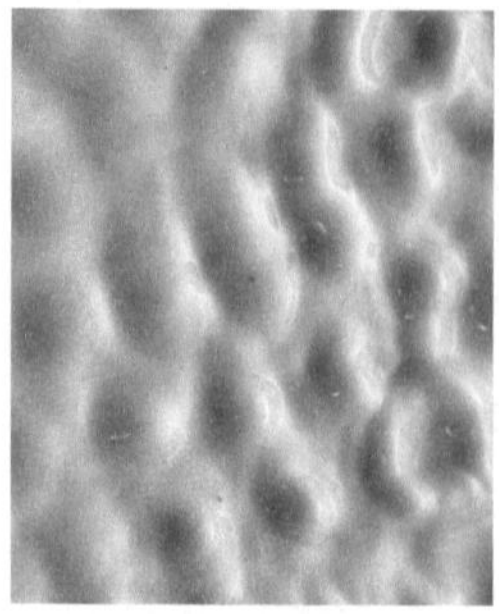

milch 25 Hz

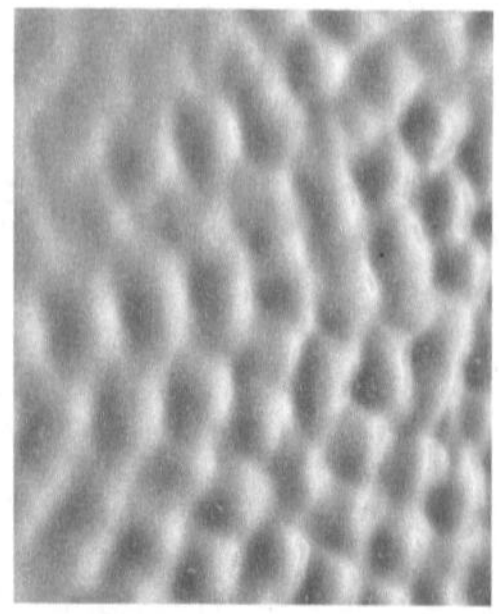

milch 40 Hz

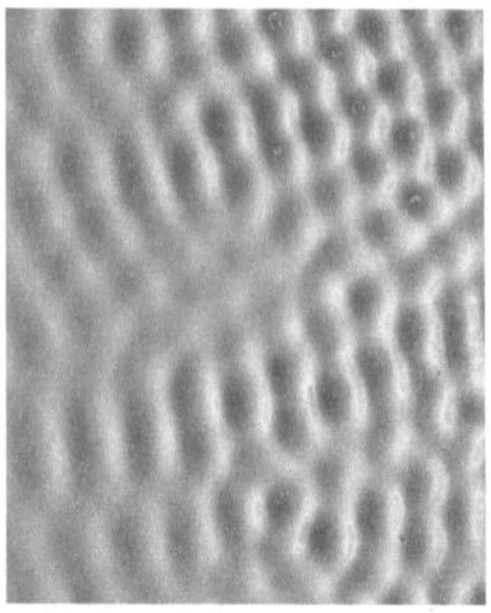

milch 50 Hz

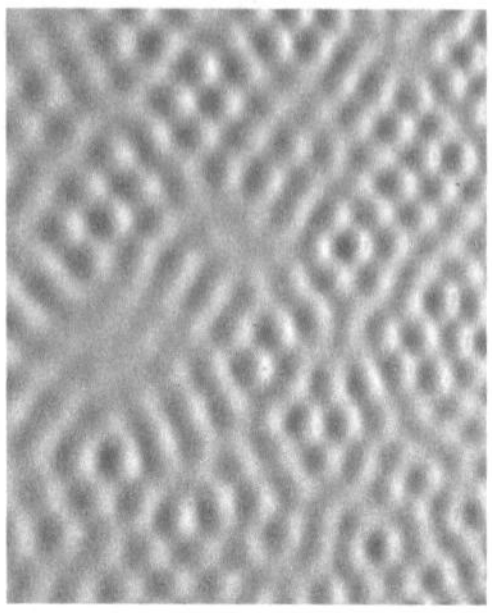

milch 75 Hz

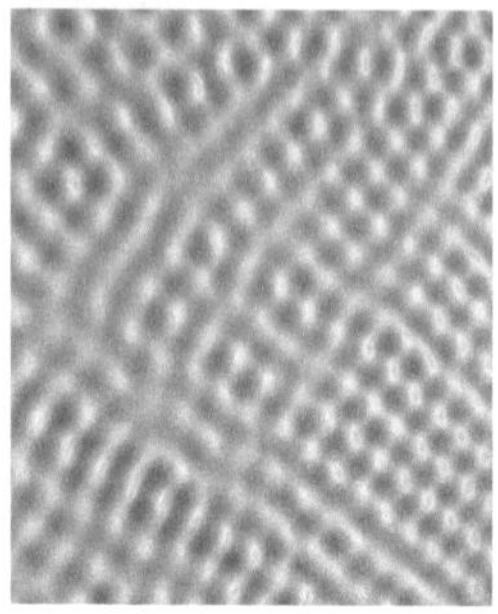

milch 80 Hz

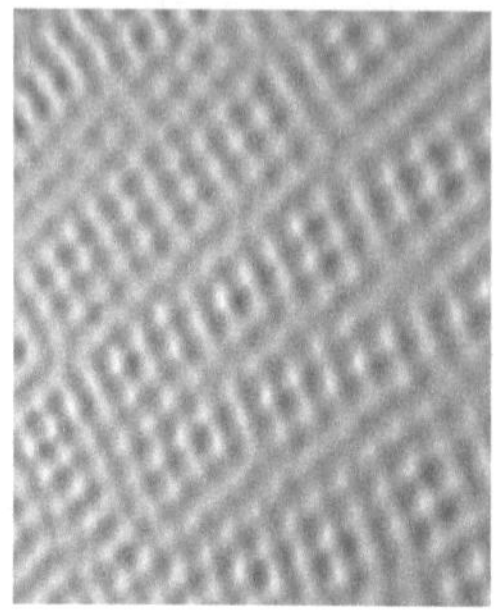

milch 95 Hz

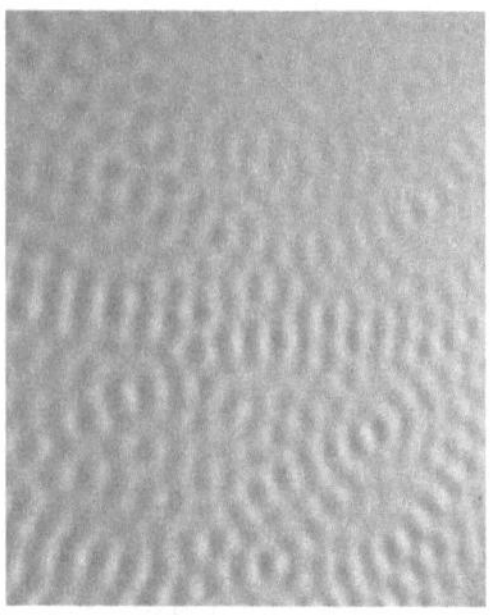

milch 110 Hz

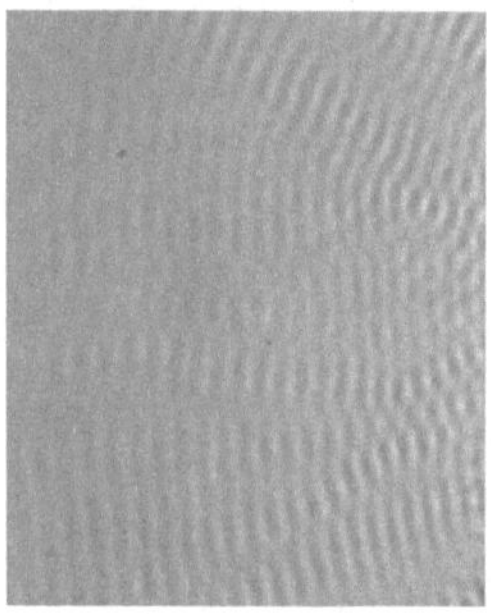

milch 140 Hz

milch, 2000

Lambda-Print auf Aluminium (Serie, 10-teilig)
je 80 x 66 cm
Auflage 5

milch geht aus einer Reihe von Experimenten hervor, die die Wirkung von tiefen Frequenztönen auf Flüssigkeiten untersuchen. In einer Testreihe wurde Milch mit Sinusschwingungen im Bereich von 10 bis 150 Hz bespielt. Der Ton, der akustisch kaum wahrnehmbar ist, lässt in dieser Versuchsreihe eine permanent bewegte, visuelle Struktur entstehen. Ähnlich wie bei den Klangfiguren, bei denen Ernst Florens Friedrich Chladni Ende des 18. Jahrhunderts Schallwellen in ornamentale Bilder überführte, entstehen auch hier komplexe Interferenzmuster, die Chaos und eine gewisse mathematische Ordnung gleichermaßen erzeugen. Die Fotografien der *milch*-Serie dokumentieren verschiedene Phasen des Experiments. Aus dem stets changierenden Prozess der Transformation von akustischen in optische Phänomene halten die Fotografien Momente fest und schaffen auf diese Weise eine Visualisierung von Schallwellen in Form abstrakter Gebilde.

milch, 2000

Lambda print on aluminum (series of 10)
80 x 66 cm each
Edition of 5

milch evolved from a series of experiments that explored the influence of low frequency sound on fluids. In the test series, milk was exposed to sinus waves ranging from 10 to 150 Hz. Sound, almost imperceptible to the ear, appears in this test series as a permanently moving, visual structure. Similar to the sound figures of Ernst Florens Friedrich Chladni, who in the 18th century transformed sound waves into ornamental images, here, too, complex interference patterns are emerging, generating both chaos as well as a certain mathematical order. The photographs of the *milch* series are documenting various phases of the experiment. From the ever changing process of transformation of acoustic into visual phenomena, the photographs freeze moments, thereby demonstrating the visualization of sound waves as abstract forms.

9

Die Muster der Natur

Nature's Patterns

Ian Stewart

1

2

Mathematiker suchen nach allgemeinen Prinzipien. Ein Dreieck, dessen Winkelsumme 180° ergibt, beeindruckt sie nicht sonderlich, doch finden sie es bemerkenswert, dass das Gleiche für jedes beliebige Dreieck gilt. Auf der Suche nach allgemeinen Prinzipien ist es hilfreich, zahlreiche Beispiele zu sammeln, denn erst dann kann man vergleichen und gegenüberstellen, um so den wesentlichen Kern der jeweiligen Fragestellung herauszuarbeiten. Eine Schneeflocke ist ein Beispiel für Musterbildung in der Natur, doch ohne eine breit angelegte Suche wird ihr Geheimnis wohl kaum zu lüften sein. || Für die Suche nach Mustern eignet sich die Erde ganz hervorragend. Während die anderen Planeten unseres Sonnensystems nur Felsoberflächen aufweisen (heiße oder kalte, aber kaum je »richtig temperierte«), gibt es auf der Erde Pflanzen und Tiere. Sowohl organische als auch anorganische Welten entfalten eindrucksvolle Muster – Schmetterlingsflügel und Regenbogen zum Beispiel zeigen das gleiche Farbspektrum –, doch offenbaren Pflanzen und Tiere ein atemberaubend vielfältiges Farben- und Formengemisch, mit dem die anorganische Welt Mühe hat mitzuhalten. | Streifen zählen zu den häufigsten Mustern in der Körperzeichnung der Tiere. Mitunter sind sie sehr regelmäßig, von quasi mathematischer Anmutung – parallele Linien wechselnder Farben, schwarz und weiß, purpurrot und gelb. Tropische Fische oder Meeresschnecken beispielsweise tragen oft mehr oder weniger regelmäßige Streifen – wenngleich bei tropischen Fischen eigentlich so gut wie alle Muster

Abb.1 und 2 Streifen zählen zu den bevorzugtesten Mustern der Natur. Es gibt sie auf Zebras und auf Fischen.

Fig.1 and 2 One of nature's favourite patterns is the stripe. There are stripes on zebras and stripes on fish.

3

Mathematicians seek generalities. They're unimpressed by one triangle whose angles add up to 180°, but they find it striking that the same is true of any triangle. When seeking generalities, it helps to have a lot of examples because then you can compare and contrast and try to extract the essence of whatever it is you are seeking. A snowflake is one example of pattern formation in nature; but the secret of the snowflake is unlikely to emerge from concealment without a wide-ranging search. || Earth is the right place to be when looking for patterns. Where the rest of the solar system has rocks (hot or cold but hardly ever "just right"), planet Earth has plants and animals. Both organic and inorganic worlds exhibit striking patterns – butterflies and rainbows show the same range of colours, for instance – but plants and animals display an apparently effortless riot of colour and form that the inorganic world struggles to match. | One of the commonest patterns in animal markings is stripes. Sometimes the stripes are very regular, the kind of thing you would associate with mathematics – parallel lines of alternating colours, black and white, purple and yellow. More or less regular stripes are common in tropical fish – for that matter, almost any pattern is common in tropical fish – and in seashells. The emperor angelfish deserves its name for its imperial dress: brilliant gold and purple, ringed with ermine, and dominated by narrow black-and-white stripes that run most of the length of its body. They're not completely regular, sometimes they fork like a Y or glitch sideways, but the regularities

Abb.3 Sogar im Meer finden sich Streifen, die wir Wellen nennen. Grundsätzlich sind Streifen ein sich wiederholendes Muster aus Reihen paralleler Linien in gleichmäßigem Abstand. Deutet die geometrische Ähnlichkeit zwischen verschiedenen Streifenformen auf eine versteckte mathematische Einheit oder handelt es sich nur um Zufall?

Fig.3 There are even stripes in the ocean, which we call waves. Stripes repeat roughly the same patterns in a series of equally spaced parallel lines. Does the geometric similarity between different instances of stripes indicate a hidden mathematical unity, or is it just coincidence?

vorkommen. Der Imperatorkaiserfisch hat seinen Namen von seinem kaiserlichen Gewand, das golden und purpurfarben leuchtet, einen Hauch von Hermelin hat und von feinen schwarzweißen, längs über den ganzen Körper verlaufenden Streifen durchzogen ist. Die Streifen sind nicht ganz regelmäßig, sondern gabeln sich hier und da wie ein Y oder scheren nach der Seite aus, doch die Regelmäßigkeiten dominieren. Streifen auf Schneckenschalen kommen in zweierlei Varianten vor: Meistens verlaufen sie entweder in die gleiche Richtung wie die Spirale der Schale oder aber im rechten Winkel dazu. Andere Tiere, wie etwa der Waschbär, tragen markante breite Streifen rund um den Schwanz. | Bei gestreiften Tieren denkt man natürlich sofort an Zebras oder Tiger. Die Streifen der Zebras sind grob und breit, keinesfalls perfekt parallel und jedenfalls deutlich variabler, als man es von einem mathematischen Muster erwarten würde. Besonders dort, wo die Beine oder der Schwanz in den eigentlichen Körper übergehen, zeichnen die Streifen bizarre Muster. Jede der drei Hauptarten von Zebras – Burchell-, Grevy- und Bergzebra – weist ihr eigenes charakteristisches Streifenmuster auf. Die Streifen eines Tigers sind noch subtiler und sehen aus wie mit dem Pinsel über die Flanke der großen Katze gemalt – ein Meisterwerk lebendiger Kalligrafie. | William Blake schrieb einst von der »Furcht erregenden Symmetrie« des Tigers, womit er vermutlich auf dessen elegante und kraftvolle Gestalt anspielte, für den Mathematiker jedoch gehören zur Symmetrie des Tigers auch die Streifen, und zwar nicht nur im übertragenen Sinne. | Auch die anorganische Welt hat ihre Streifenmuster. Wellen branden normalerweise in langen, parallelen Reihen an den Strand – Wellenberg und Wellental statt schwarz und weiß. Die einfachsten Muster der Sanddünen in der Wüste sind Quer- oder Transversaldünen, die sich senkrecht zur vorherrschenden Windrichtung als Streifen formieren, und Längsdünen, die sich in einem Winkel mit wechselnden Windrichtungen aufbauen. Streifenmuster gibt es auch auf Steinen. In Australien befindet sich die weltweit einzige Ablagerung von »Zebragestein«, das aussieht wie gestreifte Zuckerbonbons (und bedauerlicherweise bald von einem Stausee überflutet sein wird). Doch Streifen in Gesteinen stellen ein Archiv ihrer Geschichte dar, da sie sich über lange Zeiträume am Grund von Flussmündungen oder seichtem Wasser Schicht für Schicht ablagerten. Im Gegensatz dazu bilden sich die Streifenmuster von Wellen und Dünen auf einer weit weniger gemächlichen Zeitskala. Besonders die Wellen des Meeres spiegeln die Dynamik der Gegenwart wider. | Sind die Streifen auf Fellen, auf Häuten, im Sand und im Wasser wirklich so verschiedenartig, wie sie scheinen? Ist die Ähnlichkeit zwischen ihnen nichts weiter als ein visueller Gag? Oder gibt es eine versteckte Einheitlichkeit? Existiert vielleicht ein universeller Mechanismus, der Streifenmuster kreiert? Damit es mir gelingt, das Geheimnis der Schneeflocke zu lüften, muss sich eine ganze Bandbreite natürlicher Muster in einen großen Rah-

dominate. Stripes on seashells come in two kinds: mostly they run either in the same direction as the spirals of the shell or at right angles to them. Other animals such as the raccoon have rings of bold stripes round their tails. | Of course, when speaking of striped animals it is the zebra and the tiger that spring most readily to mind. The zebra's stripes are bash and bold, by no means perfectly parallel, distinctly more subtle than you might expect from anything mathematical. The stripes do funny things where the legs or tail merge into the main part of the animals's body, and each of the three main species – Burchell's zebra, Grevy's zebra and the mountain zebra – has it's own distinct pattern of stripes. The tiger's stripes are more subtle still, arrayed along the big cat's flanks like brush-strokes, an elegant exercise in living calligraphy. William Blake memorably wrote of the tiger's "fearful symmetry", probably referring to its elegant, powerful form; but to a mathematician, the tiger's "fearful symmetry", probably referring to its elegant, powerful form; but to mathematicians, the tiger's symmetry includes it's stripes, and not just metaphorically. | The inorganic world has its own stripes, too. Waves rolling up a beach are typically arranged in long, parallel rows – peak and trough in place of black and white. In the deep desert, the simplest patterns of sand dunes are transverse dunes, stripes of sand piled at right angles to the prevailing winds, and linear dunes, stripes arranged at an angle to more variable winds. You also find stripes in rocks. In Australia there is the world's only deposit of zebra rock, which looks like striped candy (and is unfortunately about to be submerged by a reservoir). But the stripes in rocks are a record of their history, as they were deposited layer by layer on the bed of an estuary or a shallow sea. In contrast, the stripes of waves and dunes are created on a far less leisurely time-scale. Ocean waves, especially, reflect the dynamics of the present. | Stripes in fur, stripes in flesh, stripes in sand, stripes in water – are they as different as they seem? Is the similarity between them no more than a visual pun? Or is there a hidden unity? Is there a universal mechanism that creates stripes? If my search for the secret of the snowflake is to succeed, there has to be a big picture into which a reasonable variety of nature's patterns fit. If I can't find unity in some portion of the world's stripes, where will I be able to find it?

Sculpted in Sand Stripes are often found in association with more complex pattern, for example wave markings left in the sand at low tide, or sand duns in the desert. Deserts, with their striking and compelling structure, are almost a laboratory of pattern formation, although we actually know remarkably little about how patterns form in sand. The physics of dunes is deceptively simple. As the wind blows across the desert, it picks up grains of sand from some places and deposits them in others. The shape of existing dunes affects the flow of air, there

4

men einfügen. Falls es mir nicht gelingt, in einem Teil der Streifenmuster der Welt eine Grundeinheit zu finden, wie soll ich sie dann anderswo aufspüren?

In Sand gemeißelt Streifen finden sich oft in Verbindung mit komplexeren Mustern, wie etwa den kleinen Rippeln im Sand bei Ebbe oder in den Sanddünen der Wüste. Wüsten sind mit ihren atemberaubenden und faszinierenden Strukturen wahre Werkstätten für Musterbildung, obgleich wir eigentlich erst recht wenig darüber wissen, wie sich Muster im Sand bilden. | Die Physik der Dünenbildung ist trügerisch einfach. Wenn der Wind über die Wüste bläst, nimmt er feine Sandkörner auf und lagert sie an anderer Stelle wieder ab. Die Form bestehender Dünen beeinflusst die Strömung der Luft und bestimmt so, wie viel Sand abgelagert oder erodiert wird und wo. Insofern besteht eine dynamische Koevolution der Windströmungsmuster und der Muster des aufgehäuften Sandes. | Die Ergebnisse dieser Koevolution sind faszinierend und weit vom Augenscheinlichen entfernt. Einfache Muster aus parallelen Streifen stellen nur den Anfang dar. Große Dünen sind mit kleinen Furchungen überzogen. Die parallelen Rücken der Querdünen nehmen manchmal selbst eine wellenartige Form an und bilden dann wiederholende Zickzackmuster, die als barchanoide Falten bezeichnet werden. Mitunter brechen die Sanddünen in unzählige mondsichelförmige Barchandünen (Barchane) auf, deren spitze Enden von der Windrichtung weg zeigen. Alternativ können

Abb.4 Selbst mit so einfachen Materialien wie Sandkörner vermag die Natur die schönsten Muster hervorzubringen. Aus dem Chaos des vom Wind herumgewirbelten Sandes erwächst die Ordnung der Dünen — riesiger Sandwellen, die sich über die Wüste ziehen, scheinbar zu Bergen und Täler erstarrt, doch in Wirklichkeit in ständiger Wanderung begriffen, Schicht für Schicht.

Fig.4 Even using materials as simple as grains of sand, nature can sculpt elegant patterns. From the chaos of whirling windborne sand emerges the order of dunes — giant waves of sand, sweeping across the desert, seemingly frozen into hills and valleys, but actually marching across the desert, rank upon rank.

5

6

is a dynamic co-evolution of the patterns of air flow and the patterns of heaped sand. | The results of this co-evolution are fascinating, and far from obvious. Simple patterns of parallel stripes are only the beginning. Big dunes are covered in small corrugations. The parallel ridges of transverse dunes can themselves become wavy, a pattern of repeated zigzags known as barchanoid ridges. The lines of heaped sand can break up altogether, forming crescent-shaped barchan dunes with the horns of the crescents trailing away from the wind. Alternatively, they can form parabolic dunes, each dome a smooth circular mound. If the wind's direction is very variable, these domes can turn into stars, speckling the desert for hundreds of kilometres. | There is astonishingly little mathematical theory to explain the myriad patterns of the desert. The physics may sound simple, but translating it into mathematics is extremely hard. Sand is composed of individual grains, unlike the finely divisible fluids of air or water, so one of the mathematician's favourite tricks – creating an idealized model of an infinitely divisible continuum – doesn't work particularly well for sand, one a genuine fluid, the other a somewhat granular one. The boundary between sand and air is part of the answer, not part of the assumptions, and it changes as time passes. And when a sandstorm surges over the desert, one of the main things that determine the shape and positions of dunes, the boundary between sand and air, becomes too fuzzy to specify. | It all looks pretty hopeless, and it is if you insist on tackling the problem directly. But,

Abb.5 Wüstensand kreiert eine große Vielfalt an Mustern, einfachen wie komplexen.

Fig.5 Desert sand creates a great variety of patterns, both simple and complex.

Abb.6 Sanddünen verhalten sich wie langsame Meereswellen und bilden genau wie diese sowohl winzige Rippeln als auch riesige Kämme und Täler aus. Die Mathematik und die Physik des Sandes stecken voller Rätsel, aber einige allgemeine Merkmale dieser Muster beginnt man gerade zu verstehen.

Fig.6 Sand dunes are like slow ocean waves – and, like water waves, they form tiny ripples as well as huge crests and troughs. The mathematics and the physics of sand are mysterious, but some general features of the patterns are starting to be understood.

sich auch Parabeldünen bilden, große halbmondförmige Sandberge mit gebogenen Kämmen, deren spitze Enden in die Richtung des Windes weisen. Wenn die Windrichtung öfter wechselt, ordnet sich die Wüste zu einer Reihe kuppelartiger Dünen, die jeweils glatte, kreisrunde Erhebungen darstellen. Sind die Winde äußerst wechselhaft, bilden diese Kuppeldünen zuweilen sternförmige Strukturen, mit denen die Wüste dann über Hunderte von Kilometern gesprenkelt ist. | Es gibt erstaunlich wenig mathematische Theorien zur Erklärung der zahllosen Muster im Wüstensand. Die Physik mag simpel klingen, aber sie in eine mathematische Theorie zu übertragen ist extrem schwierig. Sand besteht aus individuellen Körnchen, anders als die homogen verteilten Fluide wie Wasser oder Luft, so dass einer der Lieblingstricks der Mathematiker – ein idealisiertes Modell eines unendlich teilbaren Kontinuums zu schaffen – bei Sand nicht besonders gut funktioniert. Bei der Dünengeometrie haben wir es zudem mit einem Mehrphasenströmungsproblem zu tun; es schließt sowohl Luft – ein schlechtes Fluid – als auch Sand – ein eher granuläres Gemisch – ein. Die Grenzschicht zwischen Sand und Luft ist Teil der Antwort, nicht Teil der Voraussetzungen – und sie verändert sich mit der Zeit. Sobald ein Sandsturm über die Wüste braust, verliert einer der Hauptfaktoren, die Form und Position der Dünen bestimmen, nämlich eben diese Grenzschicht zwischen Sand und Luft, seine Spezifizierbarkeit. | Das Ganze scheint also ziemlich aussichtslos – und das ist es in der Tat, wenn Sie das Problem auf direktem Weg angehen wollen. Doch, wie schon gesagt, Mathematiker denken gerne in allgemeinen Prinzipien. Vielleicht gibt es ja andere, leichter handhabbare Systeme, die ähnliche Muster entfalten wie die, die sich in der Wüste finden. Wer weiß, vielleicht können uns die Wellen der Meere etwas über Querdünen sagen – oder Dünen etwas über Zebras. Oder das Geheimnis, das allen Streifenmustern gemeinsam ist, lässt sich aus einer ganz anderen Perspektive entschlüsseln. | Warum bilden sich überhaupt Muster? Keine schlechte Frage, um der Sache näher zu kommen. Über weite Wüstenabschnitte hinweg herrschen ziemlich ähnliche Windbedingungen, und selbst wenn der Wind sich ändert, variiert er überall in fast gleicher Weise. Warum streicht der über die Wüste gleitende Wind nicht einfach den Sand glatt und ebnet ihn – so wie ein Messer den Zuckerguss auf einem Kuchen glatt verteilt? Und warum stellt sich das Meer stets als eine Masse aus wogenden Wellen dar? Schon gibt es in den Fragen eine gewisse Einheitlichkeit, doch noch haben wir nicht einmal angefangen, nach Antworten zu suchen. Um es noch einmal auf den Punkt zu bringen: Jeder Tiger und jedes Zebra ist mit Haaren bedeckt, die überall auf dem Körper des Tieres die gleiche Struktur aufweisen – warum aber bilden die Pigmente dann in den Haaren farbige Bereiche mit deutlich erkennbaren Regelmäßigkeiten aus? | Warum sind Zebras nicht einfach alle einheitlich grau?

as I've already said, mathematicians like to think in generalities. Maybe there are other, more tractable systems that exhibit similar patterns to those that can be found in the desert. Maybe ocean waves can tell us something about transverse dunes. Maybe dunes can tell us about the zebra. Maybe something quite different from any of these can unlock the common secret of all the world's stripes. | One fruitful line of attack is to ask why patterns form at all. Much the same wind conditions exist across huge swathes of desert, and even if the wind varies, it varies in much the same manner everywhere. As the wind sweeps across the desert, then, why doesn't it smooth the sand out and lay it flat, like a knife smoothing icing on a cake? By the same token, why are oceans always a mass of heaving waves? Why does the water slop around so much? And when it does slop, how come it slops in recognizable patterns, rather than slopping in random muddle? Already there's a common unity of questions, and we haven't even started to look for answers. To press home the point: since the entire tiger or zebra is covered in hairs, which have much the same structure everywhere on the animals' bodies, why does the pigment in the hairs create coloured regions with visible regularities? | Why aren't zebras grey all over?

Bienenwaben Es gibt natürlich noch andere Muster in der Natur außer Streifen. Punkte oder Flecken zum Beispiel. Warum sind Tiger gestreift, Leoparden aber gefleckt? Vielleicht können die Zeichnungen von Tieren so gut wie alle Formen annehmen. Denken Sie nur einmal an das bizarre Federkleid der Paradiesvögel. Vermutlich können die Gene eines Tieres die Zellen anweisen, so ziemlich jedes Muster hervorzubringen. Das Merkwürdige ist nur, dass sie das alles in allem nicht tun. Die meisten Muster greifen auf einen Standartkatalog einfacher Formen zurück – Streifen, Punkte oder gleichmäßige Farbflächen. Häufig finden sich diese zu den seltsamsten Kombinationen zusammen, besonders bei Vögeln und Fischen. Vielleicht fügen ja die Gene die Muster zu einem Ganzen zusammen, aber etwas anderes bestimmt den Grundkatalog. | Was gibt es noch in diesem Katalog? Bienenwaben zum Beispiel. Ein Bienen- oder Honigwabenmuster sieht in der Tat mathematisch aus – Reihen von regelmäßig aneinander gefügten Sechsecken in schöner zweidimensonialer Ordnung. Bienenwaben und Schneeflocken ist die magische Zahl 6 gemeinsam – ein Zusammentreffen, das auch Kepler nicht entging. Jede Wabe in einem Bienenstock stellt eine winzige sechseckige Zellkammer dar, in der sich eine Larve entwickeln oder Honignektar gespeichert werden kann. Ich erinnere mich, wie ich als Kind gestaunt hatte, als der Mann, der früher zu Hause regelmäßig der Wespenplage den Garaus machte, ein altes Wespennest unter unserem Hausdach hervorholte, und wie ich etwas bange dieses Wunderwerk an stofflicher Baukunst auseinander nahm. Im Inneren fand ich eine wunderschöne Konstruktion aus nahezu identischen sechseckigen Kammern. Diese waren jedoch ziemlich unregelmäßig aneinander gefügt, was darauf hindeutete, dass die Wespen zur gleichen Zeit und ohne Generalbauplan mit dem Bau verschiedener Teilstücke angefangen hatten und diese dann, sobald sie aufeinander trafen, an den Enden irgendwie zusammenmontierten. | Bienen bauen ihre Honigwaben im Allgemeinen in vertikaler Richtung, wobei die sechseckigen Tunnel horizontal verlaufen; Wespen hingegen bauen sie in horizontaler Richtung und die sechseckigen Tunnel verlaufen vertikal. Wie kommt es, dass Bienen und Wespen so kluge Baumeister sind? Es handelt sich um staatenbildende Insekten, die in der Gesellschaft weit mehr zu leisten vermögen als das einzelne Individuum. Ich glaube gar nicht, dass sie so besonders klug sind. Ich denke eher, dass irgendetwas ihnen in gewisser Weise Starthilfe gibt. Bestimmte Muster sind einfach leichter zu erzeugen als andere, sie entstehen, weil das Universum im Grunde einfachen Regeln gehorcht. | So überrascht es nicht, dass Wespen und Bienen nicht die einzigen Kreaturen sind, die Wabenmuster hervorbringen. Auch manche revierbildende Fischarten tun dies. Und hier sind die Gründe für die Wabenstruktur einleuchtender. Eine winzige Fischart im Huronsee in den USA liefert

Abb. 7 Bienenwabenmuster kommen seltener vor als Streifen, sind in der natürlichen Welt aber ebenfalls weit verbreitet. Die ihnen zugrunde liegende mathematische Struktur ist offensichtlicher als die der Streifen, da Bienenwaben aus Sechsecken bestehen, und die sind aus der Geometrie wohl bekannt.

7

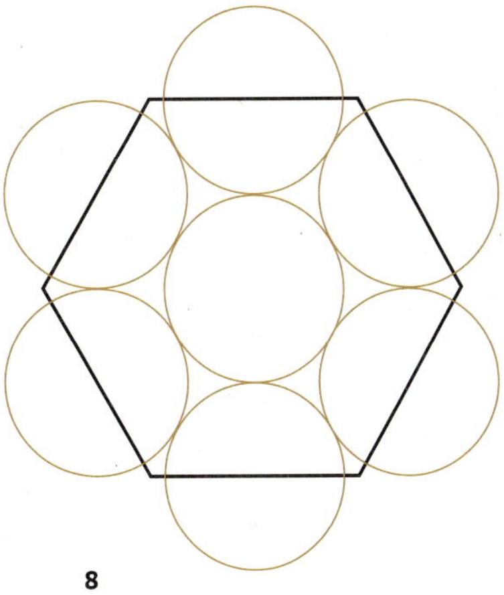

8

Honeycombs There's more to life's patterns than stripes. Spots, for a start. Why are tigers stripy but leopards spotty? Maybe animal markings can be pretty much anything they want to be. After all, look at the bizarre plumage of birds of paradise. Presumably an animal or bird's genes can instruct its cells to form pretty much any pattern. The curious thing, though, is that on the whole they don't. Most patterns come out of standard catalogue of simple forms — stripes, spots, uniform areas of colour. Often these patterns are assembled into strange combinations, especially in birds and fish, so maybe the genes put the patterns together into a whole, but something else determines the catalogue. | What else is in the catalogue? Honeycombs. A honeycomb pattern really does look like mathematics — line upon line of perfect hexagons stacked in a neat two-dimensional array. Honeycombs and snowflakes share the magic number six, a coincidence that was not lost on Kepler. In a beehive the hexagons are tiny chambers, each capable of holding a grub or some honey. I still remember my amazement when the man who dealt with our regular wasp infestations produced an old wasp's nest from inside the roof of our house and, with some trepidation, I broke open this miracle of paper engineering. The interior was a beautiful construction of nearly identical hexagonal chambers, though put together in a rather irregular fashion which suggested that the wasps started building various bits of it at the same time without

Fig.7 Honeycomb patterns are less ubiquitous than stripes, but still quite common in the natural world. Their underlying mathematical structure is more evident than that of stripes, because honeycombs are built from hexagons, and hexagons come straight out of the geometry texts.

Abb.8 Bienenwaben illustrieren zudem ein Schlüsselmerkmal der Musterbildung: die wiederholte Verwendung der immer gleichen Grundeinheit. Mehrere identische kreisähnliche Formen lassen sich in der Ebene am effizientesten in einem Bienenwabenmuser zusammenfügen und die Natur hat viele Gründe, genau dies zu tun — insofern spielen das Sechseck und das damit verbundene Bienenwabenmuster eine Schlüsselrolle in der Musterbildung, sowohl in der Physik als auch in der Biologie.

Fig.8 Honeycombs also exemplify a key feature of pattern formation: the use of the same basic unit, repeated over and again. A honeycomb is an efficient way to fit together a lot of identical, roughly circular regions, and nature has many reasons for doing just that — so the hexagon and its associated honeycomb play a key role in pattern formation, in both physics and biology.

ein typisches Beispiel. Diese Tiere besitzen ein ausgeprägtes Revierverhalten und schaffen sich schüsselartige Vertiefungen mit einem Durchmesser von etwa 30 cm als Territorien. Jeder Fisch platziert sich in der Mitte seines Territoriums und wehrt alle Eindringlinge ab. Da es sehr viele von diesen Fischen gibt, liegen die Territorien dicht gepackt aneinander. Dabei ergibt sich wie von selbst ein wabenartiges Muster. Auf den ersten Blick erscheint dies wie eine beeindruckende Ingenieurleistung, doch gibt es einen Trick. Und der heißt »dichte Verpackung«: Wenn Sie eine Menge identischer Kreise – zum Beispiel Münzen – auf einen Tisch legen und versuchen, sie so eng wie möglich zusammenzurücken, dann entsteht ein Muster ähnlich einer Bienenwabe. Es wird nicht perfekt regelmäßig sein, doch das sind Fischterritorien und Bienenwaben auch nicht. Aber annähernd regelmäßig ist es allemal. Vor etwa 100 Jahren haben Mathematiker bewiesen, dass sich Kreise in einer Ebene am dichtesten in einem Wabenmuster zusammenfügen lassen. Der Hauptgrund dafür besteht darin, dass immer genau sechs Kreise um einen anderen Kreis der gleichen Größe passen, und dieses Strukturprinzip wiederholt die Bienenwabe Kreis für Kreis. | Aus alledem lerne ich jedenfalls, dass regelmäßige Muster in großem Maßstab durch einfache lokale Regeln erzeugt werden können. Dies deutet auf ein logisches Grundprinzip hinter dem Musterkatalog der Natur. Bereits Kepler hatte das erkannt – die Regelmäßigkeit dicht zusammengefügter Kreise ist eine Schlüsselbeobachtung in seinem Buch über die Schneeflocken. Fische packen ihre Territorien dicht aneinander; Bienen und Wespen haben eine Struktur entwickelt, durch die ihre Larven eng zusammenkommen. | Was sorgt für die Packung einer Schneeflocke?

Gefrierende Kügelchen Kepler hat sich genau die gleiche Frage gestellt. Er wusste, dass Schnee aus kondensiertem Wasserdampf besteht, und er fragte sich, ob dieser wohl in einem bestimmten Muster kondensiert. »Angenommen, der Wasserdampf gefriere, sobald ihn die eindringende Kälte berührt, zu Kügelchen von bestimmter Größe … Zweitens soll angenommen werden, dass sich diese Dampfkügelchen gegenseitig in bestimmter Anordnung berührten …« Aufgrund der irrigen Annahme, die Schneeflocke sei im Grunde eher dreidimensional als flach, verrannte er sich hier zunächst, kehrte dann aber kurze Zeit später zur Geometrie der Ebene zurück und verwandelte seinen Irrtum in einen Sieg: »…dass sich hier das Sechseck entsprechend den materiellen Notwendigkeiten durch das Formvermögen ergibt, damit es keine Leere gibt und die Kristallisation des Wasserdampfes bei der Bildung des Schnees umso besser vor sich gehen kann.« | Im letzten Absatz setzt er die geometrische Regelmäßigkeit der Schneeflocken in Beziehung zur regelmäßigen Geometrie der Kristalle: »Daraus ergibt sich, dass dieses Formvermögen wahrscheinlich bei verschiedenen Flüssigkeiten

recourse to any master plan and kind of fudged the joins when the different parts met up. | Bees generally build their honeycomb in the vertical direction, with the hexagonal tunnels running horizontally; wasps build them in the horizontal direction, with the hexagonal tunnels running vertically. How can bees and wasps be smart enough to make such things? They are social insects, somehow capable of much more when they act as a collective than any individual could possibly achieve. I don't think they are that smart. I think that something gives them a helping hand, a head start. Some patterns are easier to make than others, they arise because at heart the universe obeys simple rules. | In consequence, it's no surprise that wasps and bees aren't the only creatures that make honeycombs. So do territorial fish. And here, the reasons for the honeycomb structure are more evident. A typical case is a species of tiny fish in Lake Huron, USA with strong territorial instincts, which carve themselves dish-shaped territories about 30 cm (12 in) across. Each fish stations itself in the middle and repels all invaders. There are a lot of these fish and their territories are packed closely together. As it happens, they pack in a honeycomb pattern. At first sight, that looks like an amazing engineering feat, but there's a trick. The clue is "packed closely". If you take a lot of identical circles — coins, say — lay them on a table, and jiggle them around until they're squashed as tightly as possible, then they form themselves into a honeycomb. In practice it's not perfectly regular, but the same goes for fish territories and bee's honeycombs. But it's pretty close. About a hundred years ago mathematicians proved that the honeycomb is the most efficient way to pack circles in the plane. The main reason is that six circles fit exactly around another circle of the same size, and the honeycomb repeats this structure round every circle. | At any rate, what this all tells me is that regular large-scale patterns can be created by obeying simple, local rules. This hints at a rationale behind nature's pattern book. Kepler realized this long ago — the regularity of close-packed circles is a key observation in his book about snowflakes. Fish pack their territories together; bees and wasps have evolved a structure that packs their grubs together. | What does a snowflake pack together?

Coagulated Globules Kepler asked himself the same question. He knew that snow was made of condensed water vapour, and he wondered whether it condensed in a definite pattern. "Granted then that vapour coagulates into globules of a definite size, as soon as it begins to feel the onset of cold ... Secondly, granted that these balls of vapour have a certain pattern of contact ..." This set him off on a wild-goose chase based on the mistaken belief that the snowflake is three-dimensional rather than basically being flat, but he soon reverted to the geometry of the plane and turned his error to triumph: "Six corneredness is chosen by the formative fac-

9

10

11

unterschiedlich ist. Beim Vitriol kommt häufig die kubisch-rhombische Figur vor. Beim Salpeter ist dieselbe Figur zu finden. Daher mögen die Chemiker sagen, ob im Schnee ein Salz stecke und welcher Art dieses Salz ist und wie es die Figur hervorbringt!« Keplers Suche nach dem Geheimnis der Schneeflocke führte ihn zu der allgemeineren Frage nach der Kristallstruktur und wir müssen demselben Pfad folgen. | Es mag offensichtlich erscheinen, dass die Form der Kristalle regelmäßig mathematische Muster zeigt – zum Beispiel hat ein gewöhnlicher Salzkristall die Form eines Würfels. Jedoch war die Idee, dass Kristalle mathematisch geformte Körper sind, lange umstritten. Anstatt sich über die Regelmäßigkeit der Kristallform Gedanken zu machen, dachten die Menschen darüber nach, ob diese Regelmäßigkeit real oder lediglich Illusionen sind. Im Begriff »Kristallograf« begann etwas von dem mitzuschwingen, was wir heute mit »Astrologe« oder »Ufologe« verbinden. Der französische Naturforscher Georges Louis Leclerc Comte de Buffon sagte zu Beginn des 18. Jahrhunderts, dass »Vielförmigkeit überall dort vorhanden ist, wo Einförmigkeit vermutet wird«. | Diese skeptische Denkrichtung war damals angemessener, als es heute vielleicht scheinen mag. Kristallexemplare aus der Natur haben normalerweise eine deutlich unregelmäßigere Struktur als im Labor gezüchtete. Die Kristallographie machte nur langsam Fortschritte, bis später in jenem Jahrhundert der deutsche Geologe Abraham Werner ein Klassifikationssystem für Minerale, die Oryktognosie, entwickelte, mittels der man Mi-

Abb. 9, 10 und 11 Kristalle haben etwas äußerst Faszinierendes – regelmäßige, aber nicht allzu regelmäßige Formen, wundervolle Farben, bizarre, komplizierte Formen, die sich bisweilen ergeben, wenn die Kristalle zu Clustern zusammenwachsen, und die herrliche Art, das Licht zu reflektieren.

Fig. 9, 10 und 11 There is something utterly intriguing about crystals – their regular, but not too regular forms, their wonderful colours, the bizarrely intricate shapes that they sometimes assume, the way several crystals grow together in clusters, and the wonderful way they reflect light.

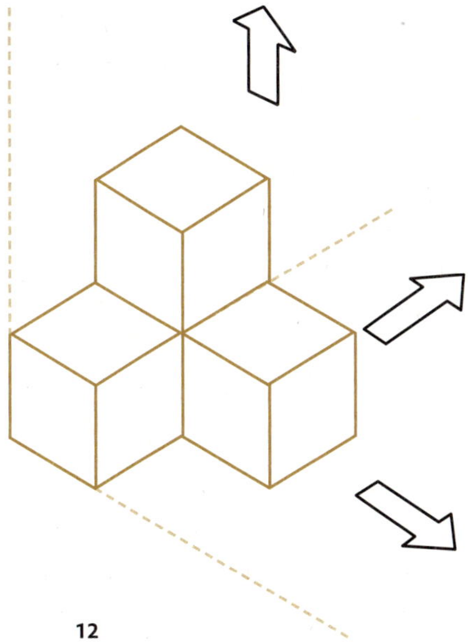

12

ulty from material necessity as well, so that no gap should be left and the gathering of vapour into formations of snow should take place more smoothly." | His final paragraph related the geometric regularity of snowflakes to the regular geometry of crystals: "It is probable, therefore, that our formative faculty varies with variations in the liquid. In sulphates of metals the rhomboid cubic shape is common, and saltpetre has its own shape. So let the chemists tell us whether there is any salt in a snowflake and what kind of salt, and what shape it assumes otherwise!" Kepler's quest for the secret of the snowflake led him to the broader question of crystal structure, and we must follow the same path. | It might seem evident that the form of crystals shows regular mathematical patterns – for example a crystal of common salt is a cube. However, the idea that crystals are mathematical used to be controversial. Instead of worrying about the regularities of crystal form, people worried about whether those regularities were real or illusory. The word "crystallographer" acquired overtones nowadays associated with "astrologer" or "ufologist". The French naturalist Count Buffon said in the early 18th century that "all the work of the crystallographers serves only to demonstrate that there is only variety everywhere they suppose uniformity". | This scepticism was more reasonable than it might seem now. Specimens of crystal found in nature are usually much less regular than those grown in the laboratory. Progress on crystals was slow until, later in the century, the German geologist Abraham Werner invented a

Abb.12 Wissenschaftler haben die Kristalle erst begriffen, als sie sich nicht mehr darauf konzentrierten, woraus Kristalle bestehen, sondern wie ihre Komponenten – welche das auch immer sein mögen – angeordnet sind. Kristalle haben eine regelmäßige Geometrie, die ihre regelmäßige atomare Struktur widerspiegelt. Sie bestehen aus identischen Einheiten, deren Muster sich entlang dreier räumlicher Achsen wiederholen.

Fig.12 Scientists started to understand crystals only when they ignored what they were made of, and concentrated on how their components – whatever they might be – were arranged. Crystals have regular geometry, which reflects their regular atomic structure. They are built from identical units, in patterns that repeat along three spatial axes.

nerale nach Farbe, Härtegrad, Dichte und so weiter bestimmen konnte. War sich ein Mineraloge sicher, dass es sich bei zwei augenscheinlich unterschiedlichen Exemplaren eines Minerals in Wirklichkeit um dasselbe Mineral handelte – oder auch nicht –, konnte er nach Regelmäßigkeiten suchen. Diese wurden schon bald in den Winkeln der ebenen Kristallflächen sichtbar. Alle Kristalle eines Gesteins, egal wie schadhaft oder unregelmäßig geformt sie sind, weisen denselben charakteristischen Winkel auf. Und nicht nur das – dieselben Winkelformen zeigen sich häufig auch bei anderen Mineralien. Wissenschaftler können Winkel messen, Zahlen ermitteln und nach den zugrunde liegenden Ursachen für die beobachteten Muster suchen. Auch bei dem Muster der Schneeflocke geht es immer wieder um Winkel: Überall gibt es Winkel von 60° und 120°. Warum? | Mathematiker suchten schon früh in dem neuen Territorium nach Mustern, noch bevor klar war, dass es überhaupt existierte. Auf Kepler folgte schon bald der englische Wissenschaftler Robert Hooke, dessen 1665 erschienenes Werk *Micrographia* Bilder enthielt, die zeigten, wie sich mit Ansammlungen eng gepackter Kreise und Kugeln kristalline Formen nachahmen lassen. Etwa ein Jahrhundert später entdeckte René Just Haüy, ein Priester und Hobbymineraloge, dass bei Zerbrechen eines Calcitkristalls alle Einzelteile immer wieder aussehen wie schief geformte Schachteln. Er schlug daher vor, die Kepplerschen und Hookeschen Kugeln durch Elemente dieser allgemeinen Form zu ersetzen. Kristallografen machten sich eifrig daran, die Natur der Grundbauteile der Kristalle aufzuspüren, stießen jedoch auf Schwierigkeiten, da diese Bauteile verschwindend klein zu sein schienen. | Mathematiker fanden einen Ausweg aus dieser Sackgasse, als sie sich entschieden, sich nicht auf die Form dieser Bauteile zu konzentrieren, sondern auf ihre Anordnung. Wie sich herausstellte, bilden die Kristalle regelmäßige Gitter – räumliche Muster, in denen sich dieselbe Grundeinheit in drei unterschiedlichen Richtungen wiederholt. Um ein einfaches Beispiel zu geben: Stellen Sie sich vor, einen Raum dicht an dicht mit Würfeln aufzufüllen – und zwar in drei Richtungen, nach Norden, Osten und nach oben. Dieser Ansatz führte zu einer Klassifikation möglicher Kristallsymmetrien, aus der sich viel später eine Antwort auf die alte Frage ergab, wie die Grundbauteile der Kristalle beschaffen sind. Sie lautet: Es sind Atome, kleinste Materieteilchen, die so winzig sind, dass sie bis vor kurzem selbst unter dem stärksten Mikroskop nicht zu sehen waren. Dies ist ein bemerkenswerter Fall, wie reine Mathematik zu einem bedeutenden Durchbruch in der Physik führt.

Spiraldrehungen Im klassischen Altertum waren die Menschen verwirrt von den Unregelmäßigkeiten natürlicher Kristalle, die häufig schadhaft oder unvollkommen waren, da sie unter veränderlichen Bedingungen gewachsen waren. Im Labor gezüchtete Kristalle sehen im Allgemeinen wie flächenbegrenzte, geomet-

classification system for mineral that he called oryctognosy, which outlined how to tell which mineral you've got by observing its colour, hardness, density and so on. Once mineralogists could be sure that two apparently different specimens of a mineral were actually the same mineral — or not — it became possible to look for regularities. These soon became apparent in the angles between the flat facets of crystals. All crystals of a given mineral, however damaged or irregular, exhibit the same characteristic list of angles. Not only that — the same list often shows up in other minerals. Scientists could measure angles, obtain numbers and seek underlying causes for their patterns. The pattern of the snowflake is all about angles: there are angles of 60° and 120° all over the place. Why? | Pattern-seeking mathematicians had already started to explore the new territory, even before it was certain that it existed. Kepler was soon followed by the English scientist Robert Hooke, whose *Micrographia* of 1665 included pictures showing how collections of close-packed circles and spheres could mimic crystalline forms. Around a century later, René Just Haüy, a priest and amateur mineralogist, noticed that when a crystal of calcite splits, it always breaks up into pieces like lopsided boxes, and he suggested replacing Kepler's and Hooke's spheres by pieces of this general shape. Crystallographers made strenuous efforts to discover the nature of the basic building blocks of crystals, but ran into difficulties because these building blocks seemed to be vanishingly small. | The impasse was broken by the mathematicians, who decided not to worry about what these building blocks were, but instead asked how they were arranged. It turned out that they were arranged in regular lattices — spatial patterns in which the same basic unit is repeated in three different directions. For a simple example think of filling space with cubes, packed in the obvious way — here the three directions are north, east and up. This approach led to a classification of the possible symmetries of crystals, which much later led to a solution of the problem that had been worrying the crystallographers all along — what is the nature of the basic building blocks of crystals? It turned out that they were atoms, particles of matter so tiny that until very recently they were invisible under even the most powerful microscope. It was a remarkable case of pure mathematics leading to a significant breakthrough in physics.

Spiral Swirls The ancients were confused by the irregularities of natural crystals, which were often damaged or imperfectly shaped because they had grown under variable conditions. Laboratory crystals typically look like flat-faced geometric polyhedrons. Such patterns are, on the whole, a bit too sharp-edged to turn up in living creatures. One of the favourite patterns of life, in fact, is based on curves — the spiral. | We use the word "spiral" in at least two senses: for a flat curve that swirls round and round, moving ever outwards as it does so; or

rische Polyeder aus. Solche Muster sind doch etwas zu scharfkantig, als dass sie in Lebewesen zum Vorschein kämen. Eines der bevorzugten Muster der Natur beruht stattdessen auf einer Kurve – der Spirale. | Das Wort »Spirale« wird in mindestens zwei Bedeutungen gebraucht: zum einen für eine ebene Kurve, die in unendlich vielen, immer weiter werdenden Windungen einen festen Punkt umläuft; zum anderen für eine sich gleichmäßig um eine Achse windende Linie im Raum (wie etwa die Wendeltreppe). Die Natur verwendet beide dieser Formen, einschließlich einer Mischung aus beiden, und nirgendwo wird dies so deutlich wie an einem Schneckengehäuse. | Spiralförmige Versteinerungen gehören zu den flachen Spiralen der Ammoniten – von denen es einst zahlreiche verschiedene Arten gegeben hat – die kennt ein jeder. Als Kind ging ich am Meerufer oft auf Ammonitenjagd und fast jedes Mal fand ich zwischen den Felsen winzige, spiralförmig gewundene Ammoniten, die die brandenden Wellen aus dem feuchten Schlamm herausspülten. Im örtlichen Museum gab es riesengroße Exemplare von einem Meter Durchmesser oder gar mehr, doch solche fand ich natürlich nie. | Die Form mancher Ammoniten ähnelt dem, was die Mathematiker als archimedische Spirale bezeichnen; diese zeichnet sich durch einen konstanten Abstand zwischen ihren Windungen aus. Die große Mehrheit der Ammoniten jedoch weist die Form einer logarithmischen Spirale auf, bei der sich der Abstand zwischen den Windungen in einem bestimmten Verhältnis vervielfacht. Der bekannteste lebende Vertreter mit einer äußeren Schale dieser Form ist der *Nautilus*. Sein Gehäuse ist erstaunlich regelmäßig und in nacheinander geordnete Kammern unterteilt. Der *Nautilus* bietet einen Schlüssel zur Spiralform solcher Gehäuse im Allgemeinen und zur logarithmischen Spirale im Besonderen. | Die Schale wird vom Körper des Weichtieres zum eigenen Schutz ausgebildet. Das Tier baut an den Rand der bestehenden Schalenteile neues Schalenmaterial an. Da der *Nautilus* mit fortschreitendem Alter an Größe zunimmt, entwächst er den bestehenden Kammern und fügt immer neue Anbauten an sein Haus an. Dabei beeinflusst die Wachstumsgeschwindigkeit des Lebewesens den Abstand der aufeinander folgenden Spiralwindungen. Wenn das Lebewesen nicht so schnell wächst, seine Schale aber trotzdem erweitert – vielleicht um ausgeschiedene Minerale aufzubrauchen, obwohl ich hier nur spekuliere –, bildet es eine archimedische Spirale aus. Wächst es hingegen exponentiell und verdoppelt seine Größe innerhalb einer bestimmten Zeit, so erzeugt es eine logarithmische Spirale. Die flachen Spiralformen der Ammoniten und des *Nautilus* resultieren also vermutlich aus einem einfachen Wachstumsmuster bei jenem Lebewesen, das die Schale um sich herum ausbildet. | An Land bauen die Schnecken ähnliche Gehäuse. Mal winden sich die Schneckengehäuse im Uhrzeigersinn, mal gegen den Uhrzeigersinn, und bei manchen Arten – doch das ist merkwürdigerweise eher die Ausnahme als die Re-

Abb. 12 b Das Gehäuse des Nautilus, eines Weichtieres aus den Tiefen des Meeres, besteht aus einer Reihe gebogener Kammern, die in immer weiter werdenden Windungen an Größe zunehmen und eine perfekte logarithmische Spirale formen. Dieses mathematische Muster zeigt, wie ein Lebewesen sein Gehäuse bildet und wie es wächst.

Fig. 12 b The shell of the nautilus, a sea-going mollusc, is made from a series of curved cells, which wind round and round, gradually increasing in size, to form a perfect logarithmic spiral. This mathematical pattern is a clue to how the creature makes its shell, and how it grows.

gel – kommen sogar beide Drehrichtungen vor. Im Jahre 1930 führte eine Gruppe von Biologen an einer Million Schnecken Züchtungsexperimente durch, um herauszufinden, was die Windungsrichtungen bestimmt. Es zeigte sich, dass diese Richtung genetisch gesteuert ist, doch nicht von den Genen der Schnecke selbst, sondern von denen der Mutter. Verblüffenderweise tragen die Mutterschnecken ein Gen, das bestimmt, in welche Richtung sich das Gehäuse der Nachkömmlinge windet, nicht aber, wie es sich innerhalb der Individualentwicklung windet. So wird sichergestellt, dass sich in einem Schneckenembryo eine bestimmte Händigkeit in der Windungsrichtung ausprägt, wenn es ein Entwicklungsstadium von nur acht Zellen erreicht hat. | Hier ist noch ein Rätsel der Natur: Schneckengehäuse sowie viele Muschelschalen winden sich oft in die dritte Dimension. Natürlich ist das Gehäuse immer dreidimensional. Was ich meine, ist, dass das »Herz« einer Spirale, die Linie, die entlang der Mitte der Kammern verläuft, nicht auf eine Ebene beschränkt bleibt, sondern sich spiralförmig in die dritte Dimension erhebt. Auch bei Fossilien finden sich derlei Schalenformen. *Turritella* beispielsweise, eine über 50 Millionen Jahre alte Gastropodenart aus dem Eozän, trägt ein konisch geformtes Gehäuse, doch findet man solche Schalen auch heute noch mit lebenden Organismen darin. Wie bei einer Wendeltreppe baut sich das *Turritella*-Gehäuse von oben nach unten auf, wobei die Stufen allmählich immer größer werden. Diese Form ist die Folge einfacher Wachstumsregeln. Das Lebewesen fügt nicht nur eine Kammer an die andere und verändert deren Größe in konstanter Regelmäßigkeit, sondern baut die jeweils neue Kammer in einem Winkel zur vorherigen.

Die Blumen des Fibonacci Im Reich der Pflanzen herrschen fast überall einige sehr charakteristische Muster vor. Zahlreiche Blüten bestehen aus vielen mehr oder weniger identischen Blütenblättern, die in einem symmetrischen Ring um einen zentralen Blütenkopf, der die Samen enthält, angeordnet sind. Ausnahmen – allen voran die Orchideen – sind oftmals links-rechts-symmetrisch. In der Erforschung von Mustern ist die Symmetrie ein mathematisches Schlüsselkonzept; sie wird auf unserer Entdeckungsreise nach der Form der Schneeflocke eine entscheidende Rolle spielen. Pflanzliches Leben entfaltet überdies einige der beeindruckendsten numerischen Muster der Natur. Es handelt sich zwar eher um Faustregeln als um letzte Wahrheiten, aber sie verkörpern einige erstaunliche Konsequenzen für die Art des Pflanzenwachstums. | Leonardo von Pisa, genannt Fibonacci, wurde im Jahre 1170 als Sohn eines Zollbeamten geboren. Als junger Mann half er seinem Vater im Zollamt, wo er das neue Zahlenschreibsystem – den Vorläufer des heutigen Dezimalsystems mit den Zeichen 0, 1, 2, 3, 4, 5, 6, 7, 8, 9 –, das von den Arabern und Hindus erfunden wurde, kennen lernte. Beeindruckt davon verfasste er das Rechenbuch *Liber Abbacci*. (»Abacus« mit einem *b* bezeich-

for a twisted curve in space like a spiral staircase. Nature uses both of these forms, including a mixture of the two, and nowhere is this more apparent than in shells. | Spiral shells appear way back in the fossil record, and are one of the most common. The flat spiral of the ammonite — of which there have been many distinct species — is widely recognized. As a child, I used to hunt ammonites on the seashore and on most days I found tiny coiled ammonites among the rocks, washed out of the clay by the action of the waves. The local museum had huge ones, a metre or more across, but of course I never found such a large specimens. | The shape of some ammonites is close to what mathematicians call an Archimedean spiral — successive coils are spaced the same distance apart. The majority of ammonites, though, form a logarithmic spiral, whose spacing multiplies by a fixed amount for each turn of the spiral. The best known modern shell with this form is the *Nautilus*, with a shape that is astonishingly regular, divided into successive compartments. The Nautilus offers a clue to the spiral shape of such shells, and to the logarithmic spiral in particular. | The shell is formed by a soft-bodied organism, for protection. New shell material is excreted on to the edge of the existing shell. As the creature's size increases with age, it outgrows its existing chamber and builds an extension on to its house. The rate of growth of the organism affects the spacing of successive coils of the spiral. At one extreme, if the organism doesn't actually grow much, but builds the extension anyway — perhaps to use excreted minerals, though I'm guessing here — then it makes an Archimedean spiral. If instead it grows exponentially, doubling in size in some fixed period of time, it makes a logarithmic spiral. So the flat spirals of ammonites and Nautilus probably result from simple patterns of growth in the creature that builds the shell itself. | On land, snails build similar shells. Sometimes snail shells coil clockwise, sometimes they coil the other way, and in some species — the exception rather than the rule, which is curious — both directions of twist occur. In 1930 a group of biologists carried out breeding experiments on a million snails and they worked out what determines the direction of coiling. The cause is genetic, but it is not in the snails genes, it is in its mother's. Amazingly there is a gene for how a snail's children's shells coil, not for how the individual's shell coils. This ensures that a particular handedness of twist is put into the snail embryo when it reaches a stage of development of just eight cells. | Here's another puzzle to add to the list. Snail shells, and indeed many seashells, often coil into the third dimension. Of course, the shape of the shell is always three-dimensional: what I mean is that the "core" of the spiral, the line that runs along the centre of the chambers, ceases to lie in a plane and starts to curl into third dimension of space. Shells like this are also seen in fossils. The cone-shaped gastropod Turritella is over 50 million years old, dating from the Eocene period, but shells just like it can be found today, with living

13

14

15

nete damals ein Rechenbrett mit Perlen, wie es auch heute noch für Grundrechenarten benutzt wird, »abbacus« mit zwei *b* hingegen bezog sich auf den Rechenvorgang an sich.) Mit diesem Werk aus dem Jahre 1202 wurden erstmals die neuen hindu-arabischen Zahlen in Europa eingeführt. | Ungefähr die Hälfte des Buches handelt von Berechnungen im Zusammenhang mit ausländischen Wechselkursen, doch unter den alltäglicheren Beispielen findet sich eine Fragestellung, die voller interessanter Mathematik steckt. Als anschauliches Beispiel dienen Kaninchen. Wir beginnen mit einem einzelnen Paar noch nicht geschlechtsreifer Kaninchen (einem Männchen und einem Weibchen). Nach einem Monat sind sie fruchtbar, die Natur nimmt ihren Lauf und am Ende des ersten Monat setzt jedes fruchtbare Paar ein neues Paar in die Welt und jedes Paar, das im Vormonat noch unfruchtbar war, zeugt im folgenden Monat, wenn es fruchtbar ist, ebenfalls ein neues Paar, das nach einem Monat fruchtbar ist. Für dieses Rechenspiel nehmen wir einmal an, dass alle Kaninchen unsterblich sind. Die Frage lautet: Wie wächst die Kaninchenpopulation? Mit ein wenig Nachdenken erschließt sich ein bestimmtes Muster. Die Zahl der Paare in aufeinander folgenden Jahren wächst wie folgt: 1, 1, 2, 3, 5, 8, 13, 21, 34, 55, 89, 144 ... und so weiter. Jede Zahl – nach den ersten beiden – ist die Summe der zwei vorhergehenden Zahlen. | Die Zahlenfolgen der Kaninchenpopulation sind heute als Fibonacci-Zahlen bekannt. Natürlich vermehren sich Kaninchen nicht nach Leonardos Regeln, und sie leben auch

Abb. 13, 14, 15 und 16
Pflanzen besitzen geheime Regelmäßigkeiten. Wenn man weiß, wonach man sucht, entdeckt man sie häufig auch dort, wo man sie gar nicht vermutet. Am besten man achtet auf Spiralmuster und bestimmte Zahlen, die mit den Spiralen in engem Zusammenhang stehen, die Fibonacci-Zahlen. Man findet sie bei zahlreichen Pflanzen, bei Blüten ebenso wie bei Tannenzapfen. Da biologisches Wachstum sehr flexibel ist, gibt es zu diesen Mustern auch Ausnahmen, doch Fibonacci-Zahlenspiele und Spiralgeometrie sind erstaunlich weit verbreitet. Sie deuten darauf hin, dass das Pflanzenwachstum einfachen, aber subtilen mathematischen Regeln gehorcht, die irgendwo im Granzbereich zwischen Dynamik, Geometrie und Arithmetik liegen.

16

organisms inside them. Turritella is like a spiral staircase made from the top down with treads that gradually get bigger and bigger. The shape is a consequence of simple rules of growth. As well as adding a new chamber on the end of the old one, and changing its size in a regular manner, the creature builds the new chamber at an angle to the plane of the previous one.

Fibonacci's Flowers Some very characteristic patterns prevail throughout much of the plant kingdom. Many flowers consist of many more or less identical petals arranged in a symmetric ring around a central seedhead. Exceptions, the most striking being orchids, are often left-right symmetric. Symmetry is a key mathematical concept in the study of patterns, and it will play a key role in our quest to discover the shape of snowflake. Plant life also exhibits some of nature's most striking numerical patterns. They are rules of thumb rather than ultimate truths, but it turns out that they embody some surprising consequences of the way plants grow. | Leonardo of Pisa, nicknamed Fibonacci, was born in 1170, the son of a customs officer. As a young man he joined his father in the customs house, where he was exposed to the new system for writing numbers invented by the Arabs and Hindus – the forerunner of today's decimal system with it's symbols 0, 1, 2, 3, 4, 5, 6, 7, 8, 9. Impressed, he wrote *Liber Abbacci*, an arithmetic text. (In those days "abacus" with one b, referred to the apparatus with beads still used to-

Fig. 13, 14, 15 and 16
There are secret regularities to plants, and once you know what to look for, you can often find these regularities in unexpected places. The things to look for are spiral patterns, and special numbers that are closely associated wit those spirals, called Fibonacci numbers. These patterns can be found in many plants, from flowers to fir cones. Because biological growth is very flexible, there are some exceptions to these patterns, but Fibonacci numerology and spiral geometry are surprisingly common. They suggest that plant growth obeys simple but subtle mathematical rules, which lie somewhere in the interface between dynamics, geometry and arithmetic.

nicht ewig. Aber verfeinerte Varianten von Leonardos Modell werden durchaus heute zur Untersuchung von Tierpopulationen verwendet. Die Fibonacci-Zahlen haben sich jedenfalls tief in die Seele der Mathematik gegraben als eine unendlich scheinende Quelle der Inspiration und Faszination. | Die Pflanzennumerologie ist durchsetzt mit Fibonacci-Zahlen. Lilien beispielsweise haben drei Blütenblätter, Butterblumen fünf, viele Rittersporne acht, Ringelblumen 13, Astern 21, Gänseblümchen und Sonnenblumen normalerweise 34, 55 oder 89 und einige größere Sonnenblumenarten sogar 144. Auch Zahlen, die keine Fibonacci-Zahlen sind, treten bei Blüten auf, allerdings seltener, zudem sind die meisten Ausnahmen entweder doppelte Fibonacci-Zahlen (für gewöhnlich als Folge von Pflanzenzuchtverfahren, welche die Anzahl der Blütenblätter verdoppeln) oder sie gehören zu der ähnlichen anomalen Reihe 1, 3, 4, 7, 11, 18, 29 ... mit der gleichen Bildungsregel, aber einem anderen Ausgangspunkt. Falls bei einer Blüte ein einziges Blatt zur Fibonacci-Zahl fehlt, kann man darauf wetten, dass es abgefallen ist. | Fibonacci-Zahlen treten außerdem in Spiralstrukturen auf, die viele Pflanzen verwenden, um ihre Samen anzuordnen. Fichtenzapfen stellen hierfür ein gutes Beispiel dar. Die Schuppen auf den Zapfen sind typischerweise in zwei Familien sich überschneidender Spiralen angeordnet; die Anzahl der Spiralen entspricht bei jeder Familie einer Fibonacci-Zahl. So laufen beispielsweise beim Fichtenzapfen fünf Schuppenspiralen in die eine Richtung und drei in die andere, während es beim Lärchenzapfen acht und fünf Spiralen sind. Im Blütenkopf der Sonnenblume zeigt sich die spiralförmige Anordnung in einem herrlich regelmäßigen Muster. Kleine Sonnenblumen haben 34 Spiralen in der einen Familie und 55 in der anderen; in größeren Arten findet man auch 55 und 89 oder sogar 89 und 144.

Tiefes Universum Das Rätsel der Muster erstreckt sich über die Grenzen des Planeten Erde hinaus. Die moderne Wissenschaft hat sich entwickelt, weil Menschen Muster am Himmel entdeckten. Und ein erstes bedeutsames mathematisches Verständnis von der Natur entfaltete sich, als wir begannen, den Kosmos zu begreifen. Schon die frühen Menschen beobachteten die Lichter am nächtlichen Himmel und fragten sich, was es mit diesen Lichtern auf sich haben mochte. So erfanden sie Kosmologien, um sie zu erklären. Die Babylonier sahen den Himmel als ein festes Gewölbe, das auf dem Meer ruht – die Sonne wohnt mit den anderen Göttern über diesem Gewölbe und tritt Tag für Tag durch eine Pforte in das Blickfeld der Menschen. Die Ägypter betrachteten den Himmel als flache Decke und die Sterne als Lampen. Die Fixsterne hingen in ihrer Vorstellung an Seilen, während die »Wandersterne« – die Planeten – von den Göttern herumgetragen wurden. Diese Vorstellungen mögen uns heute kurios erscheinen, stellen aber allesamt Versuche dar, sich mit sehr realen und sehr wichtigen Bewegungsmustern

day for arithmetical calculations, but "abbacus" with two *b's* referred to the process of calculation.) This book, which dates from 1202, was the first to introduce Hindu-Arabic numerals to Europe. | About half of it is given over to calculations related to foreign currency exchange, but in among the more prosaic examples is a problem that has spawned a huge amount of mathematics. Ostensibly it is about rabbits. Start with a pair of immature rabbits (one male, one female). After one season they mature, nature takes its course and this one pair gives rise to a new pair of immature rabbits. In successive seasons all mature pairs give rise to one new immature pair, and all pairs that had been immature for the previous season mature and give rise to one new immature pair. All rabbits are assumed to be immortal. How does the population of rabbits grow? Just a little thought reveals a pattern. In successive years the number of pairs is: 1, 1, 2, 3, 5, 8, 13, 21, 34, 55,89, 144... and so on, where after the first two, each number is obtained by adding together the previous numbers. | Leonardo's rabbit population figures came to be known as Fibonacci's numbers. Of course, rabbits don't reproduce according to Leonardo's rules and they don't live forever. Nonetheless, more sophisticated versions of Leonardo's scheme are used today in the study of animal populations. The Fibonacci numbers, though, have penetrated deep into the mathematical psyche as an apparently endless source of inspiration and wonder. | Plant numerology is riddled with Fibonacci numbers. Lilies have 3 petals, buttercups have 5, delphiniums often have 8, corn marigolds have 13, asters have 21 and daisies and sunflowers typically have 34, 55 and 89. Some bigger sunflowers have 144 petals. Numbers that are not Fibonacci numbers can occur in flowers, but they are less common; moreover, most of these exceptions are either twice Fibonacci numbers (usually a consequence of plant-breeding techniques that double the number of petals) or belong to the similar anomalous sequence 1, 3, 4, 7, 11, 18, 29... with the same rule of information but a different starting point. If you get one petal less than a Fibonacci number, than you can bet it fell off. | Fibonacci numbers also occur in the spiral structures that many plants use to arrange their seeds. Fir cones are a good example. The scales on fir cones are typically arranged in two families of intertwined spirals. Thus the Norway spruce cone has 5 spirals of scales in one direction and 3 in the other, while the larch has 8 and 5. The seedhead of a sunflower displays these spirals in a gloriously regular pattern. In small sunflowers there are 34 spirals in one family and 55 in the other; in larger ones these numbers may be 55 and 89, or even 89 and 144.

Deep Space The puzzle of patterns extends beyond the bounds of planet Earth. Modern science came into existence because people noticed patterns in the heavens, and the first significant flowering of our mathematical understanding of

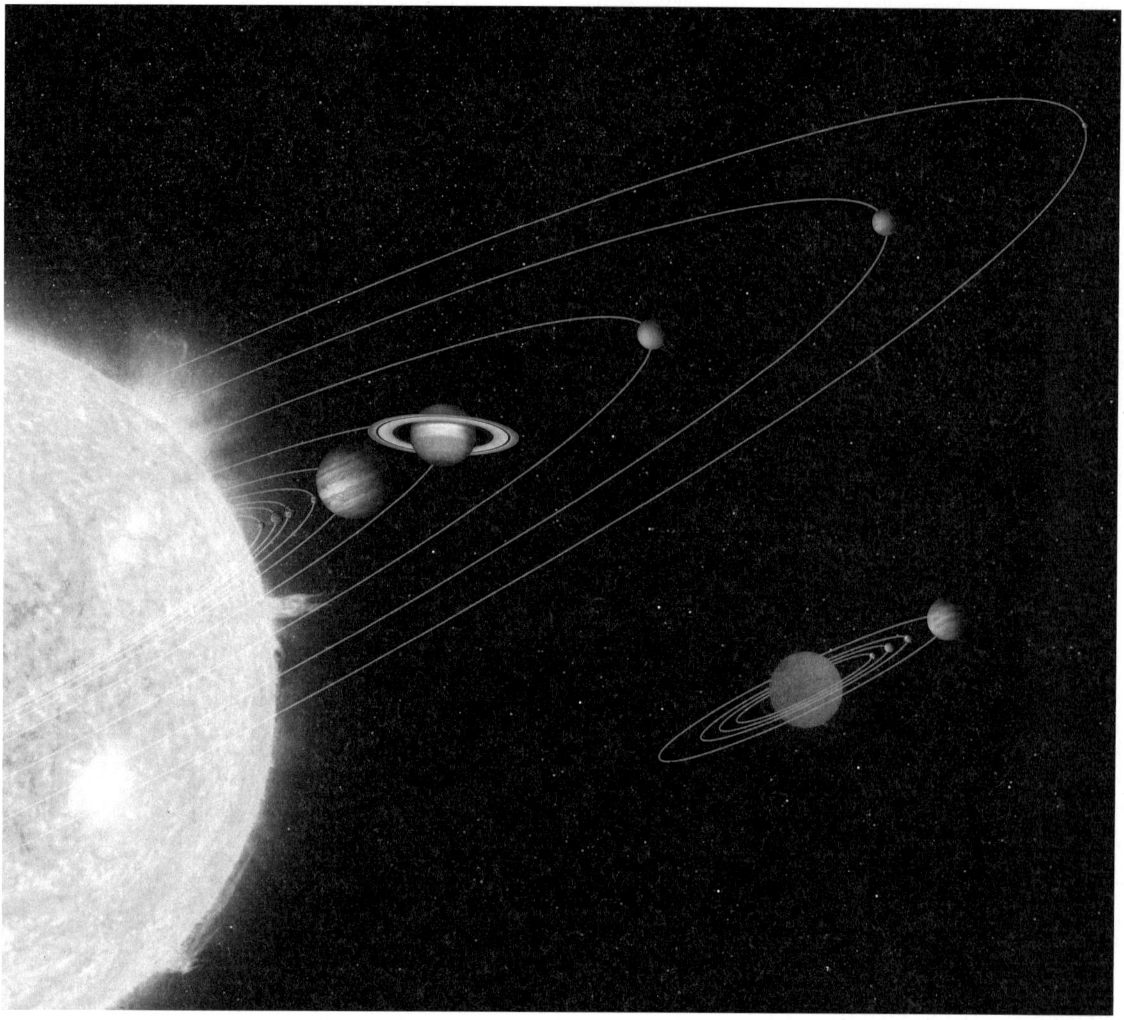

17

Abb. 17 und 18 Die Ringe des Saturn bilden nicht einfach eine flache, kreisrunde Scheibe, sondern weisen Lücken auf und haben eine sehr komplizierte Struktur. Mit einigen wenigen seltenen Ausnahmen besitzen die Einzelringe eine kreisförmige Symmetrie — zum Beispiel haben die Lücken im Kreis alle die gleiche lichte Breite. Da die Schwerkraft in alle Richtungen die gleiche Wirkung hat, ordnen sich die Eis- und Gesteinsbrocken der Ringe kreisförmig an. Aus ähnlichen Gründen ist auch unser Sonnensystem von einer annähernd zirkulären Geometrie gekennzeichnet.

18

nature occurred when we started to comprehend thr cosmos. Ancient humans observed the lights in the night sky, wondered what they were, and invented cosmologies to explain them. The Babylonians saw the heavens as a solid vault resting on the ocean — the Sun lives above the vault with the other gods and emerges every day into human sight through a doorway. The Egyptians saw the heavens as a flat ceiling, and the stars as lamps. The fixed stars hung on cords, while the "wandering stars" — the planets — were carried around by gods. These explanations are quaint to modern eyes, but they are attempts to come to grips with very real, and very important, patterns of movement. | Some cosmological patterns are more obvious than the motion of the planets. The Earth is a (slightly flattened) sphere. So are many other bodies in the solar system — the Sun, the Moon, Mars, the rest of the planets. The biggest asteroids — Ceres, Pallas, Vesta, Juno — are also round. Many of the smaller asteroids are more like giant potatoes, though — Castalia looks like a dog's bone. Saturn is surrounded by rocks and ice arranged in another mathematical shape — rings. Originally Saturn's rings were thought to form an annulus, a flat circular disc with a circular hole in the middle, like a washer. Then the disc was found to have gaps, also circular in shape. When the *Voyager* spacecraft reached Saturn they sent back images revealing that the fine texture of the rings is so intricate that it defies complete description; it is almost like grooves on an old vinyl gramophone record, only far more densely

Fig.17 and 18 Saturn's rings are not just a flat disc: they have gaps, and a very delicate structure. With a few rare exceptions, the rings have circular symmetry — for example the gaps have the same width all the way round. Because gravity has the same effect in all directions, it arranges the ice and rock of the rings into circles. Approximately circular geometry also characterizes our solar system, for similar reasons.

auseinander zu setzen. | Einige kosmologische Muster sind offensichtlicher als die Bewegungsmuster der Planeten. Die Erde ist eine (an den Polen leicht abgeflachte) Kugel – so wie andere Himmelskörper unseres Sonnensystems auch: Sonne, Mond, Mars und die restlichen Planeten. Die größten Asteroiden sehen eher wie riesige Kartoffeln aus und Castalia erinnert an einen Hundeknochen. Der Saturn ist umgeben von Eis- und Gesteinsbrocken, die in einer anderen mathematischen Form angeordnet sind – in Ringen. Ursprünglich hielt man die Saturnringe für eine einzelne flache Kreisscheibe mit einem kreisrunden Loch in der Mitte, ähnlich einer Unterlegscheibe. Dann fand man heraus, dass die Scheibe durchbrochen ist und aus mehreren getrennten Einzelringen besteht. Auf ihrem Flug durch das All lieferte die Raumsonde *Voyager* Bilder und Messdaten vom Saturn, die deutlich zeigen, dass die Detailstruktur der Ringe derart kompliziert ist, dass sie sich einer vollständigen Beschreibung widersetzt – fast wie die Rillen auf einer alten Vinylschallplatte, nur wesentlich dichter gepackt. Einige Ringe sind geflochten, andere haben Lücken, und wieder andere sind nicht einmal kreisrund. Dennoch ist die Geometrie der Saturnringe von den Kreisstrukturen dominiert. Der Saturn ist in dieser Hinsicht nicht einzigartig. Jupiter, Uranus und Neptun besitzen ebenfalls Ringe, die aber viel blasser sind, viel weniger Materie als die des Saturn enthalten und oft unvollkommen bogenförmig sind. | Die Sonne ist ein Stern, in dessen Zentrum Kernprozesse ablaufen, die eine Wasserstoffbombe wie ein Tischfeuerwerk erscheinen lassen. Die Leuchtkraft einiger Sterne ist veränderlich und folgt häufig einem sich wiederholenden Zyklus von Aufhellen und Abdunkeln. Einem solchen Muster folgt unsere Sonne nicht, obwohl auch sie nicht konstant hell leuchtet. Sie durchläuft eine andere Art von Veränderung – sie vibriert. Sie »klingelt« wie eine Glocke, wenn sie von Sternbeben erschüttert wird, so wie unser Planet Erde auf ein Erdbeben reagiert. Allerdings bedarf es hochsensibler Messgeräte, um die Vibrationen beobachten zu können. Die Messungen lassen geometrische Muster erkennen. Die Sonne ist zudem mit Sonnenflecken gespickt, gigantischen magnetischen Wirbeln, die in einem Zyklus von etwa elf Jahren kommen und gehen. Kepler entdeckte überdies subtilere Muster im Kosmos und löste damit die alte Frage nach den Wandersternen. Dabei stützte er sich auf die Hinterlassenschaften des dänischen Astronomen Tycho Brahe, der im 16. Jahrhundert über viele Jahre hinweg detaillierte und präzise Beobachtungen der Planetenbewegungen angestellt hatte. Auf der Suche nach Mustern gab sich Kepler aber nicht mit einer langen Liste von Zahlen zufrieden. Er fahndete stattdessen nach verborgenen, bis dahin unbemerkten Regelmäßigkeiten. In seinem 1609 verfassten Werk *Neue Astronomie (Astronomia Nova)* leitete er zwei wohlbegründete Muster der Planetenbewegung aus Brahes Beobachtungen ab. Nach dem ersten Keplerschen Gesetz bewegen sich die Planeten auf Ellipsen, in

packed. Some rings are braided, some have gaps and some aren't even circular. Nonetheless, the geometry of Saturn's ring is dominated by circular structures. Saturn is not unique in this respect. Jupiter, Uranus and Neptune also possess rings, but they are much fainter, contain much less material than Saturn's and are often incomplete arcs. | The Sun is a star powered by nuclear reactions that make a hydrogen bomb seem like a whisper. The light output of some stars is variable, often following a repetitive cycle of brightening and dimming. Our Sun's output isn't like that, though it's not constant either. The Sun undergoes another kind of variation – it vibrates. It "rings" like a bell when struck by starquakes, just as our own planet reacts to earthquakes, however, we need very sensitive equipment to observe the vibrations. When we do measure them, we see geometric patterns. The Sun is also speckled with sunspots, gigantic magnetic whirlpools that come and go in a cycle lasting roughly 11 years. Kepler also discovered more subtle patterns in the cosmos, resolving the old questions about wandering stars. He inherited many years of detailed and accurate observations of the movements of planets, made by the Danish astronomer Tycho Brahe in the 16th century. Being a pattern seeker, Kepler wasn't satisfied with a huge table of numbers. Instead, he looked for hidden hitherto unnoticed regularities. In his *New Astronomy (Astronomia Nova)* of 1609 he extracted two valid patterns of planetary motion from Brahe's data. That the orbits of planets are ellipses, with the Sun at one focus, and the line joining the planet to the Sun sweeps out equal areas in equal periods of time. In 1619 in *The Harmony of the World (Harmonice Mundi)* he stated a third: the square of the orbital period (the time it takes a planet to complete one orbit) is proportional to the cube of the planet's distance from the Sun. The wanderings of the planets against the backdrops of "fixed" stars, an aspect of the heavens that had previously looked capricious, turned out to have hidden patterns. We have since found many other patterns in the heavens, some temporal and some spatial. Among the most amazing are gigantic spirals – we call them galaxies. Galaxies come in other shapes, but the most common shape has two spiral arms. A spiral galaxy is a slowly turning pinwheel made up of hundreds of billions of stars. We don't fully understand galaxies; in fact, we don't fully understand rather a lot of what we see through our telescopes. This shouldn't come as a surprise – ignorance, after all, is the human condition. The triumph of science is not yet Ultimate Knowledge, it is that every day we understand a little more about nature and the universe than we did the day before.

deren einem Brennpunkt die Sonne steht. Nach dem zweiten Gesetz überstreicht eine von der Sonne zu einem Planeten gezogene Linie in gleichen Zeiträumen gleiche Flächen. In seinem 1619 veröffentlichten Werk *Die Weltharmonik (Harmonice Mundi)* beschreibt Kepler eine dritte Gesetzmäßigkeit: Das Quadrat der Umlaufzeit (der Zeit, die ein Planet für eine Umlaufbahn braucht) ist proportional zur dritten Potenz des mittleren Abstands der Planeten von der Sonne. | Die Wanderungen der Planeten vor dem feststehenden Hintergrund der »Fixsterne« – ein Himmelsphänomen, das bis dahin willkürlich erschienen war – offenbarten auf einmal verborgene Muster. Seither hat man zahlreiche weitere Muster am Firmament entdeckt, zeitliche wie räumliche. Zu den beeindruckendsten gehören gigantische Spiralen – die Galaxien. Sie kommen zwar aus in anderen Formen vor, doch die üblichste Form besteht auch zwei Spiralarmen. Eine Spiralgalaxie ist ein sich langsam drehendes Leuchtrad, das aus Abermilliarden Sternen besteht. Noch sind unsere Kenntnisse über Galaxien recht lückenhaft. Tatsächlich verstehen wir vieles, was wir durch unsere Teleskope sehen, bislang nicht sonderlich gut. Das dürfte nicht weiter überraschen – Unkenntnis liegt schließlich in der Natur des Menschen. Der Triumph der Naturwissenschaft besteht (noch) nicht im »ultimativen Wissen«, sondern darin, dass wir mit jedem Tag ein klein wenig mehr von der Natur und dem Universum begreifen.

10

Schalentiere und Seeigel

Crustaceans and Sea Urchins

Frei Otto

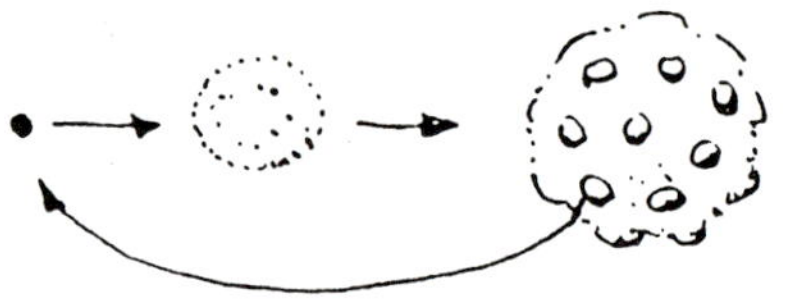

1

3

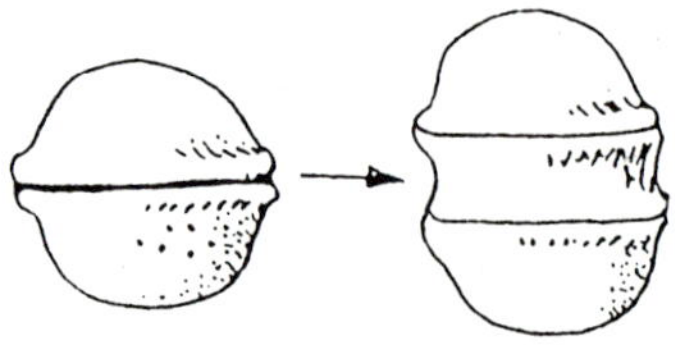

2

Wachstum trotz Erhärtungen bei Tieren Nur weiche Zellen können sich vergrößern, teilen und vermehren. Erhärtungen jeder Art erschweren diesen Prozess, besonders wenn harte Substanzen eingelagert werden, die von eigenen Organismen nicht wieder aufgelöst werden, wie Zellulose, Chitin und Kieselsäure. Lediglich die kalkreiche Knochensubstanz von Skeletttieren ist löslich und erlaubt Veränderung. Da sich die Verfestigung nie auf den ganzen Organismus erstreckt, verbleiben die Wachstumsprozesse im Bereich der Weichteile.

Gepanzerte Einzeller (Abb.1): Einige weiche Einzeller, wie Radiolarien und Diatomeen, umhüllen sich plötzlich mit einer unlöslichen Schale. Jedes weitere Wachstum, jede Teilung ist dann ausgeschlossen. Die Vermehrung erfolgt durch ausgeschiedene kleine Sporen, die wieder als weicher Pneu heranwachsen. || *Sich teilende verfestigte Einzeller* (Abb.2): Bei einigen Kieselalgen teilen sich die Schalen an vorgegebenen »Solltrennstellen«. Weiche Zwischenringe werden eingebaut und die jeweils fehlende Schalenhälfte wird in weicher Form ergänzt, die nach Fertigstellung wieder aushärtet. || *Randwachstum* (Abb.3): Schnecken und Muscheln wachsen am Rand. Mineralische Substanz wird unter einer mitwachsenden Membrane ausgeschieden. Diese Tiere wachsen kontinuierlich bis zu ihrem Tod. || *Flucht aus dem Panzer* (Abb.4): Weichtiere mit harter Schale, wie Krebse und Insekten, sind im Panzer gefangen. Sie müssen ihn sprengen und ablegen. Sie

Abb.1 gepanzerte Einzeller
Fig.1 Armoured single cell organisms

Abb.2 sich teilende verfestigte Einzeller
Fig.2 Solidified single cell organisms which divide

Abb.3 Randwachstum
Fig.3 Edge growth

4

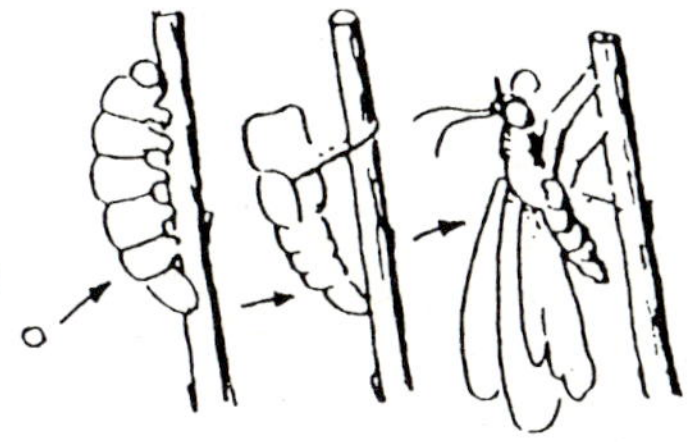

5

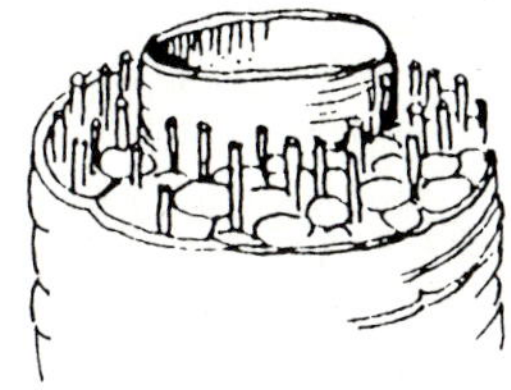

Growth in animals despite solidification Only soft cells can increase their size, divide and reproduce. Any sort of solidification will make this process more difficult especially when hard substances are formed in the soft tissue which cannot be dissolved by the organism such as cellulose, chitin and silicic acid. Only the calcium-enriched bone substance of vertebrates is soluble and allows change. Since solidification never embraces the entire organism, growth processes within the soft tissue will continue.

Armoured single cell organisms (Fig. 1): Some soft single cell organisms such as radiolaria and diatoms quite suddenly surround themselves with an insoluble shell. This excludes any further growth or division. Reproduction is performed by the shedding of small spores which in turn grow into soft pneus. || *Solidified single cell organisms which divide* (Fig. 2): In some diatoms the shells divide at certain existing »lines of division«. Soft intermediate rings are subsequently inserted and the missing half-shell is replaced by a soft structure which has the same shape and solidifies once its growth is completed. || *Edge growth* (Fig. 3): Snails and mussels grow at the edges. Mineral substance is excreted under a membrane which grows with the shell. These animals grow continuously until they die. || *Escape from the armoured shell* (Fig. 4): Molluscs with a hard shell such as crayfish and insects, are trapped in their carapace. They must destroy this carapace and get

Abb. 4 Flucht aus dem Panzer
Fig. 4 Escape from the armoured shell

Abb. 5 Knochenwachstum
Fig. 5 Bone growth

6

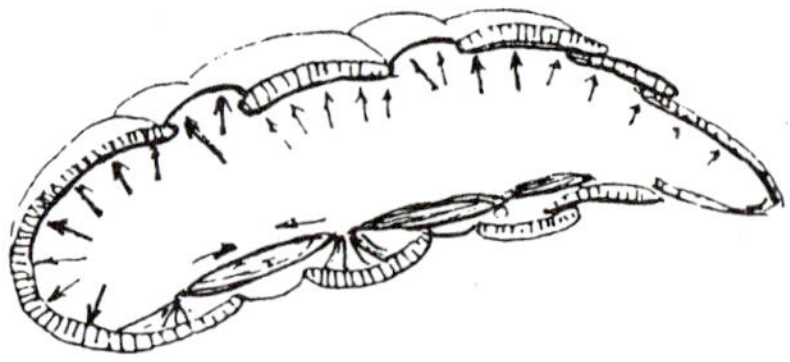

7

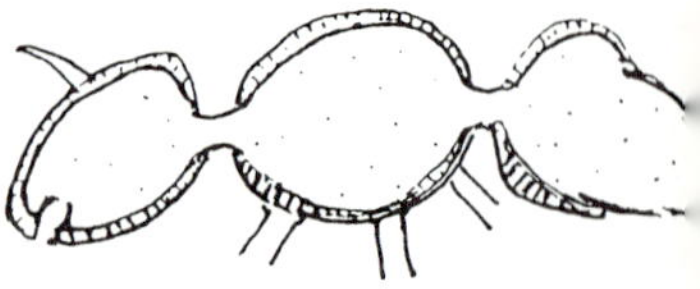

8

»fliehen«. Dann pumpen sie sich auf, bis ihre äußere Haut erhärtet. Sie haben ein stufenweises Wachstum. || *Knochenwachstum* (Abb. 5): Knochensubstanz wird zwischen den Zellen ausgeschieden, die Knochen sind allseitig von einer Membrane umschlossen. Knochen wachsen also in einem Pneu, der die Form bestimmt. Knochensubstanz wird ständig abgebaut und zugleich nachgeliefert. Kontinuierliche Vergrößerungen, Umbau und Verkleinerungen sind möglich.

Schalen Bei Tieren treten häufig äußere Erhärtungen auf. Dies führt unter anderem zu den Schalentieren. Manche kleine Schalentiere – besonders Krebse – sind eigentlich Weichtiere geblieben, die einen harten Panzer haben. Sie setzen die Muskelkraft weitgehend gegen den Körperinnendruck ab. | Mit anderen Worten: Bei Muskelkontraktionen wird nicht die Schale druckbeansprucht, sondern, wie bei Weichtieren, wirkt der Innendruck als Gegenkraft, die vom den Weichkörper durchziehenden Fasernetzwerk aufgenommen wird. Die Beweglichkeit der Schalenteilstücke wäre sonst behindert. | Funktionell kann man kleinere Schalentiere somit als gepanzerte Weichtiere bezeichnen. Da mir aber die Innendruckveränderungen bei Bewegungen dieser Schalentiere nicht bekannt sind, kann ich hier keine grundsätzliche Aussage machen. Vermutlich sind die Innendrucke bisher auch nicht ermittelt worden. | Bemerkenswert ist die begrenzte, in Zentimetern messbare Größe der auf dem Land lebenden Schalentiere. | Bei größeren Scha-

rid of it. They "escape". Subsequently they will inflate themselves until their outer skin hardens. Such as animals grow in stages. || *Bone growth* (Fig. 5): Bone substance is excreted between the cells and the bones are completly enclosed by the membrane. Bones therefore grow inside a pneu which determines the shape. Bone substance continously decays and is produced at the same time. This makes continous growth, restructuring and size reductions possible.

Shells External solidification is frequently found in animals. This results, for instance, in crustaceans. Many small crustaceans, in particular crayfish, have essentially remained molluscs which have a hard carapace. They largely replace muscle power with the internal pressure of the body. | In other words: When the muscles contract it is not the shell which is pressurised but, as in the case of molluscs, the internal pressure acts as a balancing force which acts on the network of fibres permeating the soft body. If this were not the case the movement of the shell segments would be impaired. | As far as their functioning is concerned, the smaller crustaceans can be described as armoured molluscs. As I have no knowledge of the changes in internal pressure taking place when these crustaceans move, I can make no generally applicable statement at this point. I suspect that so far no research has been done to determine the internal pressures of such animals. | Of particular note is the limited size of terrestrial crustaceans which can be measured in centimetres. | With larger crustaceans and thinner extremities we can observe clearly developed joints which allow the compressive forces caused by the muscles to be transmitted to the shell at least in one direction (e.g. in lobsters, langoustes and the larger crabs). | Armour supported by pneumatic pressure has tubular segments (Fig. 7). As a rule the entire structural system is internally pressurised, except the respiratory organs. The muscles are able to contract and move the skeleton. | The rigid rings cannot move in themselves. This rigidity is achieved by dense layers of bonded fibres but also by the deposition of hard substances. Despite this rigidity of the segments, the system as a whole remains mobile. | Many crustaceans are capable of moving when young, i.e. when external skin areas have not yet become rigid. In this state they are totally soft Pneus. Once the outer shells have hardened the force acting against the force of the muscles can partially be transmitted to the shell, especially if joints capable of taking compressive loads exist which, however, generally allow only movement in one plane. | Insects (Fig. 6 and Fig. 8) are capable of the most rapid movements. They are small, have a low mass and an external skeleton which is divided into segments by insections which serve as joints. This construction allows rapid movement in three axes such as growing, running or flying.

lentieren und bei dünnen Extremitäten ist zu beobachten, dass ausgeprägte Gelenke vorhanden sind, die das Absetzen der durch die Muskelkräfte hervorgerufenen Druckkräfte auf die Schale, zumindest in einer Richtung gestatten (z.B. bei Hummer, Langusten und größeren Krabben). | Die pneumatisch gestützte Panzerung hat röhrenartige Körper (Abb.7). Das gesamte konstruktive System steht in der Regel unter Innendruck, mit Ausnahme der Atmungsorgane. Die Muskeln können kontrahieren und das Skelett bewegen. | Die steifen Ringe selbst sind in sich unbeweglich. Dies wird durch dichte, verklebte Faserlagen erreicht (auch durch Einlagerung von harten Substanzen). Dennoch bleibt die Beweglichkeit des Systems als Ganzes erhalten. | Viele Schalentiere können sich im Jugendstadium bereits bewegen, wenn noch keine äußeren Hautpartien erhärtet sind. Sie sind in diesem Zustand total weiche Pneus. Sind die äußeren Schalen erhärtet, kann die der Muskelkraft entgegenwirkende Kraft teilweise auch auf die Schale abgesetzt werden, insbesondere wenn druckfeste Gelenke vorhanden sind, die aber in der Regel nur Bewegungen in einer Ebene erlauben. | Die schnellsten Bewegungen können Insekten (Abb.6 und Abb.8) durchführen. Sie sind klein, massearm und haben ein Außenskelett, das durch Einschnitte, die als Gelenke dienen, stark verengt ist. Damit sind schnelle Bewegungen in drei Achsen möglich, wie bspw. Rudern, Laufen oder Fliegen.

Muschelschale und Schnecke Viele Schneckenarten haben ein Haus. Das Haus bietet Schutz, hindert aber bei Bewegung. Die Erhärtung erfolgt am Rande, technisch gesprochen unter einem Pneuschlauch. Die Bewegungsform der Nacktschnecke bleibt aber erhalten. | Das Wachstum der Schalen bei Muscheln, Schnecken u.a. erfolgt unter einer dünnen Haut, unter der sich an die bereits versteifte Substanz Material anlagert. Auch hier hat die biegeweiche, unter Innendruck stehende Haut großen Einfluss auf die Form der Schale. | Der Wachstumsprozess ist an jeder Schneckenschale und Muschel von außen gut zu erkennen. Er hinterlässt ein netzartiges Raster oder zumindest Wachstumsspuren. | Einige Schnecken und Muscheln wachsen gleichmäßig. Sie haben glatte Schalen. Andere wachsen in Stufen. Sie bilden Rippen und Stacheln aus. | Ammonshörner und Nautilusse haben innere Kammern, die durch Wände abgeteilt werden; deren Form wird durch verspannte Membranen gebildet, die von Zeit zu Zeit entstehen und in die sich Hartsubstanz einlagert. Die Form dieser Kammerwände ist derjenigen gespannter Zelthaut ähnlich; häufig sind sie durch Druckdifferenzen pneumatisch gespannt und bilden den Weichkörper des Tieres versteift ab.

Erhärtete Einzeller Einige hochentwickelte Einzeller erhärten, insbesondere Strahlentiere (Radiolarien) (Abb.11) und Kieselalgen (Diatomeen). | Einzelne

9

10

Mussel shells and snails Many types of snail have a shell. The shell offers protection but also impairs movement. Hardening occurs at the edge of the shell or, in technical terms, inside a pneu tube. The style of movement of the slug is, however, retained. | In mussels, snails etc. the shell grows under a thin membrane under which material is deposited on the existing solid substance. In this case again the flexible, internally pressurised membrane greatly affects the shape of the shell. | Any snail or mussel shell clearly shows the process of growth. Growth leaves behind a net-like grid or at least traces of growth. | Some snails and mussels grow continually. These have smooth shells. Others grow in stages. The latter form ribs and spines. | Ammonites and nautiluses have internal chambers or compartments which are divided by partitions. Their shape is formed by braced membranes which are grown from time to time and in which hard substance is deposited. The shape of these chamber partitions resembles that of a tensioned tent canvas. Frequently they are pneumatically tensioned by pressure differences and represent the solidified fore of the soft body of the animal.

Hardened single cell organisms Some highly developed single cell organisms become hard, especially radiolaria (Fig. 11) and diatoms. | Individual elements of the flexible cell Pneu, which in most cases cannot be made visible, often harden in a matter of minutes; in the case of radiolaria sometimes in several stages. Typical

Abb. 9 Weinbergschnecke mit Haus. Deutlich ist die weich abgerundete Haut am Schalenrand zu erkennen – unter dieser wächst die Schale. Ebenfalls sind die Wachstumsringe deutlich auszumachen.

Fig. 9 Snail with shell. The softly rounded membrane at the edge of the shell is clearly visible – the shell grows under this membrane. Growth rings are also clearly visible.

Abb. 10 Nautilus, von außen
Fig. 10 Nautilus, external view

11
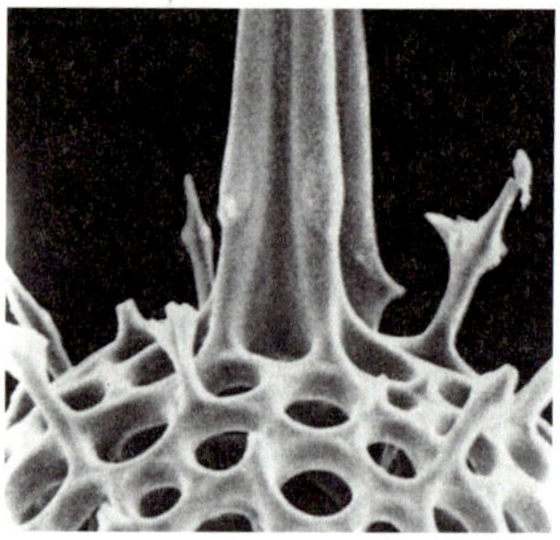

12
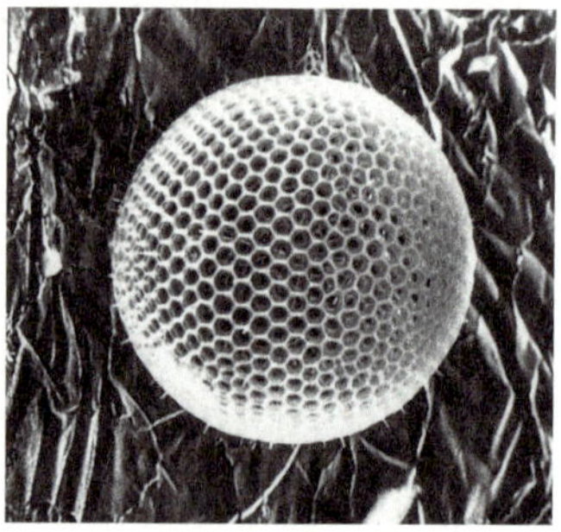

13
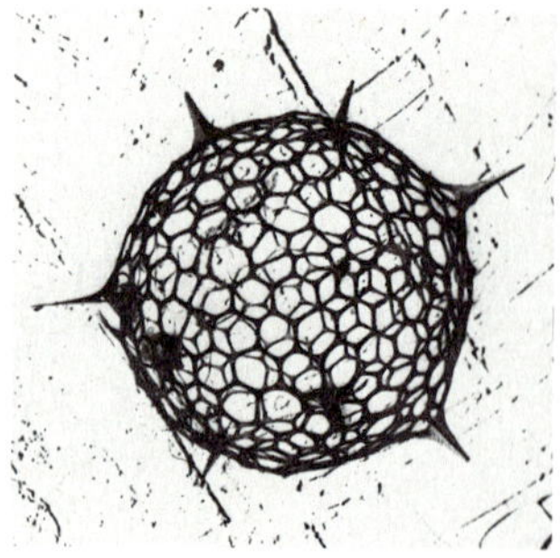

Elemente des zumeist nicht sichtbar zu machenden Pneus der flexiblen Zelle verhärten oft in Minuten, bei Radiolarien manchmal in mehreren Stadien. Man sieht typische Gitterschalen. Helmcke hatte um 1958 die Idee, dass der Hohlraum zwischen winzigen Flüssigkeitströpfchen, die auf oder in der Zellmembrane sitzen, erhärten. Modellversuche führten tatsächlich zu ähnlichen Formen. | Auch Stacheln treten auf (Abb.12), die auf Grund ihrer Form als fest gewordene Flüssigkeitsfäden angesprochen werden können. | Radiolarien und Diatomeen haben Skelette aus Kieselsäure. Sie sind glasklar und in ihrer Zusammensetzung dem Kunstprodukt Glas sehr verwandt. Jedoch sind organische Substanzen in den Skeletten nachweisbar. Dies führte zu meiner Hypothese, dass auch die Schalen dieser Organismen ein feinstes Faserskelett oder Großmolekülgitter haben und vermutlich die Synthese der glasartigen Hartsubstanz fördern oder bewirken. Ich vermute, dass das Faserskelett bereits vor der Verkieselung vorhanden sein muss, demzufolge ist dann das Kieselskelett ganz oder teilweise identisch mit dem Faserskelett. | Über die Weichteile dieser Organismen weiß man wenig. Haeckel hat jedoch solche Weichteile aufgezeichnet, die er bei mikroskopischen Beobachtungen sah. Im Rasterelektronenmikroskop sind sie (bisher) nicht zu sehen (Abb.13 und Abb.14).

Siehe hierzu auch IL 28 Diatomeen I und IL 33 Radiolarien.

Abb.11 Radiolarie. Man erkennt den Aufbau mit innerem Skelett und die äußere Haut.

Fig.11 Radiolaria. The structure consisting of internal skeleton and external membrane is clearly visible.

Abb.12 Stachel einer Radiolarie
Fig.12 Spine of a radiolarium

14

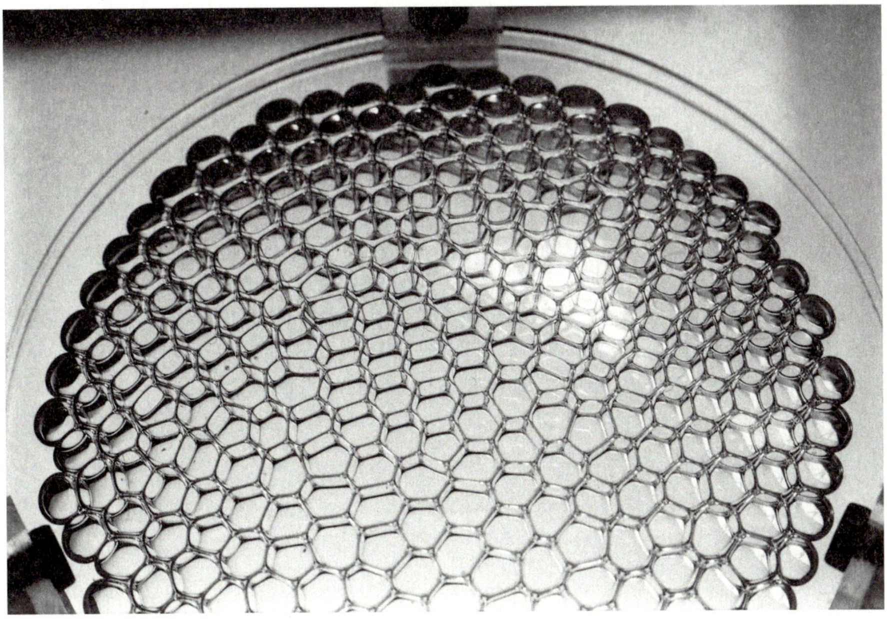

grid shells emerge. Around 1958 Helmcke had the idea that the spaces between the minute liquid droplets resting on the cell membrane will eventually harden. Experiments indeed led to similar forms. | Spines also occur (Fig. 12) which because of their shape may be regarded as solidified liquid threads. | Radiolaria and diatoms have skeletons consisting of silicic acid. They are as clear as glass and their composition is closely related to artificial glass. The presence of organic substances, however, can be shown in the skeletons. This led to my hypothesis that the shells of these organisms also have a microscopically fine fibre skeleton or large molecule grid and possibly promote or cause the synthesis of the glass-like hard substance. It further led to the assumption that the fibre skeleton pre-existed silicification and that therefore the silicic skeleton is wholly or partially identical to the fibre skeleton. | We know very little about the soft parts of these organisms. Heckel has, however, recorded such soft parts which he was able to see under his microscope. So far we have not been able to view them by using a scanning electron microscope (Fig. 13 and Fig. 14).

I refer the reader to IL 28 "Diatoms I" and IL 33 "Radiolaria".

Abb. 13 Kugelförmige Radiolarien mit Stacheln

Fig. 13 Spherical radiolaria complete with spines

Abb. 14 Experiment des IL: Seifenbläschen auf gekrümmter Unterlage. Wenn diese erhärten würden, könnte man die Schale als ein Ausschnittsteil der Kieselalge auf Abb. 12 ansprechen. Das Experiment wurde gemacht, um den möglichen Einfluss von Selbstorganisation aufzuzeigen.

Fig. 14 Experiment carried out by IL: soap bubbles on a curved base. If these bubbles solidified the shell could be likened to a segment of the diatom shown in Fig. 12. The experiment was made to show the possible effect of self-organisation.

Der Seeigel – Gedanken zu Form und Schalenkonstruktion Mit größter Wahrscheinlichkeit ist der Seeigel (Abb. 15) in seiner gesamten Lebenszeit ein Pneu. In ihm herrscht nach meiner Interpretation ständig Innendruck. Er könnte ohne diesen nicht existieren. | Die Schale entsteht dadurch, dass Platten aus Hartsubstanz in der äußeren Hülle entstehen – und zwar zwischen der äußeren und inneren Schicht der Hülle. Die Hülle (die Ballonhaut des Pneus) bleibt immer erhalten. | Die Form der Hülle und der Schale als Ganzes wird bestimmt durch innere Verspannungen und durch die Zugkraft der Schlauchfüßchen, die den Seeigel am Boden fortbewegen. | Die »Baukonstruktion« eines Seeigels ist die durch Innendruck gespannte flüssigkeitsgefüllte, flexible Hülle, in die zu irgendeiner Zeit der stammesgeschichtlichen Entwicklung Panzerplatten und Stacheln eingebaut wurden. | Die Frage, ob es im frühesten Stadium der Entwicklung schalenlose Seeigel gegeben haben könnte, ist heute nicht mehr klärbar. Es kann aber vermutet werden, dass dem so ist. Dann müsste die Form von Seeigeln aller Stämme, Familien, Arten und aller Individuen (gesund oder krank) der Formenwelt der Pneus angehören – und zwar ausnahmslos. Auch dann, wenn außergewöhnliche Ereignisse eintreten, wie die von Dafni (Eilat, Israel) angesprochene Verseuchung des Wassers, die Deformationen hervorruft, wie sie nur bei schwach gespannten Pneus vorkommen. | Der »klassische« Pneu hat die Form einer Kugel (Abb. 16). Er hat überall die gleiche Oberflächenspannung. Mit anderen Worten: Bei der Kugelform ist die Flächenspannung an jeder beliebigen Stelle der Oberfläche konstant. | Dies gilt genaugenommen für Formen, die nicht durch äußere Kräfte deformiert werden (wie Seifenblasen, Luftballone in Luft, Mikrosphären oder wassergefüllte Ballone unter Wasser). | Jede Abweichung von der Kugelform signalisiert ungleiche Flächenspannung oder das Vorhandensein anderer Kräfte. | Schon geringe Spannungsunterschiede haben großen Einfluss auf die Form. Oder umgekehrt: Große Formänderungen bewirken kleine Spannungsunterschiede. | Sehr häufig sind bei Organismen innere Verspannungen bereits im embryonisch kugeligen Zustand (Abb. 18a) angelegt und werden erst sichtbar, wenn der Pneu sich durch Innendruck vergrößert (Abb. 18b). | Alle von mir bisher betrachteten Seeigel, einschließlich äußerst flächiger, z. B. der Sanddollars, sind als weiche Pneus bildbar; und zwar die Formen Abb. 19a–19d *ohne* innere Verspannungen und die Formen (Abb. 20a, 20b) *mit* inneren Verspannungen. | Treten die inneren Verspannungen konzentriert auf, wird dies durch Vertiefungen von außen erkennbar. Diese sind bei einigen Arten sehr ausgeprägt. | Wenn äußere oder innere Kräfte wirken, kann ein kugelförmiger Flüssigkeitstropfen sich auch verformen, ohne dass sich die Spannungen ändern. | Bestes Beispiel ist der aufliegende Wasser- oder Quecksilbertropfen (beim letzteren auch unter Wasser) (Abb. 21). Die Verformung wird durch das Eigengewicht bewirkt. Flüssigkeitstropfen haben

The sea urchin – Reflections on shape and shell structure It is most likely that the sea urchin (Fig. 15) remains a pneu during its entire life. As far as I can interpret it, it is continously under internal pressure. Without this pressure it could not exist. | The shell is formed in the outer envelope by the build-up of plates or segments consisting of hard substance, i.e. between the outer and inner layers of the envelope. The envelope (the balloon skin of the pneu) is retained. | The shape of the envelope and the shell as a whole is determined by internal bracing and by the tensile force exerted by the tubular feet which enable the sea urchin to move along the floor of the sea. | The "structure" of a sea urchin is a flexible envelope which is tensioned by internal pressure and filled with liquid and which at some stage in the evolution of this species was equipped with armour plates and spines. | We can no longer ascertain whether at the earliest stage of development sea urchins existed which had no shells. We may, however, suspect that that is the case. This would mean that the shapes of all genera, families and species of sea urchins and of all individual sea urchins – whether healthy or sick – belong to the category of pneu forms – and that without exception; even when extraordinary events occur such as the pollution of water mentioned by Dafni (Eilat, Israel), which causes the sorts of deformations which occur only in low-pressure pneus. | The "classic" Pneu has the form of a sphere (Fig. 16). The surface stress is the same at all points of the surface. In other words: with a spherical shape the area unit stress is constant for all points of the surface. | To be accurate, this applies to shapes which are not deformed by external forces (such as soap bubbles, balloons suspended in air, microspheres or waterfilled balloons submerged in water). | Any deviation from the spherical shape indicates unequal surface stresses or the presence of other forces. | Even small stress differences greatly affect the form. Or conversely: large changes of form cause small changes of stress. | Quite frequently internal guying or bracing is present in organisms at the embryonic spherical stage (Fig. 18a) which do not become visible until the pneu grows as a result of internal pressure (Fig. 18b). | All the sea urchins which I have examined so far, including the extremely flat sand dollar, can be formed as soft pneus: shapes such as those shown in Fig. 19a–19d without internal bracing, and the shapes shown in (Fig. 20a, 20b) with internal bracing. | If the internal bracing is concentrated, this will show outside in the form of depressions or indentations. These are quite pronounced in some species. | Under the influence of external or internal forces a spherical droplet of liquid can also change its form without any change in the stresses. The best example is a water or mercury droplet resting on a surface (the latter also under water) (Fig. 21). The deformation is caused by the droplet's own weight. Like soap bubbles, liquid droplets have equal surface stresses. | A highly flexible spherical rubber balloon (Fig. 22) filled with concentrated

15

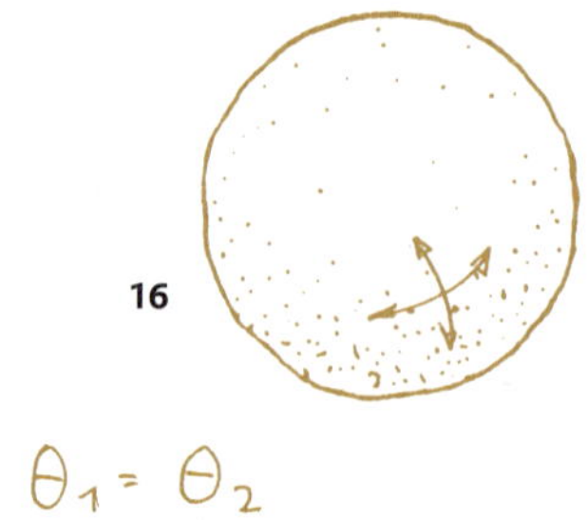

16

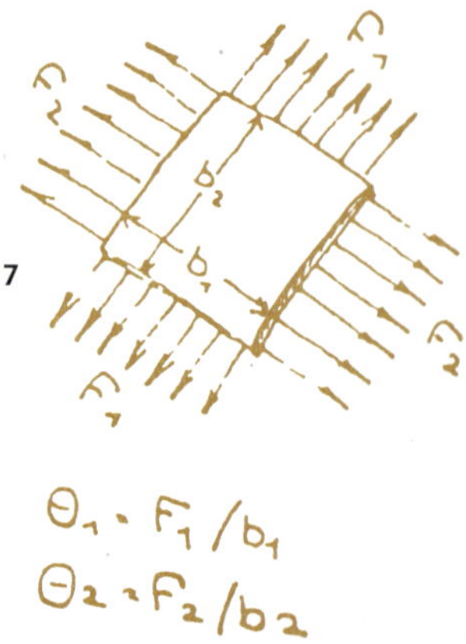

17

wie Seifenblasen gleiche Oberflächenspannungen. | Ein mit konzentrierter Salzwasserlösung gefüllter hochflexibler kugeliger Gummiballon (Abb. 22) sinkt auf den Boden und nimmt eine Tropfenform ein – auch ohne, dass innere Verspannungen da sein müssen. In diesem Fall (also bei einer Füllung, die schwerer als das umgebende Wasser ist) sind ohne innere Verspannungen alle die in Abb. 21 skizzierten Formen möglich (jeweils im Grundriss etwa kreisrund). | Solche Formen sehen wir auch bei Seeigeln. Sind diese aber bemerkenswert schwerer als Wasser? | Dafni schreibt, dass die Schlauchfüßchen, die den Pneu fortbewegen, einen Einfluss auf die Form bewirken, wenn sie an dem Pneu ziehen. Das müssen sie ja, wenn sie ihn bewegen (Abb. 23a–23c, Abb. 24). | Das stimmt, wenn man den Seeigel als echten unversteiften Pneu betrachtet. Es stimmt vermutlich auch für den Seeigel mit erhärteter äußerer Schale, dessen Hartteile ein Bestandteil der äußeren Haut sind. | Zwar wird deren Form nicht so extrem schnell reagieren wie ein weicher Pneu, wird aber im Laufe der Zeit eine dem mittleren Bewegungsverhalten entsprechende Form annehmen; d.h. selbst die Form der »harten« Schale wird auf den Zug der Füßchen reagieren. | Im ganz extremen Fall übernehmen die Füßchen *alle* Kräfte der oberen Membrane OM. Die Bodenmembrane BM ist dann völlig entlastet (Abb. 25). Sie kann beliebig wirre Falten haben oder sich dem Boden anpassen oder sogar frei schweben. Ein solcher »Seeigel« wäre dann konstruktiv gut verankert, besonders gegen Wasserströmungen im Brandungsbereich.

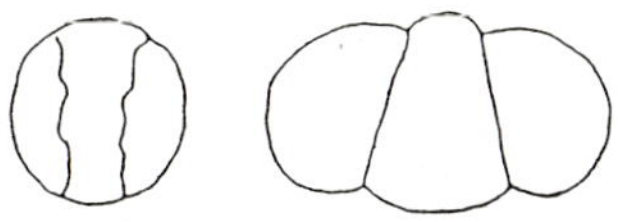

a b

18

a b

20

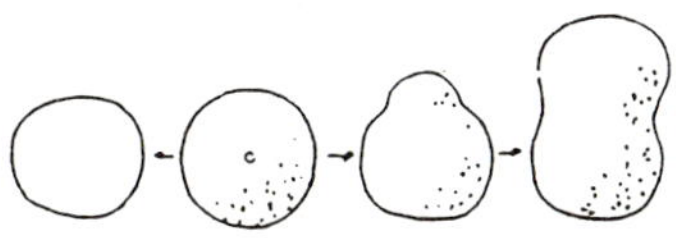

a b c d

19

21

saline sinks to the bottom and assumes the shape of a drop without any internal bracing being present. In this case (i.e. with a filling which has a greater density then the surrounding water), all the shapes sketched in Fig. 21 are possible without internal bracing (roughly circular in plan view). We observe such forms also in sea urchins. Are these, however, noticeably heavier than water? | Dafni writes that the tubular feet which move the pneu affect its shape when they are pulling at the pneu. This pulling is necessary to move it (Fig. 23a–23c, 24). | This is correct if the sea urchin is regarded as a genuine i.e. non-rigid pneu. It is probably also correct for a sea urchin with hardened external shell whose hard parts are part of the external skin. | It is true that their shape will not react as extremely rapidly as a soft pneu, but it will in the cause of time assume a shape which corresponds to the average motive behaviour; i.e. even the form of the "hard" shell will react to the pull exerted by the feet. | In an extreme case the feet will transmit *all* forces of the upper membrane OM. In this case the bottom membrane BM is totally relieved of any load (Fig. 25). It can have any number of random wrinkles or adopt the shape of the sea floor or it may even float freely. Such a "sea urchin" would be well anchored especially against the strong water currents found in areas of surf. In such areas a sea urchin needs very strong feet with large suckers and a solid base to cling to. | A "soft" sea urchin, i.e. one without hard parts, can have any of the shapes which we regard as pneu and it can move in the same way as real sea

22

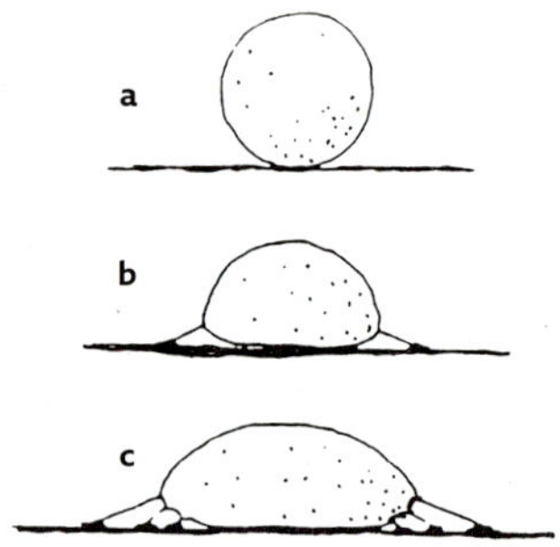

23

24

Er braucht dann sehr kräftige Füßchen mit großem Saugnapf und einen festen Untergrund. | Ein »weicher« Seeigel – also ohne Hartteile – kann alle Formen haben, die wir von den Pneus kennen, und er kann sich so, wie Seeigel es tun, bewegen. Er hat aber eine Schwierigkeit (Abb. 26). Die Zugkraft F zieht immer etwas schräg nach unten. Dies bewirkt eine Gegenkraft in Form eines erhöhten Anpressdruckes P der Bodenhaut. | Ein solcher Pneu hat es also schwer, seine Bodenreibung zu überwinden; dagegen ist er – wie bereits gesagt – bei äußeren, seitlich wirkenden oder nach oben saugenden Kräften (Abb. 27) gut verankert, und zwar umso mehr, je rauher (noppiger, stacheliger) die Bodenpartie ist. | Seeigel bewegen sich sehr langsam. Wir können sie als statische Konstruktionen betrachten. Auch wenn es eine Utopie sein mag, möchte ich kurz vom Seeigel sprechen, der leichter als das umliegende Wasser wäre, der also einen Auftrieb hätte (Abb. 28a). Der Auftrieb A könnte die Schrägkraft der Füßchen kompensieren. Die Reibung geht nach Null. Ohne Bodenreibung wäre die Bewegung sehr effektiv. | Wenn jedoch eine kräftige Strömung kommt, dann kann die Bodenreibung durch die zusammenziehbaren Füßchen sofort hergestellt werden (Abb. 28b). | Wie wir dies vom Nest der Wasserspinne und von technischen Anwendungen kennen, könnte beim Seeigel, wenn er leichter als Wasser ist, die Bodenmembrane sogar völlig berührungsfrei über dem Boden schweben (Abb. 29). | Es ist nun nicht meine Aufgabe, einen extrem mobilen Seeigel zu konstruieren, den es vielleicht gar nicht

urchins move. It will, however, face one difficulty (Fig. 26). The pull force F will always act obliquely downward. This causes a balancing force in the form of an increased contact pressure P between the bottom membrane and the ground. | It is therefore more difficult for such a pneu to overcome ground friction. On the other hand it is, as I have said before, well anchored against external lateral or uplift forces (Fig. 27), the more so the rougher (nobbly or spinous) its bottom part. | Sea urchins move very slowly. We may regard them as static structures. Even though it may be a utopia, I would like to express briefly a few thoughts on a sea urchin which is lighter than the surrounding water and which therefore would have buoyancy (Fig. 28a). The buoyancy A could compensate the oblique force exerted by the feet. In this case the friction would approximate zero. Without ground friction the movement would be very efficient. | If, however, this theoretical sea urchin met a strong current, it could produce the necessary friction with the ground by using its contracting feet (Fig. 28b). | As the nest of the water spider and technical applications show, the bottom membrane of a sea urchin which is lighter than water, could float freely without having any contact with the sea floor (Fig. 29). | It is not my task to design an extremely mobile sea urchin which perhaps does not exist at all only because it cannot make itself lighter than water for the time during which it moves. We observe, on the contrary, that sea urchins collect weight in the form of stones, manganese nodules etc. (according to Reif), presumably to increase the anchorage when the animal is at rest.

25

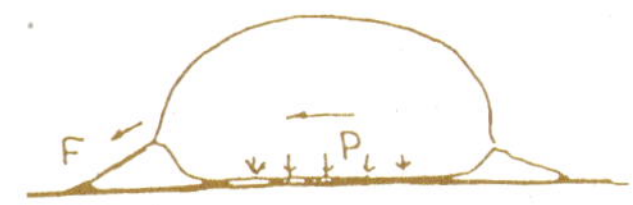

26

27

There are several questions which in my view remain to be answered: How much heavier than the surrounding water are sea urchins living on the sea floor? We are dealing here with parts per thousand. What is the degree of salinity inside their bodies? Do they have (controlable) gas bladders inside their bodies? etc. etc. | In addition the friction of the sea floor could be overcome for short periods during motion by water being ejected from the underside of a sea urchin which is heavier than the surrounding water (Fig. 30).

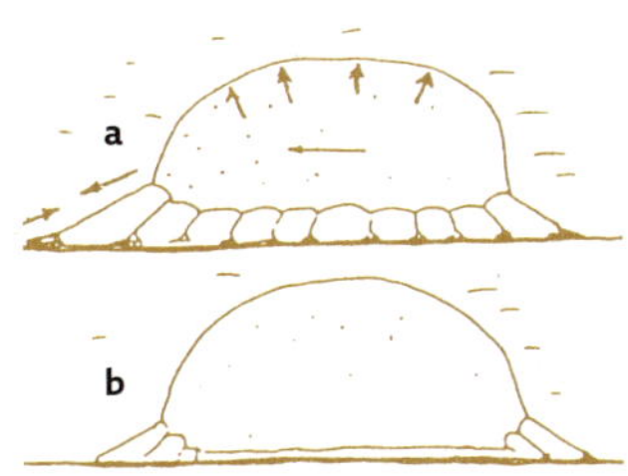

a

b

28

Now we have to ask: What is the purpose of the shell and spinous processes? The hard shell is undoubtedly a good armour against the sharp teeth of predators, since flexible membranes are not capable of withstanding loads concentrated in a point. Almost all efficient killing instruments, including those used by man, are pointed like the canines of animals or like arrows, spears and projectiles. | Hard shells are less efficient in withstanding distributed loads such as falling stones (Fig. 31) or the feet of animals (Fig. 32), than pneus with a thicker leatherlike skin. (In my view these types of load are not a particularly important selection criterion even though treatises have been written on the loadbearing capability of shells which in

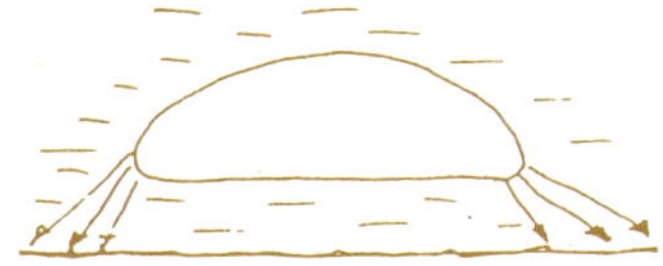

29

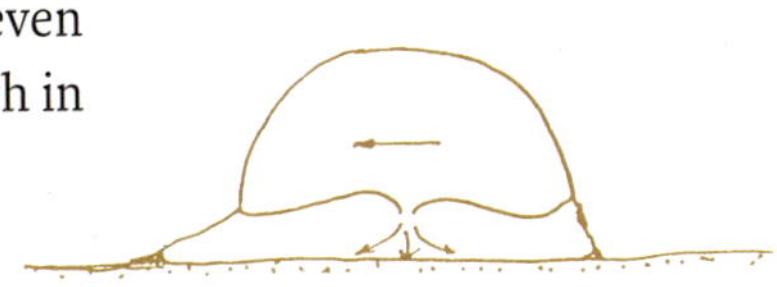

30

gibt; vielleicht nur deshalb, weil er sich für die Zeit, in der er sich bewegt, nicht leichter als Wasser machen kann. Es ist im Gegenteil zu beobachten, dass Seeigel Gewichte sammeln – wie Steine, Manganknöllchen usw. (nach Reif), vermutlich, um die Verankerungskräfte in Ruhelage zu erhöhen.

Für mich sind Fragen offen: Um wie viel schwerer sind denn die am Boden lebenden Seeigel als das sie umgebende Wasser? Hier zählen Promille. Wie salzig ist ihr Körperinhalt? Gibt es (regulierbare) Gasblasen in ihren Körpern? usw. usw. | Außerdem könnte die Bodenreibung auch bei einem Seeigel, der schwerer als das ihn umgebende Wasser ist, durch düsenhafte Auspressung von Wasser auf der Unterseite kurzfristig während der Bewegung überwunden werden (Abb. 30).

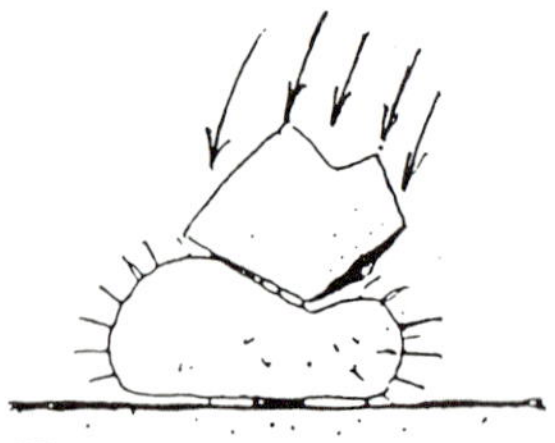

31

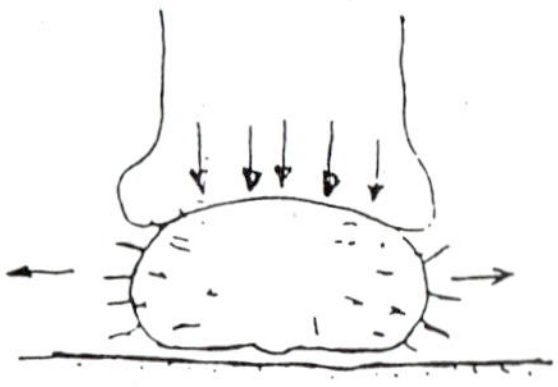

32

Nun zu der Frage: Wozu Schale und Stacheln? Die harte Schale ist unbezweifelbar ein guter Panzer gegen die spitzen Zähne von entsprechend bewaffneten Fraßfeinden, denn punktförmige Belastungen vertragen flexible Membranen nur schwerlich. Fast alle wirksamen Tötungsinstrumente – auch des Menschen – sind Spitzen oder spitzig; wie die Reißzähne von Tieren oder wie Pfeile, Speere und Geschosse. | Für Angriffe gegen Flächenlasten, z.B. herabfallende Steine (Abb. 31) und darauf tretende Tiere (Abb. 32), sind harte Schalen weniger effektiv als Pneus mit dicker lederartiger Haut (für mich sind diese Belastungsarten kein besonders wichtiges Selektionskriterium, auch wenn über die für diesen Fall einsetzende – übrigens sehr bescheidene – Schalentragwirkung Abhandlungen geschrieben wurden. Die Seeigelschale ist für mich ähnlich zerbrechlich wie die Schale eines Hühnereis). | Bei stoßartigen Belastungen bricht die Schale des Seeigels. Das mag sogar gut sein. Die stets vorhandene, durch die Bindegewebsfasern gebildete, hochzugfeste Außenhaut fängt den Stoß durch Deformation auf. Und da die Panzerplatten fest in dem Fasernetz der Haut verankert sind, wird diese selten zerstört – was vermutlich den Tod bedeuten würde. | Trotz gebrochener Schalenteile kann der Seeigel aber bei unverletzter Haut weiterleben und die Brüche reparieren. Der Innendruck im Inneren des Seeigels ist notwendig, um den Wachstumsprozess (durch »Aufpumpen«) zu steuern und um die Form zu halten. | Im Übrigen ist, wie auch bei Innenskelett-Tieren, zu vermuten: Falls der Innendruck höher als üblich ist, wächst der Seeigel schneller und wird größer als üblich. Wenn der Innendruck sich jedoch an der unteren Grenze bewegt, dann bleibt der Seeigel klein. | Sollte der Innendruck in der Wachstumsphase abnormal schnell steigen, ist es einleuchtend, dass besonders beanspruchte oder geschädigte Teile des Faserskeletts sich mehr dehnen als andere. | Es entstehen Deformationen. Sie entstehen auch dann, wenn die Erhärtung der Schale unvollkommen erfolgt. | Dann entstehen Verformungen. Vielleicht deutet die Form der Abb. 33 (Foto: Dafni) auf

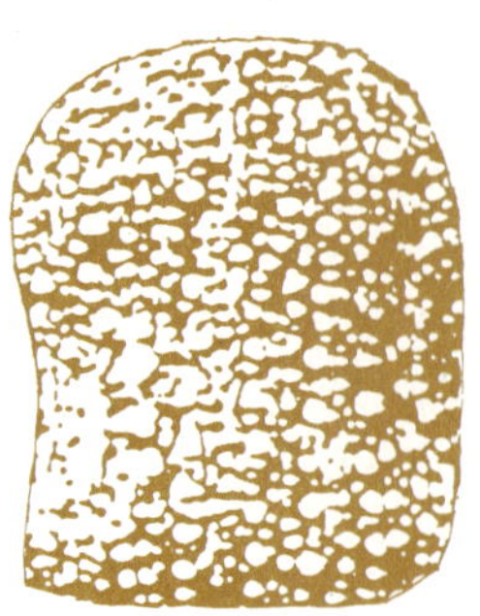

33

this case would apply and would, by the way, be very modest. In my view the shell of a sea urchin is similarly fragile as the shell of a hen egg). | The shell of the sea urchin will break under impact loading. This may even be of advantage. The highly tensile outer skin which consists of connective tissue fibres and is always present, parries the impact by undergoing a deformation. Since the armour plates are solidly anchored in the fibrous net of the membrane, these will be rarely destroyed, which would probably mean that the urchin would die. | Despite broken shell segments the sea urchin can continue to live and repair the fractures provided the membrane is undamaged. The pressure inside the sea urchin is required to control the growth process (by "inflation") and to maintain the shape. | We may assume with respect to the sea urchin what also applies to animals with internal skeletons: if the internal pressure is higher than normal the sea urchin will grow more rapidly and larger than normal. If the internal pressure is, however, close to the lower limit, the sea urchin will remain small. | It is evident that an abnormally rapid rise in internal pressure during the growth phase will cause parts of the fibrous skeleton which are under increased loading or damage, to stretch more than other parts of the skeleton. | In this way deformations are produced. They will also result if the shell hardens imperfectly. Deformations will be the result. Perhaps the shape of Fig. 33 (photograph: Dafni) may point to such a deformation. | If, on the contrary, the internal pressure is reduced or removed altogether or if it becomes persistent *vacuum*, the shell of the sea urchin will become indented or develop wrinkles (Fig. 34). | The shapes of the sea urchins affected by environmental pollution and presented by Dafni, show the typical changes of form caused by pressure changes with which we are familiar from inanimate (technical) pneus. Thus the shapes of Fig. 35 and Fig. 36 (Dafni) look like deformations caused by a vacuum. | We have studied such forms in experiments with rubber balls. Asymmetrical bulging caused by pressure or indentation caused by vacuum may even share the same cause, e.g. a shell which is excessively thin. | The armour plates of the sea urchins are so arranged that they form a particularly efficient shell, especially when the internal pressure drops. Because of its shape it is a typical "vacuum shell" designed for the drops in internal pressure which probably last only a short time in each case. | I must also assume that a sea urchin will be able to tolerate, or even actively produce, a considerable vacuum inside its body without suffering any damage. This vacuum can, however, persist only for a short time as it will inhibit the growth and form-stabilising process and slowly cause indentation. | It is evident that a sea urchin can live with an internal pressure for a short time. Could, on the other hand, a sea urchin exist with a permanent vacuum inside? In my opinion this is not possible. | A sea urchin which is always under internal pressure (if such a creature should exist), requires, like an airhouse, locks to ingest food and fresh water and

34

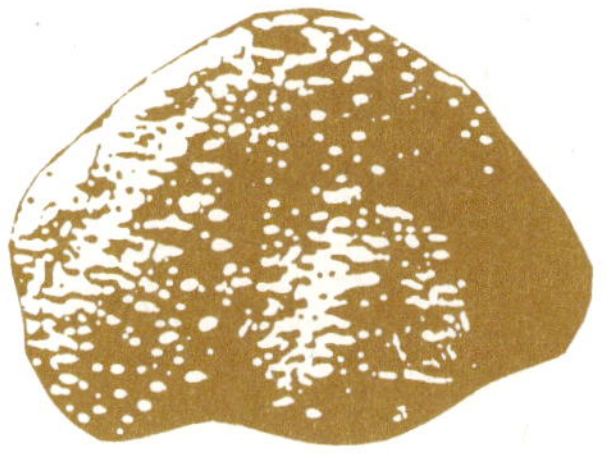

35

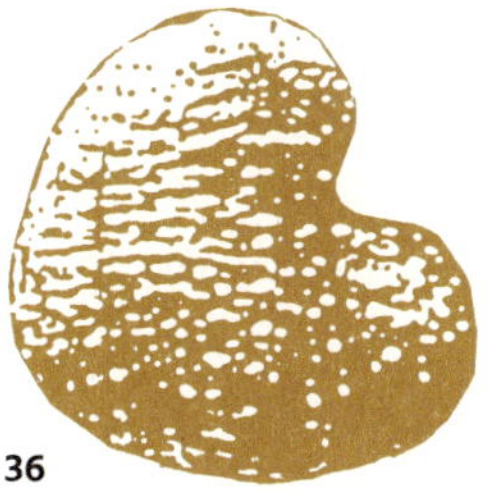

36

eine solche Deformation hin. | Wenn der Innendruck im Gegensatz dazu kleiner wird oder sogar wegfällt, ja zu einem länger haltenden *Unterdruck* wird, dann muss die Schale des Seeigels nach innen beulen oder faltig werden (Abb. 34). | Die Formen der durch Umweltverschmutzung krank gewordenen Seeigel, die Dafni vorführt, zeigen die durch Druckeinflüsse hervorgerufenen typischen Formänderungen, wie wir sie auch bei nicht lebenden (bzw. technischen) Pneus kennen. So sehen auch die Formen der Abb. 35 und Abb. 36 (Dafni) nach Deformationen aus, die durch Unterdruck hervorgerufen werden. | Wir haben solche Formen in Versuchen an Gummibällen studiert. Asymmetrische Ausbeulungen durch Überdruck oder Einbeulungen durch Unterdruck können sogar die gleiche Ursache haben – z. B. eine zu dünne Schale. | Die Platten des Seeigels sind so angeordnet, dass sie eine besonders effektive Schale bilden, besonders dann, wenn der Innendruck abfällt. Sie ist für den vermutlich nur jeweils kurzfristigen Abfall des Innendrucks aufgrund ihrer Form eine typische »Unterdruckschale«. | Ich muss auch annehmen, dass ein Seeigel schadlos auch einen beträchtlichen Unterdruck im Körperinneren aushalten oder sogar aktiv erzeugen kann, doch kann der Unterdruck nur kurzzeitig wirken, denn bei Unterdruck ist der Wachstums- und Formstabilisierungsprozess lahmgelegt – und das langsame Beulen beginnt. | Dass ein Seeigel existieren kann, der kurzzeitig Überdruck im Inneren hat, ist einleuchtend. Könnte aber ein Seeigel existieren, der ausschließlich Unterdruck hat? Ich meine: Kaum. | Ein Nur-Überdruck-Seeigel (sofern es den geben sollte) braucht, wie eine Traglufthalle, Schleusen, um Nahrungsmittel und Frischwasser ein- und verbrauchtes Wasser auszustoßen. Das wäre denkbar. Der Mund vieler Tiere ist eine solche Schleuse (auch der des Menschen). | Ein ›Kurzzeit-Unterdruck-Seeigel‹ braucht solche Schleusen nicht, wenn er eine drucksteife Schale besitzt, die das Kollabieren eines Pneus bei kurzzeitigem Einsaugen verhindert. Das Organ, um Unterdrücke zu erzeugen, ist da (Abb. 37). | Ich meine also: Seeigel sind sowohl Pneus (also Überdruckorganismen), als auch ›Kurzzeit-Unterdruck-Schalentiere‹. Ein Muskel bewegt den Mund des Seeigels nach innen. Er erzeugt da Überdruck. Ein anderer Muskel zieht ihn nach außen; er erzeugt Unterdruck (vorausgesetzt, ich interpretiere Dafnis Zeichnung (Abb. 5) richtig). Der Seeigel kann danach sein Druckgeschehen selbst steuern, jedoch nur, solange seine Muskeln nicht geschädigt sind – er also »bei Kräften« ist.

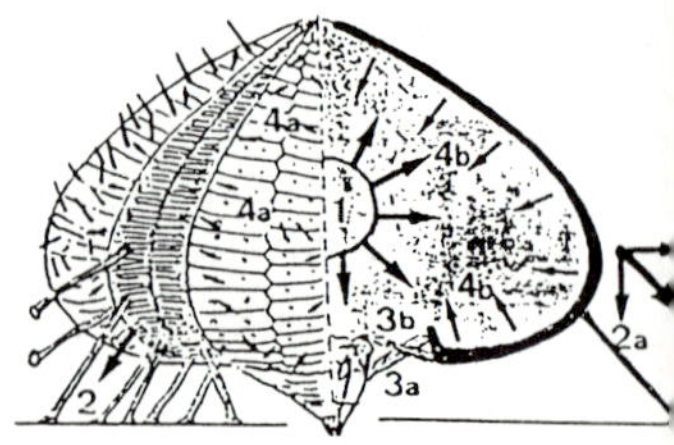

37

38 a

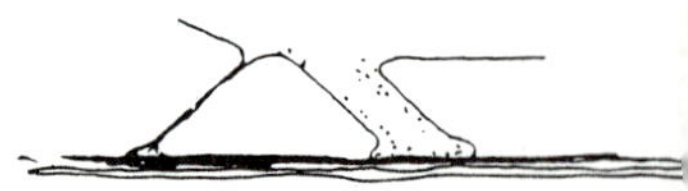
38 b

Zu den Stacheln: Die Stacheln wehren Feinde ab, bringen ihnen mit ihren Spitzen möglicherweise Verletzungen bei und machen Seeigel für Feinde weniger »schmackhaft«. | Die Stacheln sind beweglich – wie auch beim landwohnenden Säugetier gleichen Namens. Soweit ich Dafni verstehe, können sie jedoch durch einen raffinierten Einrastverschluss (durch Muskelzug) starr werden. Erstarrt sind

39 a

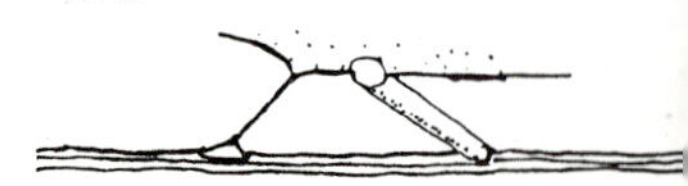
39 b

to dispose of used water. This is quite possible. The mouth of many animals (including man) functions like such a lock. | A "short-term vacuum" sea urchin does not need such locks if it is equipped with a rigid shell which prevents the collapse of the pneu during the short ingestion or suction phase. The organ needed to produce vacuum exists (Fig. 37). | I therefore have the following theory: sea urchins are pneus (i.e. internally pressurised organisms) as well as crustaceans which experience short-term vacuum phases. The mouth of the sea urchin is moved inward by a muscle. | In this way it produces internal pressure. Another muscle pulls the mouth outwards; in this way an internal vacuum is produced (provided that I interpret Dafni's sketch (Fig. 37) correctly). Accordingly the sea urchin is capable of controlling its pressure as long as its muscles are not damaged, i.e. as long as it has the necessary strength.

Some thoughts on the spinous: The spines serve to protect the animal against predators, to injure the predator and to make the sea urchin less "palatable" to potential predators. | The spines move like those of a hedgehog. As far as I understand Dafni, they can be held rigid by a clever locking mechanism (activated by muscle tension). In this rigid position the spines represent an effective defence against predators. | Could these spines have an additional function? Why can they move? I see one reason for this, i.e. efficient movement. | "Soft" sea urchins (if they existed) could considerably reduce the ground friction caused by the pull of the feet if they had very flexible pushing feet which in contrast to the pulling feet would have to have a certain plumpness or an articulated joint (Fig. 38a, 38b). In this case the bottom of the sea urchin would no longer slide over the sea floor. | The resulting motion would produce little friction and require little energy. If a sea urchin had articulated spines or if such pushing feet had turned into spines, the latter would take care of balancing the compressive force. Such a sea urchin would virtually walk on "stilts". At the end of each step the spines, just like stilts, would have to swing laterally to make contact with the ground again. Individual tubular feet would also have to find a new position while the remaining feet would take care of anchorage and propulsion (Fig. 39a, 39b). | Movement would be assisted even if the spines were passive during rapid movement and formed trailing skids (like skis) (Fig. 40).

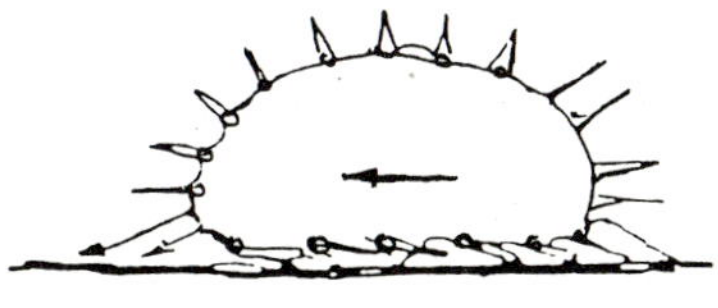

40

sie eine wirksame Feindabwehr. | Blitzschnell angreifende Feinde können meines Wissens jedoch durchaus bei Seeigeln bis zur Schale vordringen, wenn sie die Stacheln auseinander biegen, bevor sie eingerastet sind. | Könnten die Stacheln noch eine andere Funktion haben? Warum sind sie beweglich? Ich sehe einen Grund: nämlich für effektive Bewegung. | »Weiche« Seeigel (sofern es die gäbe) könnten die durch den Zug der Füßchen hervorgerufene Bodenreibung dann bedeutend verringern, wenn ein sehr flexibles Druckfüßchen vorhanden wäre, das im Gegensatz zum Zugfüßchen eine gewisse Plumpheit oder ein Gelenk haben müsste (Abb. 38a und 38b). Die Bodenpartie gleitet dann nicht mehr über den Boden. | Die Fortbewegung ist reibungs- und energiearm. Wenn ein Seeigel gelenkige Stacheln hat oder solche Druckfüßchen zu Stacheln geworden sind, dann übernehmen diese den Druckkraftausgleich. Er läuft quasi »auf Stelzen«. | Am Ende eines Schrittes müssen die Stacheln – wie Stelzen – seitlich herumschwenken und neuen Bodenkontakt finden; auch einzelne Schlauchfüßchen müssen eine neue Position suchen, während die übrigen Füßchen Halt und Vortrieb übernehmen (Abb. 39a und 39b). | Selbst wenn die Stacheln sich bei schneller Bewegung passiv verhalten und nachschleppende Kufen – wie Skier – bilden würden, könnte dies die Bewegung unterstützen (Abb. 40).

11

Seifenblasen

Forming Bubbles

Frei Otto

Dreidimensionale, offene und geschlossene Minimalwegesysteme Räumliche Wegesysteme entstehen, wenn vier oder mehr Punkte einer Konfiguration nicht mehr in einer Ebene liegen. Sie haben im Prinzip die gleichen Bildungsgesetze wie ebene Minimalwegesysteme, sind aber im Gegensatz zu diesen bedeutend weniger erforscht. | Räumliche, offene Minimalwegesysteme mit dreiarmigen Knoten sind nicht mehr durch Experimente mit Seifenlamellen zu bilden. Sie können hingegen in mechanischen Modellexperimenten (mit gespannten Fäden und 120° Röllchen, Feder-Ring-Modellen u. ä.) mit ausreichender Genauigkeit ermittelt werden. Da sich die Anzahl der Knoten bei diesen Methoden nicht mehr selbst bildend einstellt, muss diese in Teilexperimenten festgelegt werden. Ungleich komplexer sind geschlossene räumliche Netze mit vierarmigen Knoten. Da regelmäßige Dodekaeder räumlich nicht packbar sind, behelfen sich die Mathematiker damit, einen Packungspolyeder angenäherter Geometrie zu erzeugen. | Als ideales Raumnetz kann man die Plateauzonen von so genanntem Polyederschaum (s. Schäume), gebildet aus gleich großen Blasen, betrachten. Wie im Kap. Schäume näher erläutert, treffen jeweils drei Lamellen mit 120° Winkeln in einer flüssigen Kante und maximal vier Linien und sechs Lamellen in einem Knotenpunkt zusammen. Der räumliche Winkel an einem solchen vierarmigen Knoten beträgt jeweils 109° 28" und 16'. | In der Natur finden sich dreidimensionale offene und geschlossene Netze mit drei- und vierarmigen Knoten in besonders großer Anzahl. In der Technik sind Netze mit vierarmigen Knoten die häufigsten.

1

2

Abb.1 und 2 Modelle offener räumlicher Netze mit dreiarmigen Knoten aus gespannten Fäden und Rollen. Gut sichtbar ist, dass jeweils drei Arme in einer Ebene liegen und am Knotenzentrum die Verdrehung in eine andere Ebene erfolgt.

Fig.1 and 2 Models of open three-dimensional networks having three-armed nodes formed by tensioned threads and rollers. It is clearly visible that three arms share the same plane and that the rotation into a different plane takes place at the nodal centre.

Three-dimensional open and closed minimal way systems Three-dimensional way systems are created when four or more points of a configuration no longer lie in the same plane. Basically they are subject to the same principles of formation as flat minimal way systems, but have been studied to a much lesser extent than the flat minimal way systems. | Three-dimensional open minimal way systems having three-armed nodes cannot be formed by experiments using soap films. They can, however, be determined with sufficient accuracy in mechanical model experiments such as tensioned threads and 120° rollers, spring/ring models etc. Since with these methods the number of nodes no longer determines itself, it must be determined in secondary experiments. Of a much more complex form are closed three-dimensional networks with four-armed nodes. Since it is not possible to cluster dodecahedrons, mathematicians use the trick of generating a cluster polyhedron having an approximated geometry. | The liquid edges of polyhedral foam (cf. Foams) which is formed of bubbles of equal size, may be regarded as an ideal 3-dimensional network. As will be explained in more detail in the chapter "Foams", groups of three lamellae will meet at angles of 120° in a liquid edge; also no more than four lines and six lamellae can meet in any one node. The spatial angles at such a four-armed node is 109° 28" and 16' in every case. | Nature provides a particularly large number of three-dimensional open and closed networks having three and four-armed nodes. In engineering it is the networks with four-armed nodes that are found most frequently.

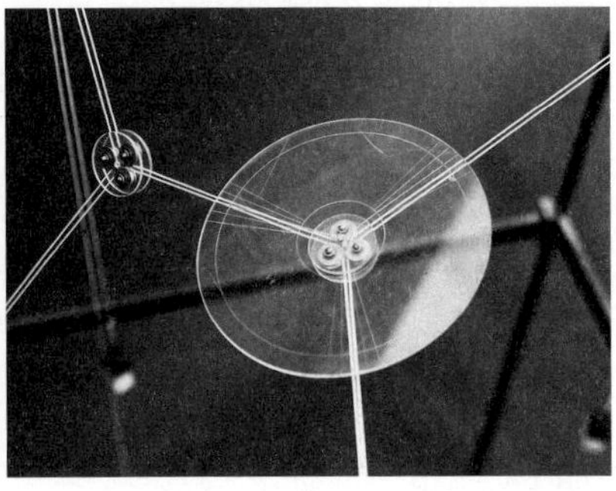

3 / 4

Abb.3 und 4 Zwei Prototypen von Knotenrollen mit drei in einer Ebene im Winkel von 120° montierten Röllchen zwischen zwei parallelen Führungsscheiben. Hieraus wurde ein kleinerer und leichterer Rollensatz, nicht größer als 10 mm, entwickelt.

Fig.3 and 4 Two prototypes of nodal rollers, with three rollers mounted in the same plane at angles of 120° between parallel guide discs. From this a smaller and lighter set of rollers not larger than 10 mm was developed.

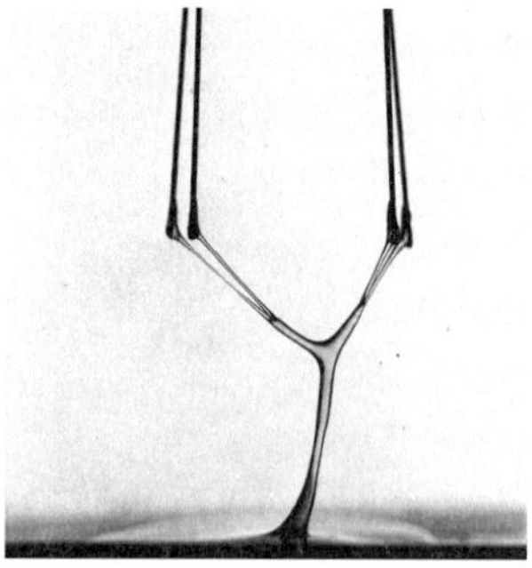

1 / 2

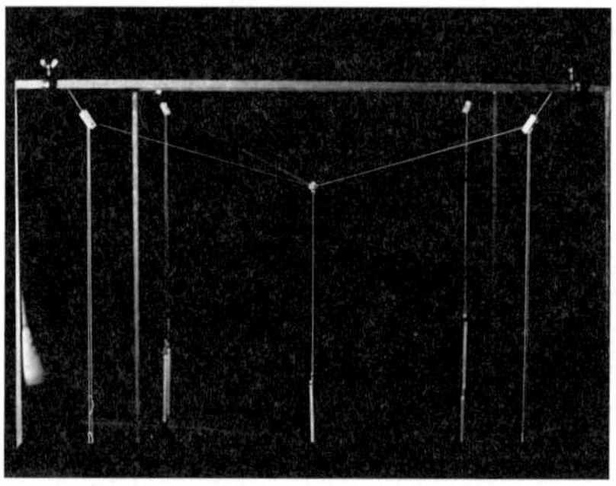

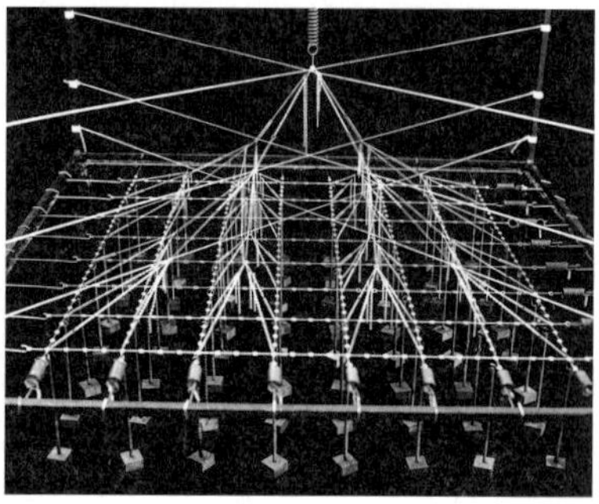

3 / 4

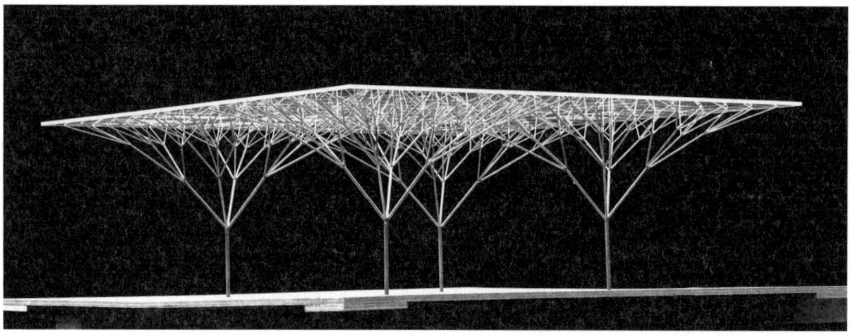

5

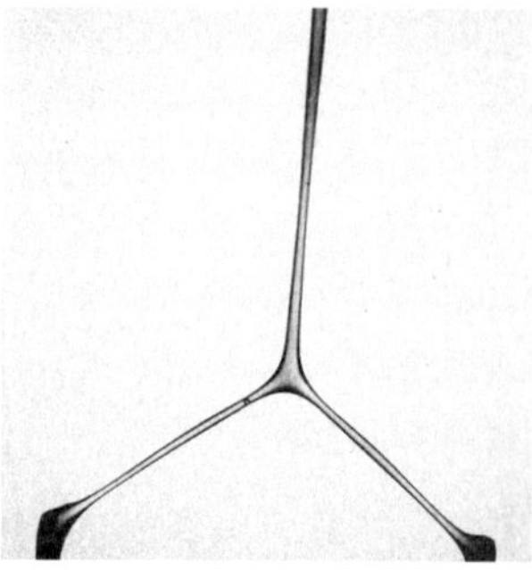

6

Abb. 1 und 2 Aus hochviskosen Flüssigkeiten lassen sich, im Gegensatz zu Seifenlamellen, verzweigte Fäden ziehen. Je nach Zahl der Anheftungspunkte entstehen vielarmige Figuren, die zu dreiarmigen flüssigen Knoten tendieren.

Abb. 3 Räumliches Knotenmodell mit fünf Armen.

Abb. 4 Zugbeanspruchtes Hängemodell einer Baumstütze mit Ketten und Federn zur Findung der optimalen Verzweigungswinkel und Stablängen. Verzweigung der Stütze in drei »Generationen« (im Verhältnis 1 zu 4) zur flächigen Deckenplatte.

Abb. 5 Modell einer Baumstütze aus Stahlrohr. Die über eine Fläche verteilten Kräfte werden gebündelt und in drei Generationen auf vier Stützen übertragen.

Abb. 6 Minimalkonfiguration in der Ebene zwischen drei Punkten mit dreiarmigem Knoten. Alle Arme liegen in einer Ebene.

Abb. 7 Räumlicher Minimalweg zwischen vier Punkten. Jeweils ein dreiarmiger Knoten liegt in einer Ebene.

Fig. 1 and 2 Unlike soap films, highly viscous liquids allow branched threads to be drawn. Depending on the number of points of adhesion multi-armed nodal figures are produced with a tendency to form three-armed liquid nodes.

Fig. 3 Three-dimensional node model having five arms.

Fig. 4 Tensioned suspended model of a tree structure consisting of chains and springs for determining the optimum branching angles and rod lengths. The supports branch in three "generations" (at a ratio of 1 to 4) towards the ceiling.

Fig. 5 Model of a tree structure made of steel tube. The forces distributed over a surface are focussed and transferred to four supports in three generations.

Fig. 6 Minimal configuration between three points having three-armed nodes. All arms share the same plane.

Fig. 7 Spatial minimal way between four points. Each plane contains a three-armed node.

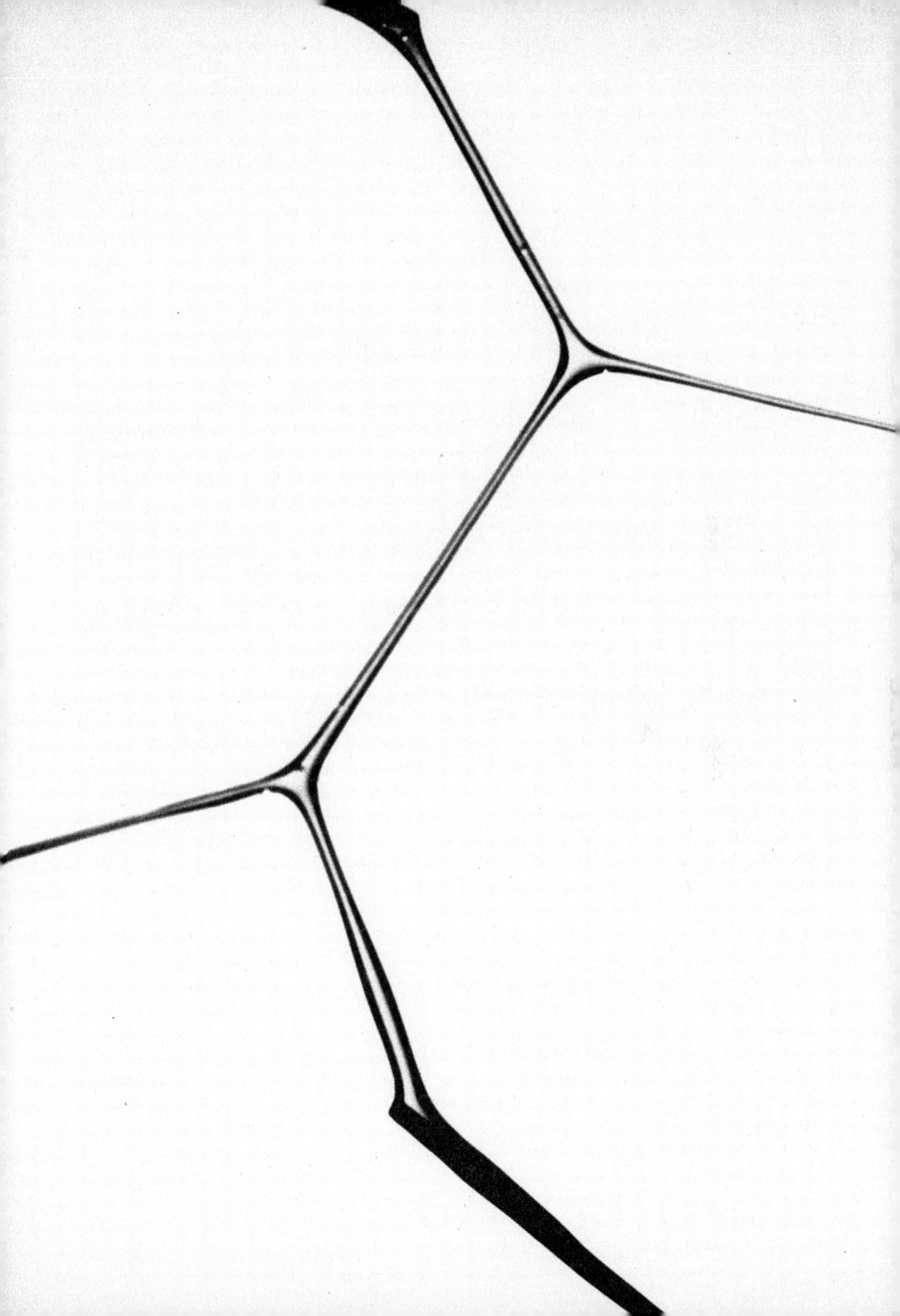

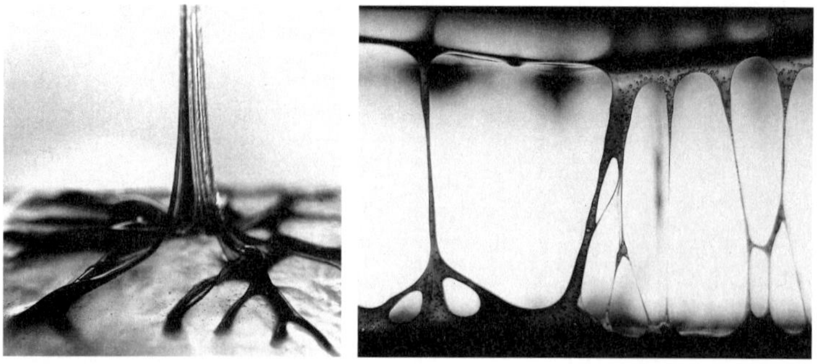

1 / 2

3

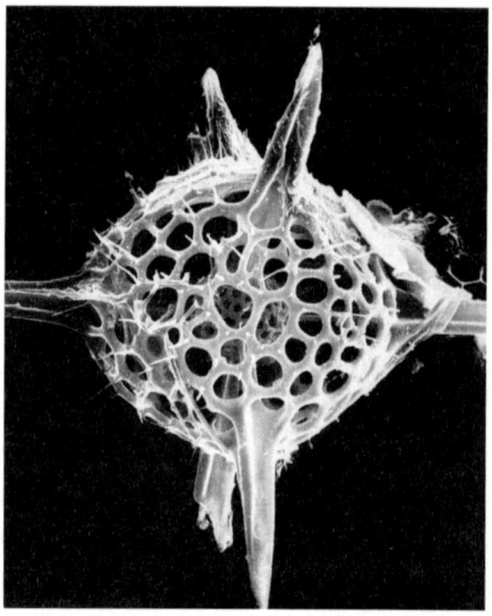

4

Zähflüssige Substanzen wie Honig, Kleber oder Kieselsäure bei Diatomeen und Radiolarien, (s. Lit.: IL18) können offene oder geschlossene räumliche Faden- und Verzweigungssysteme mit den bei Seifenlamellen bekannten Gesetzmäßigkeiten bilden. Nicht exakte 120° Winkel sind bei ihnen mit Inkonsistenz der Gemenge oder vorzeitiger Erstarrung zu erklären. Diese Systeme entstehen nicht nach rein geometrischen Bildungsgesetzen, sondern sind das Ergebnis von Randpunkten und Kraftverteilungen in den linearen Elementen. Alle sind während ihrer Formbildung nur zugbeansprucht.

Abb.1, 2, 5 Stachel-, Faden- und Verzweigungssysteme hochviskoser Substanzen.

Abb.3 Zusammenklebende Fäden bei Spinnweben.

Abb.4 In der Analogieforschung kann bei ähnlichen Formen auch auf ähnliche Bildungsprozesse geschlossen werden. Formelemente bei Radiolarienskeletten zeigen Entsprechungen zu 1 und 2.

Viscous substances such as adhesives, honey, or silica in diatoms and radiolaria, (cf. bibl. IL18), can form open and closed spatial thread and branch systems based on the principles known from soap lamellae. The fact that the 120° angles are not attained precisely may be explained by inconsistencies in the mix of these substances or premature solidification. These systems are not created on the basis of purely geometrical laws of formation but are the results of boundary points and the distribution of forces in the linear elements. During the shaping process all these systems are exclusively subjected to tensile stress.

Fig.1, 2, 5 Spike, thread and branching systems of viscous substances.

Fig.3 Viscous spiders web threads adhering to each other.

Fig.4 In analogous research similar forms suggest similar formation processes. Form elements of radiolaria skeletons have their formal counterparts in Fig.1 and 2.

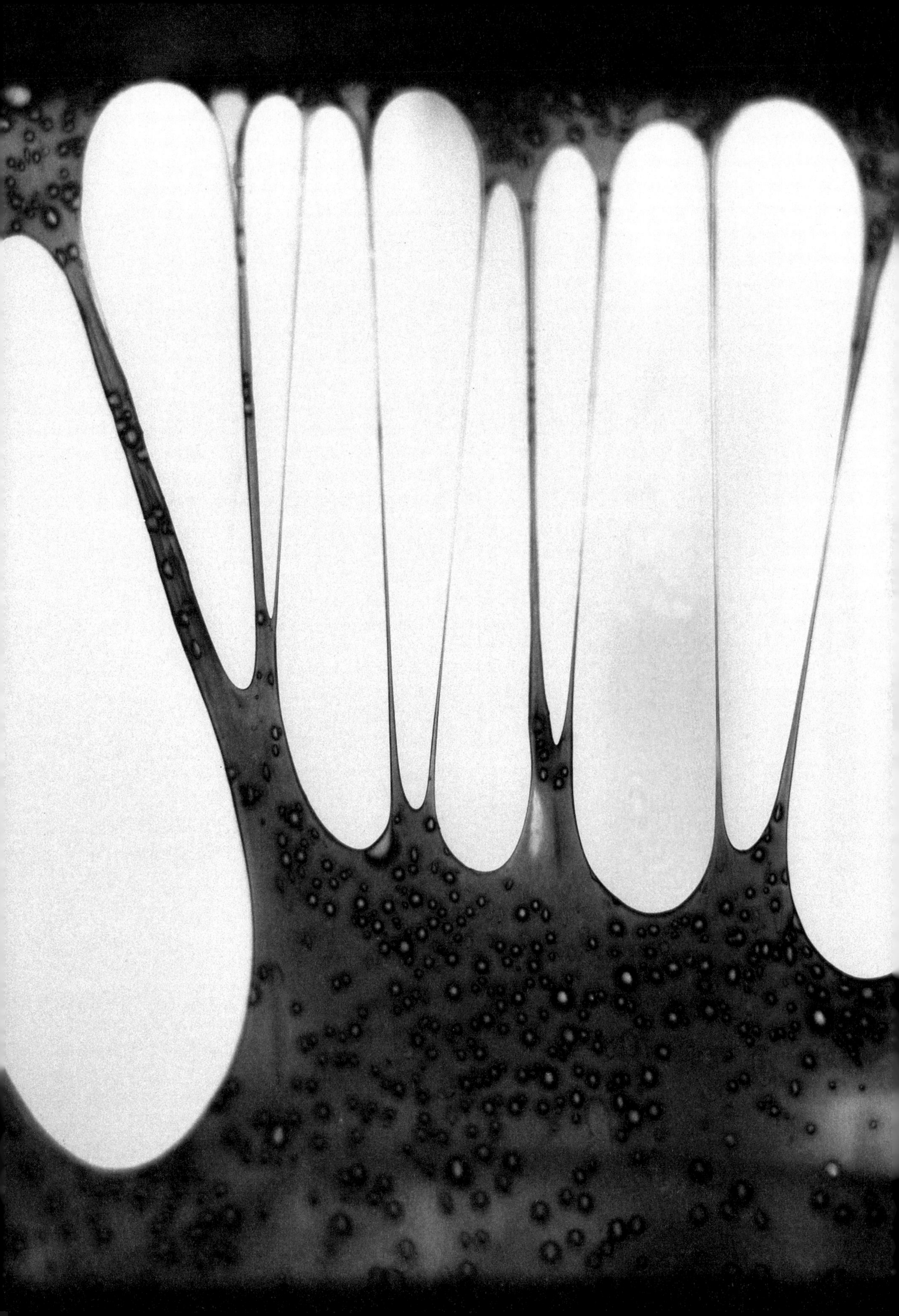

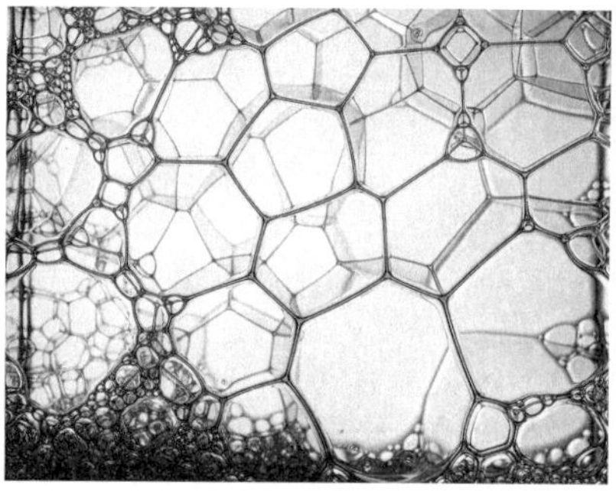
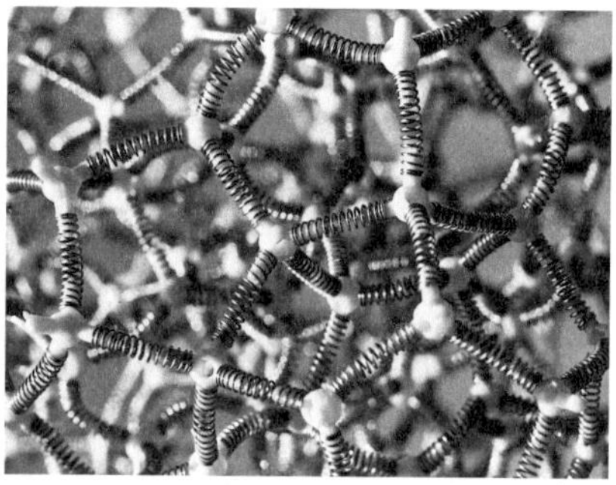

1 / 2

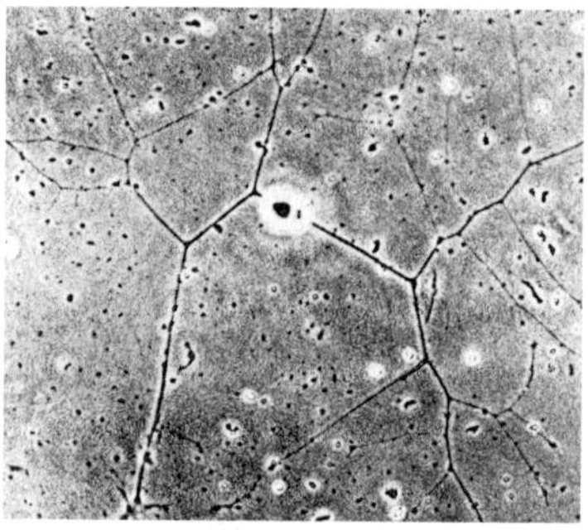
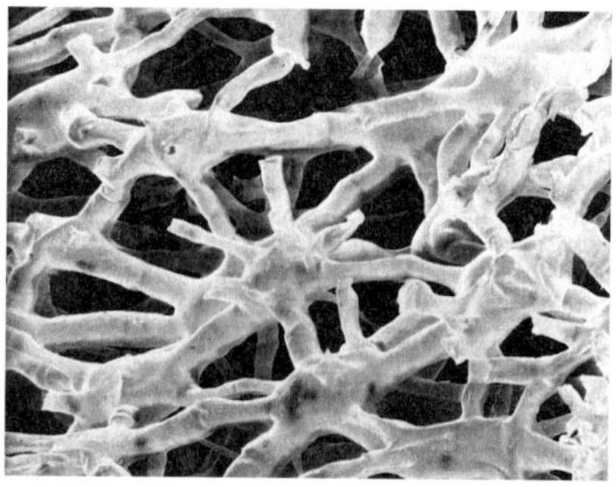

3 / 4

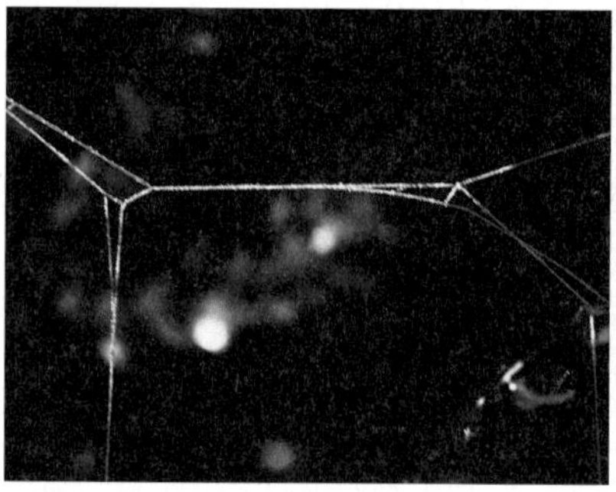

5 / 6

Abb.1 Flüssige Kanten einer Schaumpackung mit unterschiedlichen Polyederformen und -größen.

Abb.2 Federmodell mit vierarmigen 109° Knoten. Schon nach wenigen Packungsgenerationen entstehen Zwängungen, die durch Federbiegung aufgenommen werden.

Abb.3 Offenes, räumliches Algennetz.

Abb.4 Stängelgewebe von Juncus mit vernetzten, mehrarmigen Knoten.

Abb.5 und 6 Dreidimensionale Netzdetails bei Abspannungen von Spinnennetzen.

Fig.1 Liquid edge of a foam cluster containing various forms and sizes of polyhedrons.

Fig.2 Spring model with four armed 109° nodes. After only a few cluster generations constrictions are produced which are compensated by the curving of the springs.

Fig.3 Open spatial network of algae.

Fig.4 Stem tissue of Juncus having cross-linked multi-armed nodes.

Fig.5 and 6 Three-dimensional network details of spiders webs.

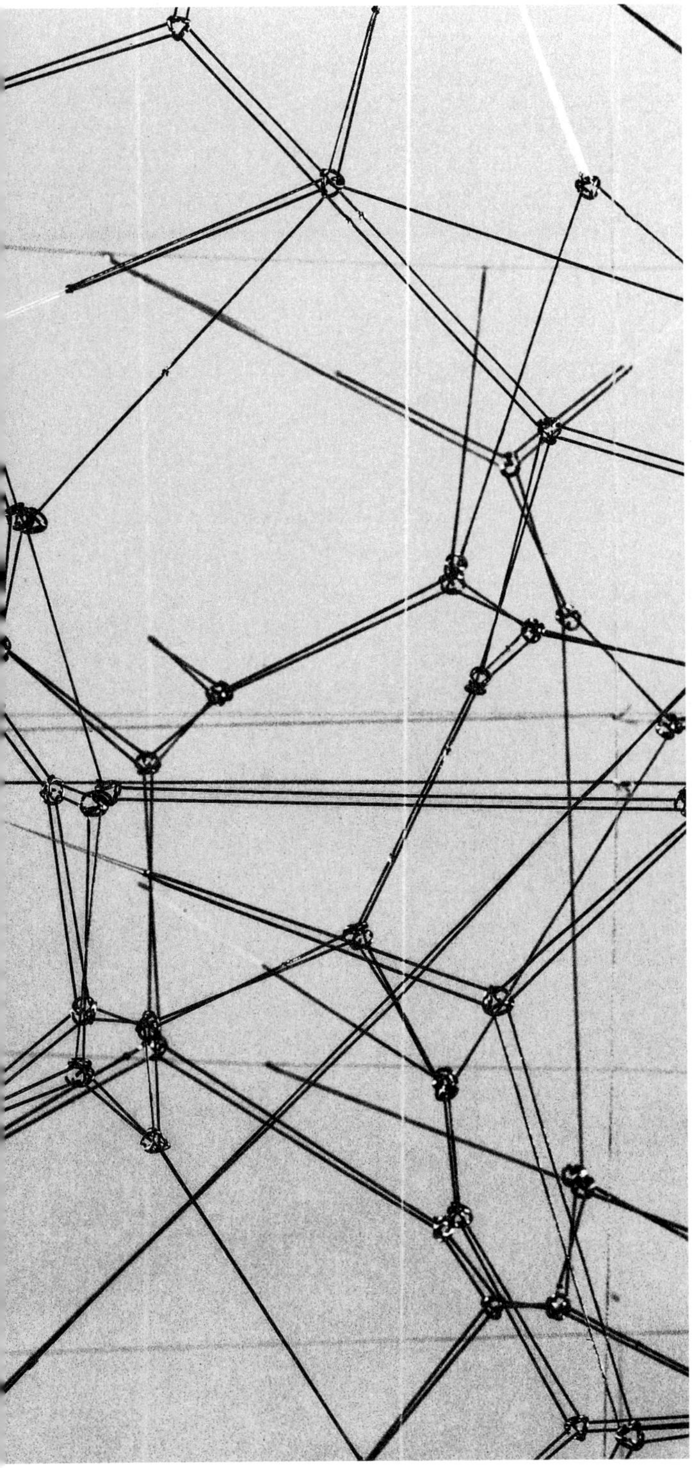

7

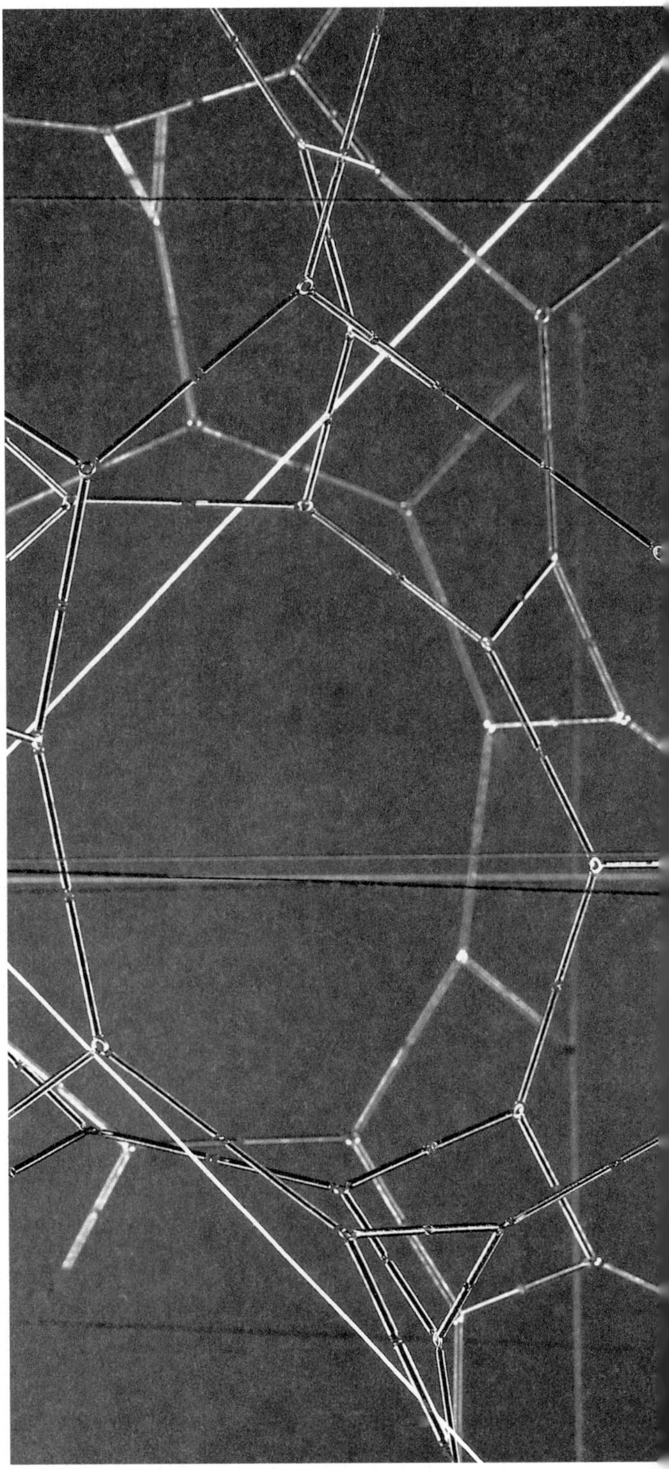

8

Abb.7 Offenes räumliches Netz mit verschiebbaren Knoten. Entsprechende Bauprinzipien finden sich auch bei natürlichen Konstruktionen oder bei Tierbauten.

Abb.8 Geschlossenes Netz mit verdrehbaren Knoten. Hier können Winkelfehler ausgeglichen werden.

Fig.7 Closed network having nodes which allow angular errors to be compensated.

Fig.8 Open spatial network having movable nodes. Corresponding principles of construction can also be found in natural structures or in animal buildings.

Einzelne Minimalflächen in einem Rahmen mit biegesteifem Rand
In jedem beliebig geformten Rahmen, der aus ebenen, geknickten oder räumlich gekrümmten biegesteifen Teilen besteht und einen geschlossenen Linienzug bildet, lässt sich *eine* Seifenhaut spannen, die stets eine zusammenhängende Fläche bildet. In ebenen Rahmen ist auch die Seifenmembran eben, wenn sie nicht von äußeren Kräften oder dem Eigengewicht verformt wird. | Der belgische Physiker *Joseph A. F. Plateau* (1801 – 1883) machte Seifenblasen zum Objekt experimenteller Beobachtungen und naturwissenschaftlicher Studien. Seifenblasen wurden zum Ausdrucksmittel physikalischer Gesetzmäßigkeiten. *Plateau* tauchte unterschiedlich geformte Drahtbügel in Seifenlauge und konnte für jeden gegebenen Rand eine Minimalfläche finden, da der Seifenfilm aufgrund seiner Oberflächenspannung stets eine kleinstmögliche Oberfläche zwischen vorgegebenen Rändern einnimmt. | Zu den den Mathematikern schon bekannten Minimalflächen des *Katenoids* und des *Helikoids* fand Plateau so eine Fülle weiterer sattelförmiger, also gegensinnig (antiklastisch) gekrümmter Minimalflächen, die allesamt die Bedingung verschwindender mittlerer Krümmung erfüllen, d.h. an jedem Punkt der Fläche ist die Summe aus positiver und negativer Krümmung gleich Null (s. Lit.: Hildebrandt, Tromba). | Eigentlich sind diese Minimalflächen nur »Stücke von Minimalflächen« und Mathematiker fragen nach jenen »vollständigen« Flächen, welche sich nach allen Seiten ins Unendliche fortsetzen lassen, so dass sie keine Ränder mehr haben (s. Lit.: Ferus). Bei komplexeren Rahmenformen ist manchmal die Eintauchrichtung entscheidend für die entstehende Form, d.h. bei gleicher Rahmenform können jeweils andere Minimalflächen entstehen.

A single minimal surface within a frame having a rigid boundary

In a frame of any shape which consists of flat, kinked or spatially curved rigid components and forms a contiguous set of lines, a *single* soap film forming a continuous surface may be tensioned. In a flat frame the soap film is flat as well provided it is not deformed by external forces or its own weight. | The Belgian physicist *Joseph A. F. Plateau* (1801–1883) made soap bubbles the subject of experimental observations and scientific study. Soap bubbles became the means of expressing physical laws. *Plateau* immersed wire loops of various shapes in a soap solution and was able to find a minimal surface for each given boundary because the soap film will always form a minimal surface due to its surface tension. | In addition to the minimal surfaces of the *catenoid* and the *helicoid*, which had been already known to the mathematicians. *Plateau* found a large number of additional saddle-shaped minimal surfaces, i.e. anticlastically curved minimal surfaces which all fulfil the condition of a disappearing mean curvature, which means that in each point of the area the sum of positive and negative curvatures is zero (cf. bibliography: Hildebrandt, Tromba). | In reality such minimal surfaces are only "parts of minimal surfaces" and the mathematicians are trying to solve the problem of those "complete" surfaces which can be continued on all sides ad infinitum so that they no longer have boundaries (cf. bibliography: Ferus). With more complex frame forms it is sometimes the direction of immersion which is decisive for the form created, i.e. the same shape of frame may produce various minimal surfaces.

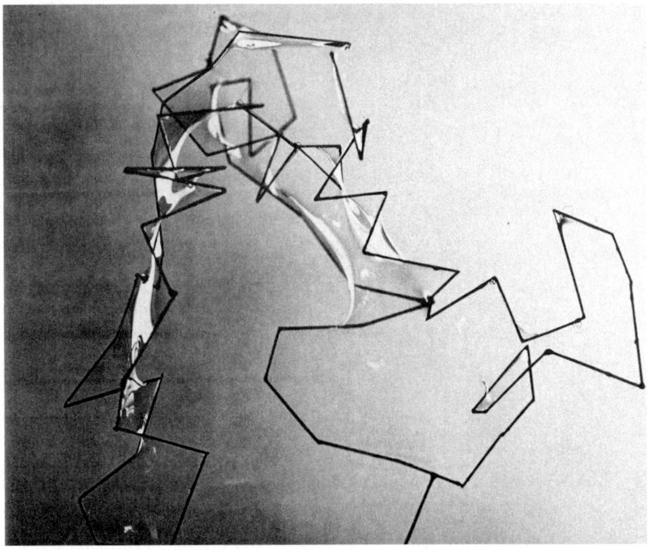

Abb.1 Minimalfläche in einem geschlossenen geknickten Rahmen beliebiger Form.

Fig.1 Minimal surface in a closed and kinked frame of random shape.

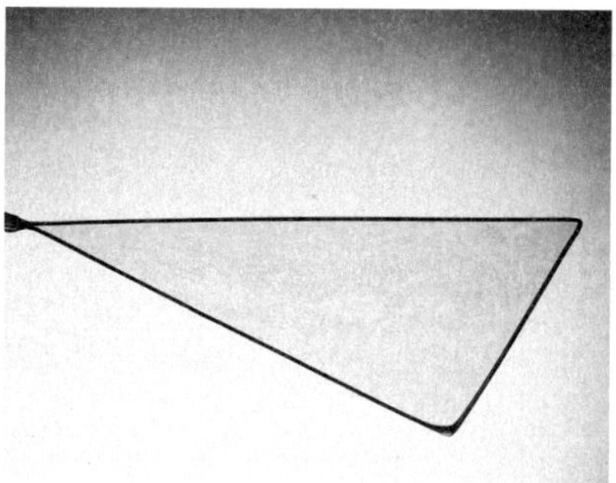

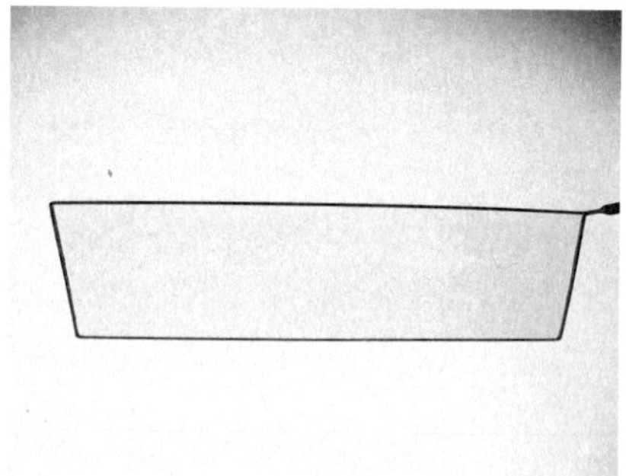

1 / 2

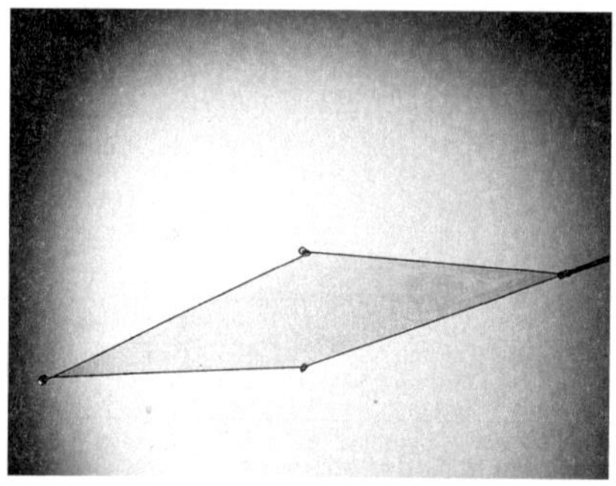

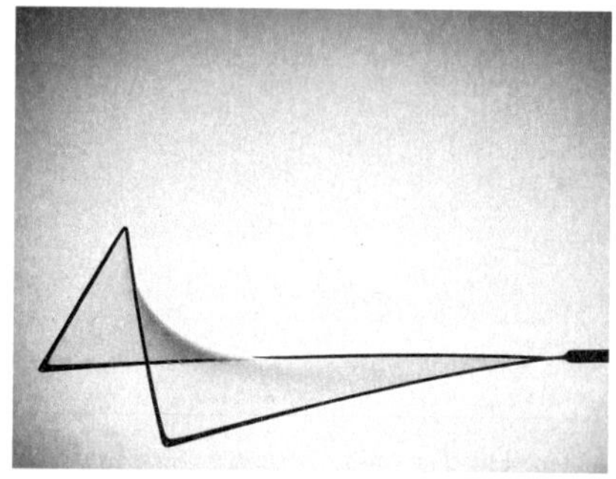

3 / 4

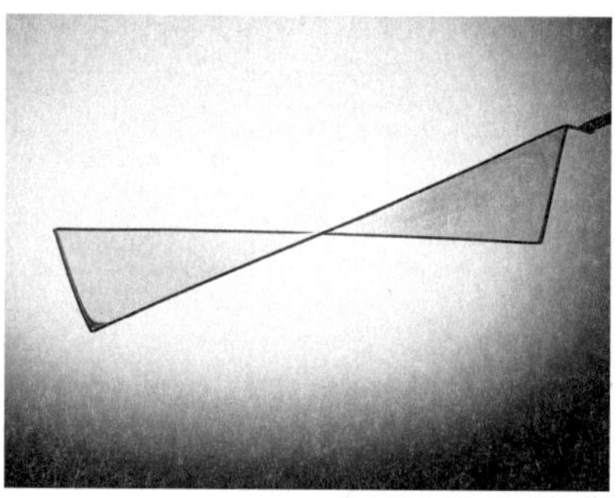

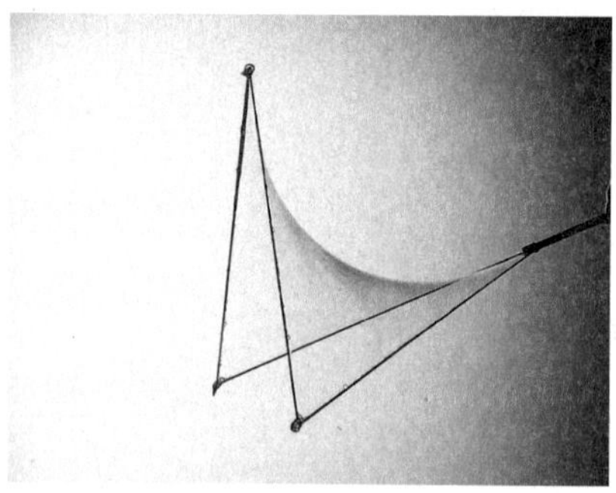

5 / 6

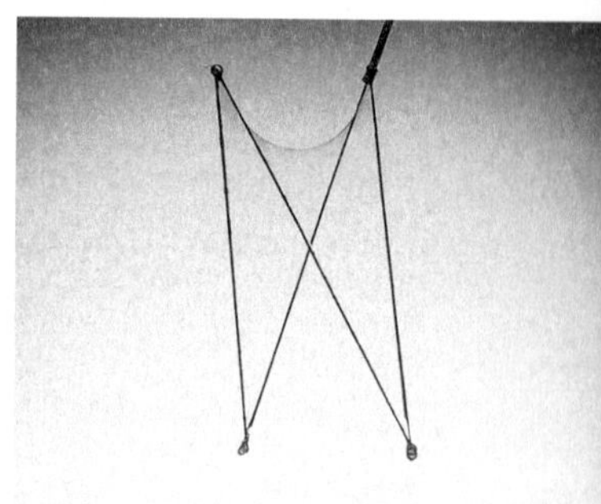

7

Abb.1 In ebenen Rahmen bilden sich, unabhängig von der Eckenzahl, auf Grund der Membranspannung der Seifenlamelle, stets ebene Minimalflächen.

Abb.2 und 4 Wird eine Rahmenseite geknickt oder der ganze Rahmen räumlich verwunden, entstehen doppelt gekrümmte sattelförmige oder antiklastische Minimalflachen.

Abb.5 - 7 In gelenkigen Rahmen kann die Membranspannung der Lamelle größer als die Steifigkeit sein.

Fig.1 Due to the surface tension of the soap lamella a flat frame will always lead to the formation of a flat minimal surface irrespective of the number of frame corners.

Fig.2 and 4 If one side of the frame is kinked or the whole frame spatially curved, doubly curved saddle-shaped or anticlastic minimal surfaces are produced.

Fig.5 to 7 In a jointed frame the tension of the lamella can be larger than the rigidity of the frame.

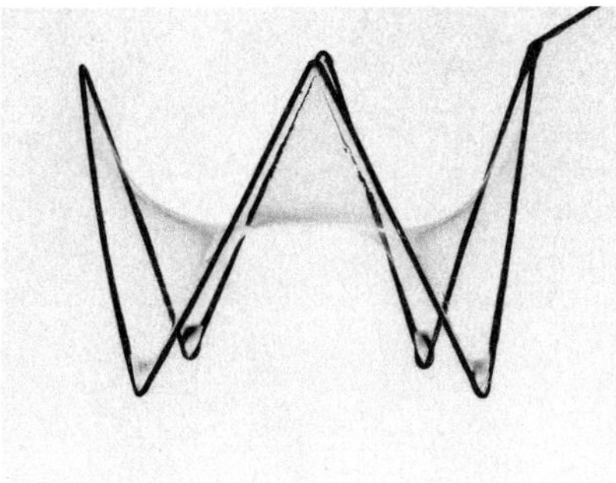

8 / 9

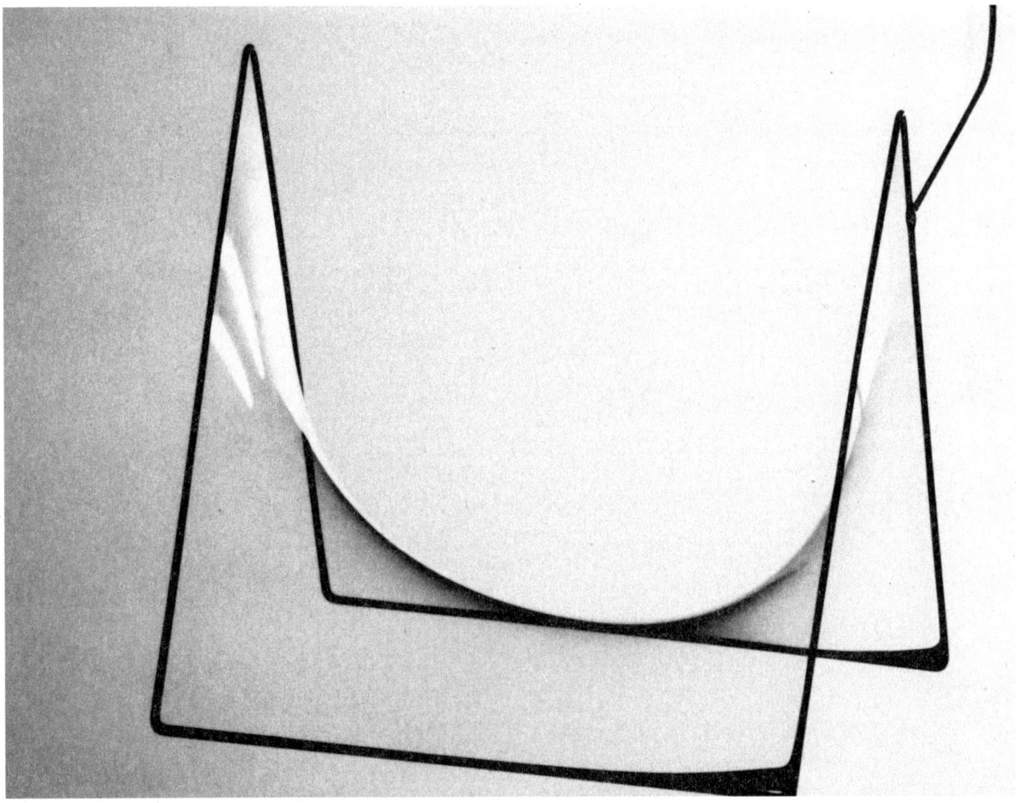

10

Abb.8 Minimalflache in geknicktem Rahmen mit vier Hoch- und vier Tiefpunkten.

Abb.9 Membranflächen zwischen biegesteifen, geknickten Rahmen. Ontario-Pavillon auf der Weltausstellung Expo '67 in Montreal.

Abb.10 Typische Minimalfläche in Sattelform in einem doppelt geknickten Rahmen.

Fig.8 Minimal surfaces in a kinked frame having four high and four low points.

Fig.9 Film surfaces between rigid, kinked frames. Ontario pavilion at the World Exhibition Expo '67 in Montreal.

Fig.10 Typical minimal surface having a saddle shape in a double kinked frame.

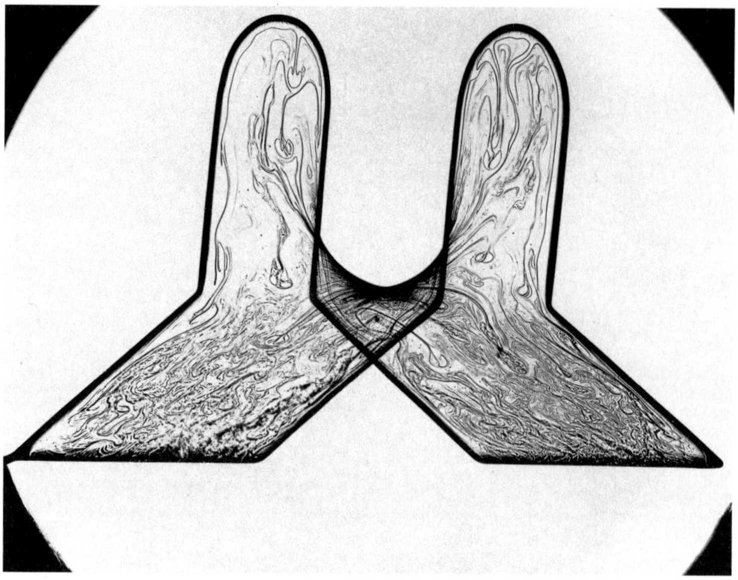

1

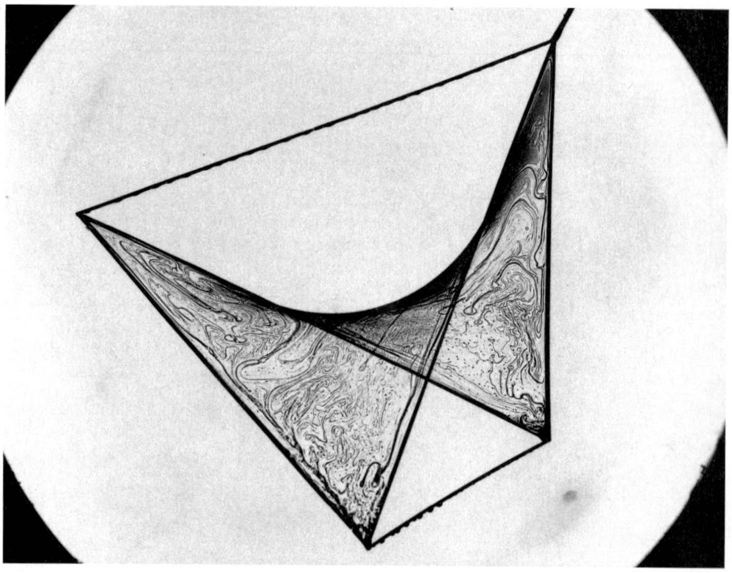

2

Drei Beispiele für sattelförmige Einzelflächen in räumlich verformten Rahmen, aufgenommen in Photogrammtechnik im parallelen Licht.

Abb. 1 Im Rahmen aus geknickten und gebogenen Stäben beeinflussen die seitlichen, senkrechten Lamellenteile die Gesamtverformung der Sattelfläche.

Abb. 2 Komplexer Tetraeder-Rahmen mit sechs Stäben, bei dem nur vier benachbarte und geknickt zueinander stehende Stäbe zum Einhängen einer Minimalfläche benutzt wurden. Die Sattelfläche ist formidentisch mit Abb. 6, S. 248. Bei Benutzung aller Randstäbe entstünde eine komplexe Sechs-Lamellen-Figur.

Abb. 3 Minimalfläche mit Hauptausdehnung in Richtung der Raumdiagonalen in einem vom Würfel abgeleiteten Rahmenmodell aus sechs benachbarten und geknickt zueinander stehenden Stäben

Three examples of saddle-shaped individual surfaces in a spatially distorted frame; photograms taken in parallel light.

Fig. 1 In a frame consisting of kinked and curved sides the lateral, vertical parts of the lamella affect the overall deformation of the saddle surface

Fig. 2 Complex tetrahedron frame consisting of six elements in which only four adjacent and angled elements are used for supporting a minimal surface. The saddle surface is identical in shape with Figure 6, p. 248. If all edge rods were used, a complex figure consisting of six lamellae would be created.

Fig. 3 Minimal surface having its main extension in the direction of the space diagonal in a frame model derived from a cube, consisting of six adjacent and angled edges

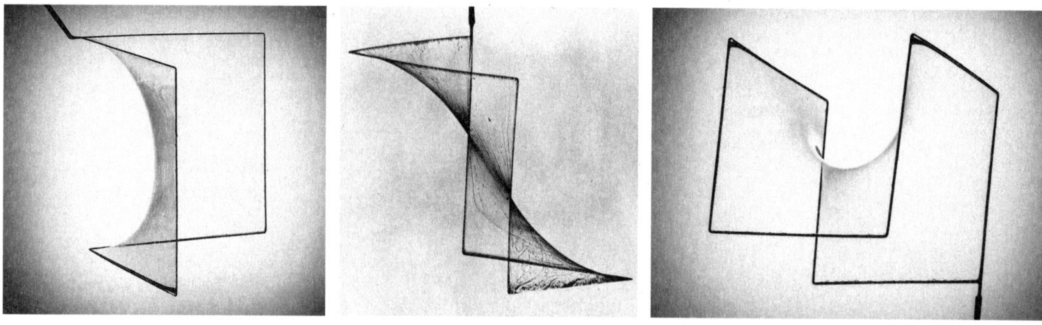

1 - 3

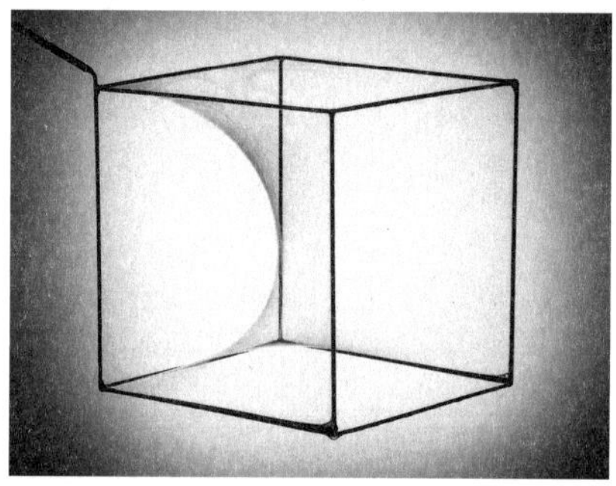

4

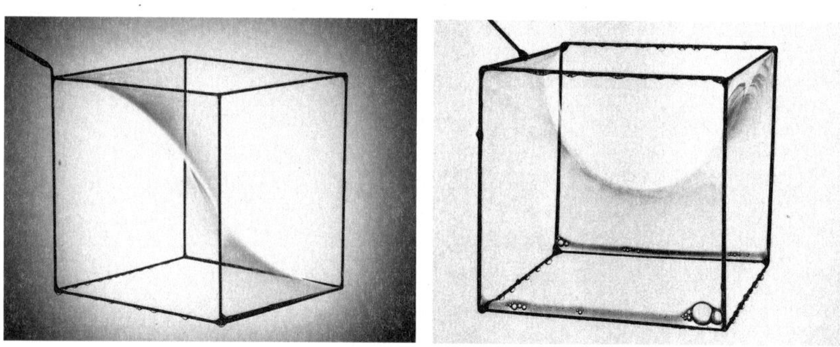

5 / 6

In komplexen Rahmen wie denen von Tetraedern (Abb.2, S.250) oder Kuben lassen sich einzelne Minimalflächen bilden, wenn vorhandene Stäbe nicht benutzt werden. Nach dem Tauchen werden gezielt alle überzähligen Lamellen zerstört. Zum besseren Verständnis sind hier die gleichen Flächenformen jeweils in Teil- und Kubusrahmen gegenübergestellt.

Abb.1 und 4 zeigen den einfachen Sattel zwischen sechs Stäben.

Abb.2 und 5 Minimalfläche in Richtung der Raumdiagonalen des Kubus zwischen sechs Stäben.

Abb.3 und 7 Sattelfläche im Doppel-U-Rahmen aus acht Stäben, fotografiert in Durchlicht- und Photogrammtechnik

Abb.6 Sattelfläche zwischen acht Stäben im Kubusrahmen

In complex frames such as tetrahedrons (Fig.2, p.250) or cubes, individual minimal surfaces may be formed if existing frame edges are not used. After immersion all superfluous lamellae are destroyed. For better clarification the same surface shapes are presented side by side in a partial and a cubic frame.

Fig.1 and 4 show a simple saddle between six rods.

Fig.2 and 5 Minimal surface in the direction of the space diagonal of the cube between six edges.

Fig.3 and 7 Saddle surface in a double "U" frame consisting of eight edges, photographed using the backlighting and photogram techniques

Fig.6 Saddle surface between eight edges in a cubic frame

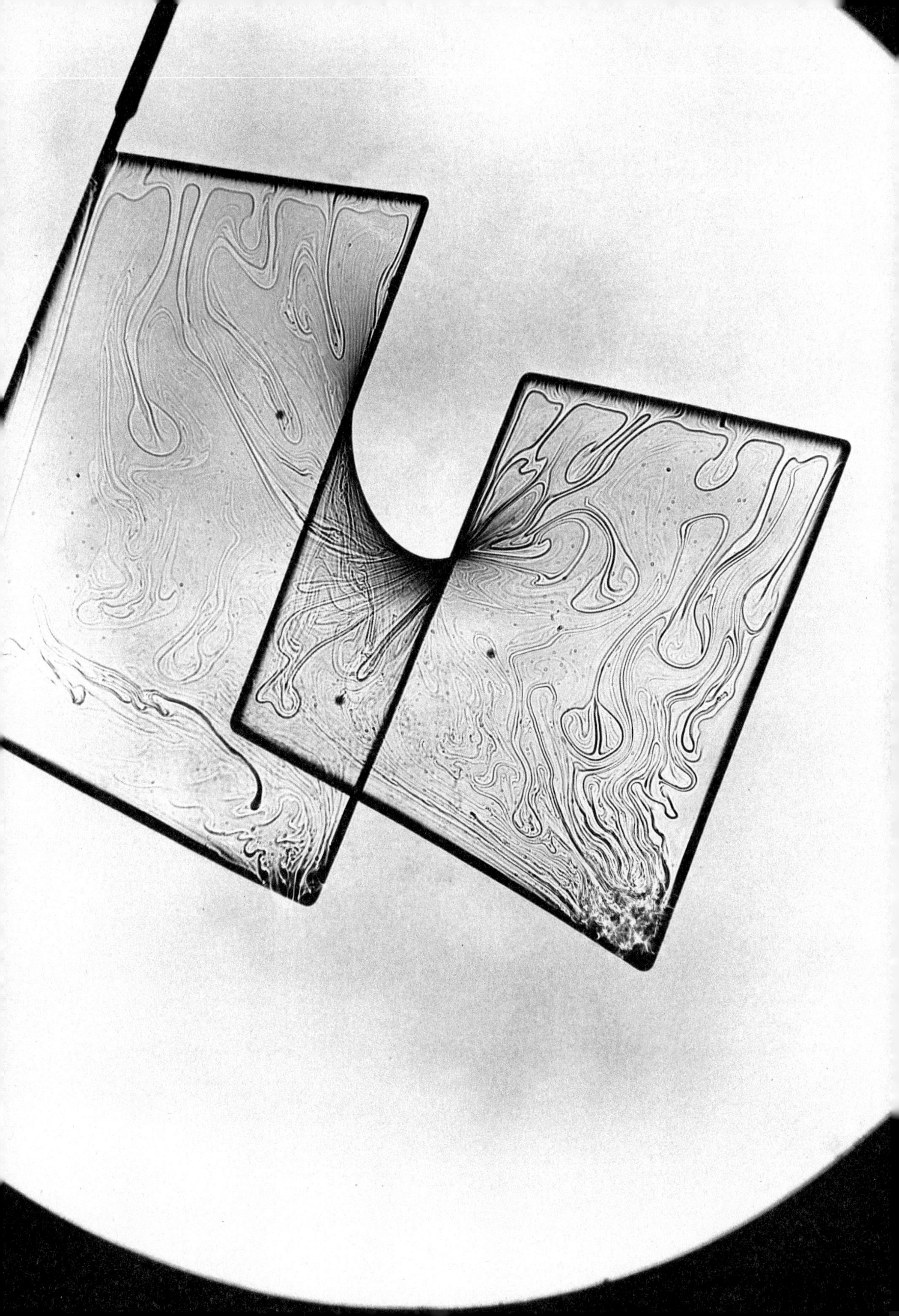

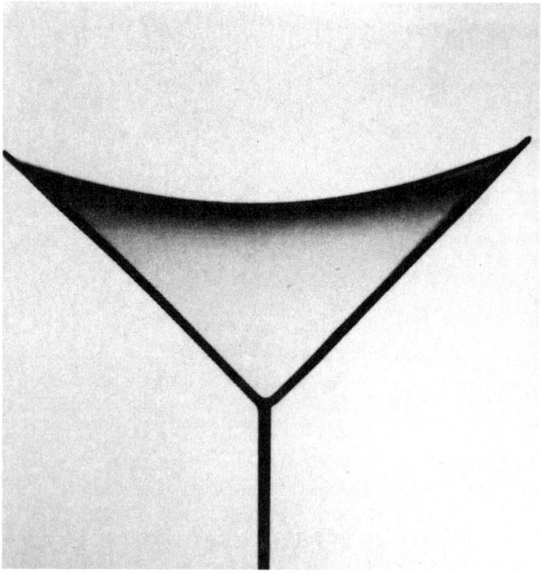
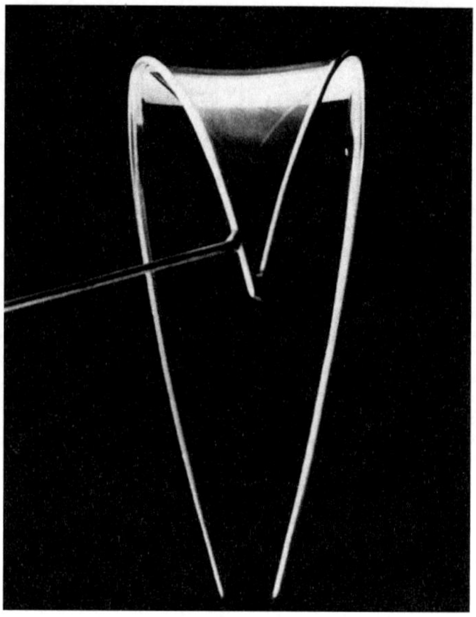

1 / 2

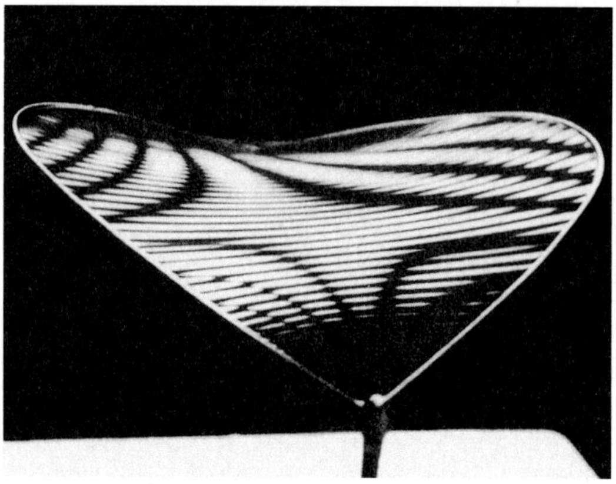

3

Abb.1 und 2 Zwei gleichgroße halbkreisförmige Rahmen im rechten und im spitzen Winkel miteinander verbunden

Abb.3 Seifenhaut-Formfindungsmodell mit aufgespiegelten Rasterlinien

Abb.4 Kirche Bremen-Grolland, ein Seilnetzbau, bei dem Dach- und Wandseilnetze von zwei halbelliptischen Bögen unterstützt sind

Fig.1 and 2 Two semicircular frames of equal size linked at right and acute angles

Fig.3 Soap film form-finding model with projected grid lines

Fig.4 The church in Bremen-Grolland; a cable net structure in which the roof and wail cable nets are supported by two semi-elliptical arches

4

5

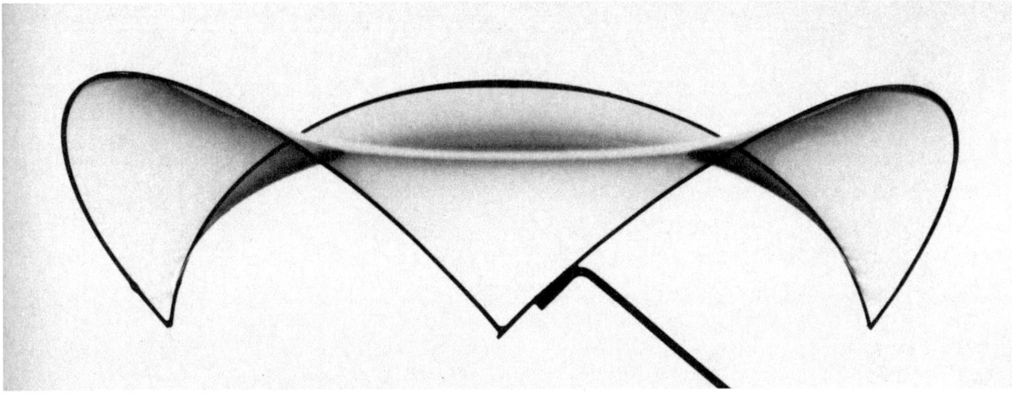

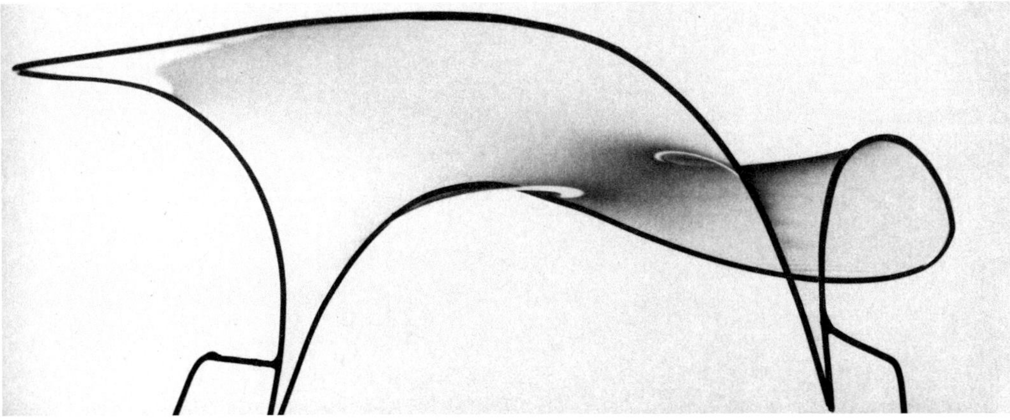

6 / 7

Abb.5 Sonnenschutzdächer aus biegesteifen, gebogenen Rahmen und eingehängten Membranen

Abb.6 und 7 Minimalflächen in Rahmen mit gebogenen Rändern. Bei Abb.6 sind die Bögen im Winkel von 90°, bei Abb.7 im spitzen Winkel zusammengesetzt.

Fig.5 Sun shades made from rigid bent frames enclosing a membrane

Fig.6 and 7 Minimal surfaces in a frame with curved edges. In Figure 6 the arches form angles of 90°, whereas in Figure 7 they form acute angles.

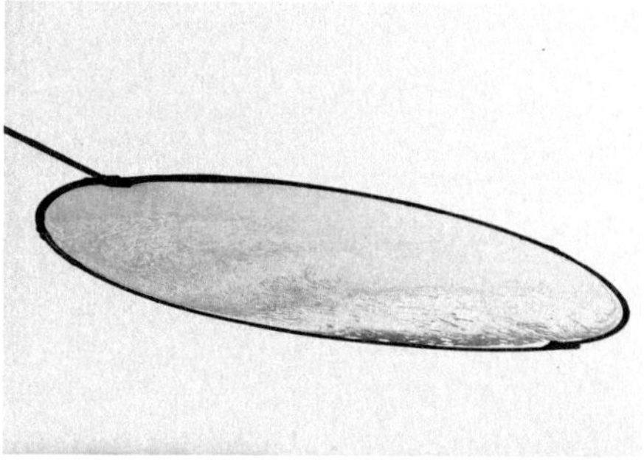

1

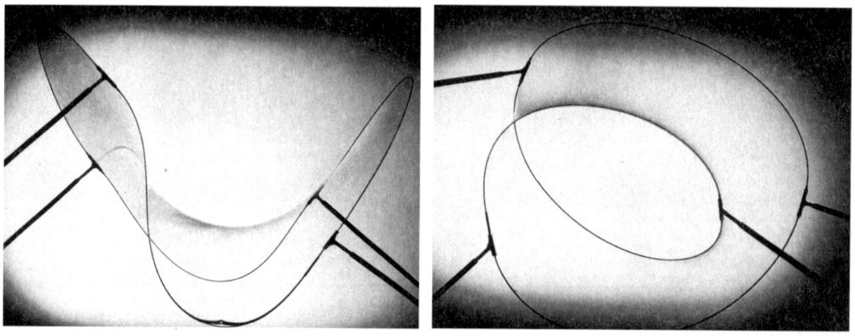

2 / 3

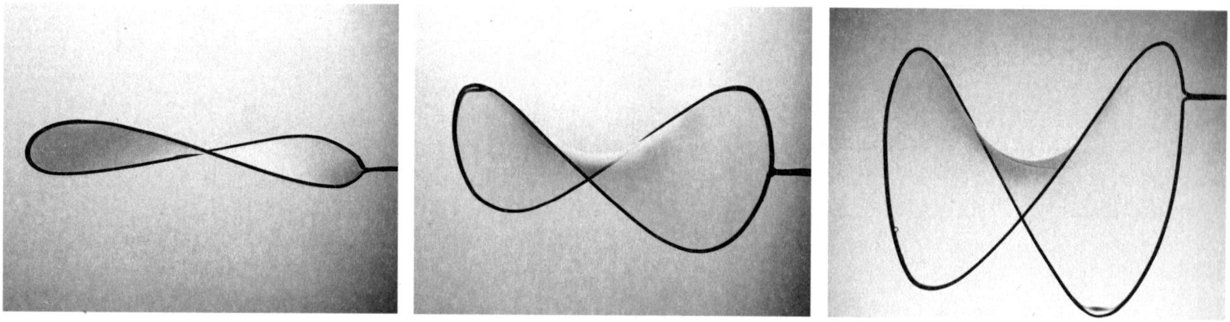

4 - 6

Abb. 1 Biegesteifer kreisrunder Ringrahmen mit ebener Lamelle.

Abb. 2 und 3 Ein Ringrahmen aus sehr dünnem, federndem Stahldraht ist manuell so verformbar, dass zunächst eine Sattelfläche, dann ein Möbiusring einstellbar ist.

Abb. 4 - 6 Drei räumlich gebogene Ringrahmen mit zunehmender Sinuskrümmung

Abb. 7 In diesem Fall bildet sich in einem nicht geschlossenen Rahmen dennoch eine Seifenhaut. Es entstehen zwei gegeneinander geneigte Schraubenflächen, die zwischen den zwei Durchstichpunkten des Rahmens eine bogenförmige flüssige Kante bilden.

Fig. 1 Rigid circular ring frame producing a flat lamella.

Fig. 2 and 3 This ring frame consisting of very thin spring steel wire can be deformed by hand in such a way that it first produces a saddle surface and then a Möbius ring.

Fig. 4 - 6 Three spatially curved ring frames with increasing sinusoidal curvature

Fig. 7 In this case a soap film is created in spite of the frame not being closed. Two helicoids are produced which are inclined towards each other forming a bow shaped liquid edge between the two points of penetration of the frame.

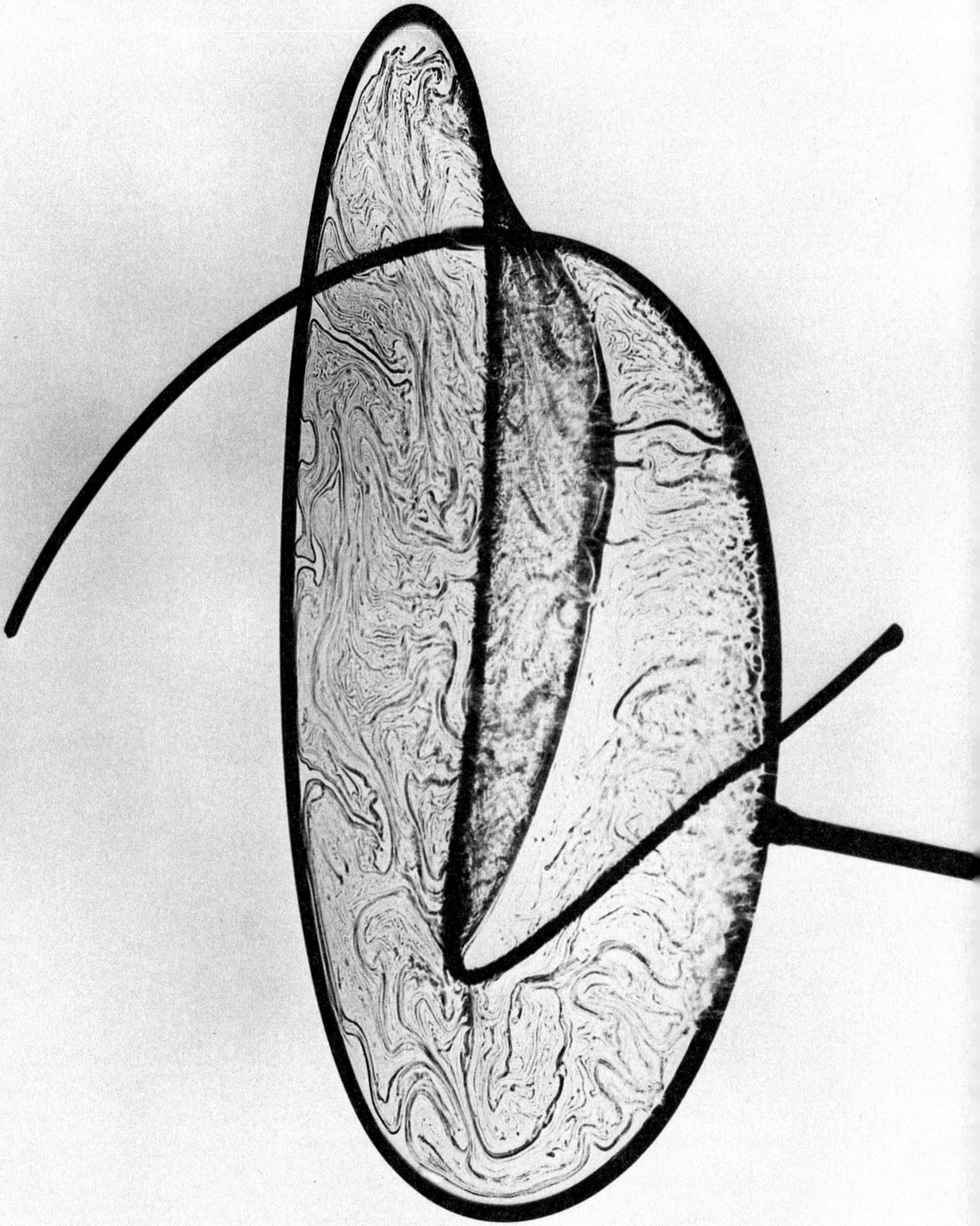

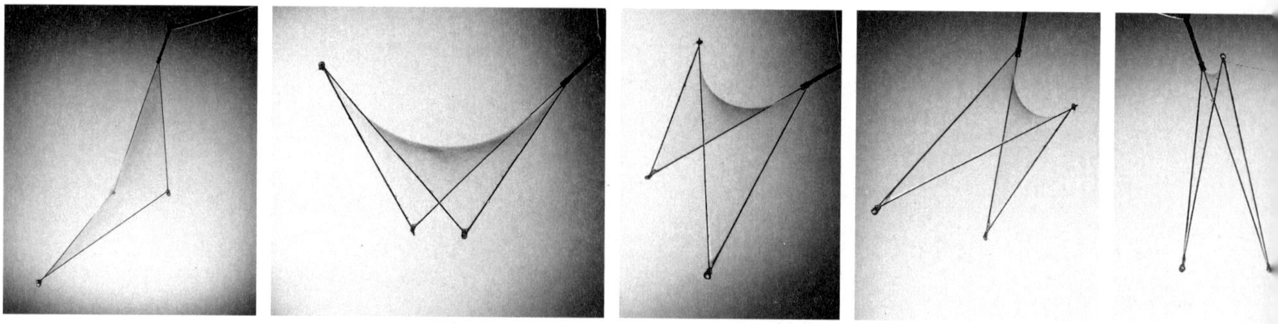

1-5

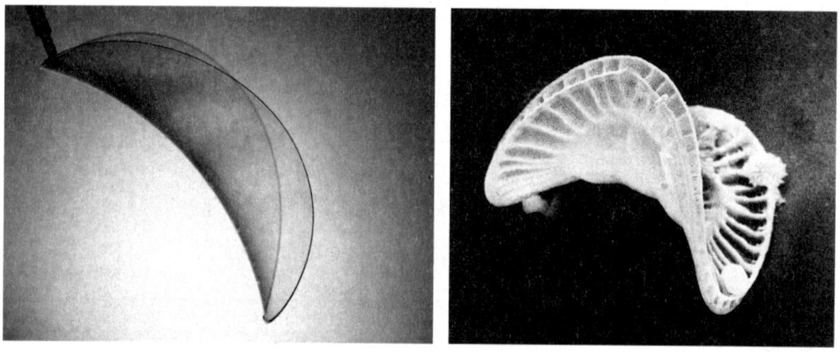

6 / 7

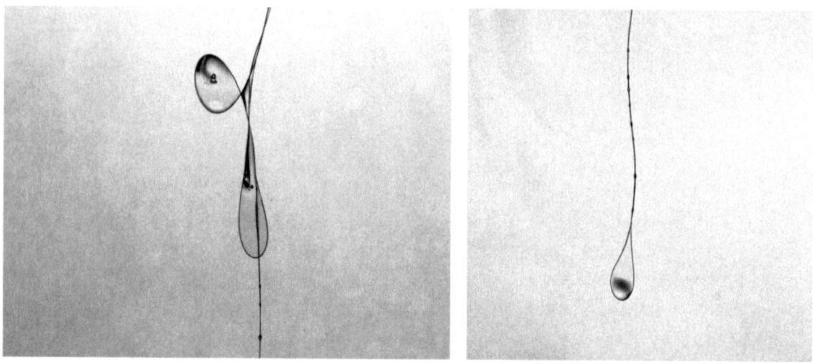

8 / 9

Abb.1-5 Sechs Phasen des Zusammenfaltens eines rhombusförmigen Rahmens mit Eckgelenken unter Einfluss der Lamellenspannung

Abb.6 Ein ebener Kreisringrahmen verformt sich unter Einfluss der Lamellenspannung räumlich. Aus der Ebene wird eine sattelförmige Minimalfläche.

Abb.7 Bei der sattelförmigen Schale der Diatomee Campylodiscus clypeus könnte im weichen Zustand ein zu Abb.6 analoger Umformungsprozess aus der Ebene stattgefunden haben.

Abb.8 und 9 Benetzte biegeunsteife Fadenschlaufen legen sich bis auf kleinste Restschlaufen aneinander, die Aufschluss geben über das Verhältnis von Oberflächenspannung zur Steifigkeit.

Fig.1-5 Six stages in the folding of a rhombic frame with corner joints subject to the lamella tension

Fig.6 A flat circular ring frame is deformed spatially when subjected to the lamella tension. The plane becomes a saddle-shaped minimal surface.

Fig.7 The saddle-shaped shell of the diatom Campylodiscus clypeus could, when still soft, have been transformed from a plane by a process analogous to Fig.6.

Fig.8 and 9 Wetted non-rigid loops will be attracted to each other right down to the smallest "residual loops", giving an insight into the relationship between surface tension and rigidity.

10

Abb.10 Mantelfläche einer Spirale mit gebogener Achse ähnlich einem Verdeck

Fig.10 The envelope surface generated by a helix having a curved axis is similar to the folding soft top of a car.

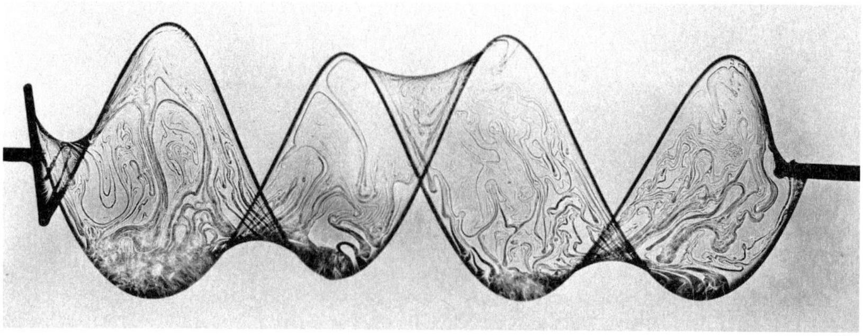

1

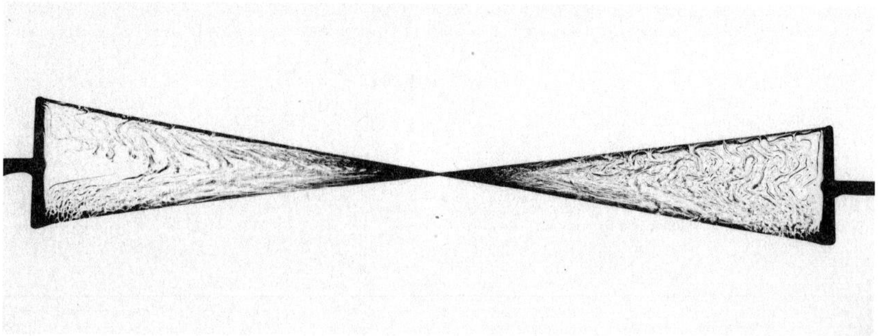

2

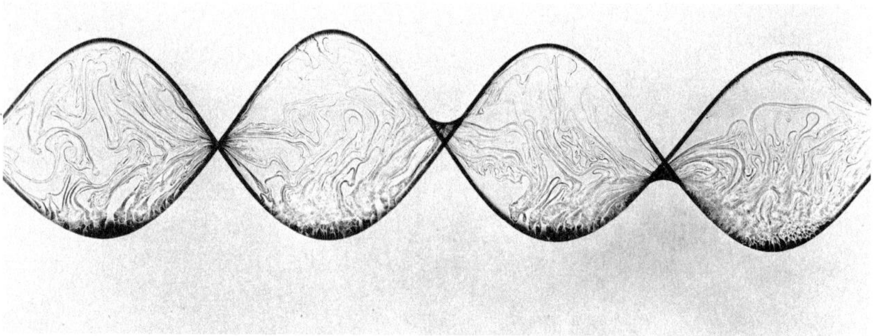

3

Schraubenflächen (Helikoidflächen)
Schraubenflächen können als eine Fläche innerhalb eines geschlossenen Randes angesehen werden. Je nach Steigung zweier um 180° gegeneinander verdrehter Spiralränder bildet sich eine Lamelle als Mantel- oder als Spiralfläche aus. Schraubenflächen sind doppelt gekrümmte Flächen, die durch Schraubung einer Geraden längs einer Senkrechten entstehen.

Abb. 1 Mantelflache einer Doppelspirale

Abb. 2 Entstehung der Schraubenfläche aus einer windschiefen Fläche mit 180° gegeneinander verdrehten Endpunkten. In der Seitenansicht schneiden sich Lamelle und Ränder genau in einem mittleren Punkt.

Helicoid surfaces
Helicoid surfaces can be regarded as surfaces within a closed boundary. Depending on the pitch of two helical boundaries offset to each other by 180°, the lamella generated is either an envelope or spiral surface. Helicoid surfaces are doubly curved surfaces generated by the spiral movement of a straight line along an axis.

Fig. 1 Envelope of a double helix

Fig. 2 Generation of a helicoid surface from a skewed surface whose end points are offset against each other by 180°. In the lateral view lamella and boundaries intersect exactly in the middle.

4

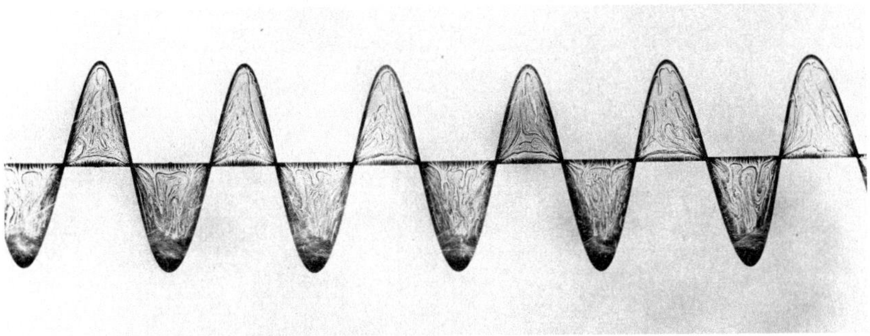

5

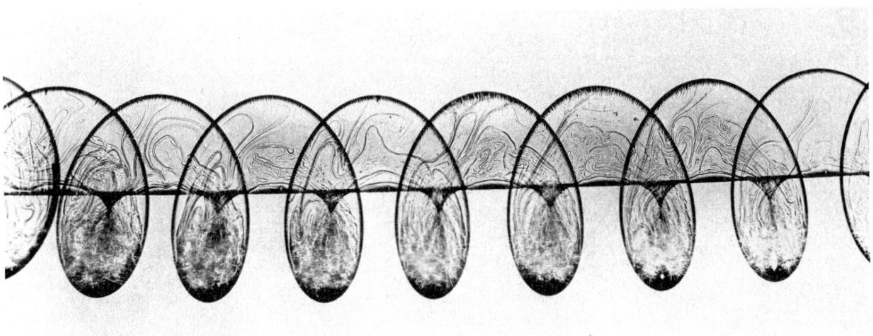

6

Abb.3 und 4 Schraubenfläche in Seiten- und Schrägansicht. Modellimperfektionen bei Abb.3 ergeben nur einen exakten Schnittpunkt.

Abb.5 und 6 Bei Einführung einer biegesteifen Achse entsteht die Helikoidform zwischen dieser und der Randspirale. Seiten- und Schrägansicht.

Fig.3 and 4 Side and oblique views of a helicoid surface. Because of imperfections of the model in 3 only one precise point of intersection results.

Fig.5 and 6 When introducing a rigid axis the helicoid shape is formed between the axis and the boundary helix. Lateral and oblique view

Schraubenflächen

Abb.1 Zwischen Spiralbügel und Acrylrohr verlaufende Lamelle einer Schraubenfläche. Die sich am Rohr einstellende flüssige Kante des »freien« Randes – die Verbindungslinie aller flächenerzeugenden Normalen – schneidet sich in der Seitenansicht exakt mit dem äußeren Bügel.

Helicoid surfaces

Fig.1 A helicoid surface generated between a helical wire and an acrylic tube. The liquid edge of the free boundary forming on the surface of the tube, which is the connecting line of all perpendicular axes generating the surface, intersects in the lateral view exactly with the outer boundary.

Abb.2 Kombination einer biegesteifen äußeren Spirale (druckbeansprucht) und eines biegeunsteifen Fadens (zugbeansprucht). Aufgrund der Membranspannung des Seifenfilms nimmt der Faden ebenfalls eine Spiralform ein.

Fig.2 Combination of a rigid outer helix (subjected to compressive stress) with a non-rigid thread (under tensile stress). Because of the tension of the soap film the thread also adopts a helical shape.

Linear addierte Blasen — Gekammerte Pneus Einzelblasen, die sich frei auf Oberflächen bewegen können, bilden zunächst locker zusammenhängende Blasenflöße. Die einzelnen Blasen bauen sich zu Kammern einer zusammenhängenden Struktur um. Ihre Zwischenwände folgen den Bildungsgesetzen der Minimalwege. | Lineare Blasenadditionen sind nur dann stabil, wenn ihre Reihung durch vorgegebene feste Ränder oder durch beidseitige Anheftungen fixiert wird. Könnern wie dem TV-bekannten Tom Noddy oder Eiffel Plasterer aus den USA gelingt es, in kurzer Zeit Konfigurationen, die aus bis zu 20 linear oder räumlich addierten Seifenblasen bestehen, zu bauen (s. Lit. GEO 5/87). | Die Krümmungen der Zwischenwände geben Aufschluss über die Druckverhältnisse in den addierten Kammern. Bei gleich großen Blasen (gleicher Innendruck) bleiben die Kammerwände eben. Bei ungleichen Blasen (ungleiche Innendrücke) beulen die Wände kleinerer Blasen in Richtung der größeren aus. | Bei linear addierten Blasen treffen an den flüssigen Kanten jeweils drei Seifenlamellen aufeinander. Die Winkel zwischen den Kammerwänden und den Blasen betragen dreimal 120°. | Die Kammerwand steht bei gleichgroßen Blasen als Resultierende genau senkrecht in der Winkelhalbierenden zur Aufstandslinie des Randes. Bei unterschiedlich großen Blasen verläuft sie schräg in Richtung der kleineren Blase.

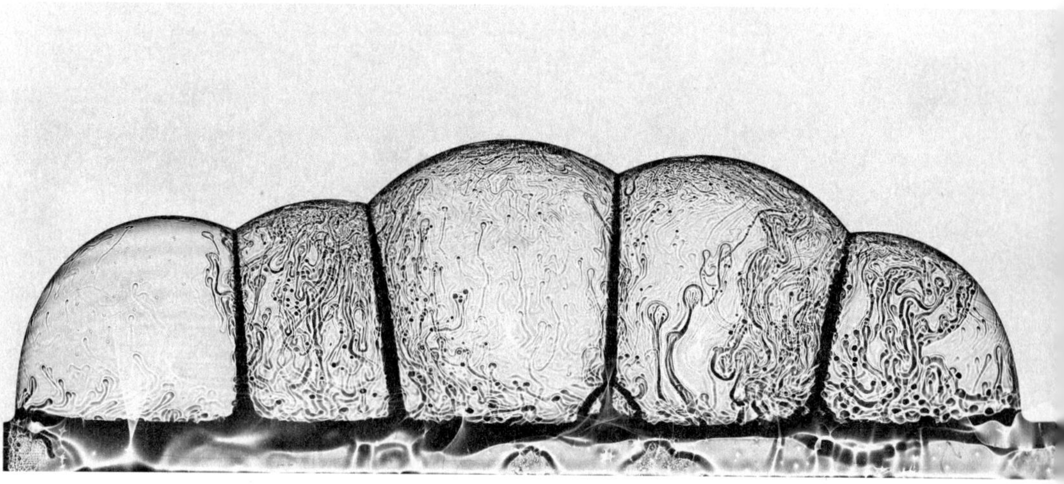

1

Linear addition of bubbles — chambered pneus Individual bubbles which can move freely on surfaces, initially form loose bubble rafts. The individual bubbles become chambers in a coherent structure. The partitions between the chambers are subject to the same laws of formation as minimal ways. | Linear additions of bubbles can only be stable if they are fixed by given solid boundaries or if they are affixed on both sides. Artists appearing on TV such as Tom Noddy or Eiffel Plasterer from the USA are capable of building within a short time configurations consisting of linear or three-dimensional additions of up to 20 soap bubbles (cf. bibliography: GEO 5/87). | The curvatures of the partitions indicate the different pressures in the added chambers. If bubbles of equal size are added, the partition walls will remain flat (equal internal pressure). With bubbles of unequal size the wails of smaller bubbles will bulge towards the larger bubble (unequal internal pressures). In the case of linear bubble additions three soap films meet at the liquid edges. The angles enclosed between the partition wall and the bubble surfaces are three times 120°. | With bubbles of equal size the chamber wall is exactly perpendicular to the line bisecting the angle between the boundary lines. With bubbles of different sizes the partition wall runs obliquely towards the smaller bubble.

2

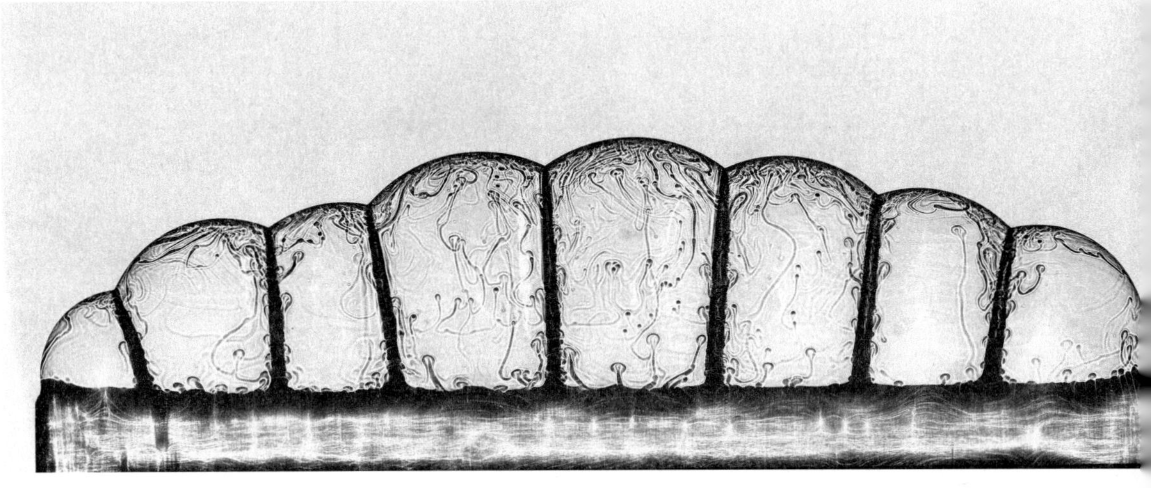

Abb.1 und 2 Lineare Addition von fünf und acht Blasen verschiedener Größen auf festem Rand. Bei gereihten Blasen ungleicher Größe sind die Kammerwände leicht gekrümmt und schräg in Richtung der Resultierenden geneigt. Die Winkel zwischen Wänden und den Blasenflächen betragen jeweils 120°.

Fig.1 and 2 Linear addition of five and eight bubbles of different sizes on a fixed boundary. In adjacent bubbles of unequal size the chamber walls are slightly curved and inclined towards the resultant. The angles enclosed by the partition walls and the bubble surfaces are 120° in each case.

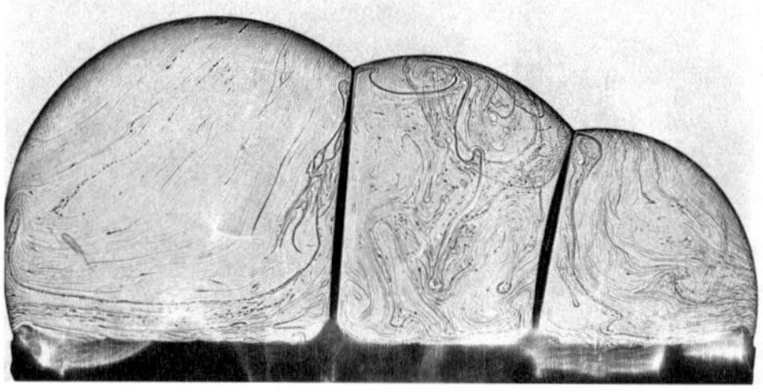

3 / 4

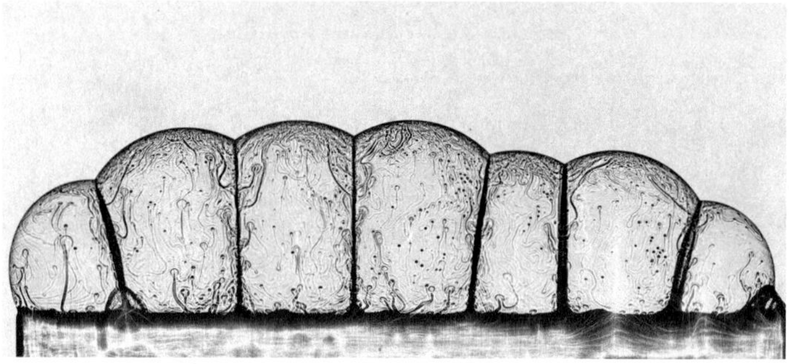

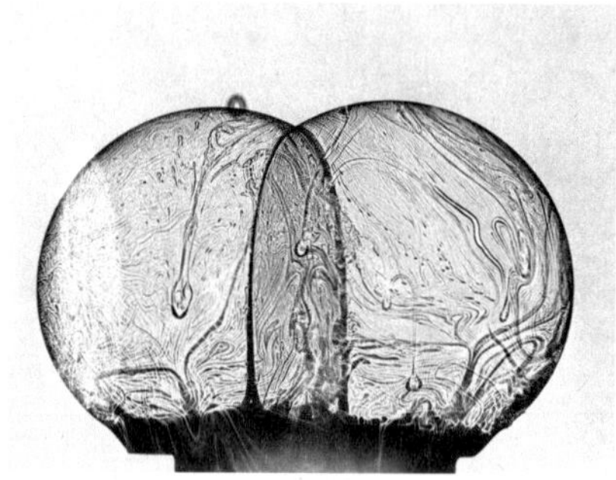

5 / 6

7

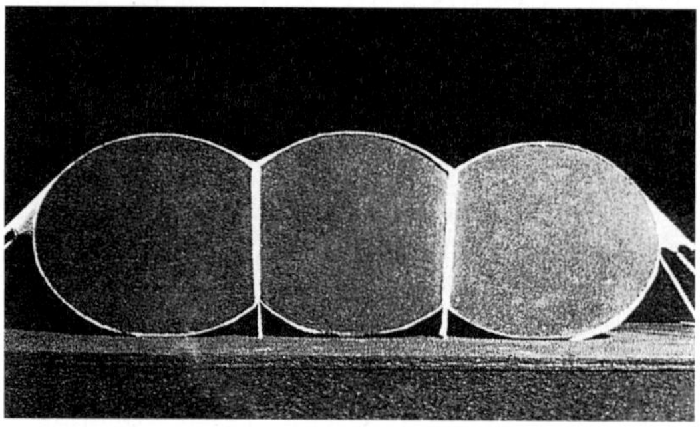

8

Abb.3 und 6 Doppelblase aus gleichgroßen Einzelblasen in Seiten- und Schrägansicht. Die ebene Kammerwand steht senkrecht zur Aufstandslinie.

Abb.4, 5 und 7 Blasenreihen mit drei und sieben Kammern, gesehen in zwei Richtungen. Die Kammerwände verlaufen als Resultierende jeweils schräg zu den kleineren Blasen. Zwischen den Wänden und den Tangenten an die Blasen stellen sich 120° Winkel ein.

Abb.8 Gipsmodell eines Dreikammerschlauches im Querschnitt. Projekt für einen Notauffangbehälter für freiwerdende Gase in chemischen Prozessen. Planung: Atelier Frei Otto, Warmbronn, 1973.

Fig.3 and 6 Twin bubble consisting of two bubbles of equal size shown in lateral and oblique view. The chamber wail is perpendicular to the line of contact.

Fig.4, 5 and 7 Rows of bubbles with three and seven chambers viewed in two directions. The chamber walls are inclined towards the smaller bubbles. Between the walls and the tangents to the bubbles 120° angles are formed.

Fig.8 Cross section of a plaster model of a three-chamber tube. Project for an emergency receiver for gases released in chemical processes. Planning: Atelier Frei Otto, Warmbronn, 1973.

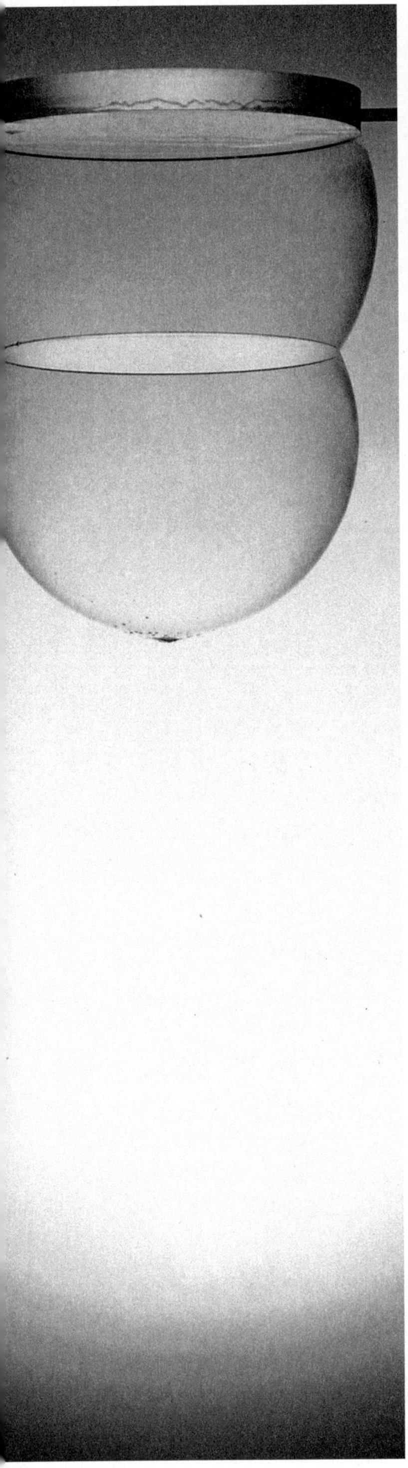

9

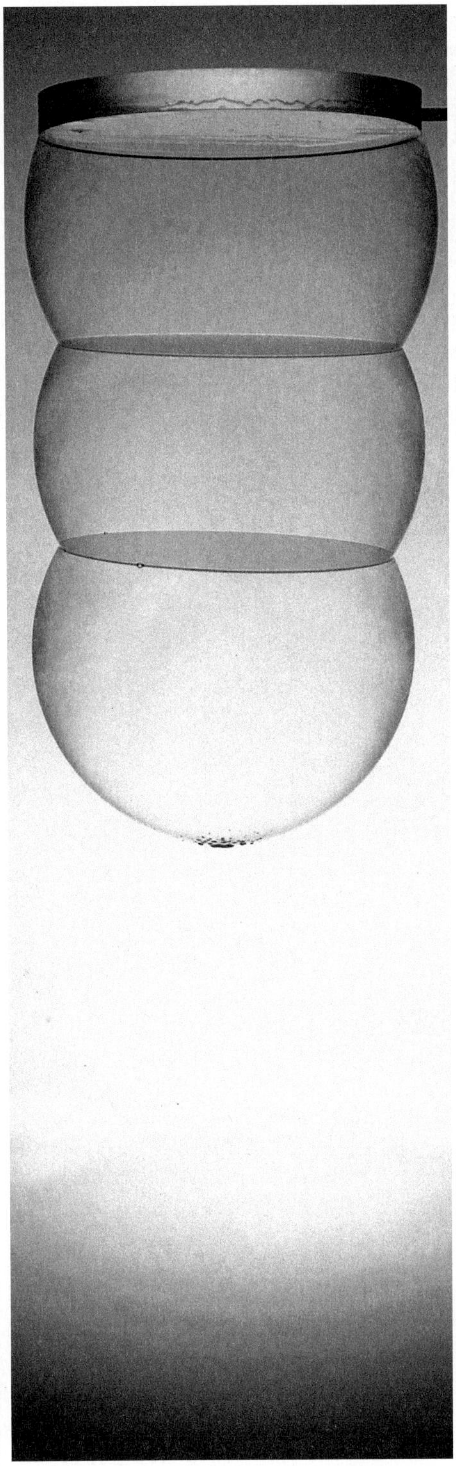

10

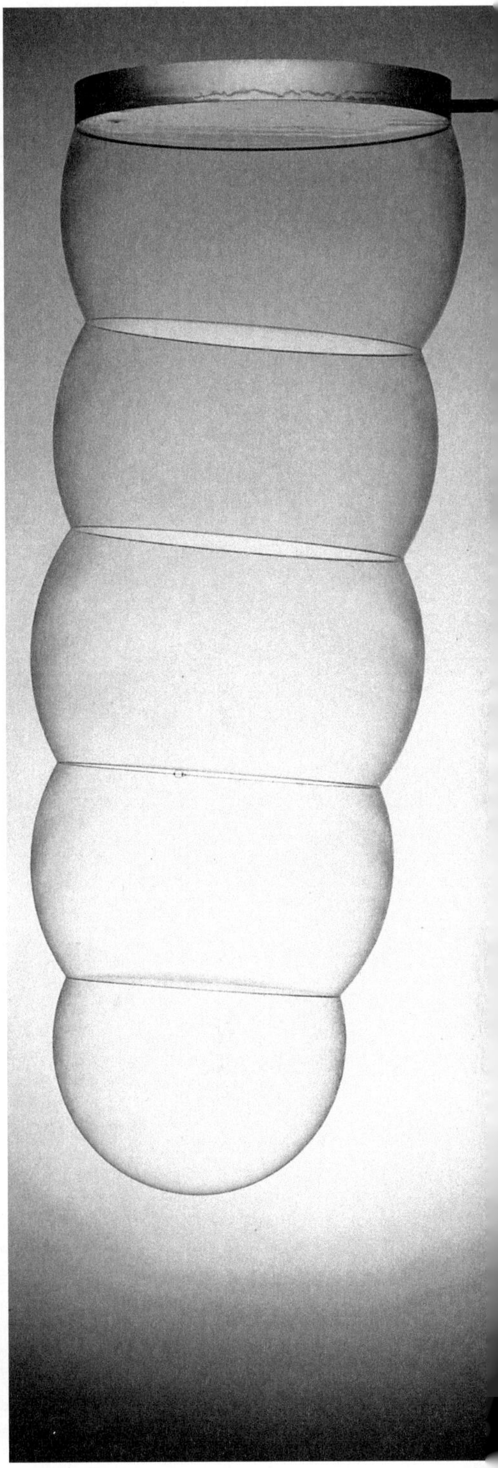

11

Abb.9-11 An ebene Scheibe angehängte Blasenreihen mit zwei (Abb.9), drei (Abb.10) und fünf (Abb.11) Kammern.

Fig.9-11 Rows of bubbles suspended from a flat disc with two (Fig.9), three (Fig.10) and five (Fig.11) chambers.

Dreidimensional addierte Blasen — Pneupackungen Polyederschäume
Frei bewegliche kugelige Blasen ordnen sich, sobald sie sich berühren, selbstbildend in dichten Packungen, so genannten Polyederschäumen, an. Polyederschäume können aus wässrigen Lösungen oder aushärtenden Flüssigkeiten gebildet werden. Polyeder entstehen nur bei allseitiger Packung, die Randkörper stellen Mischformen aus Kugelblasen und Polyedern dar. Schäume bauen sich so um, dass die Gesamtfläche ihrer Blasenlamellen ein energetisches Minimum einnimmt. Die flüssigen Kanten der Blasen können als geschlossenes, räumliches Minimalwegenetz (Abb. 8, S. 245) angesehen werden. Auch hier gilt, dass niemals mehr als drei Lamellen unter 120° Winkeln in einer Kante und maximal vier Kanten und sechs Lamellen in einem räumlichen Knotenpunkt unter Winkeln von 109° 28" 16' zusammentreffen (s. Lit.: Hildebrandt). Im anderen Falle ist die Konfiguration instabil. Sie baut sich sofort zu einer stabilen Regelform um. Bei erhärteten Schäumen, z. B. solchen aus Kunststoffen, herrschen andere Verhältnisse. Bei ihnen können sich instabile Konfigurationen verfestigen. | Polyederschäume sind sehr formstabil, sie können Kräfte übertragen. Je kleiner ihre Blasen werden, desto größer wird die Festigkeit des Schaums (Sahne, Rasierschaum). Geometrisch sind Polyederschäume bislang durch identische Polyeder nicht darstellbar. Mit dem bekannten Regelflächner Pentagondodekaeder ist eine absolut dichte Packung nicht zu erreichen. Schon nach wenigen Packungsschritten gibt es Lücken oder Zwängungen (Abb. 2, S. 244). Eine realistische geometrische Näherung bietet aber ein vom Dodekaeder abgeleiteter unregelmäßiger Polyeder (s. Lit.: Aubert u. a.). | Bei Schäumen ist zu beobachten, dass jede Blase von der Regelform abweicht. Weder sind die flüssigen Kanten exakt gerade, noch die Trennflächen eben. Jede Blase hat eine individuelle Form. Es besteht eine grundsätzlich andere Struktur als in Kristallen.

Three-dimensional added bubbles – pneu clusters polyhydral foams

Freely mobile spherical bubbles automatically form clusters as soon as they touch, forming so-called polyhedral foams. Polyhedral foams may be formed from aqueous solutions or liquids which may be hardened. Polyhedrons are formed only if there is clustering on all sides, the peripheral bodies exhibiting mixed forms consisting of spherical bubbles and polyhedrons. Foams will always re-form themselves until the total surface of their films reaches an energy minimum. The liquid edges of the bubbles can be regarded as a closed three-dimensional minimal way network (Fig.8, p.245). In this case the same law applies, i.e. that no more than three films can meet in a liquid edge and no more than four edges and six films can meet in a nodal point forming angles of 109° 28" 16' (cf. bibl. Hildebrandt). In the other case the configuration will be unstable in the short term. It will, however, immediately re-form itself into a stable shape. This does not apply to hardened foams such as plastic foams. They can form unstable configurations when they harden and set. | Polyhedral foams are stable and can transfer forces. The smaller their bubbles are the greater the rigidity of the foam (cream, shaving foam). Polyhedral foams can so far not be represented geometrically by identical polyhedrons. Using the known regular body pentagondodecahedron, absolutely tight packing cannot be achieved. After only a few clustering stages gaps or squeezings appear (Fig.2, p.244). A realistic geometrical approximation is, however, provided by an irregular polyhedron derived from the dodecahedron (cf. bibliography: Aubert et. al.). | In foams one can observe that each bubble deviates from the regular shape. Neither are the liquid edges precisely straight nor are the separating surfaces flat. Each bubble has its individual form. The structure is fundamentally different from that of crystals.

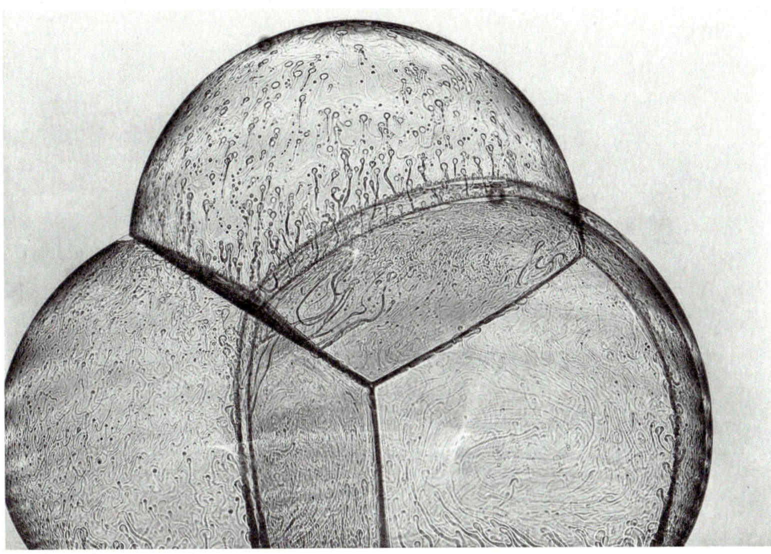

Abb.1 Schaumstruktur dicht gepackter, räumlich addierter Blasen. In der Projektion des Photograms sind drei Kammerwände, die im 120° Winkel zueinanderstehen, gut zu erkennen. Der räumliche Winkel an der Verschneidungszone beträgt dagegen 109°.

Fig.1 Foam structure of densely packed and threedimensionally clustered bubbles. The photogram projection clearly shows three chamber walls meeting at 120° angles. The three-dimensional angle at the intersection zone is 109°.

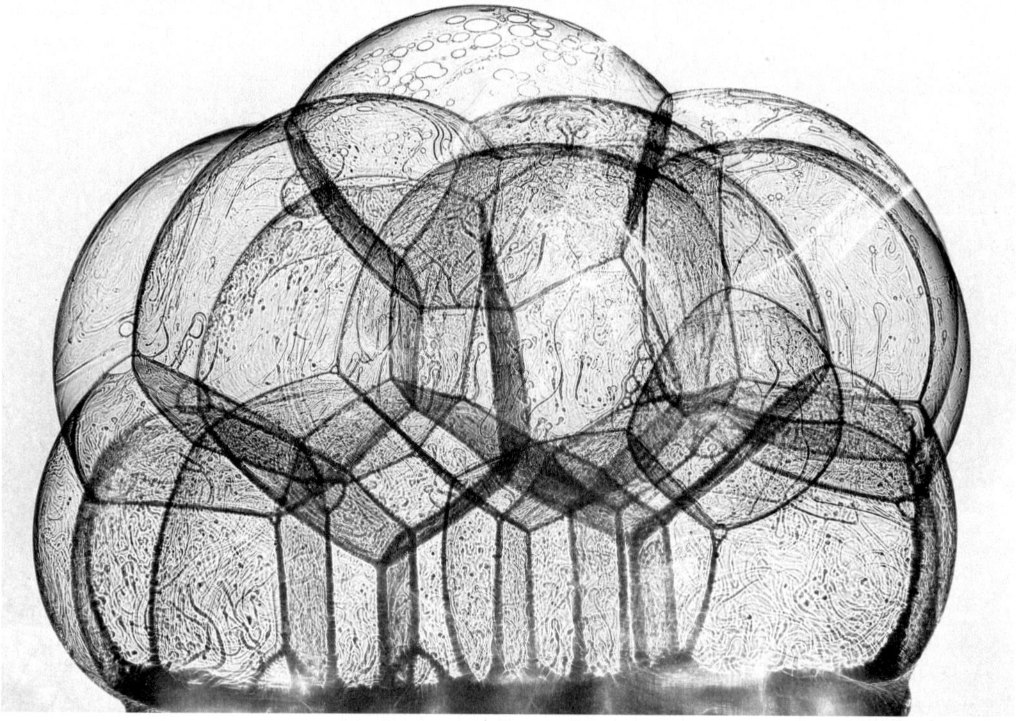

2

3

Abb.2 Im Inneren einer Blasenpackung ist die Umbildung zu Polyedern erkennbar. Die polygonalen Zwischenwände wirken wie zusammenhaltende Verspannungen.

Abb.3 Lufthalle mit kugelig eingeschnürter Oberfläche zum Formvergleich. Die Funktion der Zwischenwände übernehmen zentrale Verspannungen und äußere Einschnürungen. Ausstellungspavillon Expo 64, New York. Arch. und Hersteller: Walter Bird, Birdair Structures, USA (Foto: George Cserna).

Fig.2 The transformation of the bubbles into polyhedrons inside the bubble cluster can be seen. The polygonal partition walls act like braces holding the whole structure together.

Fig.3 Airhall with spherically constricted surface shown for comparison. In this example central braces and external constrictions replace the partition walls. Exposition pavilion Expo 64, New York. Architects and manufacturer: Walter Bird, Birdair Structures, USA (Photo: George Cserna).

4 / 5

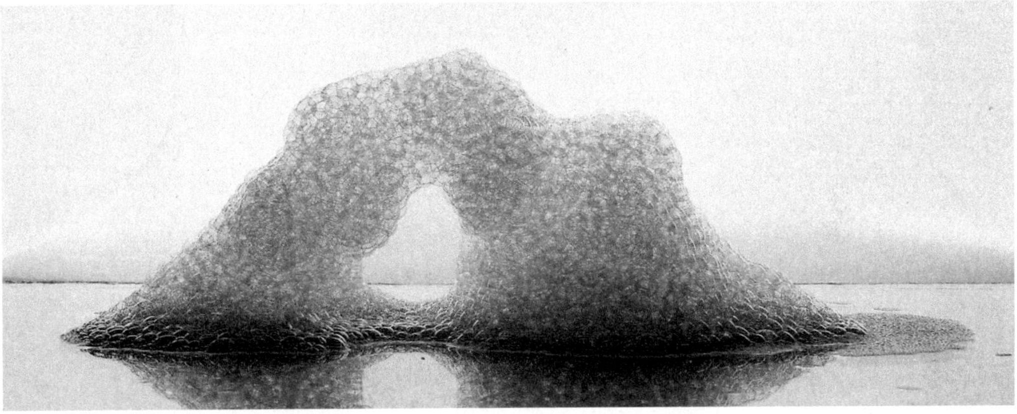

6

Abb. 4 - 6 Schäume bilden relativ beständige, tragfähige Konstruktionen. Flacher (Abb. 4) und steiler Schaumkegel (Abb. 5) und Schaumbogen (Abb. 6) aus gleich groben und selbstbildend zu Polyedern in dichter Packung umgeformten Blasen.

Fig. 4 - 6 Foams form relatively durable loadbearing structures. The photographs show a flat (Fig. 4) and steep (Fig. 5) foam cone and a foam arch (Fig. 6) formed from bubbles of equal size which automatically have transformed themselves into densely clustered polyhedrons.

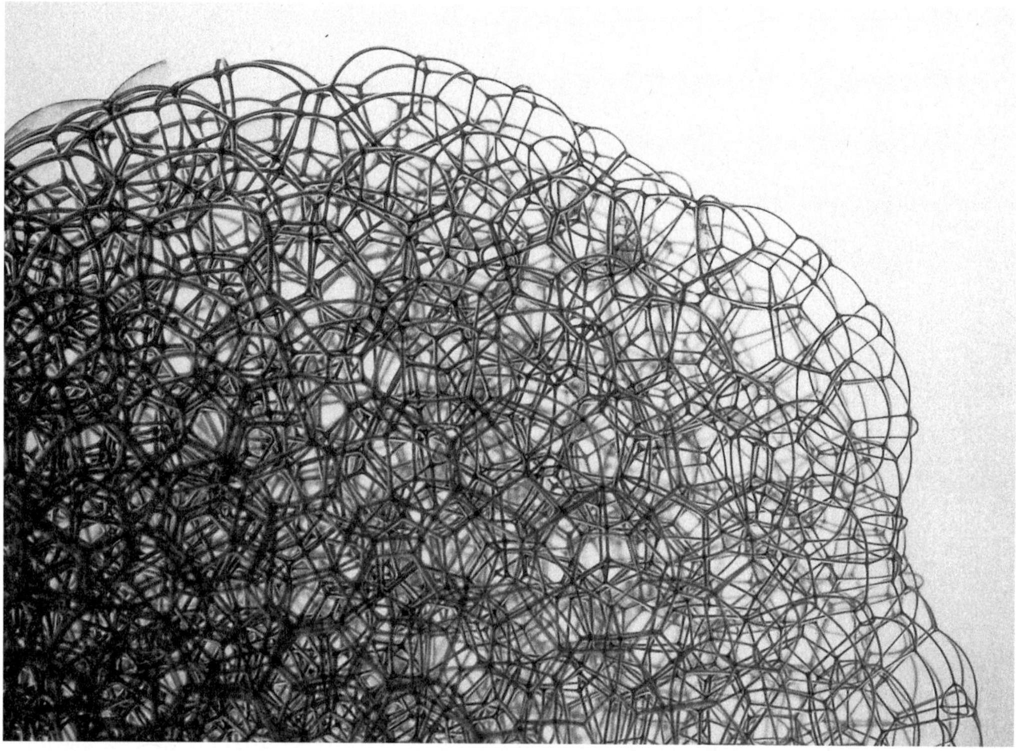

7

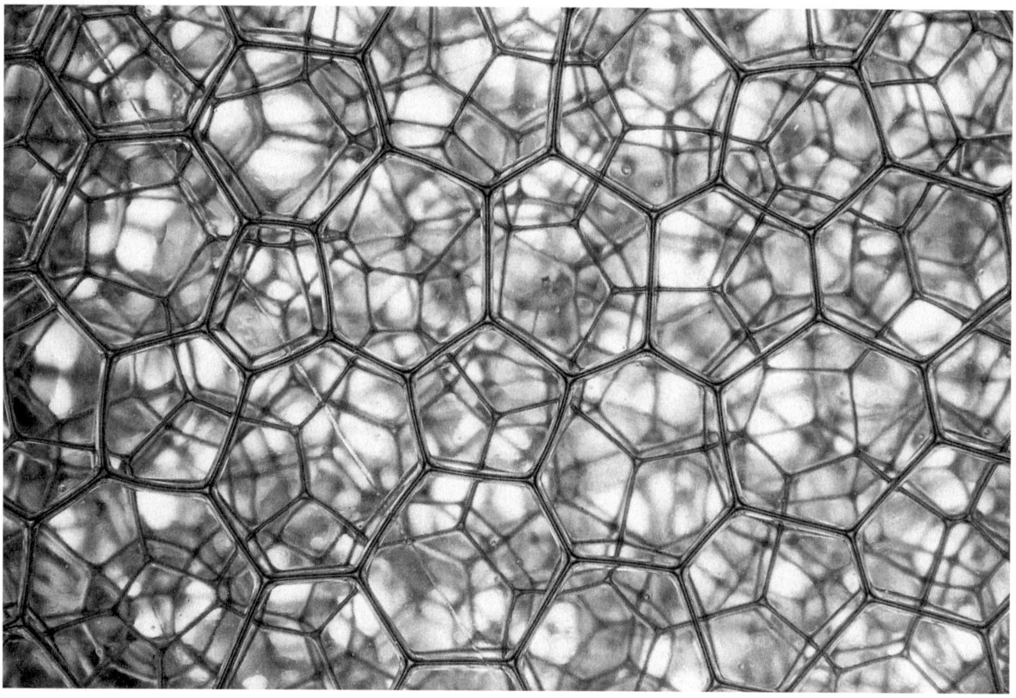

8

Abb.7 Seitenansicht eines Randausschnitts von Polyederschaurn. Bei den meisten äußeren Blasen sind allein die flüssigen Kanten, die Linien der zusammentreffenden Seifenhäute, nicht mehr die Lamellen erkennbar (schwarze Bereiche!). Makrofoto, Vergrößerung rund 10 zu 1.

Abb.8 Räumlicher Polyederschaum zwischen planparallelen Platten. An den Außenseiten überwiegen Sechseckflächen, in der Mitte sind flüssige Kanten von Polyedern mit Fünfeckseiten erkennbar. Makrofoto, Vergrößerung rund 20 zu 1. Polyederschaum aus Rasierschaum.

Fig.7 Lateral view of a peripheral section of polyhedral foam. In the majority of outer bubbles only the liquid edges of the contact lines between the soap lamellae are visible, not the soap films themselves (black areas!). Macroscopic photograph, magnification approx. 10 times.

Fig.8 Polyhedral foam between parallel plates. The outer sides show predominantly hexagonal surfaces, whereas in the middle we can see liquid edges of polyhedrons with pentagonal sides. Macroscopic photograph, magnification approx. 20 times. Polyhedral shaving foams.

12

Wohnbauten der Naturvölker

Domestic architecture of the native people

Günther Hartmann

Zunächst einige Vorbemerkungen zum Begriff Naturvolk. Dem Begriff Naturvolk steht die Bezeichnung Kulturvolk gegenüber. Objektiv betrachtet gibt es aber keine Völker ohne Kultur und die Dichotomie Naturvolk-Kulturvolk ist irreführend und falsch. Beide sind nicht getrennt durch Wesensunterschiede, sondern durch Gradunterschiede. | Die wissenschaftsgeschichtlich wichtige Trennung in schreibende und schriftlose Völker – wobei diese mit den Naturvölkern identifiziert werden – ist zwar frei von irreführenden Assoziationen, kann jedoch aus mancherlei Gründen nicht als Einteilungsgrundlage dienen. Auch die Kennzeichnung als »Primitive« ist, im üblichen, herabsetzenden Gebrauchssinn angewandt, entstellend und höchstens angebracht im Sinne des zeitlich Früheren. Überraschen doch viele Naturvölker durch eine verwirrende Fülle des Formenreichtums der Kunst und Sprache, sowie teilweise durch höchst komplizierte soziale Organisationsformen. | Naturvölker sind vielmehr »Völker geringerer Naturbeherrschung«, die versehen sind mit verhältnismäßig einfachen Werkzeugen, Geräten, Kenntnissen und Fertigkeiten, die sich z.T. jedoch durch einen hohen Grad von Anpassung an die sie umgebende Umwelt auszeichnen. Das gilt in ganz besonderem Maße auch für die Bauten der Naturvölker. | Im Folgenden sollen nach einigen allgemeinen Bemerkungen Wohnbauten von Naturvölkern aus verschiedenen Klimagebieten der Erde beispielhaft dargestellt werden. | Bei menschlichen Gruppen mit einfacher Kultur machte häufig ihre Wirtschaftsform jede wirkliche Sesshaftigkeit und damit feste Bauten unmöglich. Der Mensch hatte den Tieren, die er jagte oder erlegte, zu folgen. Ein regelmäßiger Jahresturnus bei diesen Nomaden oder Wandervölkern ist ein Kennzeichen ihrer Kulturstufe. Eine sichere Grundlage für eine feste Niederlassung existierte nicht. Sie besteht erst dann, wenn ein gewisser Anbau von Nahrungspflanzen, zuweilen auch ein Halbackerbau betrieben wird. Der Anbau oder Halbackerbau ist bis zu einem gewissen Grade immer noch von der Jagd und den Gewohnheiten der Tiere, die man nebenher jagt, abhängig. Die bebaute Erde erschöpft sich aber nach wenigen Jahren, so dass die Rodungen bzw. Pflanzungen in bestimmten Zwischenräumen neu angelegt und verlegt werden müssen. Eine Siedlung muss sich aber auch jeweils den geographischen Umständen anpassen. | Als Beispiele dafür nenne ich die Wintersiedlungen der Eskimo, die sich um Fjorde mit ebener Eisschicht erstrecken, wo der Fang am Atemloch betrieben werden kann oder um Stromschnellen, deren offenes Wasser z.B. die Tiere anlockt, während die Sommerlager der Eskimo an Fischplätzen und an Wanderwegen der Rentiere gesucht werden müssen. | Im subtropischen Neu-Mexiko und Arizona liegen Äcker und Dörfer der Pueblo-Indianer oft kilometerweit voneinander entfernt. Die Felder wurden dort angelegt, wo die Möglichkeit der Bewässerung besteht; menschliche Siedlungen hingegen lehnten die Pueblo-Völker an flache, manchmal steilwandige Felsköpfe an, die gleichzeitig Schutz ge-

Firstly some preliminary remarks on the concept primitive people. The term primitive people (Naturvolk) is the opposite of the term civilised people. Objectively speaking, however, there are no peoples without civilisation, and the distinction between primitive and civilised people is misleading and incorrect. The two are separated not by differences in fact, but by differences of degree. | The separation, important in terms of scientific history, into literate and illiterate peoples – whereby the latter are identified with primitive peoples – is free of misleading associations, but for many reasons cannot serve as a basis for classification. Also the characterisation "primitive" (Primitive) tends to be used in a disparaging sense, distortingly and mainly applied in the sense of early in time. However, many primitive peoples surprisingly have a confusing abundance of wealth in art form and spoken language, as well as sometimes highly complicated forms of social organisation. | Primitive peoples are much more "peoples with less ability to control nature", who have relatively simple tools, implements, knowledge and skills, but who sometimes excel through a high degree of adaptation to the surrounding environment. This applies especially to their architecture. | In the following paragraphs, after some general observations, examples of the domestic architecture of primitive peoples from different climatic zones are given. | In the case of human groups with a simple culture their way of domestic life frequently made any proper settlement and thereby fixed buildings impossible. Man had to follow the animals which he hunted or killed. A regular annual rotation in these nomadic or wandering peoples is the characteristic sign of their stage in civilisation. There was no safe basis for a settled colony. That only comes about if specific cultivation of plants for nourishment, sometimes even semi-agriculture, can be carried on. Cultivation or semi-agriculture is to a certain degree still dependent on hunting and on the habits of the animals which are also hunted. The cultivated earth is exhausted after a few years however, so that the clearings and/or plantations must be laid out anew and transferrred at certain intervals. | A settlement must however also adapt to the geographical conditions. | As an example consider the winter settlements of the eskimo, which are stretched out along fiords covered with a smooth layer of ice, where the prey can be driven to an airhole, or around rapids, whose open water attracts animals for example, while the summer encampments of the eskimo must be sought at fishing places and on the wandering paths of the reindeer. | In subtropical New Mexico and Arizona fields and villages of the Pueblo Indians often lie kilometers away from each other. The fields are laid out where there are watering facilities; on the other hand the Pueblo Indians lean their human settlements against smooth, often steep walled, tops of crags, which at the same time offer protection against enemy tribes fighting around them. | Three main demands are made of a dwelling: it should offer protection from above (rain/sun) by means of

gen umherstreifende feindliche Stämme boten. | An eine Behausung werden drei Hauptforderungen gestellt: Sie soll Schutz nach oben (Regen / Sonne) durch das Dach, nach den Seiten (Wind / Kälte) durch die Wände und nach unten (Feuchtigkeit / Kälte des Erdbodens) durch den Boden bieten. | Die Urform der Hütte stellt die zu einem Schutzdach verflochtenen Zweige eines belaubten Busches dar, wie es z. B. die Buschmänner in Südafrika praktiziert haben. Bricht man die Zweige ab, steckt sie im Kreis in die Erde und verflicht sie oben, so entsteht das Gerippe der Kuppelhütte; stehen nur starre Stangen zur Verfügung, entsteht das Zelt. Legt man zwei Windschirme gegeneinander und dichtet First und Giebel ab, so entsteht das Giebeldach. | Hinsichtlich des Bau- und Deckmaterials bestanden in pflanzenreichen Gebieten (Tropen, Subtropen) keine Schwierigkeiten bei der Auswahl. In nördlichen Gebieten, in Steppen und Wüsten, wo pflanzliches Material seltener vorhanden ist, hat der Mensch zum Leder gegriffen und den Filz erfunden oder aber sich dem Erdbau zugewandt. Der arktische Mensch schließlich hat sogar den Schnee als Baumaterial heranziehen müssen.

Arktis — die Eskimo Die Kultur der Eskimo hat sich weitgehend den Bedingungen des Lebens an den Gestaden des Eismeeres angepasst. Wir wissen heute nicht, welches der historische Antrieb gewesen sein mag; es steht heute lediglich fest, dass die Vorväter der Eskimo ihr Leben den Erfordernissen der arktischen Umwelt in vollendeter Geschicklichkeit angepasst haben. Ein Beispiel dafür ist das Winterhaus und das Schneehaus (Iglu). | Die Eskimo kannten in den Hauptjahreszeiten unterschiedliche Behausungen. Im Sommer diente das kuppelförmige Zelt aus Karibu-, Robben- oder Walrossfellen als Unterkunft. Dieses kuppelförmige Zelt kann als Vorläufer der gerüstlosen Schneehütte angesehen werden. — Kegelförmige Zelte sind erst in jüngerer Zeit von benachbarten Indianergruppen übernommen worden. | Für den Winter und auch als vorübergehende Unterkunft auf Jagdzügen diente den Eskimo die Schneehütte (Iglu). Sie kam nur in den zentralen Siedlungsgebiet des nordamerikanischen Küstensaumes und teilweise auch in Labrador vor. | Zum Bau der Schneehütte mussten große rechteckige Blöcke mit Hilfe eines Messers aus Walknochen oder Geweih aus festem Schneeboden ausgeschnitten und in einer ansteigenden Spirale aufeinander gelegt werden. | Die Spirale wurde nach und nach enger, da die Blöcke so gelegt waren, dass sie immer stärker einwärts sich neigten. Innerhalb von wenigen Stunden entstand auf diese Art ein regelmäßiger Kuppelbau. Die Fugen zwischen den einzelnen Blöcken wurden mit Schnee ausgestopft, ein durchsichtiger Eisblock zuweilen als Fenster eingebaut. | Das Innere der Hütte erhielt eine Sitz- und Schlafbank aus Schneeblöcken, die mit Matten aus Heidekraut und Fellen ausgekleidet waren. Abgesehen von der Erwärmung des Raumes durch die sich dort aufhaltenden Menschen

the roof, from the side (wind / cold) by means of the walls, and from below (dampness / coldness of the earth) by means of the floor. | The original form of the hut is the leafy branches of a shrub woven into a protective roof, as practised by the Bushmen in South Africa. If one breaks off the branches, sticks them in a circle in the earth and weaves them above, one obtains the framework of a domed hut; if only rigid stakes are available, a tent is constructed. If one lays two windbreakers against each other and seals off ridge and gable, a gabled roof is formed. | As far construction and covering materials are concerned there were no difficulties in choice in areas rich in plants (tropics, subtropics). In northern regions, in the savannah and deserts, where plant material is not so often available, man turned to leather and felt, or to earth structures. Arctic man finally had to even use snow as a building material.

Arctic — the Eskimo The civilisation of the Eskimo has been extensively adapted to the conditions of life on the shores of the Arctic Sea. We do not know today what the historical impulse may have been. Today it is only certain that the fore-fathers of the Eskimo adapted their life to suit the demands of the Arctic environment with a high degree of skill. An example of this is the winter house and the snow house (igloo). | The Eskimo had different dwellings for the two main seasons. In summer the dome-shaped tent made of caribou, seal or walrus skins served as shelter. This dome-shaped tent can be seen as a predecessor of the frame-free snow hut. Cone shaped tents were also taken over in recent times by neighbouring groups of Indians. | For the winter and also as temporary shelter on hunting trips the snow hut (igloo) was used by the Eskimo. It occurred only in the central settlement areas of the north American coastal belt and sometimes also in Labrador. | In the construction of the snow hut large rectangular blocks had to be cut out of the firm snow, with the aid of a knife made of whalebone or horn, and laid one on top of the other in a rising spiral. The spiral became narrower and narrower as the blocks were laid in such a way that they inclined inwards more and more. | In this way a uniform dome structure arose within a few hours. The joints between the individual blocks were packed with snow; a transparent block of ice was sometimes put in as a window. The interior of the hut contained a bench for sitting and sleeping made out of blocks of snow, which are covered with heather and skin matting. Apart from the heating of the room by the bodies of people staying there, an oil lamp brought the temperature inside the hut to a few degrees above freezing point. | In front of the main structure, but connected with it, there were often some small snow huts which served as shelters for the dogs or as provision stores. These small constructions had also the function of a cold trap, as the floor of the main house lay higher than that of the provision stores so that the

1

2

brachte es die Tranlampe mit sich, dass einige Grade über dem Gefrierpunkt im Hütteninneren erreicht wurden. | Vor dem Hauptbau, aber mit ihm verbunden, entstanden oft einige kleine Schneehütten, die als Hundehütten oder Vorratskammern dienten. Diese kleineren Behausungen hatten gleichzeitig die Funktion einer Kältefalle, da der Boden des Haupthauses höher lag als die Vorratskammern, so dass die kalte Außenluft nicht sofort in das Haupthaus dringen konnte. | Siedelten mehrere Familien zusammen, so kam es auch vor, dass man mehrere Iglus verband, die häufig nur einen gemeinsamen Eingang aufwiesen. | In anderen Siedlungsgebieten verwendeten die Eskimo auch andere Baumaterialien. So war es üblich, dass in Alaska meist vollständig die Behausungen aus Holz und/oder aus Grassoden bestanden. Im Gebiet der Beringstraße benutzte man Pfahlhäuser.

Subtropen – der vorkolumbische Südwesten der heutigen Vereinigten Staaten von Amerika: In der so genannten Pueblo/III-Phase erreichten die Völkerschaften des Südwestens der heutigen USA einen bemerkenswerten wirtschaftlichen, politischen und künstlerischen Höhepunkt. Dank der Baumringdaten kann diese Phase zwischen 1100 und 1300 n. Chr. festgelegt werden. In dieser Zeit entstehen große Klippensiedlungen (Cliff Dwellings) im Gebiet des heutigen Bundesstaates Colorado. Die größte bisher bekannt gewordene Wohneinheit ist der so genannte Cliff Palace. Er verfügt über mehr als 200 Einzelräume und 23

Abb. 1 Bau eines Iglu. Eskimo
Fig. 1 Building of an igloo. Eskimo

Abb. 2 Noch heute bewohntes Pueblo Taos, Neumexiko, USA

Fig. 2 Still used Pueblo Taos, New Mexico, USA

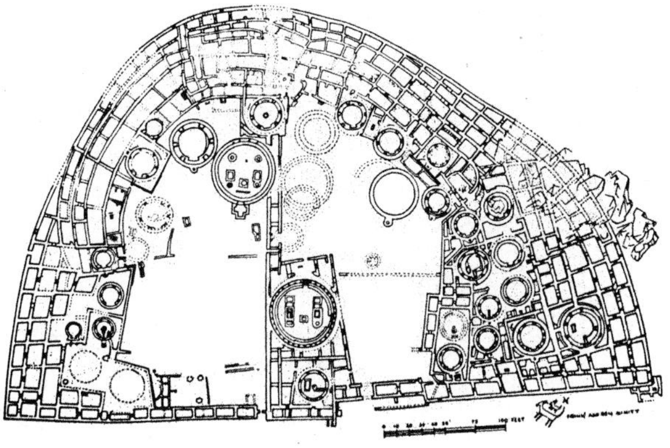

3

cold outside air could not penetrate immediately into the main house. | If several families settled together, then several igloos were sometimes joined together and frequently had only one common entrance. | In other settlement areas the Eskimos used other construction materials. Thus it was usual in Alaska for the houses to consist almost exclusively of wood and/or sods of grass. In the area of the Bering Straits houses were erected on piles.

Subtropics – The south west of the present United States of America in pre-Columbian times. In the so-called Pueblo/III phase the tribes in the south west of the present USA achieved a noteworthy economic, political and artistic peak. Thanks to tree ring dating this phase can be established as being between 1100 and 1300 AD. In this period large cliff dwellings occur in the area of the present State of Colorado. The largest living unit known so far is the socalled Cliff Palace. It possesses more than 200 individual rooms and 23 round ceremonial chambers, situated half under the earth, called kivas. | In the neighbouring Chaco Canyon area twelve such large settlements have been discovered so far. Here also lies the largest building of that time in North America, called Pueblo Bonito. | It is laid out in the shape of a D and comprises more than 800 adjoining rooms which tower one above the other to a height of four storeys. The walls of the living accomodation were erected out of layers of finely chiseled sandstone blocks,

Abb. 3 Pueblo Bonito in Chaco Canyon, Neumexiko, USA (Grundriss)

Fig. 3 Pueblo Bonito at Chaco Canyon, New Mexico, USA (Floor plan)

4

runde, halbunterirdische Zeremonialkammern, die Kivas genannt werden. | Im benachbarten Chaco-Canyon-Gebiet sind 12 solcher Großsiedlungen bisher entdeckt worden. Hier liegt auch das größte Bauwerk jener Zeit in Nordamerika, Pueblo Bonito genannt. | Es ist D-förmig angelegt worden und setzt sich aus mehr als 800 aneinander grenzenden Räumen zusammen, die bis zu vier Stockwerke übereinander emporragten. Die Mauern der Wohnanlage wurden aus Schichten fein behauener Sandsteinblöcke errichtet, die durch Lagen kleinerer Steine voneinander getrennt waren. Das Verputzen der Wände erfolgte mit Lehm. Der heute so charakteristische Trockenlehmziegelbau ist für diese Periode noch nicht belegt. Die Decken bestanden aus viereckig behauenen großen Holzbalken; darüber lag eine Schicht dünner Zweige und Stämme. Diese Schicht war mit einer 15–20 cm dicken gestampften Lehmschicht abgedeckt. | Große, leicht trapezförmige Türöffnungen verbanden die Räume miteinander. Die rückwärts gelegenen Räume dienten als Vorratskammern, während die nach außen liegenden hellen Räume die Wohnungen waren. | Um in die Wohnanlage zu gelangen, bedurfte es des Einstieges über eine lukenförmige Öffnung im Dach mit Hilfe einer einziehbaren Leiter. Da die Stockwerke treppenartig zurückgesetzt gebaut waren, konnte man mit den Leitern das jeweils höhere Stockwerk erreichen. Im Verteidigungsfalle konnten die Leitern heraufgezogen werden und machten so die gesamte Anlage fast uneinnehmbar. Aufgrund archäologischer Untersuchungen ist zu sagen, dass

Abb. 4 Keet Seel Pueblo, Kayenta Region, Arizona, USA

Fig. 4 Keet Seel Pueblo, Region Kayenta, Arizona, USA

5

which were separated from each other by layers of smaller stones. The walls were then plastered with clay. The dry clay brick construction which is so characteristic today is not yet in evidence in this period. The roofs consisted of square-hewn, large wooden beams: over this lay a layer of thin branches and twigs. This layer was covered with a 15–20 cm thick layer of temped clay. | Large, slightly trapezoid door openings connected the rooms with each other. The rooms towards the back served as storage rooms, while the rooms towards the outside with more daylight were the living accommodation. | In order to reach the living quarters, one needed to climb up through a dormer type window opening in the roof by means of a rectractable ladder. As the storeys were built set back like a staircase, one could reach the storey above by means of ladders. When they needed to defend themselves the ladders could be drawn up and thus the whole quarters were made almost impregnable. On the basis of archeological investigations it can be stated that the construction of Pueblo Bonito was begun in 920 and finished in 1067. For reasons unknown to us the settlement was abandoned in 1300. | The Pueblo Indians of the Southwest still live today in similar box-shaped one house villages, built adjacent to and on top of each other and made out of stones and mortar or air dried clay bricks (adobes), whose walls are plastered thick with clay. | Exactly as in the pre-Columbian dwellings, the rooms are connected with each other by door openings, there are ladders made of notched treetrunks to each storey, and entry

Abb. 5 Der Cliff Palace im Winter, Mesa Verde, Colorado, USA

Fig. 5 Cliff Palace in Winter. Mesa Verde, Colorado, USA

mit dem Bau des Pueblo-Bonito um 920 begonnen und 1067 der Bau abgeschlossen wurde. Aus uns bekannten Gründen ist die Siedlung um 1300 aufgegeben worden. | Die Pueblo-Indianer des Südwestens leben auch heute noch in gleichartigen kastenförmigen, neben- und übereinander gebauten Einhausdörfern aus Steinen und Mörtel bzw. luftgetrockneten Lehmziegeln (Adobes), deren Mauern dick mit Lehm verputzt sind. | Genau wie bei den vorkolumbischen Anlagen sind die Räume untereinander durch Türöffnungen verbunden, kennt man Leitern aus gekerbten Baumstämmen zu jedem Stockwerk und erfolgte der Einstieg ursprünglich über Dachluken. Ein solches Pueblo gleicht dem Zellenbau von Bienen. Die Anlage dieser Pueblos ist architektonisch in Nordamerika einmalig. | Wenn auch der Zellenbau keine architektonische Ordnung verrät, so gliedert sich doch das Dorf für den Indianer nach einem strengen Muster. So teilt sich jedes Pueblo in Teile auf, die den Weltvierteln entsprechen. Die Fächerung passt nicht zur Topographie. Diese in Gedanken der Menschen lebende Gliederung wandelte die Siedlung in ein Sinnbild des Kosmos. Nach unserem heutigen Wissen prägte diese Einteilung die gesamte Kosmologie und auch Soziologie der Pueblo-Völker.

Tropen — Indianergruppen Nordwestbrasiliens, Süd-Venezuelas. Bei tropischen Urwaldstämmen in Nordwest-Brasilien bzw. Süd-Venezuela existieren Hütten mit elliptischem Grundriss, Kegeldachhäuser, Satteldachhäuser, z. T. sogar mit Anbauten. Bei einigen Gruppen dominieren Einhaussiedlungen (Malokas), wobei das Dorf meist nur aus einem, selten aus zwei bis drei derartigen Bauwerken besteht. | Bei letzteren handelt es sich um Satteldachhäuser mit beträchtlichen Ausmaßen. Sie können eine Länge bis zu 28 m, eine Breite von 21 m und eine Höhe bis zu 10 m aufweisen. Im Inneren des Wohnhauses befinden sich Pfostenreihen, die sich zu den Seiten hin abstufen und untereinander durch Längsbalken verbunden sind. Die Pfosten der beiden innersten Reihen werden ebenfalls durch Querbalken zusammengefügt, so dass in der Mitte des Hauses ein großer freier Längsraum entsteht, der als Durchgang und Verkehrsraum, als Festsaal und Tanzplatz dient. In den Seitenräumen befinden sich die Wohnstätten der einzelnen Familien, die durch niedrige Mattenwände voneinander getrennt sind. — Als Verschluss des Hauses existiert eine Art Falltür, die von oben nach unten klappt und nachts geschlossen bleibt. — Ein Rauchfang fehlt; der Rauch entweicht durch die lockere Palmblattbekleidung des oberen Teiles der beiden Giebelwände. | Bemerkenswert ist die Tatsache, dass bei diesen Bauten die aus mächtigen Baumstämmen hergestellten Pfosten und Querbalken ohne Beschläge und Nägel nur durch Bänder von Schlingpflanzen zusammengehalten werden. | Bei sehr vielen Indianergruppen im tropischen Südamerika wird das Wohnhaus (Maloka), mit dem Kosmos gleichgesetzt und der Bau des Hauses entspricht in der Verklei-

is initially by means of dormer-type windows. Such a Pueblo is similar to the cell structures made by bees. The layout of these Pueblos is architecturally unique in North America. | Even if the cell construction reveals no architectural arrangement, the village is still formed for the Indians according to a strict pattern. Thus each Pueblo is divided into parts which correspond to the quarters of the world (NSEW). This division which lives in the mind of man transforms the settlement into a symbol of the cosmos. According to our present knowledge this arrangement affected the whole cosmology and sociology of the Pueblo peoples.

Tropics – Indian tribes in northwest Brazil and south Venezuela. In the tropical primeval forests in northwest Brazil and south Venezuela the tribes have huts with elliptical ground plans, houses with conical or saddle shaped roofs, sometimes even with outhouses. In some tribes one house settlements (malocas) are dominant, whereby the village consists of usually just one, more rarely of two or three such structures. | In the latter case these are saddle-roofed houses of considerable proportions. They can have a length of up to 28 m, a width of 21 m, and a height of up to 10 m. Inside the house there are rows of posts which are stepped down towards the side of the hut and are connected to each other by longitudinal beams. The posts of the two innermost rows are also joined together by transverse beams, so that in the middle of the house there is a large, clear room and dance hall. In the rooms at the side are the living quarters of the individual families, which are separated from each other by low matting screens. To close off the house there is a kind of trap door, which flaps down from above and is kept closed at night. There is no chimney; the smoke escapes through the loose palm leaf covering at the upper part of the two gabled walls. | It is worth noting that in these structures the posts and transverse beams made of huge treetrunks are held together only by bundles of creepers without the use of fixings or nails. | For very many Indian tribes in tropical South America the living accommodation (Maloka) is equated with the cosmos and the construction of the house corresponds in its diminuation to their conception of the universe. The dwelling is not only a residential space and banqueting area, but also an area in which the inhabitants can find protection from evil spirits and influences from outside. In a Maloka there are separate male and female zones. Lack of time prevents me from going into further details here.

Africa The African architecture is also in the main part adapted to the landscape and the economic requirements of the people. Numerous mass migrations have brought forth a multiplicity of types of dwelling in the course of its development. The building materials were dependant on the natural environment. In the for-

6

7

nerung der Weltvorstellung. Wohnung ist nicht nur Aufenthalts- und Festraum, sondern auch ein Areal, in dem die Bewohner vor bösartigen Geistern und Einwirkungen von außen Schutz finden können. Es werden in einer Maloka männliche und weibliche Zonen unterschieden. Auf weitere Einzelheiten kann ich an dieser Stelle nicht eingehen.

Afrika Auch die afrikanische Architektur ist im Großen und Ganzen der Landschaft und den wirtschaftlichen Bedürfnissen der Menschen angepasst. Zahlreiche Völkerwanderungen haben im Laufe der Entwicklung eine Vielfalt von Wohnformen hervorgebracht. | Die Baumaterialien waren abhängig von der natürlichen Umgebung. In Waldgebieten dienten Bambus, Palmblattrippen, Baumrinden, Blätter, Gräser und geflochtene Matten zum Hausbau, in Steppengebieten vor allem Lehm- und harte Termitenerde. Eine Verwendung von Steinen war recht selten; gebundener Bast ersetzte Nägel und Zapfen. Die einfachsten Formen der Haustypen waren Kuppel- oder Bienenkorbhütten, die aus den Windschirmen der Pygmäen hervorgegangen sind. | Aus Platzmangel soll im Folgenden nur noch kurz auf die Lehm-Architektur des Westsudan hingewiesen werden. Sie war in hohem Maße durch alle mediterrane, orientalische, später auch islamische Impulse beeinflusst. Ganze Siedlungen, Paläste, Burgen, Moscheen bestanden aus Lehm oder waren aus luftgetrockneten rechteckigen Ziegeln zusammengesetzt. Hölzer-

Abb. 6 Maloka der Kana, R. Aiary, NW-Brasilien

Fig. 6 Maloka of the Kana, R. Aiary, NW-Brazil

Abb. 7 Struktur der Maloka. Kána, R. Aiary, NW-Brasilien

8

est areas bamboos, the ribs of palm leaves, tree bark, leaves, grasses and plaited mates were used in house building; in the savannah regions it was mainly clay and hard termite soil. Stone was very seldom used; bound tree fibre replaced nails and pins. The most simple forms of dwellings were dome-shaped or beehive huts, which were derived from the windbreaks of the pygmies. | Through lack of space I will give just a short summary of the clay architecture of the west Sudan. It was influenced to a large extent by old Mediterrenean, Oriental, and later by Islamic impluses. Entire settlements, palaces, castles and mosques were made of clay, or were put together from air-dried, rectangular bricks. Wooden support beams are raised up from the masonry and are typical decorative features of the so-called mosque-style, which with its spiky towers, jagged walls, and moulded inner walls of clay is of moorish influence. The walls, sleeping places etc. are also made of clay. | A further peak of African clay architecture is found in the region south of the Chad Lake at Logone. The external walls of the domed structures were covered with decorative patterns, from which a smooth door arch is cut out. | To summarise, the dwellings of primitive peoples were extensively dependent on geographical and economic factors in the surrounding landscape and were to a large extent adapted to the requirements of the people.

Fig. 7 · Structure of the Maloka, Kána, R. Aiary, NW-Brazil

Abb. 8 Moschee von Mopti (Mali), Westafrika, Lehmarchitektur

Fig. 8 Mosque of Mopti (Mali), Western Africa, Adobe Construction

ne Tragbalken heben sich vom Gemäuer ab und sind typisches dekoratives Merkmal des so genannten Moscheestils, der mit seinen Igeltürmen, Mauerzacken und reliefierten Tonwänden maurischen Ursprungs ist. Aus Lehm bestehen auch die Wände, Schlafstellen usw. | Einen weiteren Höhepunkt afrikanischen Lehmbaus findet man im Gebiet südlich des Tschad-Sees am Logone. Die Außenmauern von Kuppelbauten wurden mit dekorativen Mustern überzogen, aus denen sich ein glatter Türbogen abhebt. | Zusammenfassend lässt sich sagen, dass die Wohnbauten der Naturvölker weitgehend von den geographischen und wirtschaftlichen Gegebenheiten der jeweiligen Landschaft abhängig und in hohem Maße den Bedürfnissen der Menschen angepasst waren.

9

Abb. 9 Wohnstätten u. Getreidespeicher aus Lehm der Musgu, Tschad, W. Afrika

Fig. 9 Adobe dwellings and grainery silo of the Mousgoum, Chad, W. Africa

13

Bauweise der lebenden Natur

Building methods of living Nature

Eda Schaur

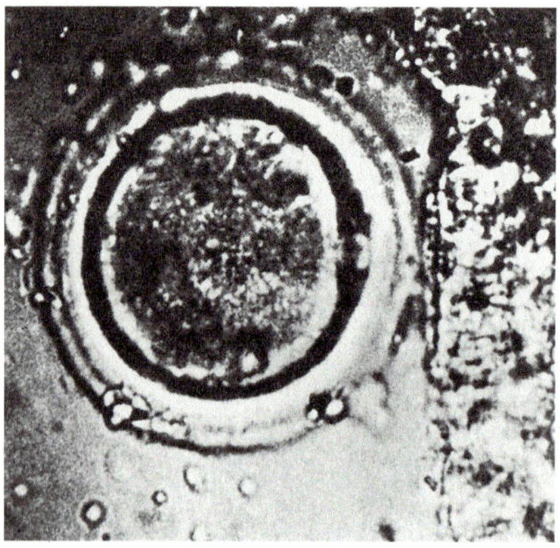

1

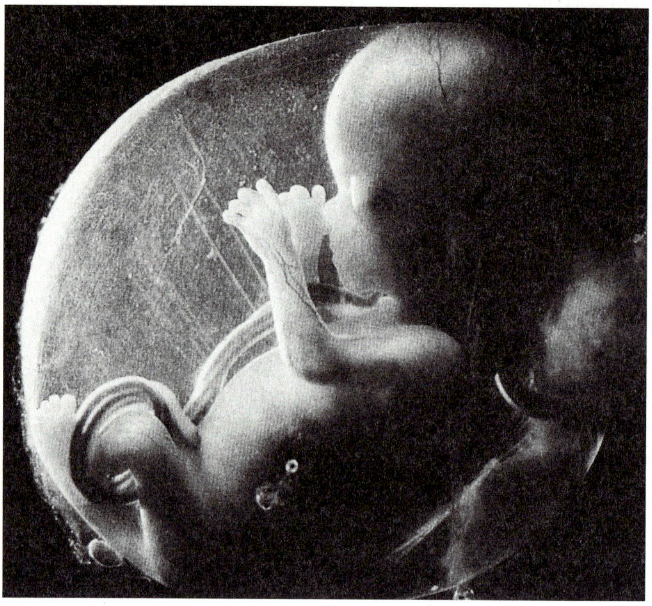

2

In den vergangenen Jahren haben wir in Zusammenarbeit mit Biologen die *Bauweise* der lebenden Natur untersucht. Die dabei gewonnenen Erkenntnisse, über die wir in IL 9 »Pneus in Natur und Technik« und in IL 19 »Wachsende und sich teilende Pneus« ausführlich berichtet haben, werde ich im Folgenden kurz darstellen. Sie können uns auf der Suche nach *natürlichem Bauen* zeigen, wie die Anwendung nur weniger Mittel zu vielfältigen, aber gleichwertigen Lösungen führen kann. || Man kann die Körper der Pflanzen, der Tiere und des Menschen als sich ständig im Wandel befindende, sich selbst bildende und selbst aufbauende komplexe Konstruktionssysteme ansehen. Sowohl die Entwicklung der Arten in der Evolution wie die Entwicklung jedes einzelnen Individuums vollziehen sich in einem ständigen Wandel der äußeren Form und des konstruktiven Aufbaus. | Jedes Lebewesen fängt seine individuelle Entwicklung mit der winzig kleinen Keimzelle an. Dieses weiche, fast kugelförmige Gebilde entwickelt sich in einer unendlichen Reihe ablaufender Formbildungsprozesse zu dem jeweiligen komplexen Organismus (Abb.1 und Abb.2). In der genetischen Kodierung ist nur der Bauplan – die *Erfahrung der Evolution* – verankert. Der Aufbau kann sich jedoch nur durch das Befolgen der physikalischen Gesetze, der Formbildung in den Grenzen der Materialeigenschaften vollziehen. || Vor wenigen Jahren machten wir eine erstaunliche Entdeckung. Die fast unübersehbare Vielfalt der Formen der lebenden Natur, der Pflanzen, der Tiere und des Menschen – ihre unterschiedlichen Lebensprozesse

Abb.1 Das Menschenei kurz nach der Befruchtung

Fig.1 The human egg shortly after fertilization

Abb.2 Embryo, drei Monate alt
Fig.2 Embryo in the third month

In past years in collaboration with biologists we have investigated the *building methods* of living nature. In the following I will give brief details of knowledge gained, which has been reported in IL9 "Pneus in Nature and Technics" and in IL19 "Growing and Dividing Pneus". In the search for "natural building" this can show how the use of only a few resources can lead to varied but equivalent solutions. || One can consider the bodies of plants, animals and human beings as complex structural systems which are constantly changing, are self-forming and self-constructing. The development of species in evolution, like the development of every single individual, takes place in a constant mutation of external form and structural arrangement. Every living object begins its individual development with a tiny germ cell. This soft, almost spherical formation develops in an endless series of continuous formation processes to the actual complex organism (Fig.1 and Fig.2). In genetic coding only the arrangement – the *experience of evolution* – is fixed. The construction can however only be brought about through compliance with physical laws, with formation within the parameters of the material properties. || A few years ago we made an astonishing discovery. The almost incalculable variety of forms of living nature, of plants, animals and human beings – their various life processes and conditions – hardly allowed one to suppose that they were all developed and built up through countless variations of one single construction principle, namely with the structural principle of the pneu. || "Pneu" we call a system in which a tension-resistant, flexible envelope surrounds a filling. The envelope and the filling together form a Ioadbearing structure. || Examples of pneus in technics are children's air balloons (Fig.3), car tyres, hot air balloons, sails, rubber dinghies, milk cartons, air supported halls. | The sphere could be described as the ideal shape for a pneu. It is the result of equal distribution of stresses in the envelope. Spherical pneus, however, occur seldom. They are formed almost exclusively from the smallest drops of liquids or water pneus floating in water – such as the first stage of the living cell of the microsphere. || The world of farm with regard to pneus is extremely varied. The forms of pneus are practically determined by the form and elasticity of the envelope and the quantity and composition of the filling. Along with the strength of the envelope they are basic to the loadbearing behaviour of the pneu. || At every change in the acting forces the pneu reacts with a change in form, if several pneus are packed closely together, they form typical compound forms. || The great form variability of the pneu can be further increased with additional elements. Compression-resistant, one-dimensional elements cause struttings of the envelope (Fig.4). | Tension-resistant, one-dimensional elements such as ties and threads can bind up the pneus (Fig.5) or, when they are found inside, form flat pneus with navel-like indentations. If these fibres that construct the pneus are connected to nets, then the net determines the

3

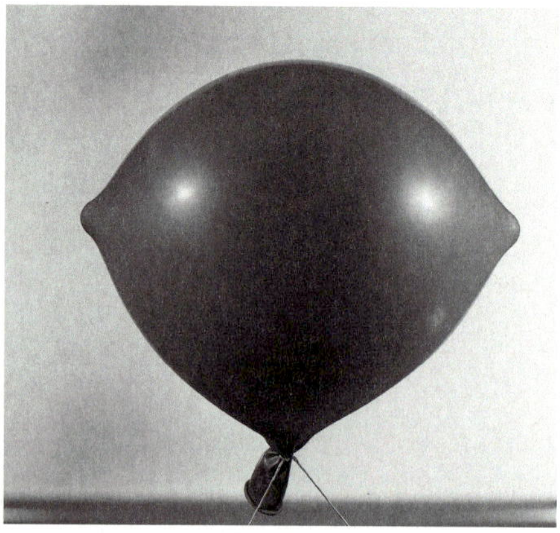

4

und Bedingungen — ließen es kaum vermuten, dass sie sich alle durch unzählige Variationen eines einzigen Konstruktionsprinzips entwickeln und aufbauen: nämlich mit dem Konstruktionsprinzip des Pneus. | »Pneu« nennen wir ein System, bei dem eine zugfeste, biegeweiche Hülle eine Füllung umschließt. Die Hülle und die Füllung bilden zusammen eine tragfähige Konstruktion. | Pneus in der Technik sind beispielsweise Kinderluftballons (Abb.3), Autoreifen, Heißluftballons, Segel, Schlauchboote, Milchtüten, Traglufthallen. Man könnte die Kugel als die Idealform eines Pneus bezeichnen. Sie ist das Resultat gleichmäßiger Spannungsverteilung in der Hülle. Kugelförmige Pneus kommen jedoch selten vor. Sie werden fast ausschließlich von kleinsten Flüssigkeitstropfen oder im Wasser schwebenden Wasserpneus — wie z.B. der Vorstufe der lebenden Zelle — der Mikrosphäre gebildet. | Die Formenwelt der Pneus ist aber äußerst vielfältig. Die Formen der Pneus werden im Wesentlichen von der Form und der Elastizität der Hülle und der Menge und Beschaffenheit der Füllung bestimmt. Sie sind neben der Festigkeit der Hülle für das Tragverhalten des Pneus maßgebend. | Auf jede Veränderung der einwirkenden Kräfte reagiert der Pneu mit einer Formveränderung. Werden mehrere Pneus eng zusammengepackt, bilden sie typische Additionsformen. | Die große Formvariabilität des Pneus wird mit zusätzlichen Elementen noch wesentlich erweitert. Druckfeste eindimensionale Elemente bewirken Verstrebungen der Hülle (Abb.4). Zugfeste eindimensionale Elemente wie Bänder und Fasern können

Abb.3 - 7 Runder Kinderluftballon, mit Druckverstrebungen. (Abb.4) mit Bändern, (Abb.5) mit Bändern und Netzen (Abb.6) und (Abb.7) eingeschnürt

Fig.3 - 7 Spherical balloon, with inner strutting (Fig.4), constructions with tapes (Fig.5) and with nets (Fig.6), (Fig.7)

5

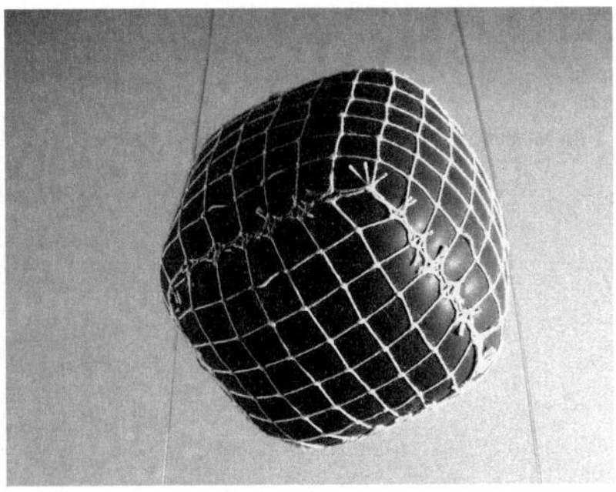

6

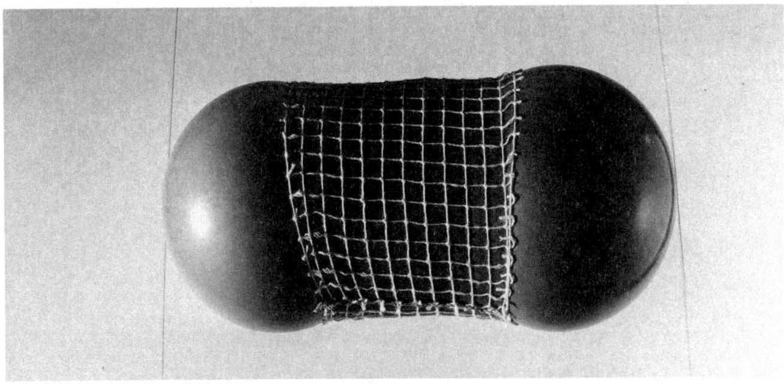

7

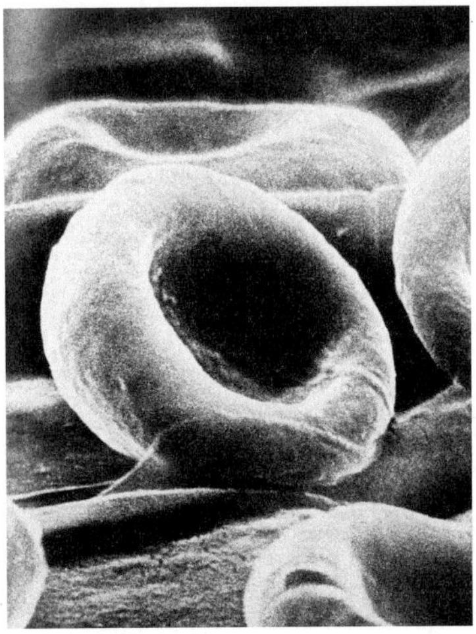

8

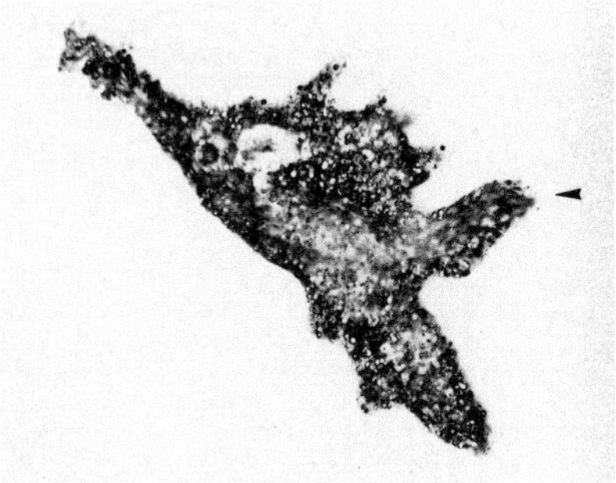

9

Abb. 8 Rote Blutkörperchen

Fig. 8 Erythrocytes

Abb. 9 Ameoba proteus

Fig. 9 Amoeba proteus

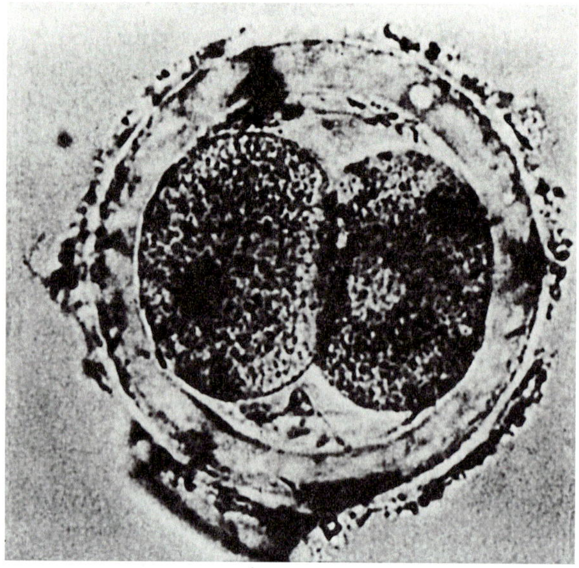

10

Pneus einschnüren (Abb.5), oder aber, wenn sie sich im Innern befinden, flache Pneus mit nabelförmigen Vertiefungen bilden. Werden diese den Pneu einschnürenden Fasern zu Netzen verknüpft, so bestimmt das Netz maßgebend die Form des Pneus (Abb.6, Abb.7). Bestehen diese Netze aus relativ zugfesten Elementen, erhöhen sie entscheidend die Belastbarkeit des Pneus. | Mit diesen Eigenschaften ist der Pneu ein konstruktives System, das lebenswichtige Prozesse wie Wachstum, Formveränderung und aktive Bewegung gewährleisten kann. Zugleich bildet aber dieses tragfähige System auch erweiterungsfähige geschlossene Räume, in denen Lebensprozesse ungestört ablaufen können. || Bereits die kleinste Einheit des Lebens, die Zelle, ist als Konstruktionssystem ein Pneu. Die Hülle, die relativ zugfeste Zellmembrane, umschließt die dickflüssige Füllung – das Zytoplasma. Vermutlich werden die unterschiedlichsten Formen der Zellen (Abb.8 und Abb.9), aber auch ihre kurz- oder langfristigen Veränderungen von in der Zellmembrane liegenden Netzen kontrolliert. Jedoch nicht nur die einzelnen Zellen sind Pneus. | Schon in der Teilung der Eizelle (Abb.10) wird das wesentliche Konstruktionsprinzip der mehrzelligen Organismen erkennbar. Nämlich das Prinzip der Pneus im Pneu. Eine äußere Haut umhüllt die im Innern sich teilenden Zellen. Bei dem Prinzip der Pneus im Pneu bilden Ansammlungen von kleineren Pneus die Füllung des größeren Pneus. Dabei ist die Zugfestigkeit der äußeren Hülle für die Festigkeit des gesamten Systems maßgebend. Mit dem Prinzip der

Abb.10 Menschenei nach der ersten Zellteilung

Fig.10 The human egg after first division

form of the pneu (Fig.6, Fig.7). If these nets consist of relatively tension-resistant elements, they decisively increase the load capacity of the pneu. || With these properties the pneu is a structural system that can guarantee vitally important processes such as growth, form changes, and active movement. At the same time this loadbearing system forms closed spaces capable of expansion, in which life processes can continue undisturbed. || In its structural system even the smallest unit of life, the cell, is a pneu. The envelope, the relatively tension resistant cell membrane, surrounds the viscous filling – the cytoplasm. Probably the very varied forms of cells (Fig.8 and Fig.9), and also their short or longterm changes are controlled by nets lying in the cell-membrane. However, not only the individual cells are pneus. || The important structural principle of the multi-cellular organism is evident in the division of the egg cell (Fig.10). An external skin surrounds the dividing cells inside. In the principle of the pneu within a pneu collections of smaller pneus act as a filling of the larger pneu. Thus the strength of the external envelope determines the strength of the whole system. Many organs and whole individuals are made upon the principle of pneus within a pneu. || Through differentiation and specialisation of cells various tissues are created which, apart from the varied biological functions, also fulfill the structural tasks of the envelopes and the fillings of larger pneus. || Organs such as the frog's sound bladder (Fig.11), the air bladder of a fish, urine bladder, gall bladder or even the intestinal tract (Fig.12) or blood vessel system (Fig.13), are gas or liquid-filled pneus surrounded by tension-resistant tissues. || The fruits of plants (Fig.14) or organs such as the liver, kidneys, brain are constructed as pneus within a pneu. The inner, relatively sensitive cell tissue is again enclosed in a relatively tension-resistant final skin (envelope), and thereby determined in its form, and protected. | The lower organisms such as worms, snails (Fig.15), caterpillars, are constructed as whole individuals exclusively as pneus within a pneu. The bodies of these animals are not stiffened by any sort of compression-resistant or bending-resistant elements. The external skin envelope the pneus of the organs and of the muscle system. The pneus of the organs support the muscle system which is necessary for mobility. || Certain forms, sizes and types of movements of living organisms are, however, only possible through various combinations of flexible pneus and support systems resistant to compression bending. Even the compression-resistant and bending-resistant support systems are created by the soft pneus of the body. They are formed by secretion and accretion of stiffening substances.

- *Plants form their reinforcement by accretion of cellulose in the originally soft pneus of the cells.*
- *The external skeletons of the animals arise through the accretion of chitin or calcium on the soft pneus of the body.*

11

12

13

14

15

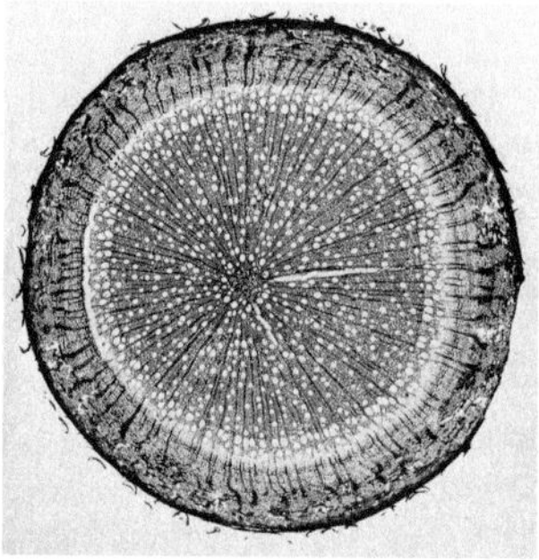

16

Abb.11 Wasserfrosch

Abb.12 Innenansicht eines Schweinedarms

Abb.13 Blutgefäßsystem der Niere

Abb.14 Quergeschnittene Orange

Abb.15 Nacktschnecke

Abb.16 Querschnitt durch die Wurzel eines Birnbaums

Fig.11 Water frog

Fig.12 Inner view of a pig colon

Fig.13 Blood-vessel-system of a kidney

Fig.14 Orange cut across

Fig.15 Slug

Fig.16 Roof of a pear-tree, cross section

17

- *The internal skeleton of the higher animals and of man arises through the accretion of calcium between the flexible bone cells and the periosteum. The bones surrounding cavities, such as the skull, are formed from the flexible pneus of the surrounding organs and the periosteum.*

The growth processes of all reinforced structures can, however, only be carried out with the aid of flexible pneus.

- *For example, the thickness growth of trees occurs in a layer of flexible, self-dividing cells, which lays itself around the hardened wood from the previous year and is protected from outside by the tension-resistant final skin – the inner fibrous bark (Fig.16).*
- *The shells of mussels and snails grow larger by calcium deposits which occur under a membrane surrounding the edge of the shell (Fig.17).*
- *Crabs grow by inflating and thus bursting and throwing off their shells which completely surrounded the body. Through the secretion of hardening substances, a new larger shell arises around the greater enlarged pneu of the soft body. The shells formed by the pneus of the body protect the soft body and make possible better mobility by supporting the muscle system.*
- *The influence of soft pneus on the molding and growth of the inner skeleton can be clearly seen in the skull. The pneus of the brain and the eyeball formed the skull during its growth. For example, if a growing individual loses an eye, the skull is deformed on one side because of the lack of eye pressure (Fig.18).*

Abb.17 Weinbergschnecke **Fig.17** Edible snail

Pneus im Pneu sind viele Organe und ganze Individuen aufgebaut. | Durch Differenzierung und Spezialisierung der Zellen entstehen nämlich unterschiedliche Gewebe, die außer den vielfältigen biologischen Funktionen auch die konstruktiven Aufgaben der Hüllen, aber auch der Füllungen größerer biologischer Pneus erfüllen. | Organe wie die Schallblase des Frosches (Abb. 11), Schwimmblase des Fisches, Harn- oder Gallenblase, aber auch Darmtrakt (Abb. 12) oder Blutgefäßsysteme (Abb. 13) sind gas- oder flüssigkeitsgefüllte von relativ zugfesten Geweben umhüllte Pneus. || Die Früchte der Pflanzen (Abb. 14) und Organe wie Leber, Niere, Hirn sind als Pneus im Pneu aufgebaut. Das innere, relativ empfindliche Zellgewebe wird wiederum von einer relativ zugfesten Abschlusshaut (Hülle) umschlossen, dadurch in ihrer Form geprägt und geschützt. | Die niederen Lebewesen wie Würmer, Schnecken (Abb. 15), Raupen sind als ganze Individuen ausschließlich als Pneus im Pneu aufgebaut. Die Körper dieser Tiere werden durch keinerlei druck- und biegesteife Elemente ausgesteift. Die äußere Haut umhüllt die Pneus der Organe und des Muskelsystems. Die für die Beweglichkeit notwendige Stützung des Muskelsystems bilden die Pneus der Organe. | Gewisse Formen, Größen und Bewegungsarten der lebenden Organismen werden jedoch erst durch vielfältige Kombinationen der weichen Pneus und durch druck- und biegesteife Stützsysteme möglich. Auch die druck- und biegesteifen Stützsysteme werden von den weichen Pneus des Körpers gebildet. Sie entstehen durch Ausscheidung und Anlagerung versteifender Substanzen.

- *Pflanzen bilden ihre Versteifungen durch Anlagerung von Zellulose in den ursprünglich weichen Pneus der Zellen.*
- *Die Außenskelette der Tiere entstehen durch die Anlagerung vor Chitin oder Kalk auf den weichen Pneus des Körpers.*
- *Die Innenskelette der höheren Tiere und des Menschen entstehen durch die Anlagerung von Kalk zwischen den weichen Knochenzellen und der Knochenhaut. Die Hohlräume umschließenden Knochen wie der Schädel, werden von den weichen Pneus der umliegenden Organe und der Knochenhaut geformt.*

Die Wachstumsprozesse aller versteiften Strukturen können sich jedoch nur mit Hilfe der weichen Pneus vollziehen.

- *So erfolgt zum Beispiel das Dickenwachstum der Bäume in einer Schicht weicher, sich teilender Zellen, die sich um das versteifte Vorjahresholz legen und nach außen hin von der zugfesten Abschlusshaut – dem Bast – geschützt werden (Abb. 16).*
- *Die Schalen der Muscheln und Schnecken werden vergrößert durch Kalkanlagerung, die unter einer den Schalenrand umhüllenden Membrane erfolgt (Abb. 17).*
- *Krebse wachsen, indem sie ihre den Körper völlig umschließenden Schalen durch Aufblähen sprengen und abwerfen. Um den so vergrößerten Pneu des weichen Körpers ent-*

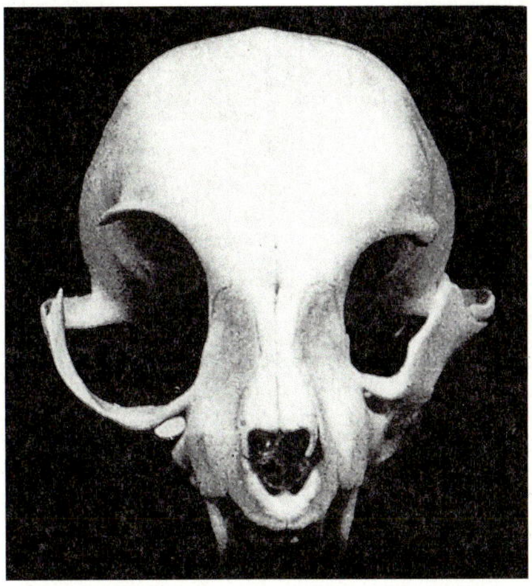

18

19

- *Hairs (Fig. 19), nails and grasses grow by a kind of ejection. Accumulations of soft cells are pushed out through nozzle-like openings in the growth zones, are thereby formed and immediately stiffened.*
- *Leaves, blossoms and insect wings develop in "buds". By means of cell division in inner spaces protected by envelopes two-dimensional, folded, soft forms arise. They are furnished with tube systems. As a whole they are not pneus, but they occur in pneus and are unfolded by the inflation of tubelike pneu systems of their supply systems Thus the dragonfly wing, for example, grows as a soft, folded form (Fig. 20), unfolds by inflation of its tube system, and in its stiffened state later forms a bending-resistant surface (Fig. 21).*

Our investigations have clearly shown that the complex bodies of plants, animals and man, as a whole as well as in their individual elements, are constructed and extensively shaped in their form by the pneu. In evolution nature has, through optimisation by means of mutation and selection, increased the capacity of the organism to integrate into the current biotope and thus increased its chance of survival. Thus evolution is not only the history of the development of self-forming structural systems. The physical laws of forming and the materials available have set restrictions. With these restrictions the viable whole — the biotope — has formed through the development of the complex variety of species and individuals.

Abb. 18 Katzenschädel
Fig. 18 Skull of a cat

Abb. 19 Kopfhaut und Haupthaar des Menschen

Fig. 19 Scalp and head hair of men

steht durch Ausscheidung erhärtender Substanzen eine neue, größere Schale. Diese von den Pneus des Körpers geformten Schalen schützen den weichen Körper und ermöglichen durch die Stützung des Muskelsystems bessere Beweglichkeit.

— *Der Einfluss der weichen Pneus auf die Formgebung und das Wachstum der Innenskelette wird am Beispiel des Schädels besonders deutlich. Die Pneus des Gehirns und der Augäpfel formen während des Wachstums den Schädel. Verliert ein sich im Wachstum befindendes Individuum zum Beispiel ein Auge, wird der Schädel wegen des fehlenden Augendrucks einseitig verformt (Abb. 18).*

— *Haare (Abb. 19), Nägel und Gräser wachsen durch eine Art Strangpressverfahren. Weiche Zellansammlungen werden durch düsenförmige Öffnungen der Wachstumszonen hinausgeschoben, damit geformt und sofort versteift.*

— *Die Blätter, Blüten und Insektenflügel entwickeln sich in Knospen. In von Hüllen geschützten Innenräumen entstehen durch Zellteilungen zweidimensionale, mit Schlauchsystemen durchzogene, zusammengefaltete, weiche Gebilde. Sie sind als Ganzes keine Pneus, entstehen aber in Pneus und werden durch Aufpumpen schlauchförmiger Pneus ihrer Versorgungssysteme entfaltet. So wächst zum Beispiel auch der Libellenflügel als weiches zusammengefaltetes Gebilde (Abb. 20), wird durch Aufpumpen des Schlauchsystems entfaltet und bildet später im versteiften Zustand eine biegesteife Fläche (Abb. 21).*

Unsere Untersuchungen haben eindeutig gezeigt, dass die komplexen Körper der Pflanzen, der Tiere und des Menschen im Ganzen wie in ihren Elementen durch den Pneu aufgebaut und in ihrer Form weitgehend geprägt sind. In der Evolution hat die Natur durch die Optimierung über die Mutation und Selektion die Integrationsfähigkeit der Organismen in das jeweilige Biotop und damit ihre Überlebenschance gesteigert. Somit ist die Evolution nicht nur die Geschichte der Entwicklung der Arten, sondern auch die Geschichte der Entwicklung sich selbst bildender Konstruktionssysteme. Die physikalischen Gesetze der Formbildung und die zur Verfügung stehenden Materialien haben Grenzen gesetzt. Innerhalb dieser Grenzen hat sich das lebensfähige Ganze – das Biotop – durch die Entwicklung der unübersehbaren Vielfalt von Arten und Individuen gebildet.

20 / 21

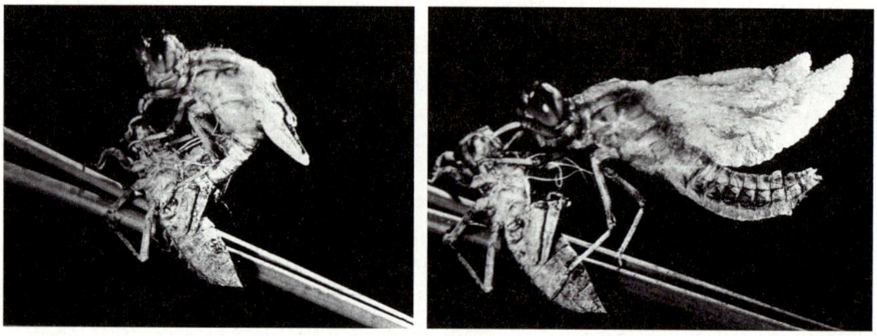

Abb. 20 Libelle kurz nach dem Schlüpfen
Fig. 20 Dragon-fly shortly after moulting

Abb. 21 Libelle nach der Entfaltung der Flügel
Fig. 21 Dragon-fly after unfolding of the wings

14

Tierbauten

Zwischen Spezialisierung und Anpassung

Animal structures

Between specialisation and adaption

Cornelius Thywissen

1

2

Die Technik der Tiere führt zu erstaunlichen Bauleistungen, Bauleistungen von oft imponierendem Umfang und bis ins Detail hinein von größter Akribie. Wir wollen hier nur ein Schlaglicht auf diese Bautätigkeiten werfen und dabei weniger die körpereigenen Hüllen, Schalen und Gehäuse betrachten, als vielmehr solche Bauten, die unabhängig vom tierischen Organismus existieren und oft mit fremden Baustoffen errichtet werden. | Freilich haben seit jeher die geometrisch präzisen und auch bizarren Gehäuse der Schnecken, Tintenfische und Ammoniten fasziniert. Das Wachstumsprinzip als Spirale taucht übrigens immer wieder in der lebenden Natur auf — nicht nur in den Gehäusen, sondern auch in der Ausbildung von Organen oder von Körperextremitäten, z.B. bei den Hörnern von manchen Ziegen und Antilopen. Schnecken, Muscheln, die bekannten und seit der Challenger-Expedition (1878) viel bewunderten Radiolarien, Diatomeen und Foraminiferen — jenen Mikroorganismen, denen moderne Forscher wie J. G. Helmcke ihr ganzes Leben gewidmet haben, die aber auch von Architekten, z.B. Buckminster Fuller, so sehr bewundert wurden — ihre Architektur besteht aus der Körperhülle, die wie auch immer veränderlich sein muss und neben dem Schutz vor äußeren Einflüssen noch manch andere Eigenschaften aufweist. Diese Hüllen sind Architektur, sind Konstruktion und in Aufbau, Tragverhalten und der Bewältigung von Aufgaben sehr komplex. | Die Hüllen der toten Tiere sinken auf den Meeresgrund. Milliarden von feinsten Einzelbauwerken bilden diese Sedimentschicht und spä-

Abb. 1 Barriere-Riff
Fig. 1 Barrier reef

Abb. 2 Nautilus-Schale
Fig. 2 Nautilus shell

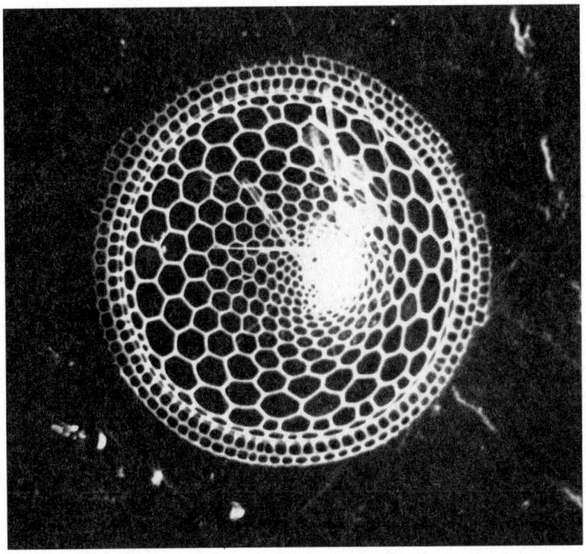

3

The technology of the animals leads to astonishing building achievements, achievements of often imposing dimensions and going into detail of the greatest accuracy. Here we wish to throw just a ray of light on these structural activities and in so doing observe not so much the hulls, shells and housings of their own bodies as those structures which exist independently of the animal organism and are often constructed with foreign material. | Indeed, since ever the geometrically precise and even bizarre housings of snails, cuttlefishes and ammonites have fascinated mankind. The growth principle of a spiral appears over and over again in living nature — not only in housings but also in the formation of organs or of bodily extremities, e.g. in the horns of many goats and antelopes. Snails, mussels, the radiolaria, diatoms and foraminifera, known and much admired since the Challenger Expedition (1878) — those micro-organisms to which modern researchers such as J.G. Helmcke have devoted their whole life, but which have also been greatly admired by architects such as Buckminster Fuller — their architecture consists of the body shell which must as always be adaptable and besides offering protection against external influences also has many other properties. These shells are architecture, are construction and are very complex in their composition, loadbearing behaviour and in the way they solve problems. | The shells of dead animals sink to the bottom of the sea. Billions of the most delicate individual structures form this layer of sediment and later the lime mountains of our earth's crust. The

Abb.3 Diatomeen-Skelett
Fig.3 Diatomes skeleton

4

5

ter die Kalkgebirge unserer Erdkruste. An diesem gigantischen Aufbau haben auch die Korallenpolypen der warmen Meere teil. Es sind in Kolonien lebende, an einem festen Standort gebundene Weichtiere, die ihre Körper in röhrenförmigen Gehäusen schützen und damit ganze Röhrenlandschaften bauen. Von Sonnenlicht, Wärme und Sauerstoff abhängig, können sie nur in ganz bestimmten Tiefenbereichen des Meeres leben. Sinkt der Untergrund ab, bauen neue Polypen auf den Gehäusen der abgestorbenen auf. So bilden sie Inseln, Riffe, neues Festland. | Die Werkmeister, die uns hier speziell interessieren, sind jene Tiere, die zu ihrem dauernden Schutz, zum Schutz ihrer Nachkommen, zum Schutz während des Schlafs oder der Fortbewegung und zum Beutefang Konstruktionen errichten, wobei sie sich von diesen Gebäuden — wenigstens kurzfristig und ohne Schaden — entfernen können. Viele Baumeister, vor allem die Larven der Insekten, manche Ameisen, Termiten und der Maulwurf haben sich dermaßen an ihre selbsterrichteten Gehäuse adaptiert, dass sie sich nicht freiwillig aus ihnen herausbewegen, da sie sonst der Umwelt wehrlos ausgeliefert wären. Diese von der Evolution vorgesehene Abhängigkeit von dem schützenden Gebäude erlaubt, Schutzorgane, Sinnesorgane, Bewegungsorgane etc. zurückzubilden oder gar nicht erst auszubilden. | Bei den Tierbauten finden sich alle Arten von Tragwerksformen: Höhlen, Stabtragwerke (viele Nester der Vögel), Membranbauten und Netzkonstruktionen (Spinnen- und Raupennetze), Schalen (die dünnwandigen Schwalbennester der

Verschiedene Tragwerksformen / Different forms of loadbearing structure:

Abb.4 Stabwerke (Laubenvogel)
Fig.4 Framework (bower bird)

Abb.5 Flechtwerke (Webervogel)
Fig.5 Plaited Work (weaver bird)

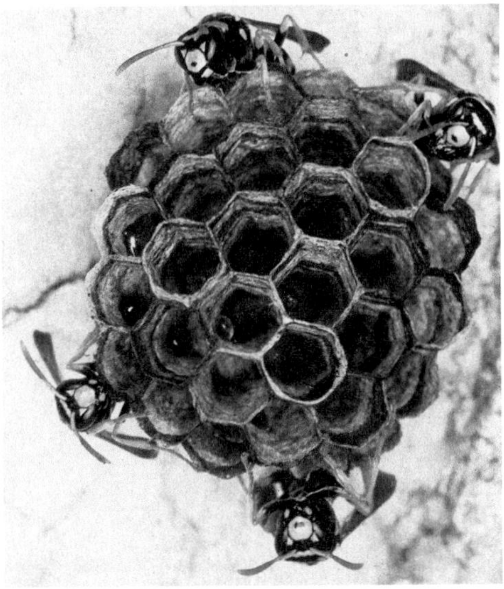

6

7

coral polyps of the warm seas also play their part in this gigantic construction. There are colonies of living molluscs connected to a fixed location, which protect their bodies in tubular shells and thus build whole tubular landscapes. Dependent on sunlight, warmth and oxygen they can only live in areas where the sea bed does not go below a certain depth. If the sea bed sinks, new polyps build up on the shells of the dead ones. Thus they form islands, reefs, new mainland. | The master craftsmen who particularly interest us here are those animals which for their permanent protection, for the protection of their descendants, for protecting during sleep or locomotion and catching prey, erect structures which they can leave without damage at least for a short time. Many master builders, above all insect larvae, many ants, termites and moles have adapted to such a degree to their self-constructed houses that they do not willingly move out of them as they are then exposed defencelessly to the environment. This evolutionary dependence on the protective housing allows protective organs, sense organs, locomotive organs etc. to reform or not to form in the first place. | Among animal structures all kinds of Ioadbearing formations are found: cavities, fram structures (many birds' nests), membrane structures and net structures (spiders' webs and caterpillar cocoons), shells (the thin-walled nests of the salanganes), folded structures (e.g. the honeycomb structures of bees), voults (e.g. the ant heaps above ground), solid structures (the cemented, cast, highly stable termite constructions). | One could classify and

Abb.6 Faltwerk (Wespe)
Fig.6 Folding structure (wasp)

Abb.7 Massivbau (Termite)
Fig.7 Solid structure (termites)

Salanganen), Falttragwerke (z.B. die Wabenbauten der Bienen), Gewölbe (z.B. die oberirdischen Ameisenhaufen), Massivbauten (die gekitteten, gegossenen, hochfesten Termitenbauten). | Man könnte Tierbauten ordnen und verstehen nach »Rangordnung« und Systematik ihrer Erbauer. Dabei fällt auf, dass die hochentwickelten Wirbeltiere — vor allem die Säuger — erstaunlich wenig entwickelte, beinahe primitive Bauten errichten, im Gegensatz dazu findet man gerade bei den Wirbellosen — den stammesgeschichtlich älteren Tieren — hochkomplizierte Bausysteme. Der Bau einer Wespe mit den umhüllenden papierfeinen Schalen um ein mehrgeschossiges räumlich ausgeklügeltes Wabensystem ist unvergleichlich raffiniert und in Statik und Funktion entwickelter als das Gang- und Höhlensystem eines Maulwurfes. Die hochentwickelten Primaten — z.B. die Menschenaffen — errichten allenfalls aus Zweigen und Blättern Schlafnester, die zwar reinlich gepflegt und ständig ausgebessert werden, aber als Architektur eher ungeschickt und willkürlich erscheinen und nicht als das Werk eines mental so überlegenen Tieres. | Tierbauten lassen sich auch vom Detail her analysieren, von der Technik des Herstellungsvorganges, von der maximalen Belastbarkeit und den tatsächlich auftretenden Kräften. An diesem Institut würde die Frage nach dem Masseaufwand einer solchen Konstruktion im Verhältnis zu den maximalen Kräften und dem Kraftweg interessieren. Wir müssen hierzu gestehen, dass uns überhaupt keine Daten bekannt sind und wir selber auch nicht in der Lage sind, solche Untersuchungen allein durchzuführen. Über das Spinnennetz freilich, über seinen Anschluss, über den Herstellungsprozess, Alterung und Belastungsverhältnisse ist vergleichsweise am meisten bekannt. Es wurden von Zoologen, z.B. von dem Spinnenforscher Ernst Kullmann umfangreiche Studien — auch aus technischer Sicht — durchgeführt. Bspw. können wir die Reißlänge (die maximale Länge eines senkrecht hängenden Fadens unter Eigengewicht) vergleichen mit technischen, hochzugfesten Materialien. Hier einige Vergleichswerte:

Material	Spinne: Wegfaden /	Kokonfaden	Wolle	Baumwolle	Viskose	Polyester	Glasfaser
Reißlänge = R (km)	-70	-20	10-20	25-50	40-75	70-95	70-120

Quelle:
IL 8, Mitteilung des Institutes für leichte Flächentragwerke Nr. 8, Stuttgart, 1975.

Wenn man die elektronenmikroskopischen Aufnahmen von Anheftungsstellen, Maschen, Randausbildung eines Spinnennetzes, die Ernst Kulimann machte, mit den Details von technischen Seilnetzkonstruktionen vergleicht, ist man über die

understand animal structures by the "order of precedence" and biological system of their builders. It would then be apparent that the highly developed vertebrates — above all the mammals — erect surprisingly meagerly-developed, almost primitive structures. In contrast to this one finds that the invertebrates — the phylogenetically older animals — have highly complicated building systems. The building constructed by a wasp with the enclosing paper-fine shell around a multi storey, ingeniously worked out honeycomb system is incomparably more refined and more developed structurally and functionally than the hollows and passages of a mole. The highly developed primates — e.g. the anthropoid apes — erect sleeping quarters out of branches and leaves, which are indeed kept clean and continually repaired, but as architecture appear rather inept and arbitrary, not like the work of an animal which is mentally so highly developed. | Animal structures can also be analysed from the details, from the technique of the production process, from the maximum loading and the forces which actually occur. At this Institute we are interested in the expenditure of mass in such a construction in relation to the mass forces and the path of those forces. Here we must admit that hardly any data at all is known to us, and we ourselves are not in a position to carry out such investigations alone. Probably most is known about the spider's web, its connections, the manufacturing processes, ageing and loading conditions. Extensive studies have been carried out — also from a technical aspect — by zoologists, such as by the spider researcher Ernst Kullmann. For example we can compare the tearing length (the maximum length of a vertically hanging thread under self-weight) with technical highly tensile materials. Here are some comparative values:

Material	Spider: traveling thread	Spider: cocon thread	Wool	Coton Wool	Viscose	Polyester	Glassfibre
Tearing length = R (km)	-70	-20	10-20	25-50	40-75	70-95	70-120

Source:
IL 8, Information of the Institute for Lightweight Structures, Nr. 8, Stuttgart, 1975.

If one compares the photos, token by Ernst Kullmann, with an electronmicroscope, of junctions, meshes, edge formations of a spider's web with the details of technical cable net constructions, one is astonished by the similarity of the solutions to the problems. Of course the spider's webs are not constructed under such constant environmental conditions as technical cable netstructures. During construction the web can be destroyed by a large insect or by the wind for example. The an-

8

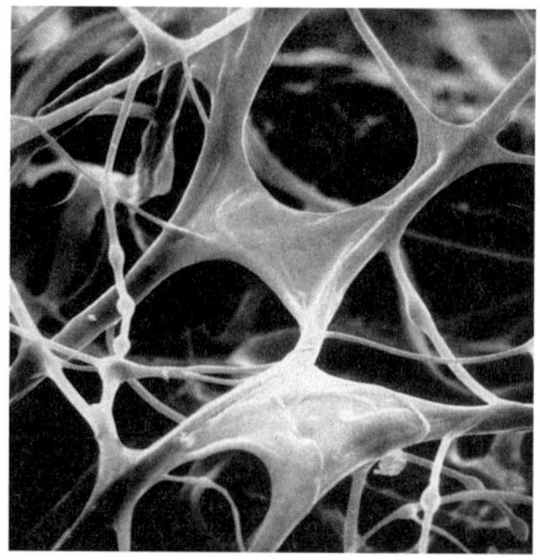

9

Ähnlichkeit der Problemlösung erstaunt. Freilich entstehen die Spinnennetze nicht unter solchen konstanten Umweltbedingungen wie technische Seilnetzkonstruktionen. Während des Baus wird bspw. das Netz von einem großen Insekt oder durch Wind zerstört. Die Ankerpunkte reißen, weil sich ein Ast heftig bewegt oder Laub in das Nest geweht wird. Trotz dieser ungleich widrigeren Verhältnisse schafft die Spinne ein Bauwerk von für ihre Art typischem Aussehen und mit immer ähnlichen Details. | Eine Bautechnik wird nicht allein von einer Tiergattung erfunden. Netzbauten aus Spinnfäden finden sich nicht nur bei den Spinnen. Im Insektenreich gibt es viele Arten, vor allem die Larven dieser Arten, die mit Hilfe von Fäden und Netzen konstruieren. Auch das typische Vogelnest wird von anderen Tierarten mit täuschender Ähnlichkeit gebaut. Spinnen, Weben, Nähen, Kleben, Zusammenfügen von Fremdmaterial, Graben, Anhäufen, Bohren, Fräsen..., es sind Techniken für spezielle Materialien, aber auch für typische Werkzeuge. Manchmal werden von stammesgeschichtlich völlig fremden Tieren ähnliche Werkzeuge für ähnliche Tätigkeiten ausgebildet, die auch für vergleichbare Bauformen eingesetzt werden. Ein bekanntes Beispiel ist der Vergleich zwischen der Grabhand der Maulwurfsgrille und des Maulwurfs. Und dennoch gilt, dass die stammesgeschichtlich älteren Tiergattungen ähnliche Bautechniken besser beherrschen, als ob die Spezialisierung längere Tradition hätte. Die Spezialisierung unterliegt ökologischen Zwängen. Fehlschläge der Evolution im Sinne

Abb.8 Spinnennest mit Tautropfen
Fig.8 Spider's web with drops of dew

Abb.9 Detailaufnahme
Fig.9 Detailed photo

chorage points tear because a branch moves violently or leaves are blown into the web. Despite these variable conditions, the spider creates a structure with a typical appearance for its species and with constantly similar details. | One building technique is not discovered by just a single animal species. Net structures made of fine viscous threads are not only found with spiders. In the insect kingdom there are many species, above all the larvae of these species, which construct with the aid of threads and nets. Even the typical bird's nest is built by other species of animals with deceptive similarity. Spinning, weaving, sewing, sticking, addition of strange material, digging, building into heaps, boring, milling ..., these are techniques for specific materials but also for typical tools. Sometimes similar tools are designed for similar activities by animals which are completely different phylogenetically and these tools are also used for comparable building forms. One known example is the comparison between the digging paw of the mole cricket and the mole. And yet it is true that the phylogenetically older species of animals master similar building techniques better, as if the specification had a longer tradition. Specialisation is subjected to ecological constraints. Failures of evolution in the sense of unsuitablility or of specialisation which is too onesided are solved by the extinction of the species. | Animals build in a suitable environment. The structures are related to the ecological conditions. Adaptability to the surroundings (not in the sense of mimicry) seems to be only necessary in a limited way. Adoption of the surroundings to the individual specialised pattern of behavoiur and needs occurs very seldom. The example of the coral which alters its surroundings to a great extent shows that animals are in reality dependent on currents, tides, light etc. The alteration of the surroundings occurs only secondarily through the growth of the polyps which is conditioned and directed by the environment. It is different in the case of the beaver; he directly adapts his surroundings to his needs, in that he only erects dams in the courses of rivers when the water supply is irregular or disturbed. Only then does he build his water castles, heaps of twigs and logs up to 4 m high with hollows to live in and entrances laid under the surface of the water. But the beaver is content with earth constructions in the slope of the bank if it is certain that the entrance to the living quarters will always remain under water. Beavers set to work so decisively in the river system that they can change whole landscapes. These water structures are unbelievably resistant. They can permanently withstand a pressure of several tons.The longest known beaver dam was 600 m long. | It seems that direct interference in the ecology of a landscape is restricted to the higher animals (in exceptional cases) and man.

von Nicht-Angepasst-Sein oder von zu einseitiger Spezialisierung werden durch Aussterben der Art gelöst. | Tiere bauen in einer ihnen angemessenen Umgebung. Die Bauten entsprechen den ökologischen Bedingungen. Anpassung an die Umgebung (nicht im Sinne von Mimicry) scheint nur bedingt erforderlich zu sein. Anpassung der Umgebung an die eigenen spezialisierten Verhaltensmuster und Bedürfnisse gibt es sehr selten. Das Beispiel der Koralle, die ihre Umgebung in größerem Ausmaße ändert, zeigt, dass die Tiere in Wirklichkeit von Strömungen, Gezeiten, Licht etc. abhängen. Die Veränderung der Umgebung geschieht erst sekundär durch das umweltbedingte und -gerichtete Wachstum der Polypen. Anders beim Biber, er passt direkt seine Umgebung seinen Bedürfnissen an, indem er an Flussläufen nur dann Dämme errichtet, wenn der Wasserhaushalt unregelmäßig oder gestört ist. Nur dann baut er auch seine Wasserburgen, bis zu 4 m hohe Reisig- und Knüppelhaufen mit Wohnhöhle und einem unter dem Wasserspiegel gelegenen Eingang. Somit begnügt sich der Biber mit Erdbauten in der Uferböschung, wenn garantiert bleibt, dass der Eingang zu der Wohnhöhle immer unter Wasser liegt. Die Biber greifen in das Flusssystem so entscheidend ein, dass sie eine ganze Landschaft verändern können. Die Wasserbauwerke sind unglaublich widerstandsfähig. Sie können permanent mehrere Tonnen Staudruck aushalten. Der längste bekannte Biberdamm war 600 m lang. | Es scheint, dass der direkte Eingriff in die Ökologie einer Landschaft den höheren Tieren (in Ausnahmefällen) und dem Menschen vorbehalten ist.

10

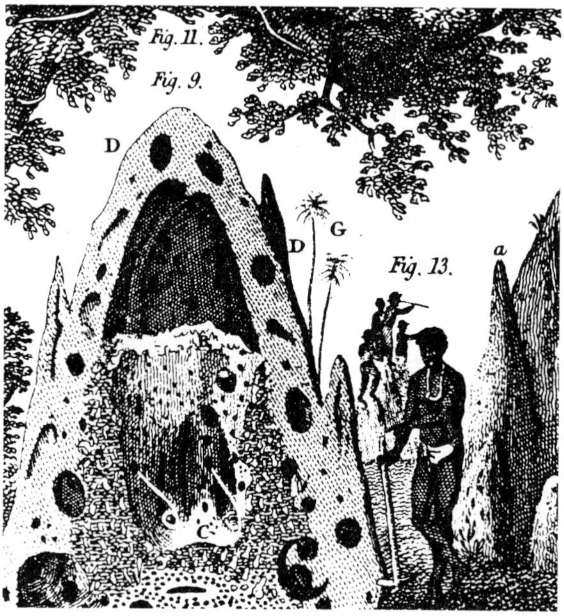

Abb. 10 Termitenbauten
Fig. 10 Termite structures

15

Konstruktionen der nicht lebenden Natur

Constructions of non-living nature

Cornelius Thywissen

1

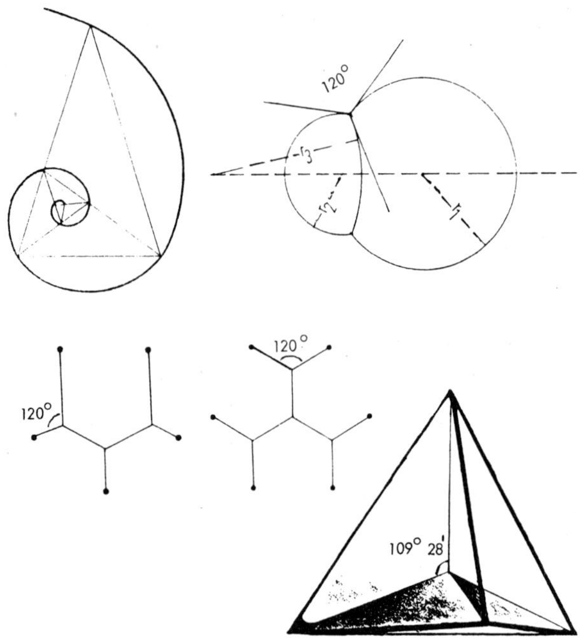

Die Erscheinungsformen in der nicht lebenden Natur: Galaxien, Wellenformen, Gletscher und Flüssigkeitstropfen und vieles mehr, sind uns als Phänotypen innerhalb eines ganz bestimmten Kanons bewusst – und werden vielleicht gerade deshalb als bekannt empfunden, da sich diese Formen unabhängig von Größe, Aggregatzustand und Zeit innerhalb eines bestimmten »Repertoires« bewegen. Um es geometrisch auszudrücken: Es gibt typische Grundmuster, die sich überall und immer manifestieren: die Spirale, die Verzweigung mit dem Winkel 120° für die Ebene und dem Knoten aus drei Strecken, für den Raum der Winkel 109° 28' und entsprechend ein Knoten aus vier Strecken, das in der Ebene immer wiederkehrende Sechseck, das Fünfeck für den Raum, die Kreisscheibe und die Kugel. In diesen geometrischen Angaben steckt die Idee der Minimierung, der kleinsten Oberfläche, der kürzesten Entfernung zwischen zwei oder mehreren Punkten, der dichtesten Packung von Masseteilchen, selbst bei expandierender Tendenz des gesamten Systems. Von daher erklärt sich auch unser Interesse an diesem Institut für die physikalischen Gesetzmäßigkeiten in der nicht lebenden Natur. | Aber solche abstrakten geometrischen Erklärungen für Abläufe in der Natur können die Wirklichkeit kaum beschreiben. Die Sprunglinien im Basalt oder das Rissesystem im Schlamm entsprechen eben nur annähernd der Idealform eines minimalen Wegenetzes von 120° Winkeln und drei Streckenknoten. Das Material erstarrt unregelmäßig auf Grund seiner Zusammensetzung. Es kann sich auch nicht wie

Geometrische Grundformen:

Abb. 1 Spirale, Kantenwinkel von 120° zwischen zwei Flächen, Eckenwinkel von 109° 28' im Raum (Seifenhautmodell eines Tetraeder)

Basic geometric forms:

Fig. 1 Spiral, edge angle of 120° between two surfaces, corner angle of 109° 28' in space (soap membrane model of a tetrahedron)

2

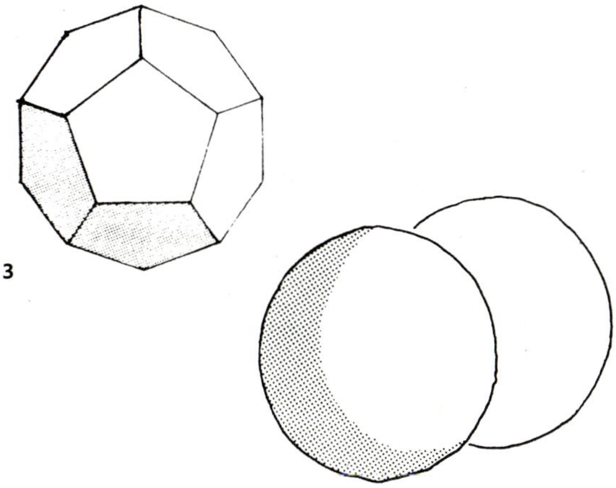

3

The manifestation of non-living nature: galaxies, wave forms, glaciers and drops of liquid and many more are known to us as phenotypes within a quite definite canon – and are perhaps felt to be known just because these forms, independent of size, physical condition and time, move within a defined "repertoire". To express it geometrically, there are typical basic patterns, which are always manifest everywhere: spirals, the ramification with the angle 120° for the plane and the node mode of three branches, three-dimensionally the angle of 109° 28' and a node of four branches, the continually recurrent hexagon in plane, the pentagon in space, the circular disc and the sphere. In these geometric statements is hidden the idea of minimisation, the smallest surface area, the shortest distance between two or several points and the densest packing of particles, even if the whole system trends toward expansion. Thus we declare our interest at this Institute for the physical procedure in non-living nature. | But such abstract geometrical explanations for developments in nature can scarcely describe the reality. The cracking lines in basalt or the fissure system in mud corresponds only approximately to the ideal form of a minimal pattern network of 120° angles and three linear nodes. The material solidifies unevenly because of its composition. It cannot behave like a liquid with biaxial equal tension. The soap bubble which floats in the air or the small falling raindrop are not ideally round like a sphere, because self weight or the force of gravity always plays a role. And yet the tendency to form certain typical

Abb.2 Das Sechs-Eck in der Ebene
Fig.2 The hexagon in plane

Abb.3 Das Fünf-Eck im Raum (Dodekaeder), die Kugel und der Kreis.

Fig.3 The pentagon in space (dodecahedron), the sphere and circle.

4

5

6

eine Flüssigkeit mit biaxialer Gleichspannung verhalten. Die in der Luft schwebende Seifenblase, der kleine fallende Regentropfen, sie sind eben nicht ideal rund wie eine Kugel, da immer noch das Eigengewicht bzw. die Gravitationskraft eine Rolle spielt. Und dennoch ist die Tendenz zur Ausbildung bestimmter, typischer Erscheinungsformen deutlich. Vergleicht man Flussläufe untereinander, so stellt sich sowohl am Wiesenbach als auch beim breiten Strom wie dem Mississippi in den gleichen Proportionen von Flussbreite zur Sinuslänge eine Mäanderbildung in ganz bestimmter Sequenz ein. || Wir betrachten hier die Erscheinungsformen der Natur in ihren materiellen Aspekten. So gesehen, sind auch Luftspiegelungen oder Blitze materiell gebundene Erscheinungen. Sie sind Objekte, die Kräfte übertragen können, und sind daher Konstruktionen mit spezifischen konstruktiven Eigenschaften. Massen in jeder Form des Aggregatzustandes bilden Systeme als mehr oder minder dichte Packungen, bis hin zum völlig ungeordneten Zustand der freibeweglichen Teile. Die Systeme bauen sich entsprechend dem Ordnungsgrad auf: Kristalle nach dem ihnen innewohnenden Gesetz von Abstoßung und Anziehung der Ionen, freibewegliche, rollende, fließfähige heterogene Teile wie Sand und Geröll zu Halden und Schüttkegeln. Große Massen schließlich – aufgrund der Massenanziehung – zu kugeligen Masseanhäufungen, zu Planeten. Wir kennen die spiralige Anhäufung als Galaxie oder Strudel, wenn die Fliehkraft gegenüber der Gravitationskraft überwiegt. Wir kennen Anhäufungen als Schüttkegel, wobei der

7

8

manifestation is clear. If one compares the courses of different rivers on a creek as well as on a wide river such as the Mississippi a meander formation in a quite definite sequence sets in the same proportions from the river width to the sine length. || We are considering here the manifestations of nature in their material aspects. Viewed in this way, mirages or lightning flashes are also material-related manifestations. They are objects which can transfer forces and are therefore manifestations with specific structural properties. Mass in any form of physical condition forms systems in the form of more or less dense packing, up to the completely disordered state of free moving particles. The systems are composed according to the degree of order: crystals in accordance with their inherent law of repulsion and attraction of tons, free moving, rolling, fusible parts such as sand and stones to banks and alluvial cones. Large masses finally – because of gravitational attraction – to spherical accumulations of matter to planets. We are familiar with the spiral accumulations as galaxies or whirlpools when the centrifugal force outweighs the force of gravity. We are familiar with accumulations as debris cones where the frictional resistance of the flowable material determines the angles of the cone. Debris cones are volcanos, high mountains and heaps of rubble. The structural form adjusts itself to the material properties and the composition of the material. || Further structural forms are linear and movement forms, and river courses and channels in the shallows, and meanders. Many geometrical and phys-

Die Hierarchie der festen Massen:

Galaxie, Planet, Gebirge, Berg, Kristall, Winkel, Kegelfaltungen, Schüttkegel, Packung.

The hierarchy of fixed matter:

Galaxy, planet, mountain, range, mountain, crystal, angel, sphere, foldings, debris cone, packing.

9

Reibungswiderstand des fließfähigen Materials den Winkel des Kegels bestimmt. Schüttkegel sind Vulkane, hohe Berge und Geröllhalden. Die Konstruktionsform stellt sich selbst ein, entsprechend den Materialeigenschaften und der Masse des Materials. || Weitere Konstruktionsformen sind die linearen und die Bewegungsformen, die Flussläufe und Prile im Wattenmeer, die Mäander. Darüber wurden viele geometrische und physikalische Studien angestellt und die moderne Geologie versucht, das Ordnungsprinzip mittels Informationstheorie und computerunterstützter Mathematik aufzudecken. Erinnert sei hier nur an die Deutung nach den Regeln des Goldenen Schnittes, nach der Fibonacci-Reihe, wonach die Anzahl der Zweige in einer bestimmten Hierarchie zunimmt. Flächige Bewegungsformen, die Wellen, hinterlassen ihre Spuren im Schlick, auf den Sanddünen, als Bewegungsformen auf dem Wasser, oder als Rauch in der Luft. || Flüssigkeiten bilden an der Grenzfläche zu einem anderen Medium eine Oberflächenspannung, die für die Form der Flüssigkeitskonstruktion maßgebend ist. Die Spannung erzwingt eine Gesamtform mit möglichst geringer Oberfläche — im schwerelosen Zustand gäbe es nur kugelige Wasserformen. An diesem Institut haben uns lange Jahre die Tropfen und Blasen aus Wasser, Öl oder Luft begeistert. Die Hauptkennzeichen dieser Konstruktionen sind: die zugbeanspruchte Hülle um eine Füllung, das Medium kann Flüssigkeit sein oder Gas. Da Formen (gemessen in Radius) und Kraft (gemessen in Membranspannung) in direkter Beziehung zueinander stehen, kann

Bewegungsformen der festen, aber fließfähigen, der flüssigen und gasförmigen Massen: Wirbel, Welle und Mäander

Movement forms of the solid, but flowable, of the fluid and of the gaseous matter: vortex, wave and meander.

Abb.9 Wasserstrudel
Fig.9 Whirlpool

10

11

ical studies have been made on this and modern geology is trying to discover the principle of order by means of information theory and computer-supported mathematics. Let us just re-mind ourselves here of the interpretation according to the rules of the "Golden Section", and according to the Fibonacci sequence, whereby the number of branches increases in a definite hierarchy. Areal movement forms, the waves, leave their traces behind in slime, on the sand dunes, as movement forms on the water, or as smoke in the air. || Liquids form a surface tension at the contact surface to another medium. This surface tension is decisive for the form of the liquid constructions. The tension gives rise to a total form with the least possible surface area — in a weightless condition there would be only spherical water forms. At this Institute for many years we have been inspired by droplets and bubbles of water, oil or air. The main characteristics of these constructions are the tension stressed skin around a filling — the medium can be liquid or gas, the form (measured by radius) and force (measured in membrane tension) are directly related to one another, one can make assertions about the infernal pressure or the stability of such constructions. || The internal pressure of very small doplets of water, for example, is enormously high and can amount to several bars. || In non-living nature the systems organise and regulate themselves according to external and internal conditions. Their structural forms are inherent in them. Yet they are in accordance with their specific properties and constantly adapt to the external environmental conditions. Apparently "dead" nature is characterised by this flowing potential, by rearrangement, construction and destruction. lt lives in its own way.

Abb.10 Riffelbildung im Sand bei Ebbe
Fig.10 Wave forms in sand at ebb tide

Abb.11 breiter Strom in Südwestfrankreich
Fig.11 Wide stream in Southwest France

man Aussagen über den Innendruck oder die Festigkeit solcher Konstruktionen machen. Der Innendruck bspw. sehr kleiner Wassertropfen ist enorm hoch und kann mehrere bar betragen. || In der nicht lebenden Natur organisieren und regeln sich die Systeme selbst, entsprechend äußeren und inneren Bedingungen. Es entstehen die Konstruktionsformen aus sich heraus, doch entsprechend ihren spezifischen Eigenschaften und ständig in Anpassung an die äußeren Umweltbedingungen. Die scheinbar »tote« Natur ist durch dieses fließende Potential, durch Umlagerung, Auf- und Abbau charakterisiert. Sie lebt auf ihre Weise.

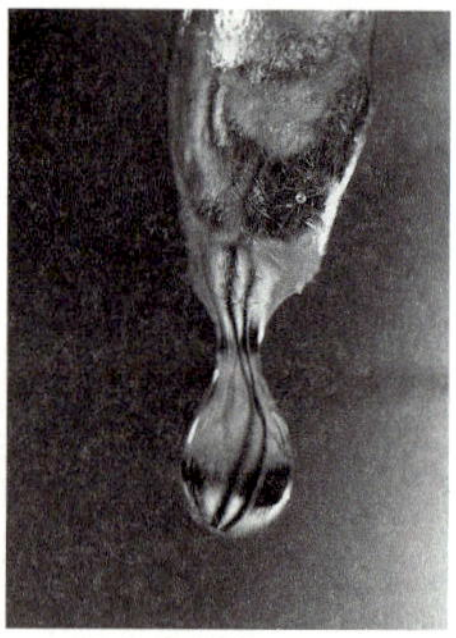

12 / 13

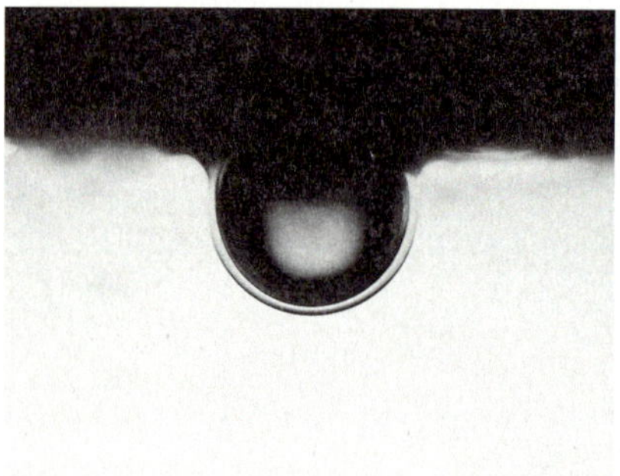
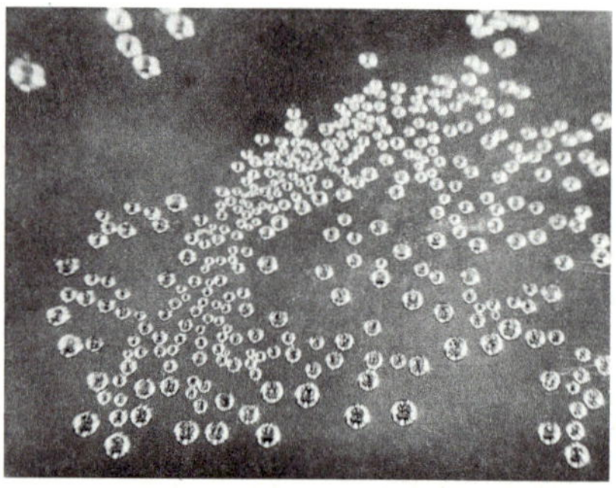

14 / 15

Wasserformen:

Abb.12 Ansammlung einzelner Wassertropfen in der Luft, die Wolke.

Abb.13 Fallender Wassertropfen

Abb.14 Mikrosphäre, eine Konstruktion aus Flüssigkeitsfüllung, doppelter Flüssigkeitshaut und umgebendem flüssigen Medium.

Abb.15 Luftblasen im Wasser

Water forms:

Fig.12 Clouds as an agglomeration of single waterdroplets in air.

Fig.13 Falling water droplet

Fig.14 The microsphere, a construction consisting of a fluid filling, a double layered fluid skin and of fluid medium.

Fig.15 Air bubbles in water

16

Wachsende und sich teilende Pneus

Growing and dividing Pneus

Frei Otto

Vorbemerkungen In der folgenden Abhandlung möchte ich aufzeigen, wie mit bekannten Mitteln der Technik die Form von Pneus geändert werden kann, wobei das Größerwerden (Wachsen) und das Teilen im Vordergrund stehen. | Es werden die wichtigsten technischen Methoden beschrieben, mit denen Formveränderungen an Pneus zu erreichen sind. Ich möchte aber auch einige Formveränderungsprozesse der Pneus in der lebenden Natur beschreiben, jedoch nur so weit, als sie mir erklärbar erscheinen. | Der folgende Text kann deshalb auf biologischem Gebiet keinesfalls als eine Darlegung neuer wissenschaftlicher Erkenntnisse gelten. Seine Hauptaufgabe ist es, zur Diskussion und vertiefenden Arbeit anzuregen. Ich bin mir dessen bewusst, dass mir viel von dem speziellen Wissen der Biologen, Zoologen, Botaniker und Zellphysiologen fehlt. | Jedoch musste ich bei häufigen Gesprächen mit Biologen feststellen, dass die Zusammenhänge von Form und Kraft nur wenig bekannt sind. Es kann daher nicht schaden, wenn dieses Thema zur Diskussion gestellt wird. || Im Sinne des vom IL betriebenen interdisziplinären Erfahrungsaustausches habe ich diesen Text folgenden Biologen, Botanikern, Zoologen, Anthropologen und Paläontologen zur Ansicht gegeben: Helmcke, Schill, Bereiter-Hahn, Gutmann, Seilacher, Nachtigall. Ihre Anmerkungen waren wertvolle Hilfe. Auf besondere Einwände habe ich hingewiesen. || Unter dieser Voraussetzung möchte ich auch einige Hypothesen aufstellen, um biologische Prozesse zu klären, die bisher offensichtlich noch nicht erklärt sind oder deren bisherige Erklärungen mir aus Gründen der Kenntnis von Form und Spannungen bei Pneus wirklichkeitsfremd erscheinen. | Seit über zwei Jahrzehnten arbeiten wir (davon 14 Jahre am IL) intensiv auf dem Gebiet der Wechselwirkung von Form und Kraft bei biegeunsteifen Membranen. Auf Grund der großen Fülle von praktischen Erfahrungen und theoretischem Wissen kann die Richtigkeit der hier gemachten Äußerungen auf technischem Gebiet — jedoch nur auf diesem — als überprüft angesehen werden. Die Frage der Übertragbarkeit von größeren technischen Strukturen auf sehr kleine Objekte (z. B. aus dem Bereich der Mikrobiologie) wurde ferner durch viele Versuche und Analogierechnungen geprüft. Es wurde von uns festgestellt, dass Vergleiche zwischen Makrostrukturen der Technik und mikroskopisch kleinen Objekten wie Zellen mit biegeweichen Wänden grundsätzlich zulässig sind. Es ist also durchaus möglich, den Formentstehungsprozess, die Spannungsverhältnisse und die Formänderung bei Belastungen an Lufthallen, Gummiballons, Seifenblasen usw. mit denselben Methoden zu untersuchen und auf biologische Objekte zu übertragen. Für ähnliche Formen unter ähnlichen Beanspruchungen gelten im Prinzip die gleichen statisch konstruktiven Gesetze. Die Gültigkeit hört im ultramikroskopischen Bereich erst dann auf, wenn die Strukturen nur noch aus einigen Tausenden von Molekülen bestehen, für die jedoch dann auch besondere — durchaus nicht völlig unbekannte — konstruktive Wechselwirkungen von Form und Kraft gelten.

Preliminary Remarks In the following article I intend to show how the shape of pneus can be changed by applying means known in technology. Growth and division will be in the forefront. The most important technical method, with which the shape of pneus can be changed, will be described. I should, however, also like to describe some changes of shape of pneus in living nature, as far as they appear to me explainable. | The following text can therefore in no way be considered as an exposition of new scientific knowledge in biology. Its principal aim is to stimulate discussion and deeper-going research. I am aware of the fact that I lack much of the specialized knowledge of biologists, zoologists, botanists, and cell physiologists. | During my frequent talks with biologists, I found, however, that the relationships between shape and force are little known. It can therefore do no harm to make this a subject for discussion. | I have, in accordance with the interdisciplinary exchange of experience, practiced by IL, submitted this text for review to the following biologists, botanists, zoologists, anthropologists, and palaeontologists: Helmcke, Schill, Bereiter-Hahn, Gutmann, Seilacher, and Nachtigall. Their observations constituted an invaluable aid. I have drawn the attention to specific remarks. || Under these conditions I would like to propose some hypotheses for explaining biological processes that have obviously not yet been explained, or whose present explanations appear to me to be wrong in view of my knowledge of the shape and stresses of pneus. For over two decodes (14 years of these at the IL) we have been working intensively on the interaction between shape and force in the case of flexible membranes. The large amount of experience and theoretical knowledge permits us to consider the statements made here as proven from the technical point of view — but only from it. The question of the transferability of results from larger technical structures to very small objects (e.g., in the field of microbiology) was investigated in many tests and analog computations. We found that comparisons between engineering macrostructures and microscopic objects like cells with flexible walls are permissible in principle. lt is therefore quite possible to investigate shape-forming process, stress relationships, and changes of shape during load applications to pneumatic halls, rubber balloons, soap bubbles, etc. with the same methods and to apply the results to biological objects. The some statical structural laws are in principle valid for similar shapes subject to similar stresses. The validity breaks down in the ultramicroscopic region only when the structures consist only of a few thousand molecules. Special, but not completely unknown structural interactions between shape and force exist for such objects.

Changes of shape in simple Pneus General: The ability of a pneu to change its shape depends on the material properties of the envelope. Its elasticity, thickness, and internal structure are of particular influence. || We shall in the following

1/2
3
4
5
6
7

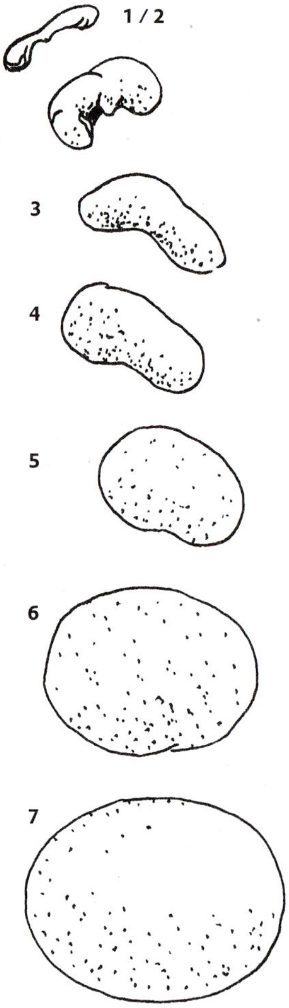

Formveränderungen bei einfachen Pneus Allgemeines: Das Vermögen eines Pneus, seine Form zu verändern, ist abhängig von den Materialeigenschaften der Hülle. Insbesondere haben deren Elastizität, Dicke und innere Struktur einen großen Einfluss. || Im Folgenden wird unter dem Begriff »Formänderung« jede Art der Änderung der Form eines Pneus verstanden: dazu gehören z. B. die vielen möglichen Formen, die ein Pneu während des Aufblasens annimmt, wie sie die Skizzen vom evakuierten Zustand zum Zustand kurz vor dem Platzen zeigen (Abb. 1–7). Ein evakuierter Pneu (Abb. 1) hat eine (kinematisch instabile) indifferente Form, ebenso der nicht völlig gefüllte Pneu (Abb. 2). Er ist bereits durch geringste äußere Kräfte verformbar. Eine deutlich analysierbare Form erhält ein Pneu erst dann, wenn im überwiegenden Bereich der Membrane feststellbare Spannungen auftreten, die die Membrane (Abb. 3) spannen. Eine völlig definierbare stabile Form wird erreicht, wenn die gesamte Haut Membranspannungen aufweist. Diese Form ist hier als »Ausgangsform« bezeichnet. Das Vermögen von Pneus, die Form zu ändern, ist von der Ausbildung und Elastizität der Hülle abhängig. || Wir kennen Pneus mit folgenden unterschiedlichen Membranen, die jeweils unterschiedliche Eigenschaften haben:

a) Flüssigkeitshaut: tritt auf bei Seifenblasen, Nebel- und Wassertropfen, Öltröpfchen im Wasser, Luftbläschen im Wasser, Antibubbles, Mikrosphären. (Abb. 8)

b) Gummihaut: also eine Haut, die in bestimmten Grenzen als äußerst elastisch bekannt ist, z. B. bei Kinderluftballonen. (Abb. 9)

c) Zugsteife, doch biegeunsteife Haut: die nur wenig elastisch ist, wie Papier, Metall- und Kunststofffolien, z. B. bei Lufthallen. (Abb. 10)

d) Plastische Haut: die sich je nach Spannung oder Temperatur wie eine Flüssigkeitshaut (nach a), wie eine Gummihaut (nach b) oder eine zugsteife Folie (nach c) verhält, z. B. bei Folienverpackungen. (Abb. 11)

e) Haut, die durch zugbeanspruchte Bänder und Netze verstärkt ist. (Abb. 12)

Formveränderung bei Flüssigkeitspneus: Bei den Wasserpneus wie Seifenblasen, Wassertropfen, Antibubbles, Mikrosphären gibt es keine Hülle konstanter Größe, die bei Evakuierung zusammenschrumpft und sich zusammenfalten kann. || Eine Seifenblase kann durch Vergrößerung des Volumens aufgebläht werden. Die Kugelform bleibt dabei erhalten (Abb. 13). Die Membranspannung bleibt gleich. Die Form bleibt geometrisch ähnlich. Die Oberfläche wächst jedoch. Bei Evakuierung ist der Vorgang umgekehrt. Die Hülle wird kleiner, bis ein luftleeres Wassertröpfchen übrig bleibt. | Das Gleiche gilt nachweisbar für eine im Wasser schwebende Antibubble (Füllung Wasser, Medium Wasser), Doppelhaut mit Luftspalt (siehe IL 9, Seite 219), die experimentell vergrößert oder verkleinert

denote by "change of shape" any kind of shape alternation of a pneu. This includes all the various possible shapes assumed by a pneu during inflation, as shown in the sketches from the evacuated pneu to the state shortly before bursting (Fig. 1–7). An evacuated pneu (Fig. 1) has a (kinematically instable) indifferent shape as does a not completely filled pneu (Fig. 2). It can be deformed by very small external forces. A pneu assumes a clearly analyzable shape only when definite stresses appear throughout the membrane, so that it is stretched (Fig. 3). A completely defined stable shape is attained when the entire skin is subject to membrane stresses. This shape will here be denoted as "initial shape". The ability of pneus to change their shape depends on the design and elasticity of the envelope.

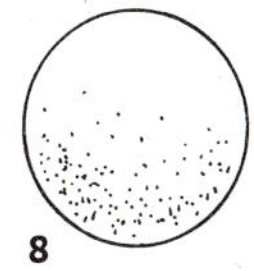
8

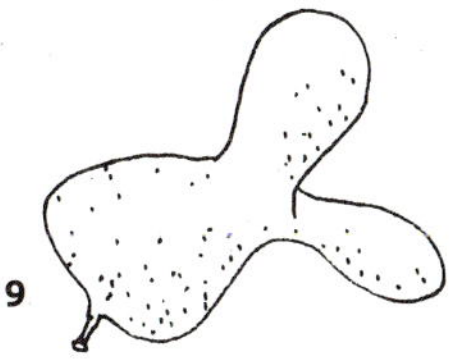
9

We know pneus with the following different membranes, having respectively differing properties:

a) Liquid skins: are present in soap bubbles, mist and water drops, oil droplets in water, air bubbles in water, antibubbles, microspheres. (Fig. 8)

b) Rubber skins: i.e., skins known to be extremely "elastic" within certain limits, e.g., air balloons for children. (Fig. 9)

c) Skins rigid in tension but flexible in bending: which are only slightly elastic, like paper, metal, and plastic foils, e.g., pneumatic halls. (Fig. 10)

d) Plastic skins: behaving like liquid skins (as under a), rubber skins (as under b), or plastic foils rigid in tension (as under c), according to the stress or temperature, e.g., plastic packagings. (Fig. 11)

e) Skins reinforced by tapes and nets subject to tension. (Fig. 12)

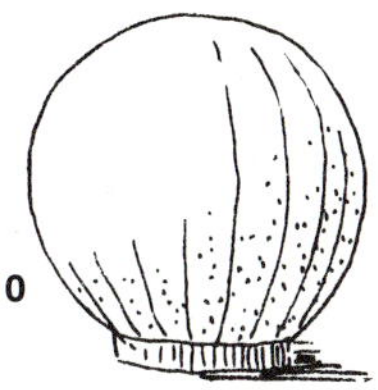
10

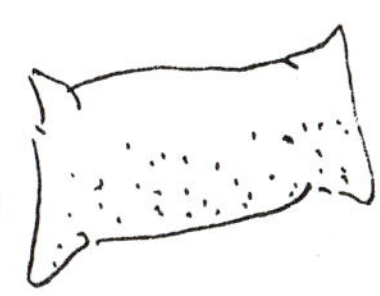
11

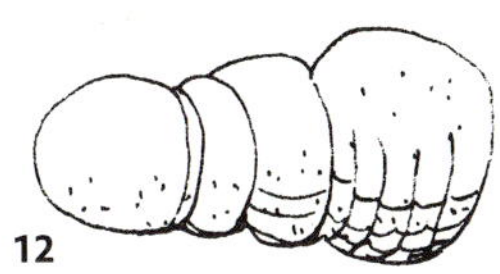
12

Changes of shape in liquid pneus: Water pneus like soap bubbles, water drops, antibubbles, and microspheres have no envelopes of constant size, which can shrink and fold together during evacuation. || A soap bubble can be inflated by increasing its volume. The spherical shape is preserved in this case (Fig. 13). The membrane stress remains constant. The shape remains geometrically similar. The surface, however, grows. The inverse process occurs during evacuation. The envelope becomes smaller, until a water drop containing no air remains. The same can be shown to hold true for an antibubble (water as filling, water as medium) floating in water and for a double envelope with air gaps (cf. IL 9, p. 219), which can be enlarged or reduced in size experimentally. This must also be true to the same extent for microspheres whose smallest diameter is approximately 10 A° (= 0.000 001 mm) and whose largest diameter may amount to more than one meter (cf. IL 9, pp. 219, 226). || A sure sign that no liquid skin is present in an otherwise unknown microscopic pneu is that it deviates noticeably from a spherical shape when inflated. It then cannot be a microsphere (antibubble). The same laws are

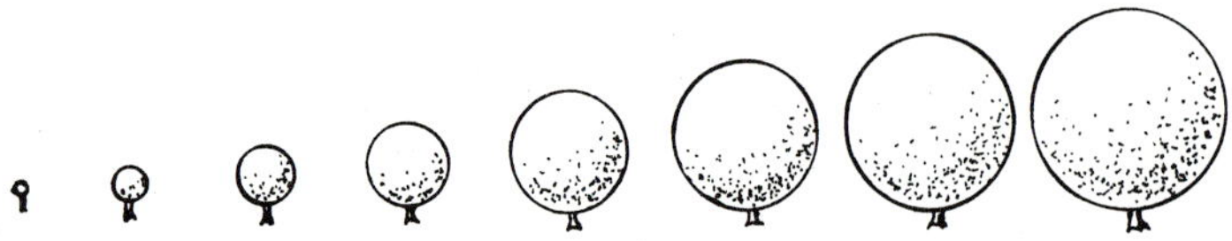

13

14

werden kann. Dies muss auch in gleichem Maße für die Mikrosphäre gelten, deren kleinster Durchmesser etwa 10 A° = 10^{-6} mm und deren größter Durchmesser über einen Meter betragen kann (IL 9, Seite 219, 226). || Wenn es gelingt, einen sonst unbekannten mikroskopisch kleinen Pneu aufzublähen, und wenn an ihm dann merkliche Abweichungen von der Kugelform festzustellen sind, so ist das ein sicheres Zeichen dafür, dass keine Flüssigkeitshaut vorhanden ist. Es kann sich also bspw. nicht um eine Mikrosphäre (Antibubble) handeln. Die gleichen Gesetze gelten auch für den frei schwebenden Nebeltropfen oder für das Öltröpfchen im Wasser. Es bleibt bei jeder Vergrößerung oder Verkleinerung kugelförmig. | Wasser- und Quecksilbertropfen, die auf einer Fläche aufliegen, sind zwischen den Dimensionen von 0,000 000 1 mm bis 1,0 mm kugelförmig, darüber erhalt das Eigengewicht der Füllung Einfluss. Eine maximale Höhe wird beim Tropfen nicht überschritten (Abb. 14).

Formveränderungen bei Gummiballonen und dergleichen: Das Verhalten eines Pneus mit Gummihaut (wie z. B. eines Kinderluftballons), der exakt kugelförmig mit exakt gleicher Wanddicke hergestellt wird, was technisch äußerst schwierig ist, gleicht weitgehend der Seifenblase: Er verändert seine Form nicht, sondern bleibt bei der Vergrößerung kugelförmig (Abb. 15 - 17). | Ein Gummiballon, der überall eine gleich dicke Haut und ferner eine Ausgangsform besitzt, die in irgend-

15 16 17

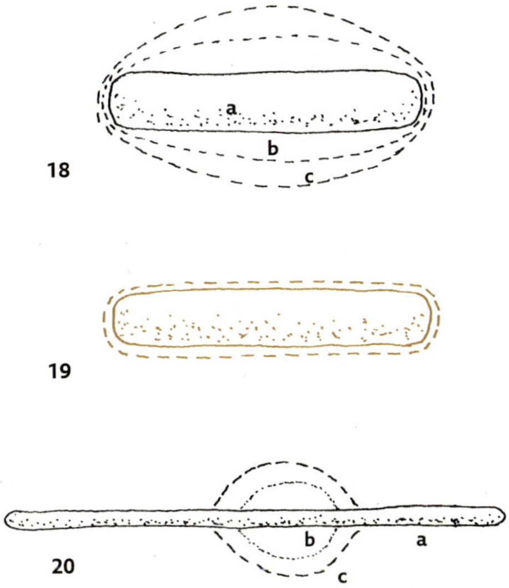

18

19

20

also valid for freely floating mist drops or oil droplets in water. They remain spherical during any enlargement or size reduction. | Drops of water or mercury lying on a surface are spherical when measured between 0.000 0001 and 0.1 mm. The dead weight of the liquid exerts an influence above this size. The height of a drop cannot exceed a maximum (Fig. 14).

Changes of shape in rubber balloons and similar objects: The behaviour of a pneu with a rubber skin (e.g., like that of an air balloon for children), which is made accurately spherical with precisely uniform wall thickness (this is technically very difficult),is very similar to that of a soap bubble: Its shape does not change, but remains spherical during enlargement (Fig. 15–17). | A rubber balloon, which everywhere has a skin of uniform thickness and an initial shape deviating from a spere, does not grow in a geometrically similar manner. Thus, a spindle-shaped body (Fig. 18a) becomes more rounded when its contents increase (Fig. 18b, 18c). The infinitely large final shape would always be a sphere if rubber could be stretched at will. Enlargement of a pneu so that it remains geometrically similar, is possible only (and even then only approximately), if the initial shape has a non uniform wall thickness (Fig. 19). || Very thin rubber hoses (Fig. 20a), made of highly elastic rubber or of elasticoplasflc material, exhibit the characteristic spherical shape known as "bulge" (Fig. 20b, 20c) already at an early stage of inflation. It often oc-

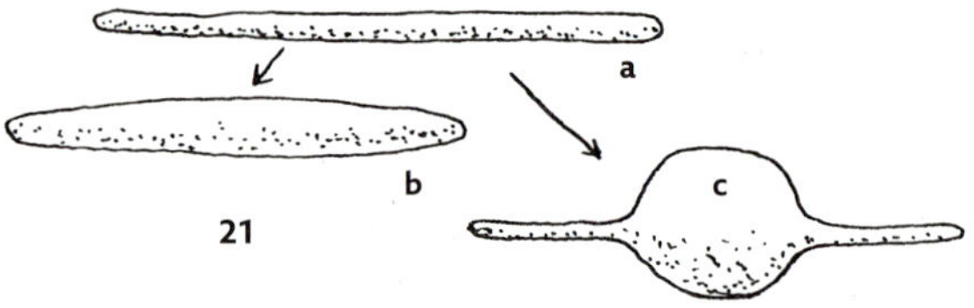

21

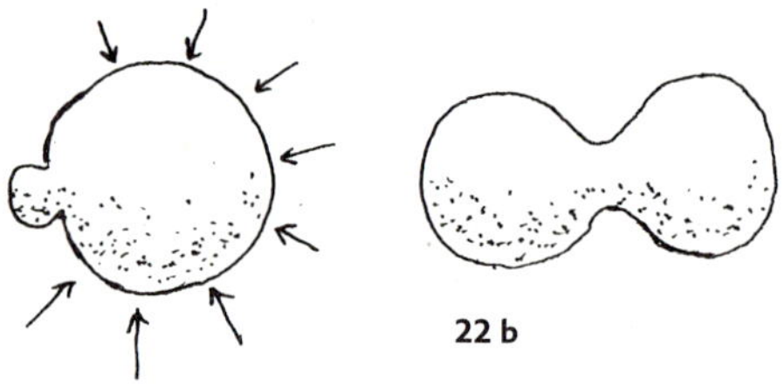

22 a

22 b

einer Weise von der Kugelform abweicht, vergrößert sich nicht geometrisch ähnlich. So wird bspw. ein spindelförmiger Körper (Abb. 18a) bei Vergrößerung des Inhaltes rundlicher (Abb. 18b, 18c). Könnte man Gummi beliebig dehnen, dann wäre die unendlich große Endform stets eine Kugel. || Soll eine Vergrößerung so erfolgen, dass der vergrößerte Pneu geometrisch ähnlich bleibt, so kann das nur – und dann auch nur angenähert – erreicht werden, wenn die Wanddicke der Ausgangsform ungleichmäßig dick ist (Abb. 19). || Sehr dünne Gummischläuche (Abb. 20a), aus äußerst elastischem Gummi oder aus elastisch-plastischem Material, zeigen beim Aufblasen bereits im frühen Stadium die typische Kugelbildung (Abb. 20b, 20c), die als »Beule« bekannt ist. Sie tritt bei Gummischläuchen, aber auch bei Därmen oft überraschend und ungewollt auf. Sie wird von Gummitechnikern und Ärzten häufig als Kennzeichen von »schwachen« Stellen angesehen. Es lässt sich jedoch leicht rechnerisch und experimentell nachweisen, dass diese Erscheinung eine stabile Form ist, die bei äußerst elastischen Membranen auftreten muss, auch wenn sie nicht geschädigt sind. Es ist jedoch einzuräumen, dass »Beulen« an geschädigten Stellen eher auftreten als an ungeschädigten. | Bei uns wurde in den letzten Jahren darüber diskutiert, ob die sog. Puffs bei Chromosomen als solche Beulen interpretiert werden können. || Die »stabile« Form einer »Beule« ist leicht zu erklären: Ein Schlauch (Abb. 21a) mit unelastischer Haut erhält bei hohem Innendruck die Form 21b. Der gleich geformte Schlauch mit elas-

curs unintentionally and suddenly in rubber hoses and also in intestines. Rubber technologists and physicians frequently consider it to indicate a "weak" point. It can, however, be easily shown by computation and experiment that this phenomenon represents a stable shape which must occur in highly elastic membranes even if they are not damaged. It must, however, be admitted that "bulges" occur at damaged points sooner than at undamaged ones. | A point debated at the IL in recent years was whether the so-called puffs in chromosomes can be interpreted as such bulges. || The "stable" shape of a "bulge" can be explained easily. A hose (Fig. 21a) with an inelastic skin assumes the shape in Fig. 21b when the internal pressure is high, while a similar hose with an elastic skin assumes the shape in Fig. 21c. The total surface area of the shape in Fig. 21c is smaller than that of the shape in at the same volume content (Fig. 21b). Less deformation energy has therefore to be expended. || A rubber balloon of spherical initial shape (Fig. 23b) becomes enlarged and changes its shape during filling if its membrane thickness is nonuniform. It thus no longer remains spherical. Balloons are very difficult to manufacture with precisely uniform skin thickness. Becoming norispherical is therefore the rule and remaining spherical the exception. Experts, however, without such analyses can make qualitative and relatively accurate quantitative statements on any shape assumed by a pneu from the initial shape until it bursts. I tshould be stressed again in this connection that the membrane stresses in the pneu with flexible skin can be deduced from the shape. The shape characterizes the design. It must, however, be measured with geodetic and mathematical means and analyzed. Pneus with membranes of materials having low elasticity (like sheet metal, paper) are in principle subject to the same laws as pneus made of rubber. Only the deformation from the initial shape up to bursting is less and "bulges" occur never or very seldom; on the other hand, foldsand buds may appear.

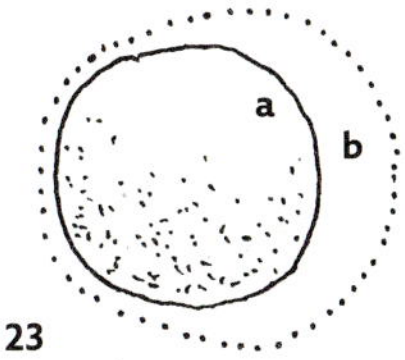

23

Budding: A so-called "bud" may occur during inflation of a rubber balloon (Fig. 24a), if its skin has a particularly thin spot there. The material of the membrane is especially stretched in the region of the bud. The bud usually bursts when the internal pressure is raised further.

Reinforcements: The shape of a rubber balloon (Fig. 25a) can be easily changed when a "patch" is glued to it (p) — e.g., a rubber foil. Further inflation (Fig. 25b) causes the patch to form a recess. The patch expands less and therefore takes up higher membrane stresses. | Such "patches" may have arbitrary shapes (Fig. 26a) and also form rings and nets (Fig. 26b). (Reticular pollen grains have a similar shape, according to Schill). No energy has to be expended in order to glue an unstretched "patch" to the inflated pneu. The energy required for deformation is only

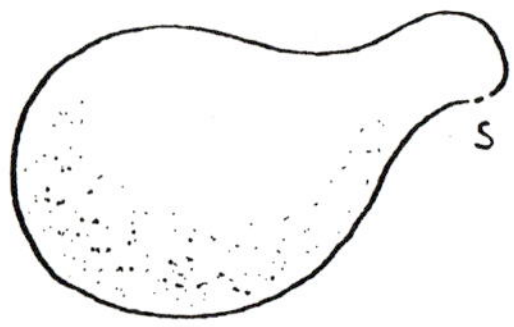

24a

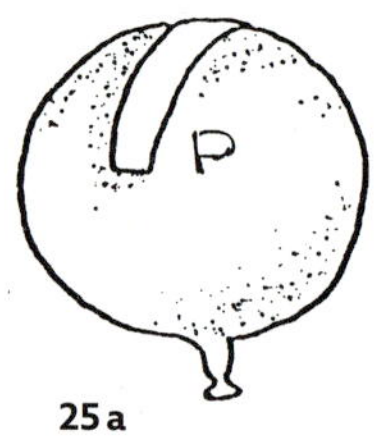

25a

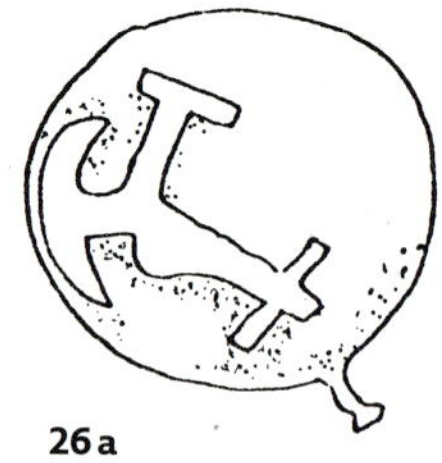

26a

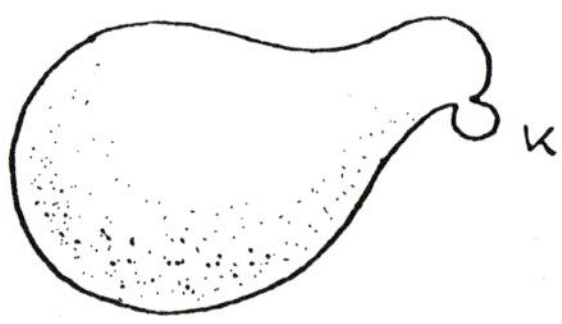

24b

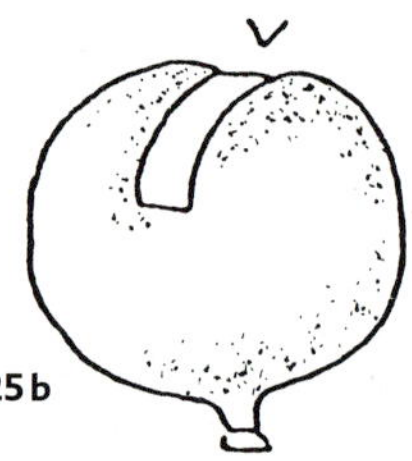

25b

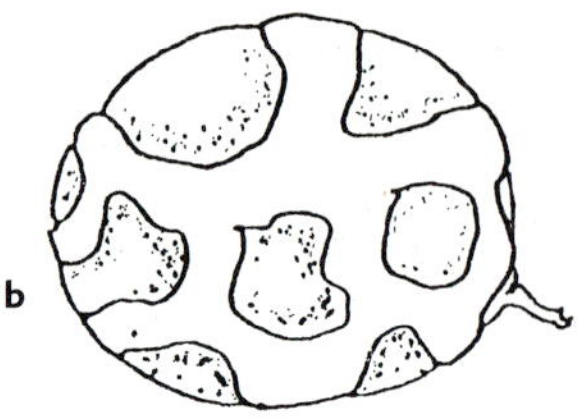

26b

tischer Haut die Form 21c. Die Gesamtoberfläche der Form 21c ist bei gleichem Volumeninhalt *kleiner* als die Form 21b. Es war weniger Verformungsenergie aufzuwenden. || Ist die Membrandicke eines Gummiballons mit kugelförmiger Ausgangsform (Abb.23a) unterschiedlich dick, dann vergrößert sich der Ballon bei Füllung unter Veränderung der Form (Abb.23b). Er bleibt also nicht mehr kugelförmig. Da sich Ballone äußerst schwierig mit exakt gleicher Dicke herstellen lassen, ist das Unrundwerden die Regel und das Kugelförmigbleiben die Ausnahme. Fachleute können aber auch ohne solche Analysen durchaus qualitative und sogar verhältnismäßig quantitative Aussagen machen, und zwar für jede der verschiedenen Formen, die beim Aufblähen eines Pneus von der Ausgangsform bis zum Platzen (Bersten) eingenommen wird. Es sei an dieser Stelle noch nachdrücklich darauf hingewiesen, dass man bei einem Pneu mit einer biegeunsteifen Haut von der Form auf die Membranspannungen schließen kann. Die Form offenbart die Konstruktion. Man muss sie jedoch mit geodätischen und mathematischen Mitteln aufmessen und analysieren. | Für Pneus mit einer Membrane aus wenig elastischem Material (wie Blech, Papier) gilt im Prinzip das Gleiche wie bei allem, was für Pneus aus Gummi gesagt wurde. Lediglich die Formänderung ist von der Ausgangsform bis zum Bersten geringer und die »Beulen« treten dafür nicht (bzw. äußerst selten) auf, dafür aber Falten und Knospen.

Abb.27 Formänderung bei einfachen Pneus. Die äußere Form eines einfachen Pneus, hier eines Luftballons, kann abhängig sein von der Füllung (Gas, Flüssigkeit, Festkörper) oder von der Beschaffenheit der Membrane.

Fig.27 Form alteration in simple pneus. The external form of a simple pneu, in this case an air balloon, can depend on the filling (gas, fluid solids) or from nature of the membrane

expended through further inflation. This fact seems to be of fundamental importance for deformation processes in living nature. We shall therefore discuss it in detail in the next section.

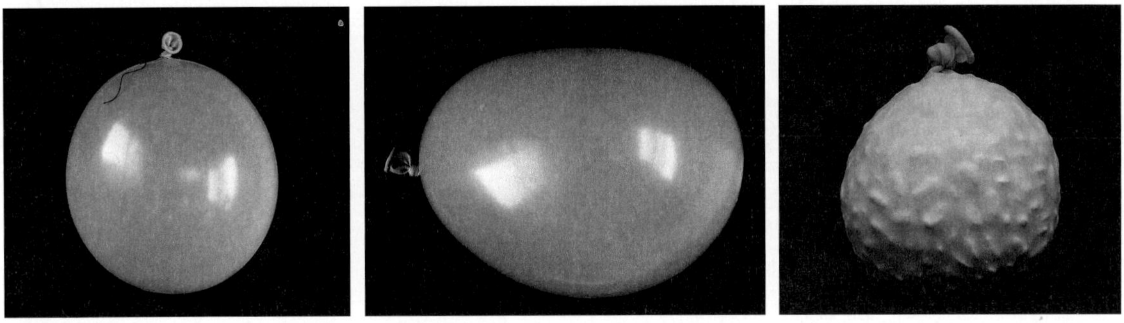

27

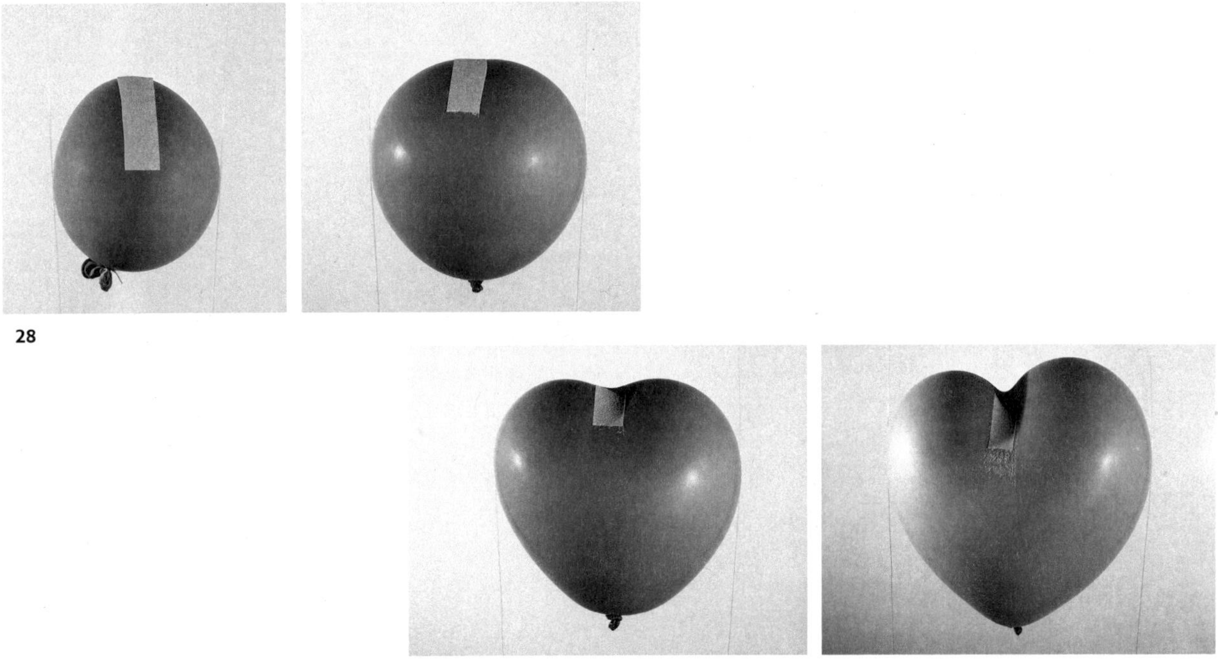

28

Abb. 28 Durch nur einfache Maßnahmen (hier Klebestreifen) wird die spätere Form eines Pneus geprägt. Der Klebestreifen entspricht einer örtlichen Verdickung der Membrane.

Fig. 28 Using only simple measure in this case adhesive strips) the later form of a pneu is decided. The adhesive strip corresponds to a local thickening of the membrane.

Knospung: Hat ein Gummiballon (Abb.24) in der Haut eine besonders dünne Stelle (s), so kann beim Aufblasen an dieser Stelle eine so genannte »Knospe« (k) auftreten. Im Bereich der Knospe wird jedoch das Membranmaterial besonders gedehnt. In der Regel platzt die Knospe bei weiterer Erhöhung des Innendrucks.

Verstärkungen: Die Form eines Gummiballons (Abb.25a) lässt sich leicht ändern, wenn ein »Pflaster« aufgeklebt wird (p) – z.B. aus Gummifolie. Bei weiterem Aufblasen (Abb.25b) wirkt sich der Einfluss des Pflasters als Vertiefung (v) aus. Das Pflaster dehnt sich weniger und zieht daher mehr Membranspannungen an sich. | Solche »Pflaster« können beliebige Formen haben und auch Ringe und Netze bilden (Abb.26a, 26b) (nach Schill haben bestimmte Pollenkörner eine ähnliche Form). Klebt man »Pflaster« spannungslos auf den aufgeblasenen Pneu auf, dann braucht man dazu keine Energie aufzuwenden. Die Energie, die zur Verformung benötigt wird, wird erst durch weiteres Aufblasen eingebracht. Diese Tatsache scheint von grundlegender Bedeutung für Formänderungsprozesse in der lebenden Natur zu sein. Deshalb wird sie im nächsten Abschnitt ausführlicher behandelt.

29

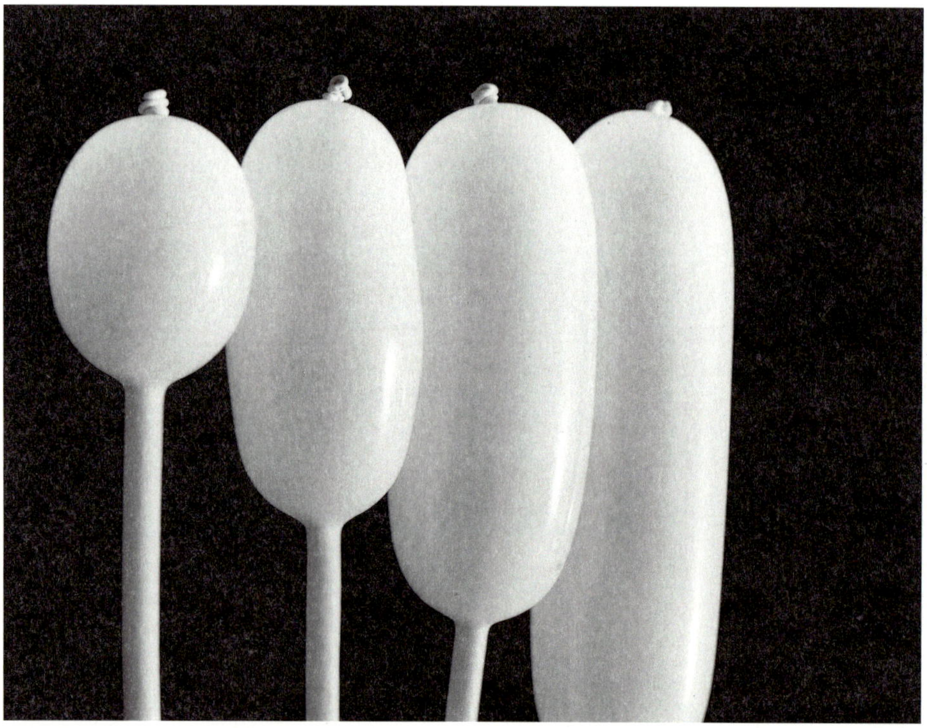

Abb.29 Das Bild zeigt die verschiedenen Stadien eines Luftballons während der Füllung. Die starken Membrankräfte in Radialrichtung zwingen bei wachsendem Innendruck die Beule in Längsrichtung.

Fig.29 The illustration shows the various stages of an air balloon during filling. The strong membrane forces the radial direction force the bulge into the longitudinal direction under growing internal pressure.

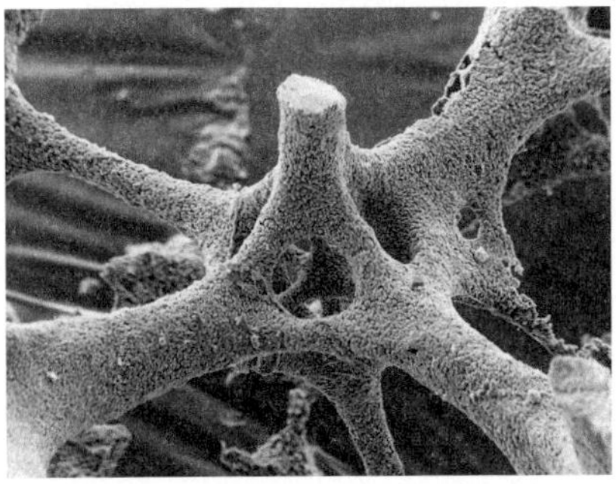

30

31

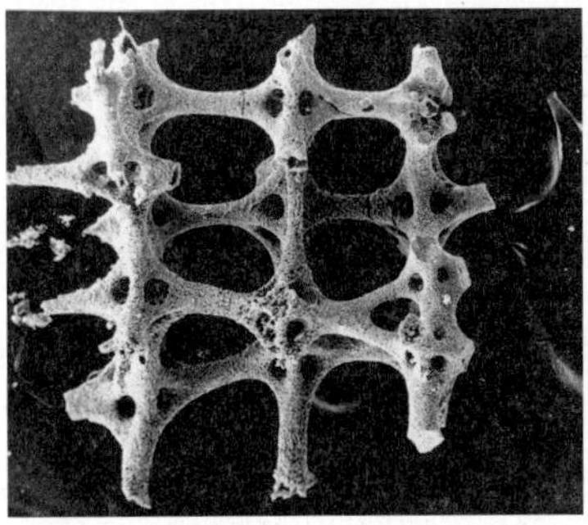

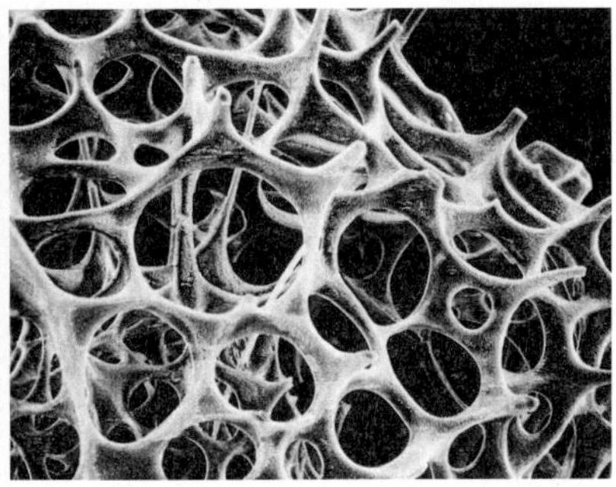

32 / 33

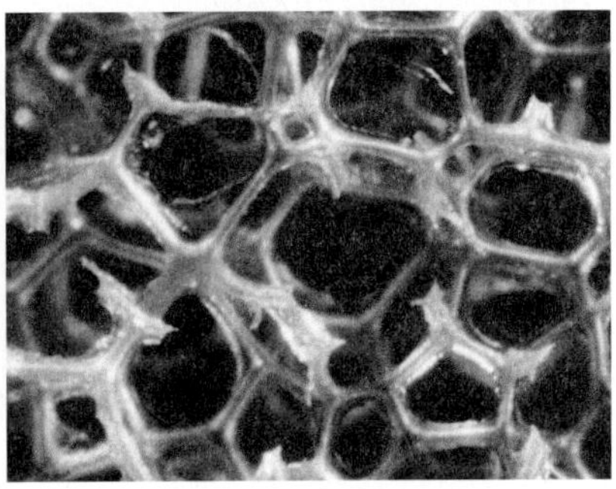

34 / 35

Schlauchnetze in Natur und Technik

Abb.30 Ausschnitt aus dem Skelett eines Kieselschwammes.

Abb.31 Studie eines Gitterträgers

Abb.32 Kieselschwamm

Abb.33 Ausschnitt aus dem Skelett einer Kieselalge 3000 x Vergrößerung

Abb.34 Studie. Der Hohlraum zwischen wassergefüllten Pneus wurde mit Gips ausgegossen.

Abb.35 PUR-Schaumzelle stark vergrößert. Nur der Zwischenraum ist verhärtet.

Tube nets in nature and technics

Fig.30 Section of the skeleton of siliceous sponge

Fig.31 Study of a grid girder

Fig.32 Flint sponge

Fig.33 Section from the skeleton of a diatom magnified 3000 times

Fig.34 Study. The hollow space between water-filled pneus was filled with plaster.

Fig.35 Polyeurethane foam cell greatly magnified. Only the intermediate space is hardened.

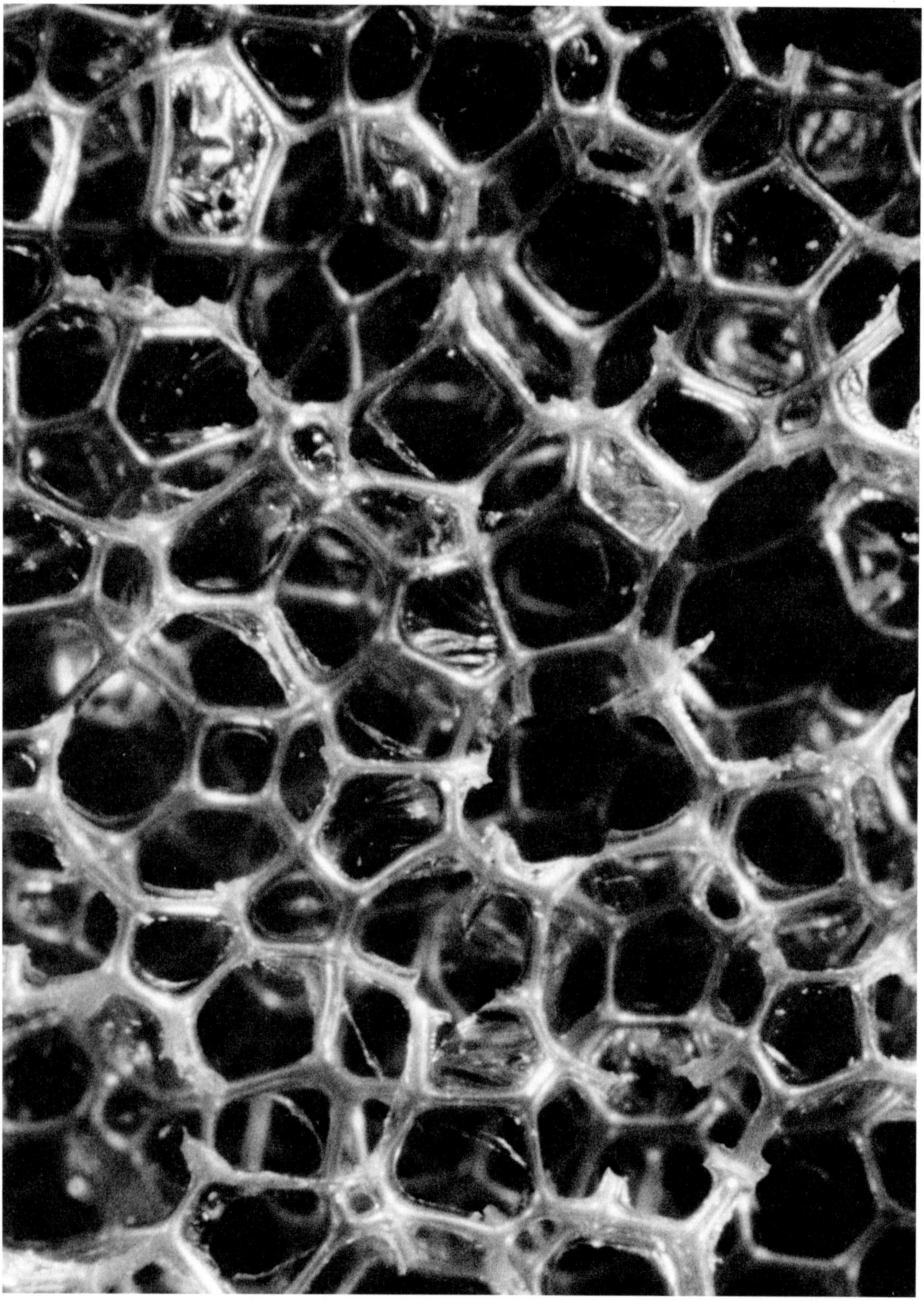

Abb.36 PUR-Schaum, vergrößert.

Fig.36 Polyeurethane foam cells, greatly magnified.

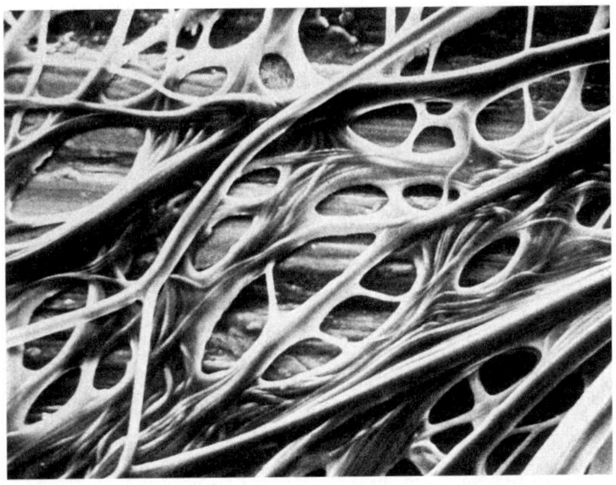

37

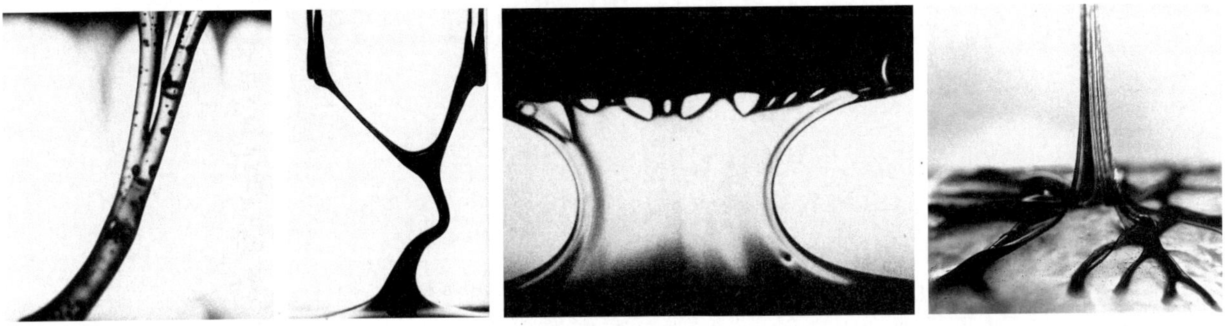

38

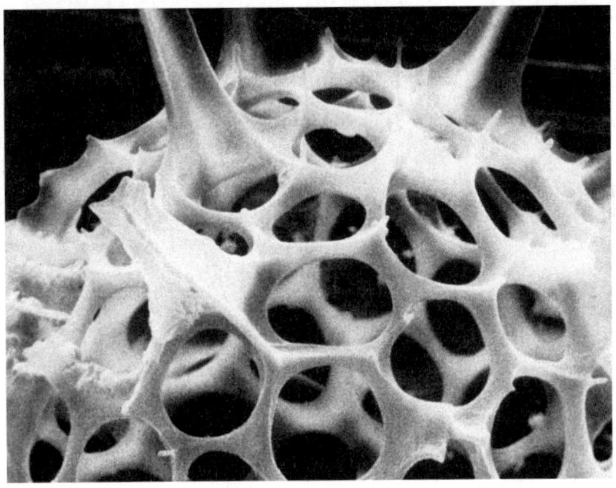

39

Flüssigkeitsfäden
scheinen pneumatische Formentstehungsprozesse zu haben.

Abb.37 Spinnfäden, Anheftungsstelle.

Abb.38 Studien mit viskosen Flüssigkeiten: Honig, Klebstoff.

Abb.39 Radiolarie, ca. 3000 fache Vergrößerung.

Formale Analogie: Radiolarienstachel - Flüssigkeitsfaden. Oder sind die Entstehungsprozesse beider Strukturen vergleichbar?

Fluid threads
appear to have pneumatic form-origin processes.

Fig.37 Spider threads, fastening pint

Fig.38 Studies with viscous fluids: honey, glue.

Fig.39 Radiolaria, approx. 3000 x magnification.

Form analogy: radiotaria barb - fluid thread. Or are the creation processes comparable?

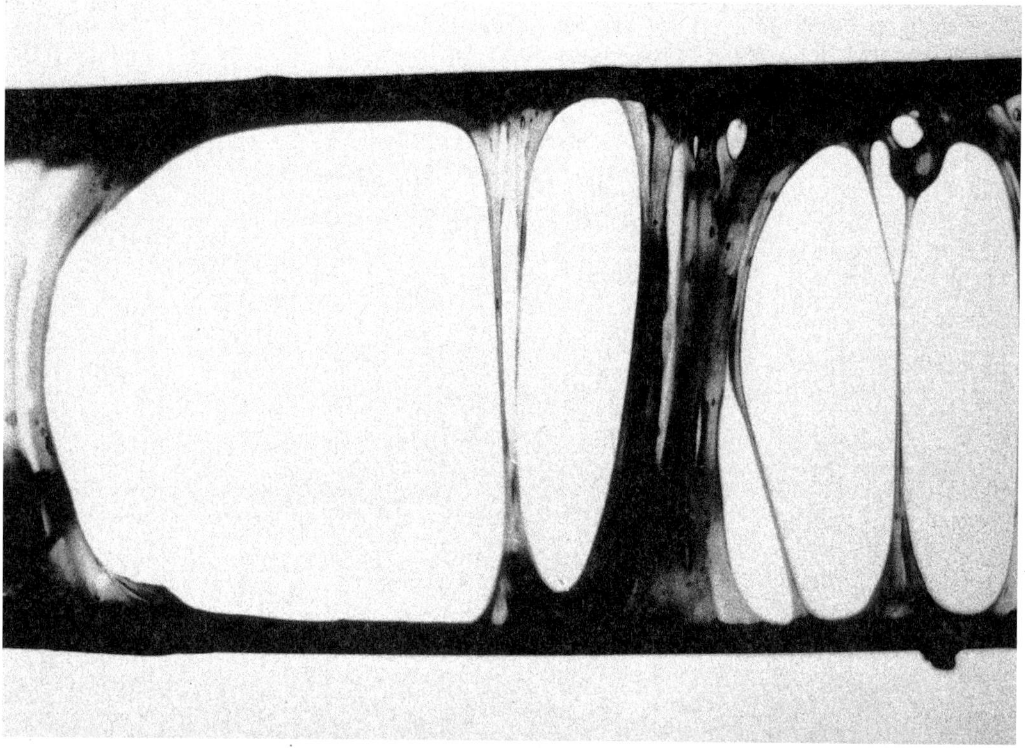

40

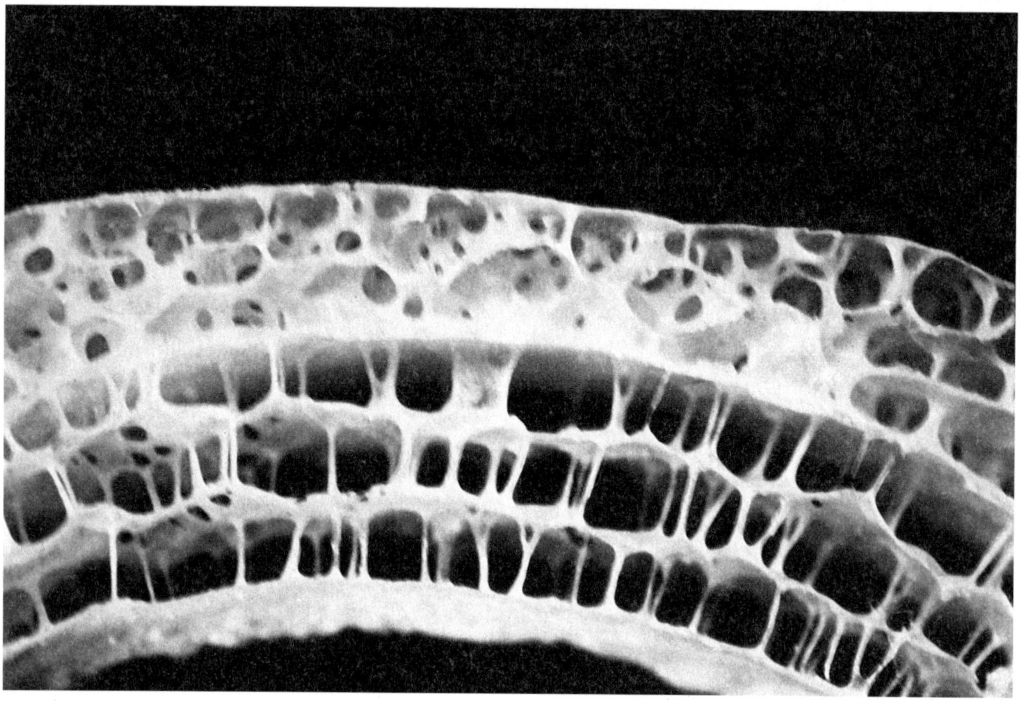

41

Abb.40 Flüssigkeitsfäden
Fig.40 Fluid threads

Abb.41 Vergrößerter Ausschnitt der Schädeldecke eines Vogels. Auch hier nur formale Analogie oder vergleichbarer Entstehungsprozess?

Fig.41 Enlarged section of the skull of a bird. Here too just a form analogy, or comparable creation process?

17

Sandwichstrukturen der Schädelkapsel verschiedener Vögel.

Zum Leichtbauprinzip bei Organismen

Sandwich structures in the skull capsules of various birds.

The principle of lightweight structures in organisms

Paul Bühler

In der Biologie wird vor allem im Zusammenhang mit Stützorganen häufig vom allgemeingültigen Prinzip der Materialersparnis gesprochen. Gibt es ein solches allgemeingültiges Prinzip? Sie erinnern sich an die Schädelkapselstruktur einer Elster, die Herr Kummer gezeigt hatte, und daran, dass er betonte, dass die Knochenstrukturen bei Vögeln sich auffallend von denen der Säugetiere unterscheiden. Die Vögel bilden zur Erhöhung der Flugfähigkeit im besonderen Maße Leichtbaustrukturen aus. Bei anderen Tierarten wie den Korallen der warmen Meere zum Beispiel, die mit ihren Skeletten ganze Felsmassive errichten, scheint das Prinzip der Materialersparnis aber keine so wesentliche Rolle zu spielen. Auffallend ist das Prinzip des Leichtbaus außer bei Vögeln auch im Skelett von Landtieren verwirklicht, die ihren Körper frei über dem Boden tragen, und zwar in Abhängigkeit von der Größe – je größer eine Art, desto ausgeprägter die Strukturen. Herr Otto hat zwar gestern darauf hingewiesen, dass die konstruktive Effektivität prinzipiell nur von der Form und den Materialeigenschaften abhinge und nicht von der Größe. Das ist nur dann richtig, wenn man die Belastung als Faktor gesondert betrachtet. Bei Konstruktionen, die durch die Erdanziehungskraft belastet werden, ist aber zu berücksichtigen, dass bei einer linearen Vergrößerung eines Objektes mit konstanter Form die Masse und damit auch die Eigenbelastung in der dritten Potenz zunimmt, während die Effektivität der tragenden Elemente von deren Querschnitten mitbestimmt wird, die nur in der zweiten Potenz zunehmen. Das Verhältnis zwischen Tragfähigkeit und Gewicht muss also, wenn Material und Proportionen unverändert bleiben, bei zunehmender Größe ungünstiger werden. Nach dieser Überlegung ist offensichtlich, weshalb ein Reh oder eine Gazelle relativ schlankere Beine hat als der größere Tapir, ein Rind oder gar ein Elefant (Abb. 1). Diese Zusammenhänge erklären auch, weshalb Landtiere nicht so groß werden können wie Meerestiere, die vom Wasser getragen werden (Elefant 6 Tonnen, Blauwal 110 Tonnen), und warum sehr große Landtiere gestreckte, säulenförmige Extremitäten ausbilden und damit auf die Fähigkeit zu springen verzichten. Das Gewichtsproblem lässt außerdem vermuten, dass die großen Landwirbeltiere in den Knochen materialsparende Strukturen ausbilden. Die (Abb. 2) und (Abb. 3) zeigen Beispiele solcher Strukturen bei Elefanten. In die Gehirnkapselwandung sind eigenartige Luftkammern eingefügt. Im Prinzip entspricht die Struktur einer Sandwichkonstruktion der Leichtbautechnik[1]: Zwei dünne, widerstandsfähige Deckplatten bilden zusammen mit einer dickeren, leichten und zerbrechlichen Zwischenschicht, die die Deckplatten gegeneinander verspannt und verstrebt, eine leichte und formhaltende Platte oder Schale. Ähnliche Strukturen findet man auch im Schädelbereich der Nashörner, bezeichnenderweise aber nicht bei den Walen, See-Elefanten und Walrossen und auch nicht bei den amphibisch lebenden Flusspferden. Merkwürdig ist die sehr grobe

1 vgl. Hertel

In biology, we often speak of the general principle of economy of materials, particularly in connection with supporting organs. Does such a general principle exist? You will remember the skull capsule structure of the magpie which Prof. Kummer showed you and his emphasis on the fact that the bone structures of birds are different from those of mammals. Birds develop lightweight structures to a considerable extent as a means of increasing the ability to fly. However, among other animal species such as the corals of warm seas, which erect great stone edifices with their skeletons, the principle of material economy appears rather unimportant. In addition to birds, the principle of lightweight construction is also noticeable in the skeletons of land animals which carry their bodies freely above the ground and indeed, independent of the size – the larger the species, the more characteristic are the structures. You will remember, Prof. Otto pointed out before that the constructive effectiveness was fundamentally dependent on only the form and material properties and not on the size. But, this is true only if the load placed upon a structure is considered as a separate factor. However, in constructions which are loaded by gravity, it must be noted that given a linear expansion of the object with constant form, the mass and thus also its own dead weight increases to the third power while the effectiveness of the supporting elements, at least partially determined by their cross-sections, increases only to the second power. Thus, the relationship between supporting capacity and weight must, if material and proportions remain the same, become less favorable with increasing size. Given this observation, it is clear why a deer or a gazelle has relatively thinner legs than the larger tapir, a cow or even an elephant (Fig.1). This relationship also explains why land animals cannot become as large as sea animals which are supported by the water (elephant: 6 tons, blue whale: 110 tons) and also why land animals develop straight, column-like extremeties thus forfeiting the capacity to jump. Moreover, the weight problem also gives rise to the presumption that large land vertebrates develop material saving structures in their bones. (Fig.2) and (Fig.3) show examples of such structures in elephants. Strange air chambers are inserted into the brain capsule walls. Basically, the structure of a sandwich construction corresponds to that used in lightweight building [1]: Two thin resistant cover layers, braced and guyed against one another by a thicker, light and fragile intermediate layer form a light and form-maintaining panel or shell. Similar structures are also found in the skull region of rhinoceroses but, interestingly enough, not in whales, sea-elephants and walruses and also not in the amphibious hippopotamouses. The very coarse chambering of the capsule walls in elephants is remarkable. In 1959, Kummer showed that a clearly recognizable relationship between the absolute size and the relative mesh width of cancellous tissue can be demonstrated in the thigh bones of mammals: the larger the animal, the finer are the meshes in

1 cf. Hertel

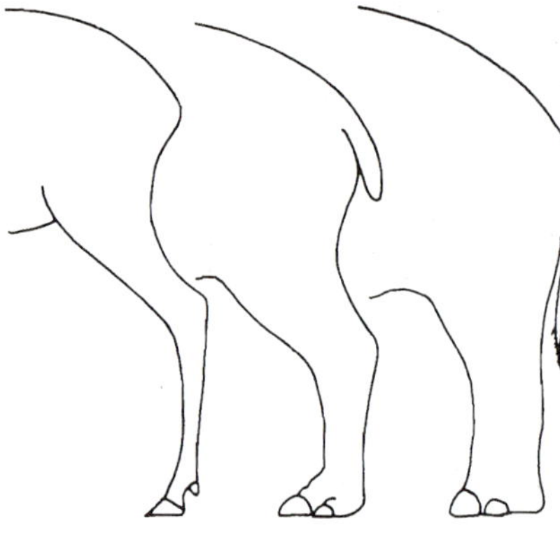

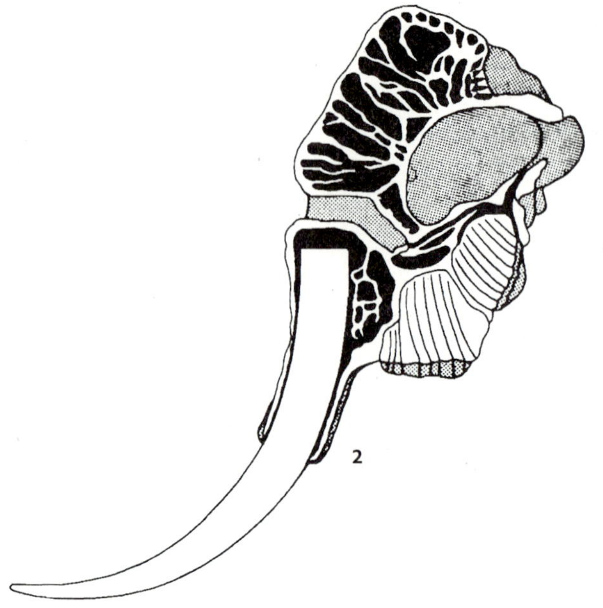

Kammerung der Kapselwandung der Elefanten, Kummer hat 1959 aufgezeigt, dass im Oberschenkelknochen der Säugetiere eine klar erkennbare Beziehung zwischen der absoluten Größe und der relativen Maschenweite einer Knochenspongiosa vorliegt: Je größer ein Tier, desto feiner die Maschen im Verhältnis zum Gesamtknochen[2]. Im Bereich des Vogelschädels hatte ich vergleichbare Verhältnisse gefunden. Für die pneumatisierten Strukturen der großen Dickhäuter scheinen aber andere Gesetzmäßigkeiten zu gelten. | Sehr weit ist das Prinzip der Sandwichstrukturen bei den Vögeln verbreitet. Es gibt allerdings auch Arten, bei denen sie im Skelettbereich unterdrückt werden. Eine Reihe von Enten zum Beispiel bildet eine einfache massive Schädelkapsel aus (Abb. 4a), und zwar auch dann, wenn z.B. Schnabelhöcker – wie bei der männlichen Trauerente (Melanitta nigra) – die Schädelform stark umgestalten; was ganz im Gegensatz zu Barnikol steht, nach dessen Ansicht starke Umformungen des Vogelschädels immer mit der Ausbildung pneumatisierter Strukturen korreliert sein müssten[3]. Um das Fehlen der Leichtbaustrukturen bei den Enten zu verstehen, muss man berücksichtigen, dass sie ihre Nahrung ertauchen und dass jedes Merkmal einer Population des Kompromissergebnis der gegeneinander gerichteten Selektionsdrucke darstellt; ist ein Selektionsdruck besonders groß – wie bei den Enten die Notwendigkeit, gut tauchen zu können –, dann werden andere Elemente der entsprechenden Merkmalausbildung unterdrückt, und zwar so stark, wie es der

2 vgl. Kummer. 1959, S.220

3 vgl. Barnikal. 1951, S. 397

Abb. 1 Umriss der Hinterextremität eines Rehs, eines Tapirs und eines Elefanten – etwa auf gleiche Größe gebracht. Mit zunehmender Körpergröße nimmt einmal der relative Durchmesser der Extremität zu. Zum anderen erhöht die Extremität eine gestrecktere, säulenähnlichere Grundform – die Sprungfähigkeit wird reduziert.

Fig. 1 Outline of the rear extremities of a deer, a tapir and an elephant – reduced approximately to the same size. As the body size increases, the relative diameter of the extremities increases, on the one hand, while the extremity becomes straighter and more column-like in basic form (thus reducing the ability to jump), on the other hand.

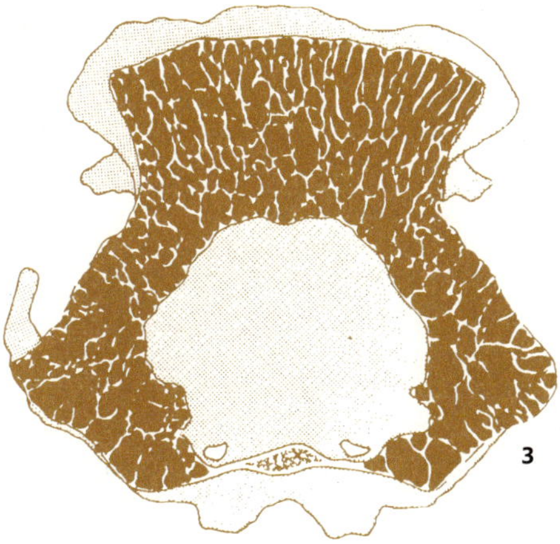

3

relation to the overall bone.[2] I have observed comparable relationships in the skull regions of birds. However, other laws appear to apply to the pneumatized structures of the large pachyderms. | The principle of sandwich structures is very widespread in birds. Nevertheless, there are species in which these structures are suppressed in the skeletal region. For example, a number of ducks develop a simple, solid skull capsule (Fig. 4a) and even in those cases where beaks, for example – as in the case in the male Common Scoter (Melanitta nigra) – severely change the skull form; this is completely contrary to Barnikol's view according to which essential alterations of the bird's skull must always be correlated with the formation of pneumatized structures[3]. In order to understand the lack of lightweight structures in ducks we must remember that they obtain their nourishment by diving and that every feature of a population represents the results of compromise with opposing selection pressures: if the selection pressure is particularly heavy – such as the necessity to be able to dive well in ducks – then other elements of the corresponding feature formation are suppressed and in fact to the extent that the counter selection pressure – in this case the necessity for flight – permits. | Now, let us return to the species which have sandwich shells as skull capsules. We may assume that the weight-reducing selection which has produced the structures stands in relation to the ability to fly in all flying species, it leads to widely differing results in the various groups. Here, we have one of the

2 cf. Kummer. 1959, p.220

3 cf. Barnikal. 1951, p. 397

Abb.2 Längsschnitt durch einen Elefantenschädel (Elephas – schematisiert nach Claus-Grobben 1910). Die Luftkammern in der Schädelwandung der Elefantenschädel sind relativ und absolut sehr groß.

Fig.2 Longitudinal section through an elephant's skull (Elephas – schematized after Claus-Grobben 1910). The air chambers in the skull wall of the elephant skull are relatively and absolutely very large.

Abb.3 Querschnitt durch einen Elefantenschädel (Loxodonta – grob schematisiert nach Bolk u. a 1936).

Fig.3 Cross section through an elephant skull (Loxodonta – roughly schematized after Bolk et al. 1936).

Gegenselektionsdruck – in unserem Falle die Notwendigkeit der Flugfähigkeit – zulässt. | Nun zurück zu den Arten, die Sandwichschalen als Schädelkapseln haben. Obwohl wir annehmen dürfen, dass die gewichtsverringernde Selektion, die diese Strukturen geschaffen hat, bei allen flugfähigen Vogelarten in Beziehung zum Flugvermögen steht, führte sie bei den einzelnen Gruppen zu recht verschiedenen Ergebnissen. Wir haben hier eines der häufigen Beispiele dafür, dass im Bereich der Organismen, ähnlich wie im Bereich der Technik, zur Lösung eines Problems verschiedene Wege eingeschlagen werden, die sich unter bestimmten Bedingungen nebeneinander bewähren können.[4] Gemeinsam ist den verschiedenen Typen, dass in der Knochenwandung ausgedehnte Hohlraum-Systeme angelegt werden, deren einzelne Abschnitte miteinander und über Mittelohr und Mundhöhle mit der Außenwelt kommunizieren. Vier Ausprägungstypen heben sich gegeneinander ab, obwohl Übergangsformen vorkommen. | Weit verbreitet ist der Spongiosa-Typ (Abb. 4b), bei dem die Zwischenschicht der Wandung von einem ungerichteten und ungeschichteten, mehr oder weniger lockeren, dreidimensionalen Balkensystem gebildet wird. Er ist bei Hühnervögeln (Meleagridinae und Phasianidae), Gänsen (Anser) und vielen anderen Gruppen ausgebildet. Abb. 9 zeigt ihn im hinteren Schädelbereich eines Bussards (Buteo buteo), Abb. 10 in der unteren Hälfte der Gehirnkapsel einer Haustaube (Columba livia). Häufig findet sich diese Struktur als Füllmaterial in voluminösen Knochenfortsätzen oder in Bereichen starker Verdickung der Kapselwandung. Ein Sonderfall des Spongiosa-Typs ist das lockere und grobmaschige System relativ langer Knochenbälkchen der Nachtschwalben (Caprimulgidae). Bei dieser Gruppe ist die Materialreduktion so weit getrieben, dass z. B. ein über 5 cm langer und über 3 cm breiter Schädel von Caprimulgus ruficollis zusammen mit dem Unterkiefer nur 0,45 g wiegt. Verantwortlich für das niedrige Gewicht ist sowohl die lockere Bälkchenstruktur als auch die geringe Dicke der Deckplatten. Die Außenlamelle der Schädelkapsel ist an manchen Stellen weniger als 1/50 mm stark vgl. Abb. 6. Die extreme Pneumatisierung dieses Schädels ist unter anderem auch die Voraussetzung für eine extreme Verbreiterung des Schädels. | Abb. 7 zeigt, dass die Schädelkapsel etwa doppelt so breit ist wie der gesamte Gehirn- und Innenohrbereich. Die Verbreitung der Gehirnkapsel ist die Voraussetzung für einen ganz gewöhnlichen Spreizmechanismus (Abb. 5 und Abb. 8), mit dem die Tiere beim nächtlichen Jagdflug ihre Mundöffnung wie einen auffaltbaren Insektenkescher in zwei Dimensionen aufspreizen können.[5] | Beim einstöckigen Typ (Abb. 4c) ist in eine relativ dünne Wandung (beim Star z. B. beträgt die Dicke der gesamten Sandwichschale nur 0,4 mm) eine einzelne Schicht verzweigter Hohlräume eingelagert. Die Wandung besteht also aus zwei Deckplatten, die über unverzweigte Bälkchen direkt miteinander verbunden sind (Abb. 10 – oberer Bereich). Man fin-

4 vgl. Bock: *multiple evolutionary pathways*. 1959

5 Bühler. 1970

Abb. 4 Schädelkapselwandungen verschiedener Vögel.
a: Massive, unpneumatisierte Wandungen (Tauchente).
b: Ungerichtete Spongiosa (weit verbreitet).
c: Einstöckiger Typ (kleine Singvögel u. a).
d: Zweistöckiger Typ (größere Singvögel).
e: Vielstöckiger Typ (Eulen).

Fig. 4 Skull capsule walls of various birds.
a: massive, non-pneumatized wall (diving ducks).
b: non-directional spongiosa (widely spread).
c: single-story type (small song birds among others).
d: two-story type (larger song birds).
e: multi-story type (owls).

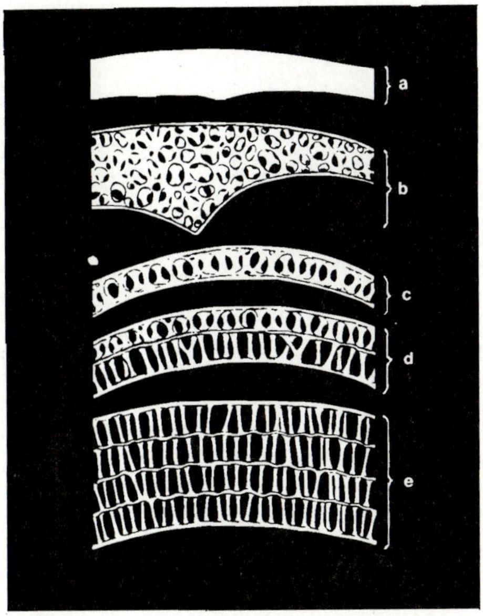

4

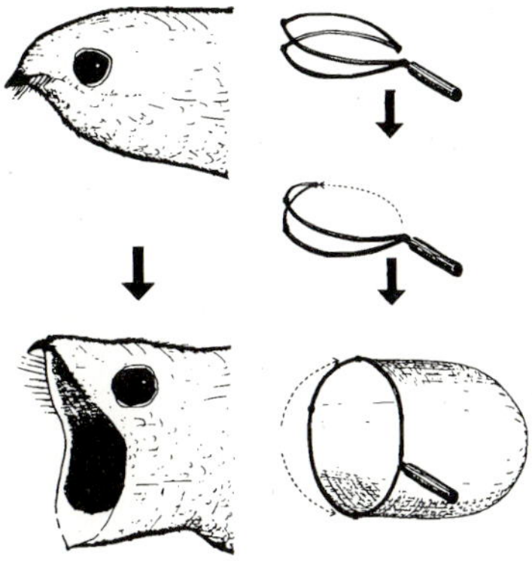

5

many examples of the fact that in organisms, as in technology, various approaches will be implemented for the solution of a problem which, under certain circumstances, can prove successful and compatible[4]. The common feature of the various types is that extended systems of caves are arranged in the bone walls, the individual sections of which communicate with one another and with the outside world by way of the middle ear and oral cavity. Four characteristic types contrast with one another although transitional forms do occur. | The spongiosa type (Fig. 4b) is widespread. In this type, the intermediate layer of the wall is formed from a non-directional and non-layered, more or less loosely 3-dimensional system of beams. It is developed in the game birds (Meleagridinae and Phasianidae), geese (Anser) and many other groups. Fig. 9 shows it in the posterior skull region of a Buzzard (buteo buteo); Fig. 10 shows it in the lower half of the brain capsule of a House Pigeon (Columba livia). Often, this structure appears as filler material in voluminous bony processes or in regions of increased thickening of the capsule wall. A special case of the spongiosa type is the loose and coarse-meshed system of relatively longer boned-beam Nightjar (Caprimulgidae). In this group, material economy is carried out to such an extent that, for example, the skull of the Caprimulgus ruficollis which is over 5 cm longer and 3 cm wider together with the lower jaw weighs only 0.45 g. The reason for the lower weight is both the loose beam structure and the reduced thickness of the cover layers. In some places the

4 cf. Bock: *multiple evolutionary pathways*. 1959

Abb. 5 Der Fangapparat einer Nachtschwalbe (vgl. Abb. 8) ist einem faltbaren Insektennetz vergleichbar, aber wirkungsvoller konstruiert, da er sich mit einer Bewegung in zwei Dimensionen öffnen lässt, während beim Netz dafür zwei Bewegungen nötig sind.

Fig. 5 The trapping apparatus of a Nightjar (cf. Fig. 8) is comparable to an insect net but more effectively constructed because it can be opened in two directions with one movement while, with a net, two movements are required.

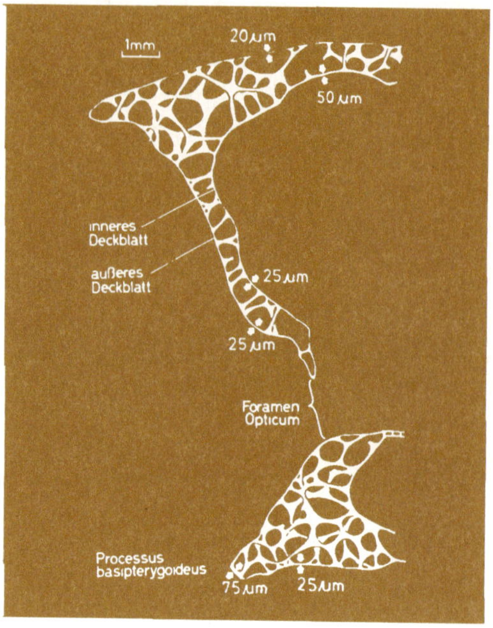

6

det die einstöckige Struktur u.a. bei Möwen (Larus), Turmfalken (Falco tinnunculus), kleinen Eulen wie dem Steinkauz (Athene noctua), Seglern (Apus) und den kleineren Singvögeln (Oscines). | Beim zweistöckigen Typ (Abb.4d) wird innerhalb des beim Nestling angelegten Sandwichs eine dritte Knochenlamelle angelegt und durch kleine Säulchen fixiert (Abb.11 und Abb.12). Gefunden habe ich den zweistöckigen Typ bei verschiedenen Rabenvögeln (Garrulus, Pica, Nucifraga, Corvus), aber er dürfte auch bei anderen mittelgroßen und großen Singvögeln verbreitet sein.[6] Gewichtsmäßig wirkt sich die Pneumatisierung bei diesen Arten etwa so aus, dass das spezifische Gewicht der Wandung nur etwa ein Viertel von dem der Trauerente mit nicht pneumatisierter Knochenwandung beträgt. | Der vielstöckige Typ (Abb.4e) ist die konsequente Weiterführung des zweistöckigen und ist durch sehr dicke und leichte Kapselwandungen vertreten. Innerhalb eines primären Sandwichs wird im Laufe der Juvenilentwicklung eine Serie von Stockwerken angelegt. Ob die auffallenden Lamellen (Abb.13, Abb.14, Abb.15) nur Stadien der Juvenilentwicklung sind, die fixiert wurden, denen aber keine besondere adaptive Bedeutung beizumessen ist, oder ob sie durch phylogenetische und ontogenetische Anpassung an trajektorielle Belastungen ihre Form erhielten, ist noch offen. Da die Lamellen vor allem bei kräftig zubeißenden Eulen ausgebildet sind, ist Anpassung wahrscheinlich. | Abb.16 zeigt – wiederum von einer Waldohreule – den hinteren Bereich der Schädelkapsel von innen. Es fällt auf,

6 vgl. Storck. 1969

Abb.6 Querschnitt durch die Schädelkapsel einer Nachtschwalbe (Caprimulgus ruficollis). Extreme Gewichtsverringerung durch die Ausbildung einer sehr lockeren Knochenspongiosa und sehr dünner Decklamellen. Überraschenderweise weist gerade die Außenlamelle die geringste Dicke auf – im Scheitelbereich stellenweise weniger als 1/50 mm. Am dicksten ist die Außenlamelle in der Gleitgelenkfläche des Basipterygoidfortsatzes (an der das bewegliche Gaumenskelett entlanggleitet).

Fig.6 Cross section through the skull capsule of a Nightjar (Caprimulgus ruficollis). Extreme weight reduction is achieved by formation of very loose cancellous tissue and very thin cover lamellas. Suprisingly enough the exterior lamellas are the thinnest – less then 1/50 mm in some places in the parietal region. The exterior lamella is thickest in the arthrodia surface of the bosiptergoid process (on which the movable palate skeleton slides).

Abb.7 Schädel einer Nachtschwalbe (Caprimulgus ruficollis) von oben. Die äußeren Knochenlamelle (a) der Sandwichschale wurde im rechten Bereich der Schädelkapsel aufgebrochen. Die innere Knochenlamelle (c, d, F) entspricht in der Form täuschend genau dem inneren Gehörapparat, dem Gehirn und dem Riechnerv, die von ihr umhüllt werden. Die Knochenbälkchen (e) verstreben und verspannen die äußere und die innere Knochenlamelle gegeneinander (vgl. Abb.6). Die Gehirnkapsel ist etwa doppelt so breit wie der Bereich (b), den Gehirn und Innenohrorgane einnehmen.

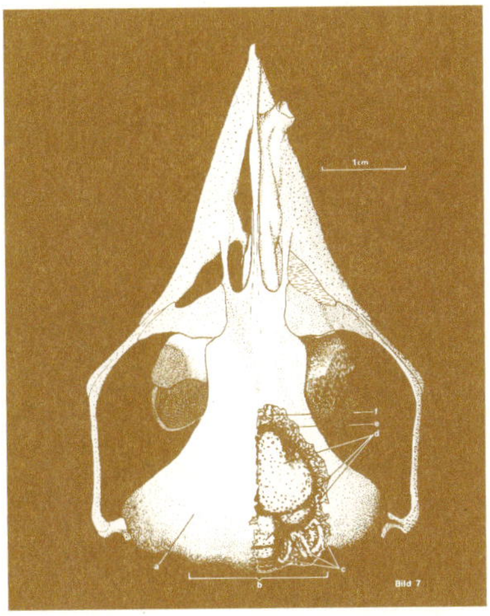

7

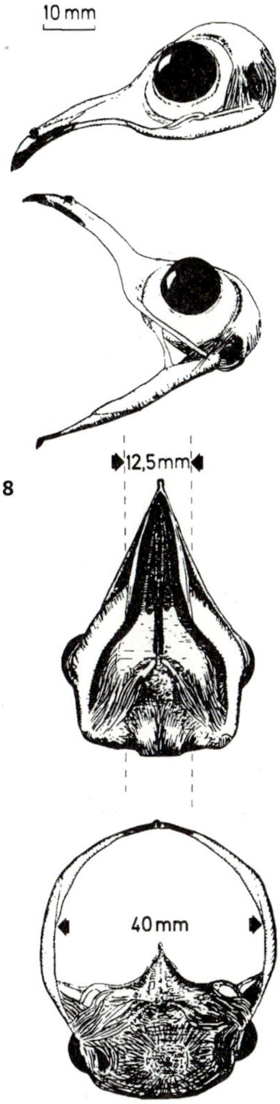

8

outer lamella of the skull capsule is less than 1/50 mm thick cf. Fig. 6. The extreme pneumatization of this skull is, among other things, also the presupposition for an extreme widening of the skull. | Fig. 7 shows that the skull capsule is approximately twice as wide as the total brain and internal ear region. The widening of the brain capsule is the presupposition for a very unusual spreading mechanism (Fig. 5 and Fig. 8) with which the bird can spread its mouth in two dimensions like a folding insect net during nocturnal hunting flights [5]. | In the single-story type (Fig. 4c) a single layer of branched caves is imbedded in a relatively thin wall (e.g. in the starling, the whole sandwich shell is only 0.4 mm thick). Thus, the wall consists of two cover panels which are directly connected to one another via branched beams (Fig. 10, upper part). The single-story structure is found, among other places, in Gulls (Larus), Kestrels (Falco tinnunculus), small owls such as the Little Owl (Athene noctua), Swifts (Apus) and the smaller song birds (Oscines). | In the two-story type (Fig. 4d) a third bony lamella is inserted in the sandwich in the nestling and is fixed by small columns (Fig. 11 and Fig. 12). I have observed the two-story type among various Corvidae (Garrulus, Pica, Nucifraga, Corvus), but it is probably also present in other middle-sized and large song birds [6]. In terms of weight, the pneumatization in these species has the effect that the specific weight of the wall represents approximately one-fourth of that in the Common Scoter with non-pneumatized bony wall. | The multi-story type (Fig. 4e) is the consequential fur-

5 Bühler. 1970

6 cf. Storck. 1969

Fig. 7 Skull of a Nightjar (Caprimulgus ruficollis) as seen from above. The exterior bony lamellas (a) of the sandwich shell were removed in the right region of the skull capsule. In terms of form, the inner bony lamellas (c, d, f) are deceptively close to the interior auditory apparatus, the brain and the olefactory nerve which it encases. The bony beams (e) brace and guy the external and interior bony lamellas against one another (cf. Fig. 6). The brain capsule is approximately twice as wide as the area (b) which the brain and inner ear organs occupy.

Abb. 8 Die Kieferspreizmechanik einer Nachtschwalbe (Caprimulgus ruficollis). Die Mundöffnung wird sowohl in der Vertikalen als auch in der Horizontalen erweitert. Eine Voraussetzung für die erstaunliche Horizontalspreizung des Unterkiefers ist die für Vögel ungewöhnliche Verbreiterung der Schädelkapsel (vgl. Abb. 7).

Fig. 8 The law spreading mechanism of the Nightar (Caprimulgus ruficollis). The oral opening can be expanded both vertically and horizontally. Cne prerequisite for the astonishing horizontal spread of the lower jaw is the widening of the skull capsule (cf. Fig. 7) which is unusual for birds.

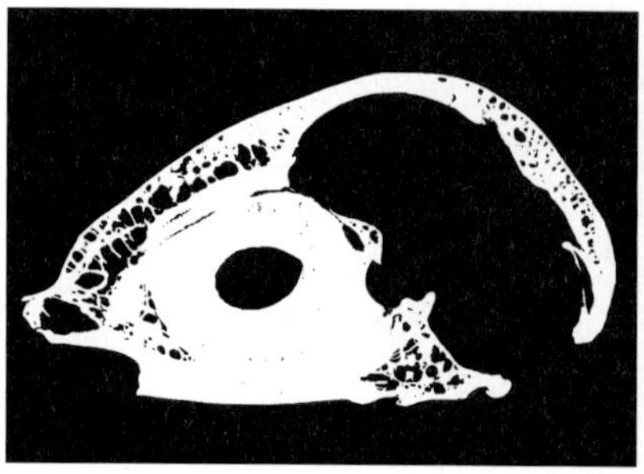

9

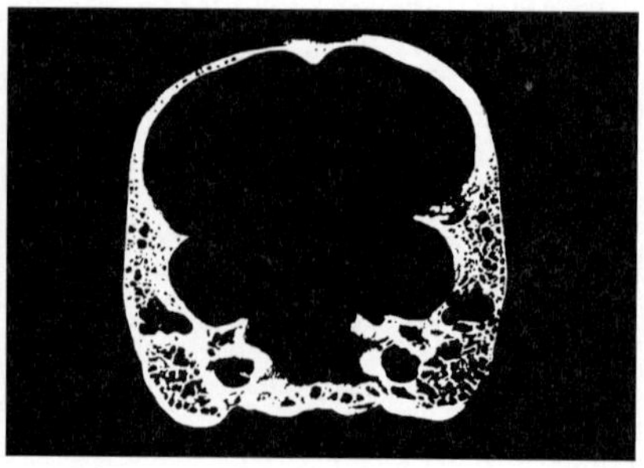

10

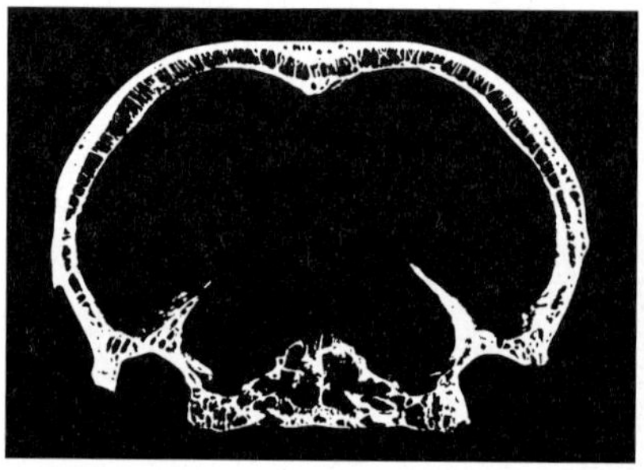

11

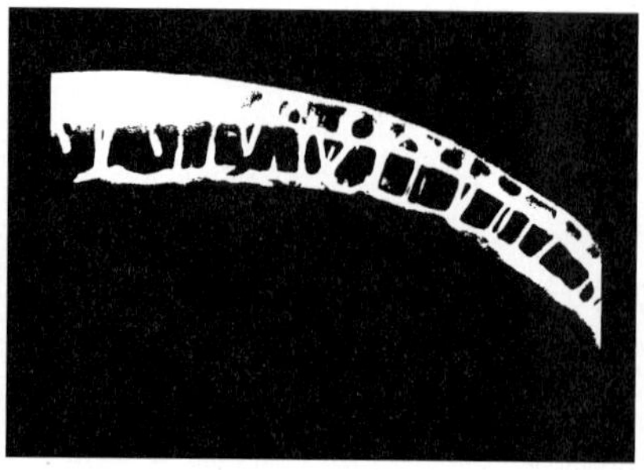

12

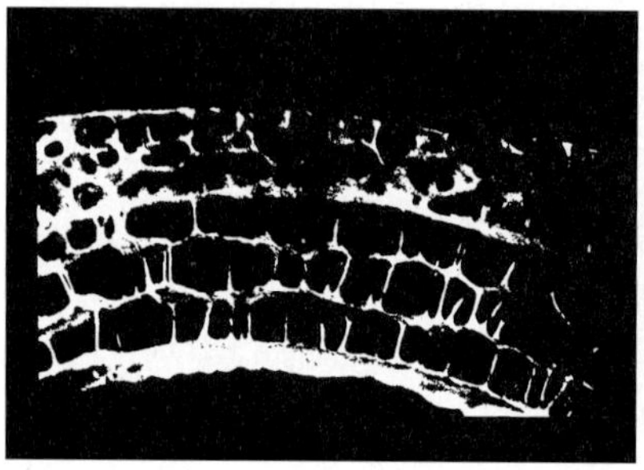

13

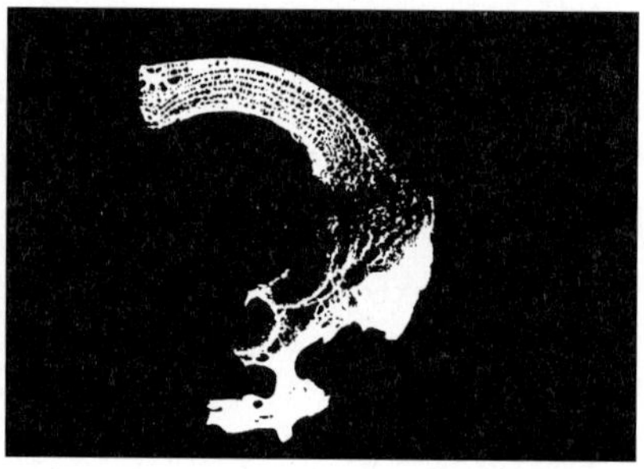

14

Abb.9 Schädelkapsel Mäusebussard, Längsschnitt (Im Gehirnkapselbereich unpneumatisierte Abschnitte und Spongiosatyp, in der Stirnregion Ausbildung mehrerer Stockwerke.

Abb.10 Schädelkapsel einer Haustaube, Querschnitt (Unterer Bereich – Spongiosa-Typ; oberer – einstöckige Struktur).

Abb.11 und 12 Zweistöckige Struktur bei Rabenkrähe und Elster.

Abb.13 Schädelkapselwandung einer Waldohreule (drei regelmäßige Stockwerke innerhalb einer Spongiosa-Struktur).

Abb.14 und 15 Querschnitte der Schädelkapsel eines Waldkauzes in der Gehirn- und Stirnregion (vielstöckiger Typ).

Fig.9 Skull capsule of a European Buzzard, longitudinal section (non-pneumatized sections and spongiosa types in the brain capsule region – cf. text: formation of several stories in the forehead region).

Fig.10 Skull capsule of a House pigeon, Cross section (lower area – spongiosa type; upper area-single story structure)

Fig.11 and 12 Two-story structures in Carrion Crows and Magpies.

Fig.13 Skull capsule wall of a Long-Eared Owl (three regular stories within a spongiosa structure).

Fig.14 and 15 Cross sectiäns of the skull capsule of a Tawny Owl in the brain and forehead region (multi-story type).

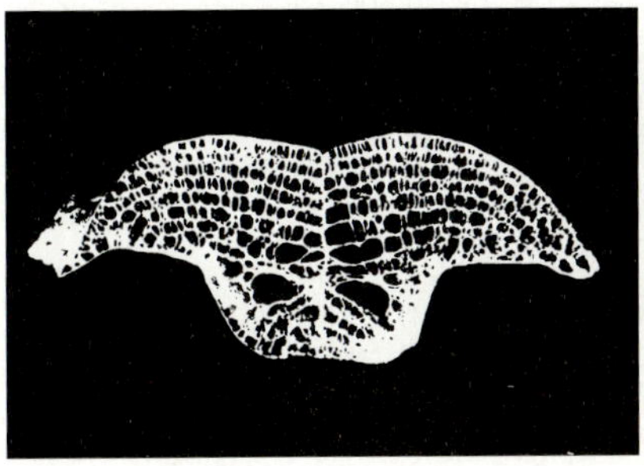

15

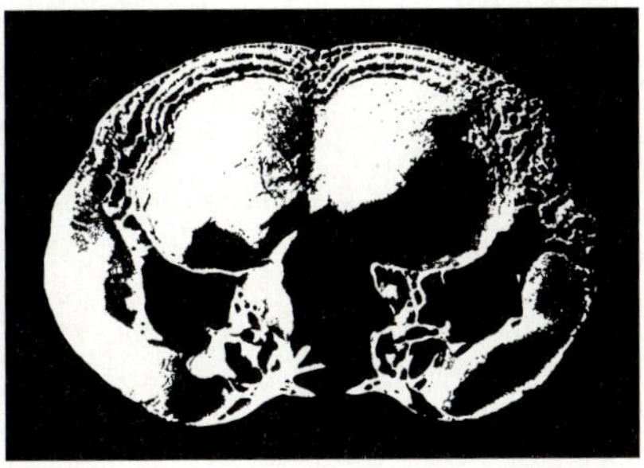

16

ther development of the two-story type and is represented by very thin and very light capsule walls. A series of stories appear within the primary sandwich during juvenile development. Whether the lamellas occurring (Fig. 13, Fig. 14 and Fig. 15) are only phases in the juvenile development which will be fixated (to which, however, no particular significance can be attached) or whether they retain their form due to phylogenetic and ontogenetic adaptation to trajectorial loads is still an open question. Since the lamellas are developed, especially in hard biting owls, adaption is the probable explanation. | Fig. 16 shows – again from a Long-Eared Owl – the posterior region of the skull capsule from inside. It is interesting to note that the exterior dimensions of the skull capsule – as is the case with the Nightjar – is much wider than the cerebral chamber inside. In order to clarify the function of this skull widening, which has nothing to do with a jawspreading mechanism in owls, allow me to make a digression. As a student I received my first Barn Owl (Tyto clba), with which I was able to perform some experiments. This owl had been raised in captivity. Among other things, I had a mouse scurry around under spread-out newspaper approximately 4 meters away form the owl on its elevated perch. Despite the fact that it was not possible to determine optically, where the rodent was, the owl – which until that point had had no experience with living prey – dove to the middle of the paper surface and caught the mouse immediately. The owl had used the rustling through the newspaper (nearly

Abb. 16 Hinterer Bereich der Schädelkapsel einer Waldohreule – Ansicht von innen (Die pilzförmige Schale im Zentrum ist das innere Deckblatt der Sandwichschale, die das Gehirn umhüllt; die nierenförmigen Schalen rechts und links sind die hinteren Knochenwandungen der äußeren Gehörgänge. Der Schädel ist viel breiter angelegt, als es die Breite des Gehirns erfordern würde. Die Verbreiterung des Schädels vergrößert das schallfangende System der Eulen.

Fig. 16 Posterior region of the skull capsule of a Long-Eared Owl – viewed from inside (the mushroom-shaped skull in the center is the internal cover of the sandwich shell which encloses the brain; the kidney-shaped shells at the right and left are the posterior bone walls of the outer auditory canals. The skull is much wider than the width of the brain would require. This widening of the skull increases the sound reception system of the owl.

dass die Außenabmessung der Schädelkapsel – ähnlich wie bei den Nachtschwalben – viel breiter ist als die Gehirnkammer im Inneren. Um die Funktion der Schädelverbreiterung, die bei den Eulen nichts mit einem Kieferspreizmechanismus zu tun hat, verständlich zu machen, eine Abschweifung: Als Student hielt ich meine erste handaufgezogene Schleiereule (Tyto alba), mit der ich auch ein wenig experimentierte. Unter anderem ließ ich in etwa vier Meter Entfernung von der erhöht sitzenden Eule eine Maus unter ausgebreitetem Zeitungspapier umherhuschen. Obwohl optisch nicht erkennbar war, wo der Nager sich befand, griff die Eule, die bis zum damaligen Zeitpunkt keinerlei Erfahrung mit lebender Beute gehabt hatte, im Sturzflug mitten in die Papierfläche hinein und fand die Maus auf Anhieb. Die Eule hatte das für den Menschen fast unhörbare Rascheln durch das Zeitungspapier hindurch für die Lokalisierung der Maus genutzt. Etwa zur gleichen Zeit hatte unabhängig davon in den USA R.S. Payne ebenfalls bei der Schleiereule – aber mit Hilfe von Dunkelkammer, Lautsprecheranlage und Infrarotfilmgerät – festgestellt, dass Eulen Objekte, die Raschelgeräusche verursachen, rein akustisch orten können, und zwar mit Winkelabweichungen, die weniger als ein Grad betragen.[7] Welche Bedeutung hat aber das akustische Orten für den Schädelbau? Die Ortung der Beute ist mit großer Wahrscheinlichkeit von der schallsammelnden Wirkung der parabolschirmartigen Federbildungen hinter den Ohröffnungen abhängig. Die fokussierende Wirkung von Parabolschirmen ist aber von der Relation Durchmesser des Schirms zur Wellenlänge Lambda abhängig. Nur wenn der Durchmesser größer als Lambda ist, kommt es überhaupt zu einer Fokussierung, und erst wenn der Durchmesser ein Mehrfaches von Lambda ist, wird die Richtcharakteristik so ausgeprägt, dass mit dem Parabolschirm eine genauere Ortung möglich ist.[8] Durch die Verbreiterung der Schädelkapsel werden aber bei den Eulen die Basis der Reflektoren und damit das gesamte, schallfangende System verbreitert. Wenn diese Überlegungen richtig sind, können wir daraus folgern, dass eine Schleiereule, deren Reflektorsystem etwa einen Durchmesser von 6 cm hat, Frequenzen unter 5600 Hz überhaupt nicht mit Hilfe der Reflektoren nutzen kann und für eine genauere Ortung Frequenzen über 11000 Hz braucht. Das stimmt aber sehr schön mit den Angaben Paynes überein, der festgestellt hat, dass seine Eulen sich weigerten, einen Lautsprecher im Dunkeln anzufliegen, wenn aus dem ausgestrahlten Laubgeraschel die Frequenzen über 5000 Hz herausgefiltert worden waren. Waren nur die Frequenzen über 8500 Hz herausgefiltert, dann versuchten die Eulen zwar den Lautsprecher zu schlagen, aber ihre Zielgenauigkeit war noch deutlich vermindert. | Zusammenfassend können wir festhalten, dass materialsparende Konstruktionen dann auftreten, wenn sie aus bestimmten Gründen einen positiven Selektionswert für eine Population haben. Aufgezeigte Beispiele für extreme Leichtbaustrukturen

7 Payne. 1962 und 1971

8 Obata und Yosida. 1930

inaudible for human ears) to locate the mouse. At approximately the some time and independently, R. S. Payne in the USA also noted in barn owls — but with the help of a darkroom, a loudspeaker system and infra-red film equipment — that they can locate objects which create rustling noises on a purely acoustic basis and in fact with angular deflections of less than one degree[7]. But, what significance does this structure of the skull have for acoustic location? It is very probable that the prey is located by means of the noise collecting effect of the parabolic, dish-like feather formations behind the ear openings. However, the focusing effect of parabolic dishes is dependent on the relationship of the diameter of the dish to the wave length Lambda. Focusing will not take place unless the diameter is greater than Lambda and then not until the diameter is a multiple of Lambda does the directional characteristic become so strong that accurate location is possible with the parabolic dish[8]. However, with the enlargement of the skull capsule in owls, the basis of the reflectors, and thus of the overall sound-reception system is enlarged. If these considerations are correct, we can conclude that a Barn Owl, whose reflector system measures approximately 6 cm in diameter, cannot use frequencies under 5600 Hz at all even with the help of reflectors and for accurate location requires frequencies over 11000 Hz. This concurs nicely with Payne's data, which indicated that his owls refused to attack a loudspeaker in the dark when the frequencies over 5000 Hz were filtered out of the leaf-rustling sound being emmitted. If only the frequencies over 8500 Hz were filtered out, the owls tried to attack the loudspeaker but their accuracy was still clearly reduced. To summarize, we can conclude that material-saving constructions occur in those cases, where, for various reasons, they represent a positive selection value for a population. The very voluminous skull capsules of Nightjars and Owls were the examples given for extremely lightweight structures. The effect of gravity which in many cases encourages the formation of lightweight structures is only one among many ecological factors which influence material deposition in organisms. Let us recall the formation of very heavy racks of antlers in certain types of deer and the different shell thicknesses between ocean and fresh-water snails. However, none of the facts observed support the thesis that there is a superordinate principle of material economy independent of environmentally-determined selection pressures.

7 Payne. 1962 and 1971

8 Obata and Yosida. 1930

sind die sehr voluminösen Schädelkapseln von Nachtschwalben und Eulen. Die Wirkung der Erdanziehung, die offensichtlich in vielen Fällen die Ausbildung von Leichtbaustrukturen fördert, ist nur eine von vielen ökologischen Gegebenheiten, die die Materialablagerung in Organismen beeinflusst. Erinnert sei an die Ausbildung sehr schwerer Geweihe bei manchen Hirscharten und an die Unterschiede der Schalendicke zwischen Meeres- und Süßwasserschnecken. Keine beobachtete Tatsache spricht aber dafür, dass ein übergeordnetes Prinzip der Materialersparnis unabhängig von umweltbedingten Selektionsdrucken wirksam ist.

Abb.17 Schnitt durch ein Schneckenhaus. Foto aus dem Fotowettbewerb Biologie und Bauen, 1968.

Fig.17 Section through a snail shell. Photo from the photo competition for Biology and Building, 1968

18

Formenkunst der Natur

Formative art of nature

Oskar Prochnow

Zur Einführung Kunst und Natur hat man oft in Gegensatz zueinander gebracht. Man sah in der Natur den starren Ausdruck ewiger, vom Menschen unabhängiger Gesetze; in der Kunst aber verehrte man die freigestaltende Kraft des schöpferischen Menschengeistes. So entgeistigte man die Natur und machte sie zum Gegenpol unseres gestaltenden Denkens – während man auf der andern Seite die Naturgebundenheit des Künstlers nicht sehen mochte. | Doch hat es auf beiden Seiten Mittler gegeben, die Natur und Geist in eins verschmelzen wollten, die wohl ahnten, dass das Schaffen des Künstlers zwar freier ist als das, was sich im »Spiel der Kräfte« offenbart, doch im Grunde von gleicher Art. | Der Naturforscher Ernst Haeckel, dessen Einstellung auf große Fragen bekannt ist, hat durch seine »Kunstformen der Natur« weiten Kreisen einen Einblick in die schönsten Bauformen der Natur, besonders im Gebiet der niederen Tiere, ermöglicht, der zugleich die Freiheit der gestaltenden Kräfte ahnen lässt. Der Künstler Karl Bloßfeldt hat durch seine »Urformen der Kunst« den Künstlern ihre Stoff- und Formgebundenheit gezeigt, indem er die Anlehnung der künstlerischen Formgebung an das Vorbild der Pflanze durch Großaufnahmen von Pflanzenteilen belegte. | Heute werden von beiden Seiten immer erneute Angriffe gegen die Grenze zwischen Natur- und Geisteswissenschaften und damit auch zwischen Natur und Kunst geführt; und wir dürfen wohl sagen, dass dem Materialismus der Biologismus folgen wird und muss: dass man die Fragen des Lebens vom Leben aus zu beantworten versuchen muss, nicht von einem künstlichen Schema aus. | In dieser Richtung liegt auch das Bestreben dieses Buches. Einmal will es Bilder zeigen, wie sie – das kann wohl gesagt werden – die meisten Menschen bisher nicht gesehen haben, Bilder von der Formungskraft der Natur, ja von ihrer Formungskunst. Darüber hinaus will es dazu führen, dass man Naturgesetz und Formgesetz als unsere Denkformen des Seienden wie des Schönen anerkenne. Ist es denn so verwunderlich, dass unsere Normen des Schönen »auch« auf die Natur anwendbar sind – die uns im Jahrmillionen langen Ringen gezwungen hat, Normen des Wahren und Schönen zu finden: sie lehrte uns, das Einfache als wahr und zugleich das Einfache als schön auffassen. | Acht Jahre lang haben wir, meine Frau und ich, Bilder aus dem Bereich der Formenkunst der Natur gesucht. Hier und da gelang es uns, in Neuland vorzustoßen und den mikroskopischen Hilfsmitteln Bilder abzutrotzen. Nicht gelang es ganz, die Wunderwelt der Radiolarien so im Bilde wiederzugeben, wie es erwünscht ist – auch nicht, als wir durch drei Zimmer hindurch mit einer Luftstrecke zwischen Linse und Mattscheibe arbeiteten, da selbst die Zweimeterkamera mit Anbau nicht ausreichen wollte. | Mit vollem Bedacht sind aus unserer um vieles größeren Sammlung von Bildern dieses Gebietes die Pflanzenformen ausgeschieden worden, da sie in Bloßfeldts Buch ausschließlich als Künder der Formschönheit der Natur auftreten. Gleichfalls

By way of introduction Art and nature have often be treated as opposites. Nature is frequently considered the rigid expression of eternal laws independent of man; while in art one revered the free and creative power of the human mind. Thus, nature was robbed of spirit and turned into the opposite pole to our creative thought – while on the other hand, eyes were blinded to the way artists are rooted in nature. There have, however, been intermediaries on both sides who sought to meld nature and mind in a single entity and evidently sensed that while an artist's creativity may be freer that what is revealed in the "interplay of forces", it is essentially similar. | In his book "Art Forms From Nature", natural scientist Ernst Haeckel, whose stance on major issues is well-known, offered a wide circle of readers an insight into the most beautiful forms erected by nature, especially in the realm of the lower creature, while likewise giving us a feel for the freedom of the forces at work creating them. In his "Art Forms in Nature", artist Karl Bloßfeldt showed how the way artists are indebted to materials and shaped by pinpointing in large-scale photos of sections of plants the way shapes in art borrow from the example set by the plant kingdom. | Today, both sides repeatedly attack the line dividing the natural sciences from the humanities and thus nature from art; and we can thus no doubt say that materialism will inevitably follow biologism, that the questions of life must be answered by life and not by an artificial scheme. | This is the thrust of this book, too. Firstly it presents images, it is fair to say, the likes of which most people have not hitherto seen, images of the creative power of nature, indeed of its artistic ability to shape things. Moreover, it attempts to persuade readers to recognize that natural laws and the laws of form constitute our way of thinking existence and the beautiful. Is it so surprising that our norms of the beautiful are "also" applicable to nature – which in a struggle lasting millions of years has forced us to identify the norms for the true and the beautiful: It teaches us to grasp the simple as true and also to see the simple as beautiful. | For eight years, my wife and I have sought images from the realm of the formative art of nature. Here and there we have succeeded in entering new terrain and to create images using microscopic means. We did not completely succeed in reproducing the miraculous world of radiolarians in images the way we wanted – even when we opted for a space between the lens and the glass screen that extended through three rooms because the two-meter camera and appendages did not suffice. | After careful consideration, we have eliminated from the far greater collection of images in the field the plant shapes, as they occur in Bloßfeldt's book solely as testimony to the formal beauty of nature. Again, quite intentionally, it is nature's art in two dimensions that is highlighted. After all, we managed to observe here a wealth of especially surprising and insightful natural forces at work here. Moreover, in particular these images may be a source of inspiration for the arts and

mit voller Absicht tritt hier die Flächenkunst der Natur stark hervor. War es uns doch möglich, hier eine Fülle besonders überraschender und aufschlussreicher Auswirkungen der Naturkräfte zu beobachten. Hinzu kommt, dass besonders diese Bilder für Kunst und Gewerbe Anregungen geben können. | Nur eine Ergänzung unserer Bildersammlung wurde vorgenommen. Die klimatischen Verhältnisse in Deutschland und die beruflichen Bindungen ermöglichten es nämlich nicht, den andern gleichwertige Bilder von Schneekristallen zu bekommen. Daher wandte ich mich an Herrn Wilson A. Bentley in Jericho U.S.A. und erhielt von ihm eine Auswahl seiner Aufnahmen, die den acht Bildern der Tafeln 43 bis 50* zugrunde liegen. Ihn, der 45 Winter hindurch Tausende von Schneekristallformen in wertvollen Abbildungen wiedergab, kann mein Dank an dieser Stelle nicht mehr erreichen. | Für bereitwillige Unterstützung meiner Arbeit darf ich den Leitungen der zoologischen und mineralogischen Sammlungen des Museums für Naturkunde in Berlin und des Naturhistorischen Museums in Lübeck danken. Die Aufnahmen, die nach den Stücken dieser Sammlungen gemacht worden sind, sind in der Übersicht durch entsprechende Zusätze gekennzeichnet. Sonst befinden sich die abgebildeten Formen – soweit sie haltbar sind – in meinem Besitz. | Ist auch für meine Mitarbeiterin und mich jedes Bild an sich reichlich Lohn und Freude, so wollen, so dürfen wir doch die Ergebnisse so vieler stiller, schöner Feierstunden der Arbeit nicht für uns behalten. Bilder sollen gesehen werden und Bilder aus der Formenkunst der Natur sollen weiterwirken.

* Anmerkung: Im Folgenden ist davon nur Tafel Nr. 49 (S. 378) abgebildet.

Berlin-Lichterfelde, am 12. April 1934. *Oskar Prochnow*

crafts. | We only supplemented our collection of images in one regard. For climatic conditions in Germany and our professional commitments prevented us from obtaining images of snow flakes on a par with the other images. For this reason I turned to Wilson A. Bentley in Jericho, U.S.A., and he kindly sent me a selection of his photographs and these form the basis for the eight images on plates 43 to 50*. Sadly, I can no longer thank him in person for his efforts – he spent all winter making invaluable photographs of thousands of different snow flake shapes. | I am most grateful to the directors of the zoological and mineralogical collections at Berlin's Museum of Nature Study in Berlin and the Lübeck Museum of Natural History in Lübeck for gladly supporting my work. The photographs that I have made of the items in these collections are labeled accordingly in the overall list. Otherwise, all the other forms presented here are in my possession, to the extent that are not perishable. | While for my staff and myself each and every image is a reward and joy in itself, we do not wish to keep the results of so many silent and enjoyable celebrations of work to ourselves. Photographs are there to be seen, and images from the formal art of nature must have a long-term effect.

* Note: In the following only plate Nr. 49 (p. 378) is pictured.

Berlin-Lichterfelde, April 12, 1934. *Oskar Prochnow*

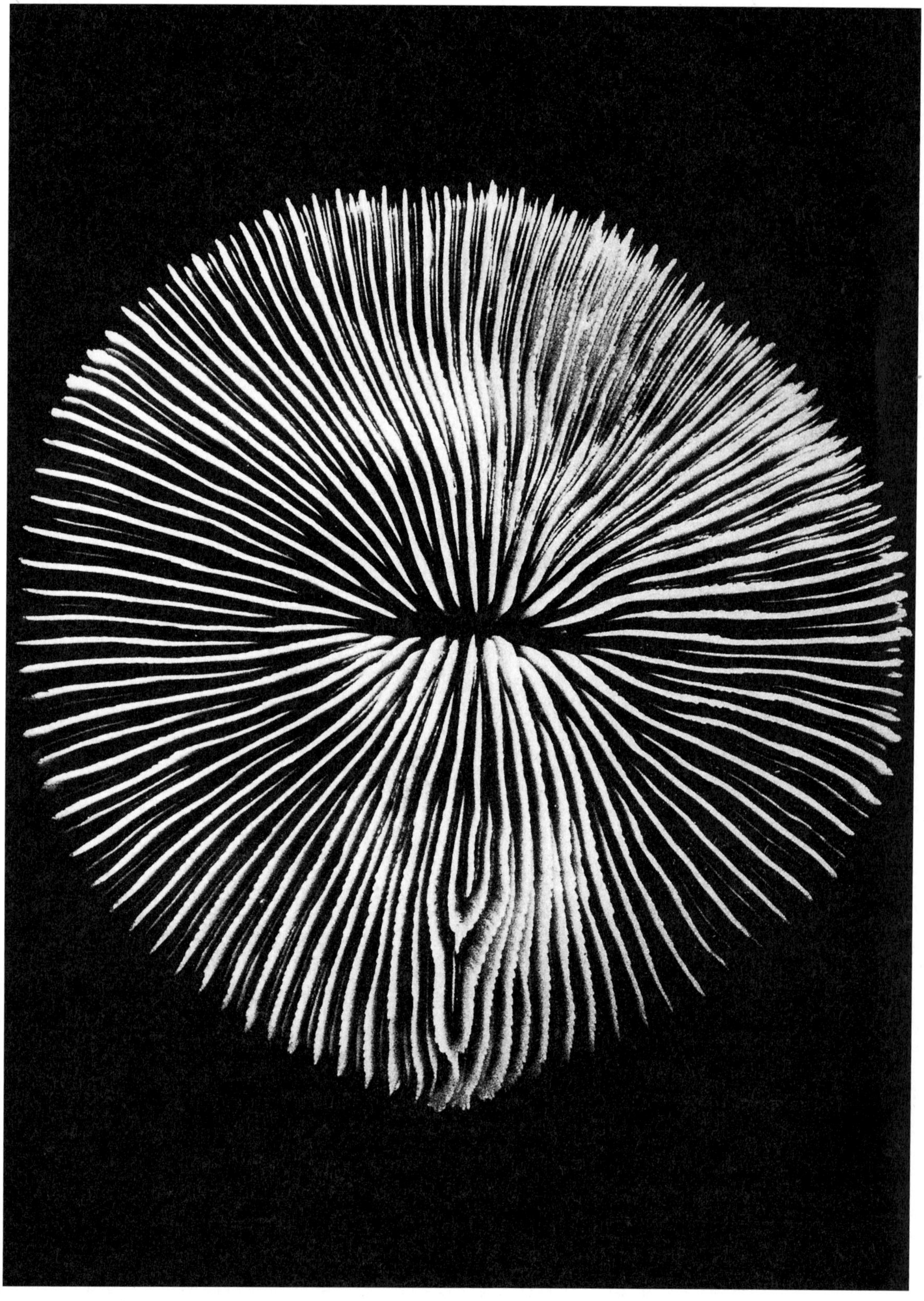

1:1

Pilzkoralle Fungia repanda.

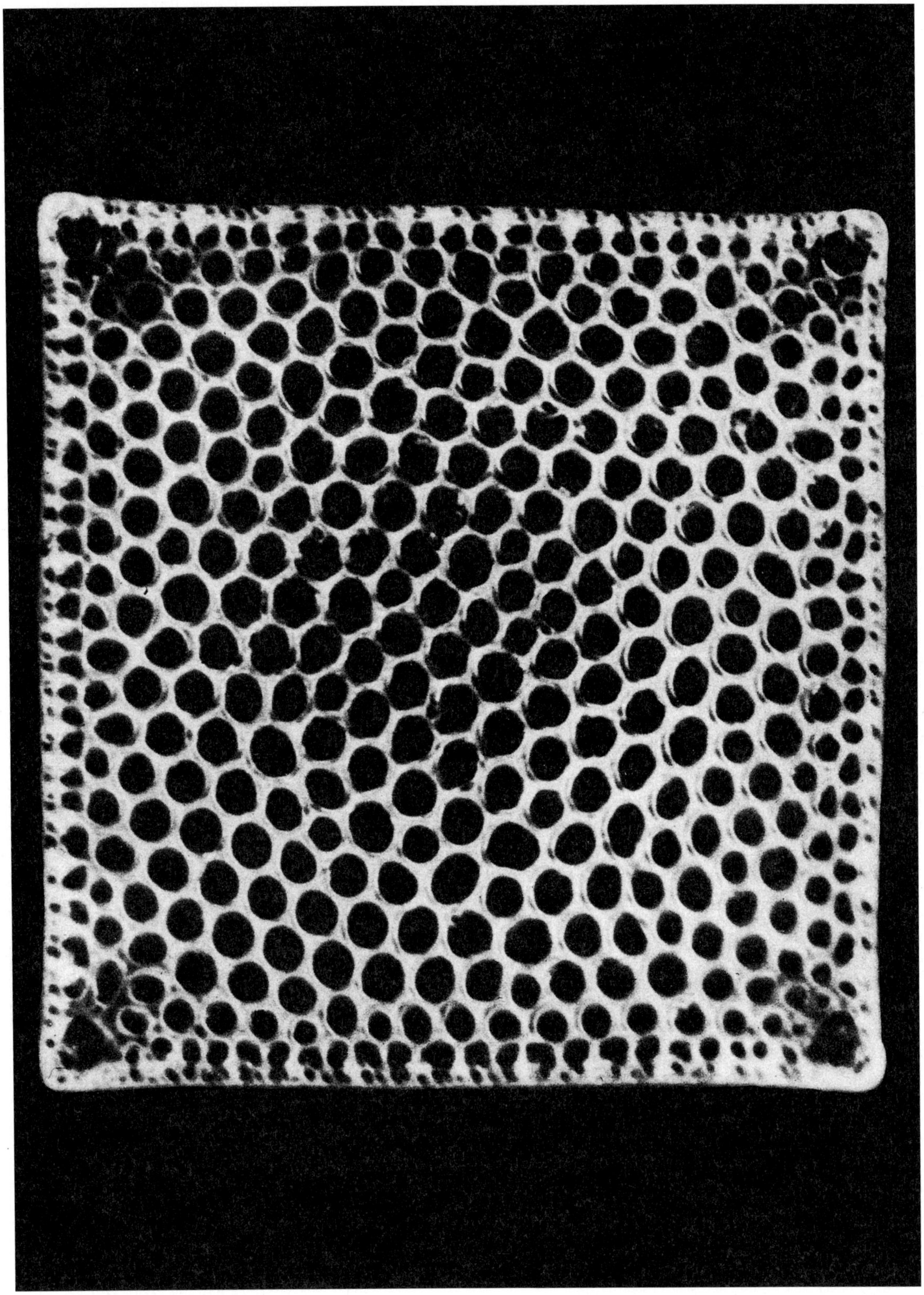

650:1

 Viereckige Kieselalge Triceratium favus.

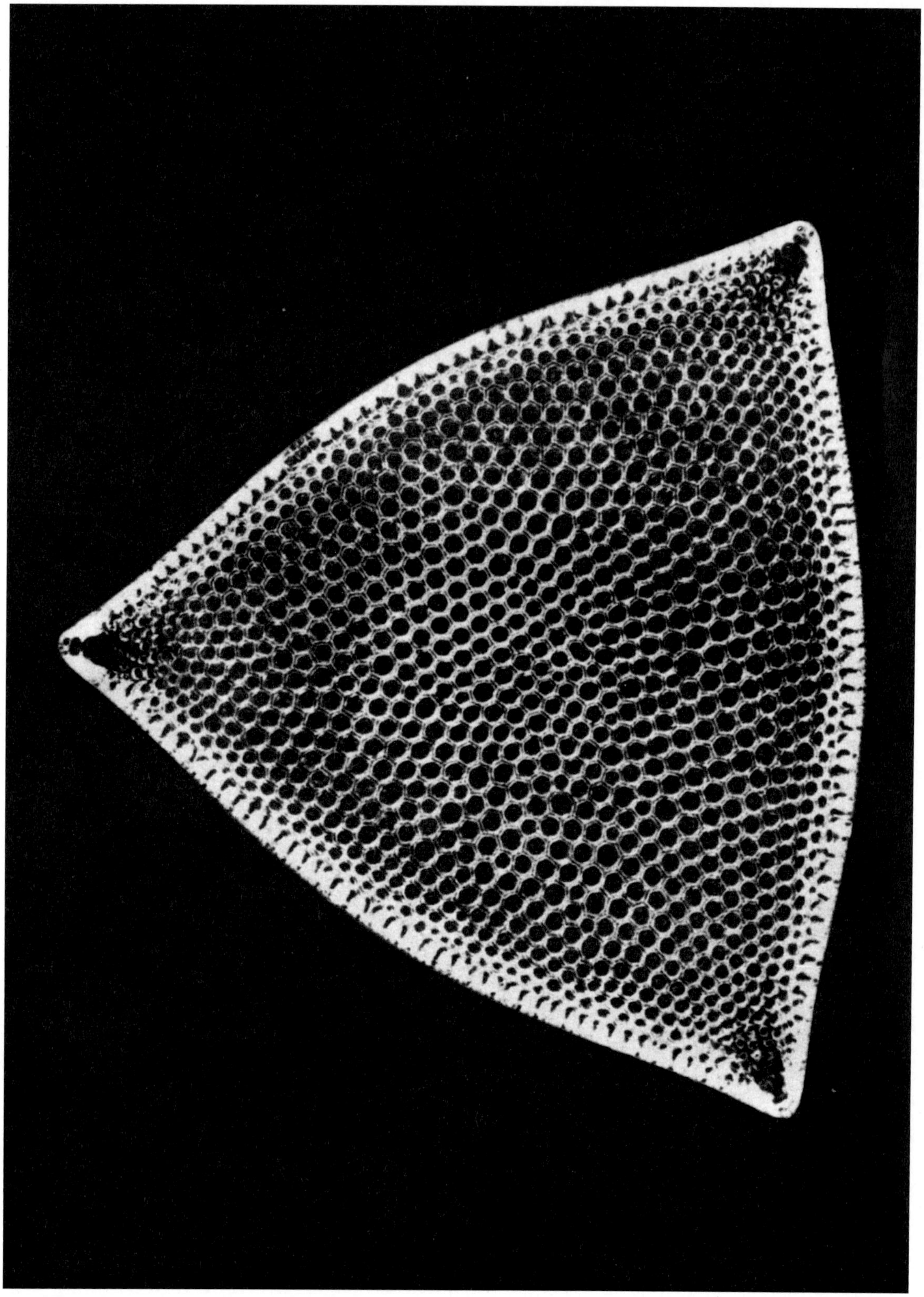

425:1

Dreieckige Kieselalge Triceratium secedens.

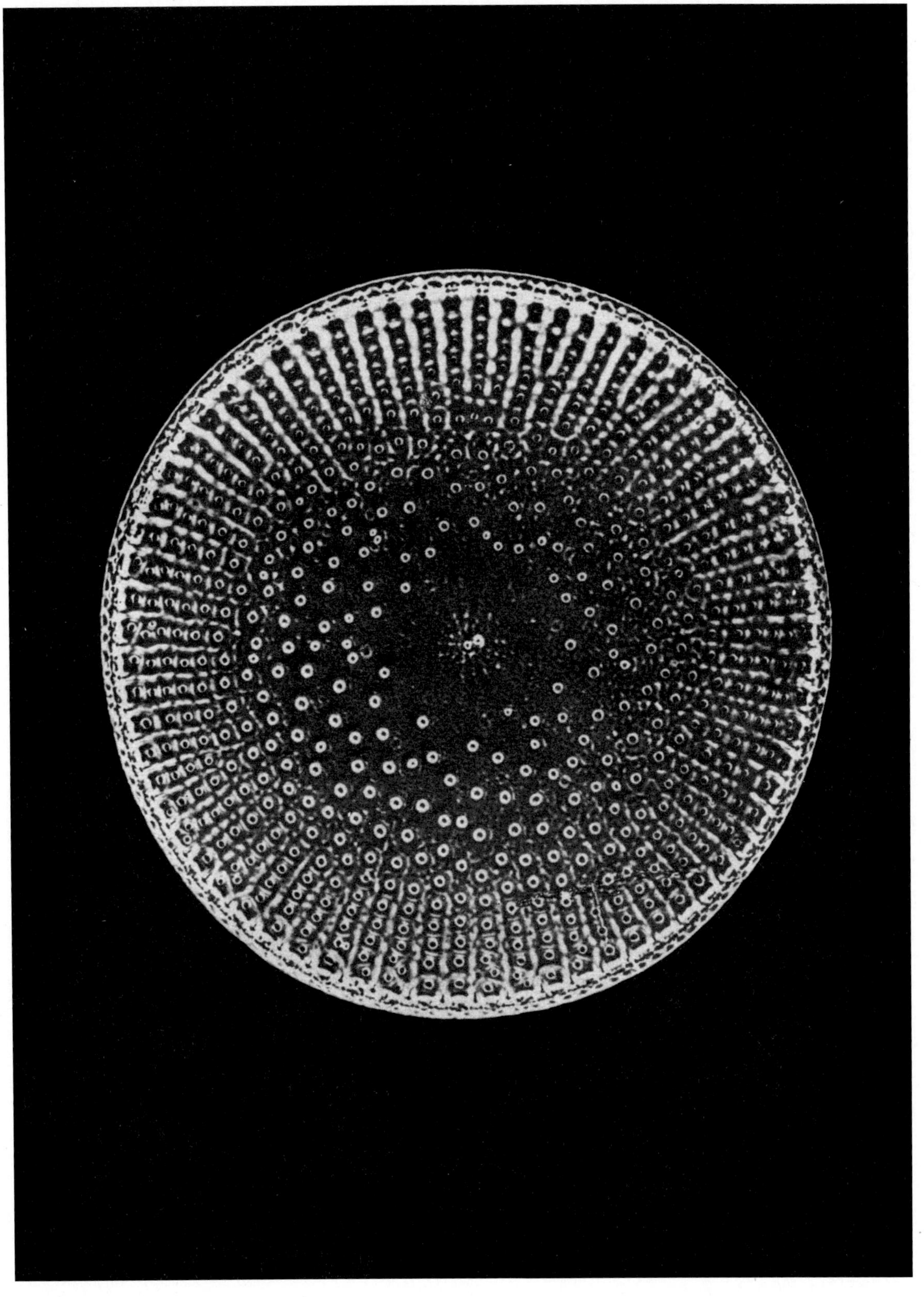

820:1

Scheibenförmige Kieselalge Stictodiscus Californicus.

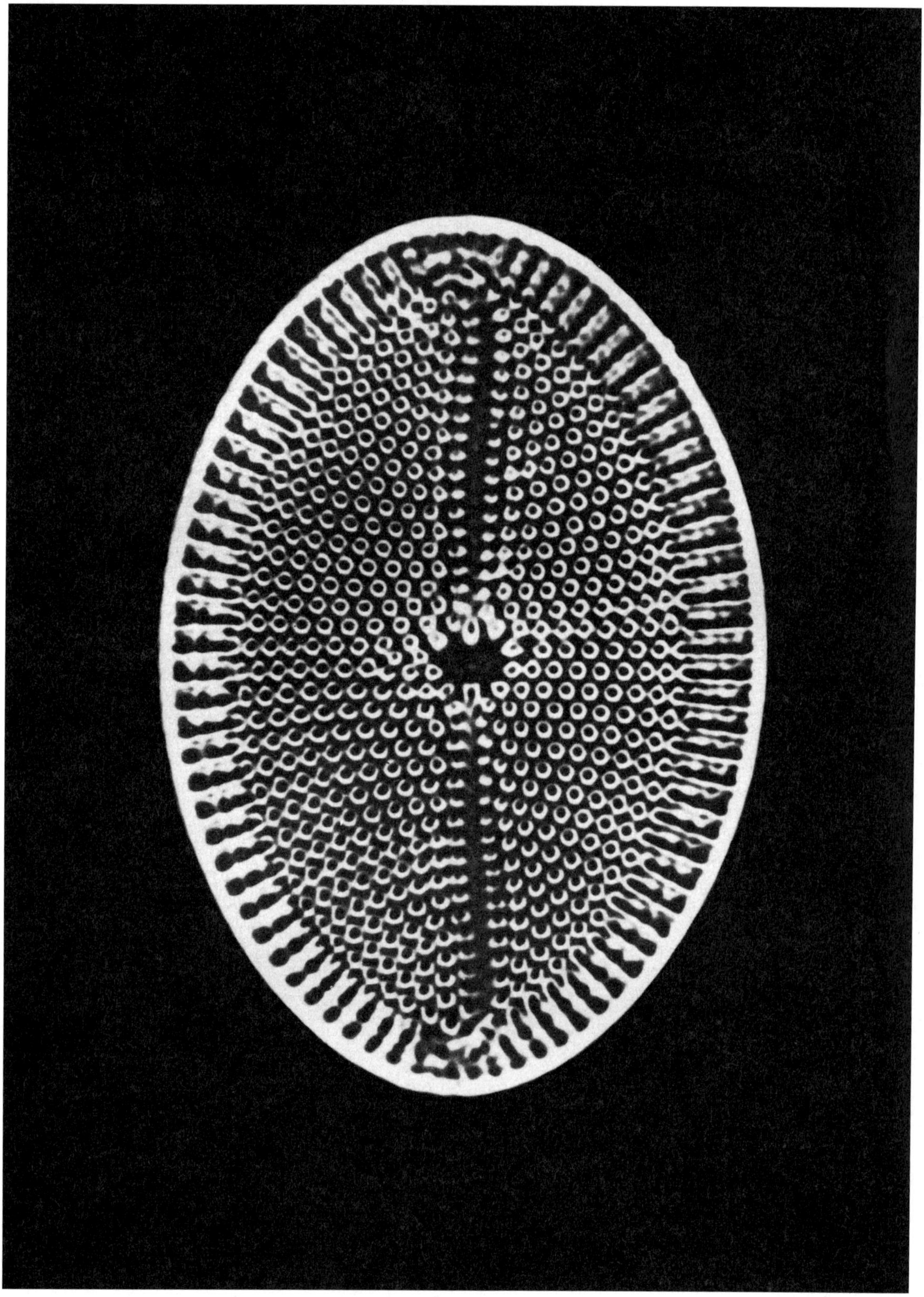

2400:1

Ellipsenförmige Kieselalge Orthoneis
splendida mit parabelförmigen Versteifungen.

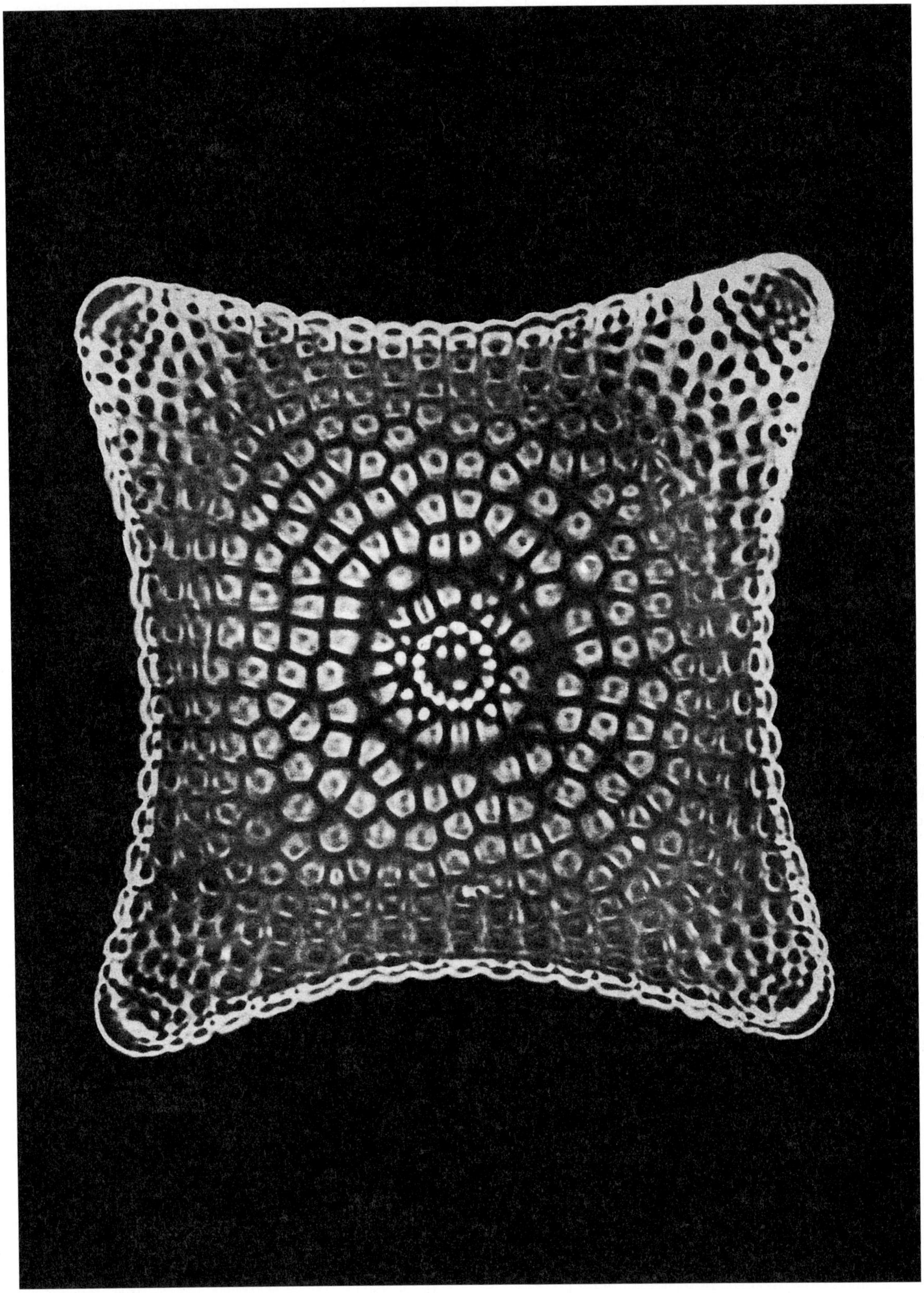

1940:1

Viereckige, kissenförmige Kieselalge
Triceratium parallelum.

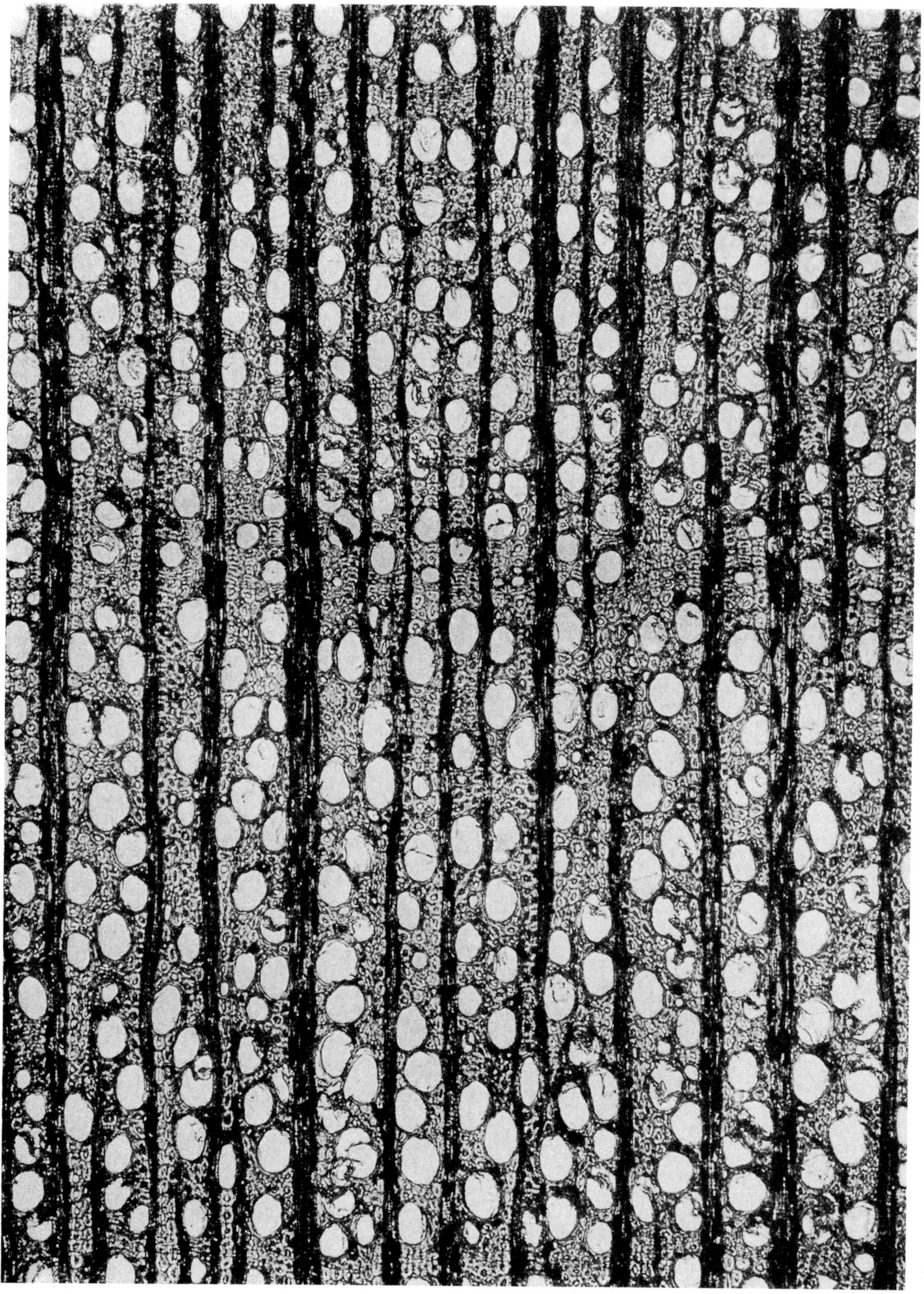

45:1

Querschnitt durch das Holz der Platane.

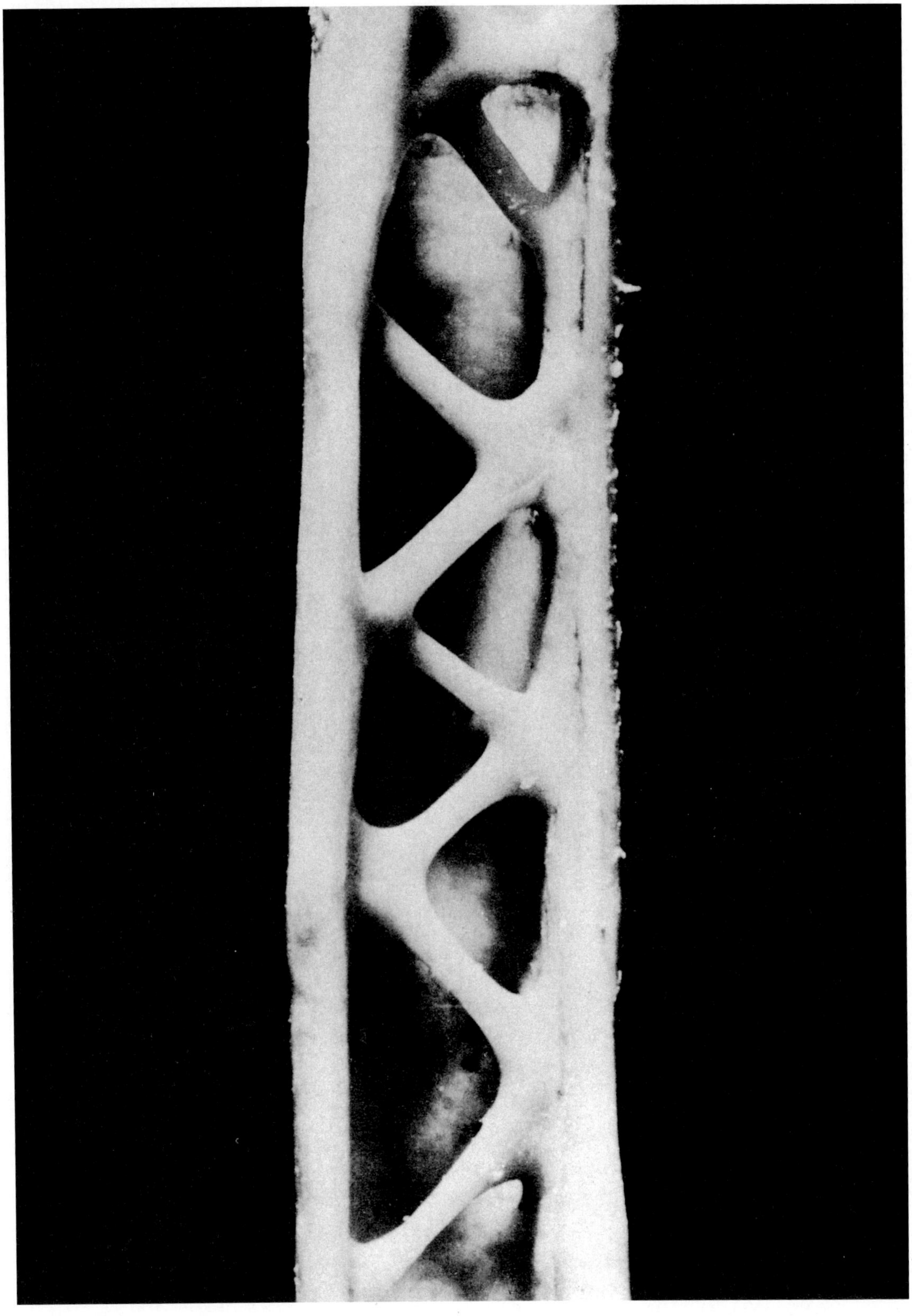

10:1

Versteifung durch V-förmige Träger in einem Mittelhandknochen des Bartgeiers.

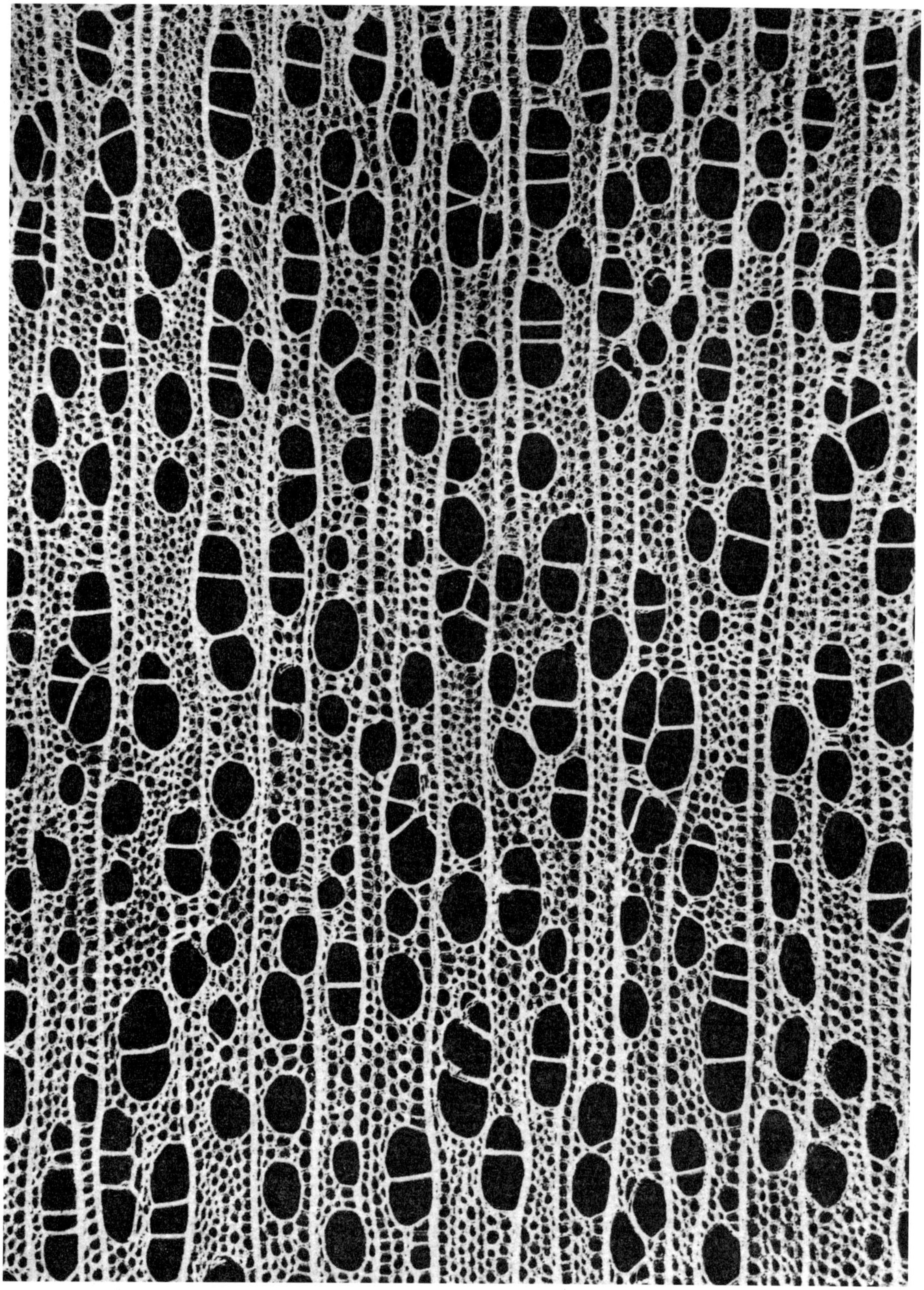

45:1

Querschnitt durch das Holz der Silberpappel.

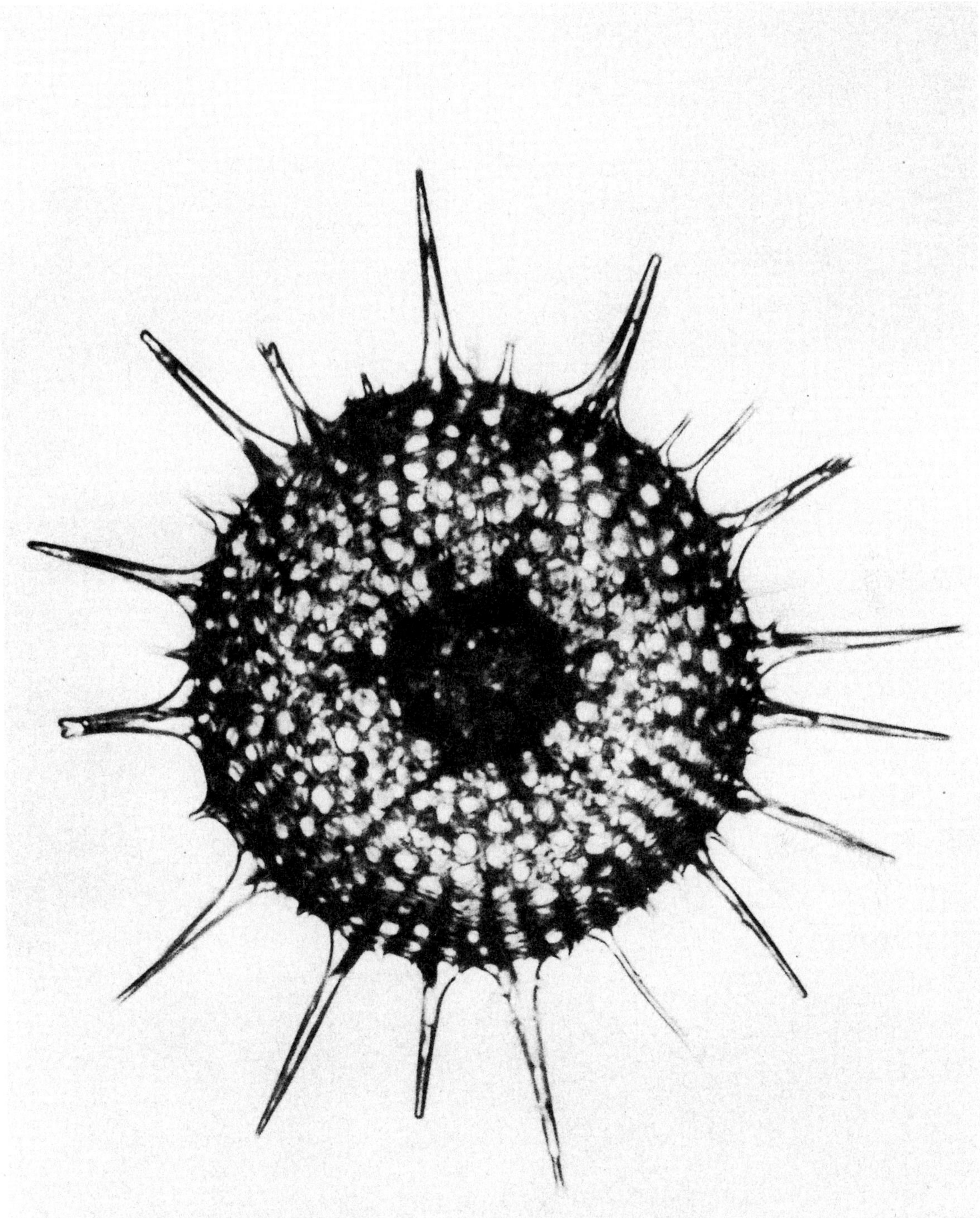

540:1

Abgeflacht-kugelförmige Radiolarie
Heliodiscus solaster.

Plättchenförmiger Schneekristall mit verzweigten Strahlen.

1,8 : 1

385 Reifsterne an einer Fensterscheibe.

24:1

Feinbau des Stützgerüstes der Koralle
Goniopora pedunculata.

19

Frank Lloyd Wright und seine organische Architektur

Frank Lloyd Wright and his organic architecture

Rudolf Finsterwalder

»Am Wege zieht uns irgendeine Blume durch ihre ungewöhnlich leuchtenden Farben oder ihre Formenschönheit an; machen wir vor ihr Halt, so nehmen wir ihre vollkommene Lieblichkeit dankbar in uns auf; wenn wir nun aber versuchen, das Geheimnis ihres Reizes zu entdecken, so finden wir, dass die Blüte, die zuerst unsere Aufmerksamkeit fesselte, in engster Beziehung zu der Textur und der Gestalt ihrer Blätter steht, wir entdecken eine seltsame Sympathie zwischen der Form der Blume und dem System, nach welchem die Blätter um den Stängel geordnet sind. Und diese Beobachtung führt uns zu einer anderen, zu der charakteristischen Gewohnheit des Wachstums und der daraus resultierenden Natur des Baus, dem sein Weg und seine Gestalt in den Wurzeln vorgeschrieben sind, die in der warmen Erde verborgen liegen und durch das beständige Bedecken mit vermoderten Blättern feucht gehalten werden. Diese Bauart geht von dem Allgemeinen zum Besonderen in ganz unvermeidlicher Weise über, bis sie schließlich die Blüte erreicht, welche in ihren Linien und Formen die Natur des Baues zeigt, welcher sie erzeugte. Es ist eine organische Einheit. Gesetz und Ordnung sind die Grundlage ihrer vollendeten Anmut und Schönheit. Anmut und Schönheit sind der Ausdruck der fundamentalen Bedingungen in Linie, Form und Farbe und ihr Vorhandensein dient dazu, sie nach irgendeinem Plane vollendet zu gestalten.«[1] | (Abb.1) Wright benutzte den Begriff *organische Architektur* und war damit einer der Ersten. Er verstand darunter einen organischen, also gewachsenen Zusammenhang der Architektur mit den verschiedenen Elementen der Kunst, der Natur und der menschlichen Lebensbereiche, also eine ganzheitliche Auffassung von Architektur. Alles sollte mitbedacht und miteinbezogen werden. | Als organische Architektur wird in der Geschichte eine Architektur bezeichnet, die die Harmonie von Gebäude und Landschaft, eine den Baumaterialien gemäße, *organisch* aus der Funktion heraus entwickelte Form, sowie eine biologische, psychologische und soziale Zweckmäßigkeit der Architektur anstrebt. | F. L.Wright hat ein tiefes, fast religiöses Verständnis von Natur. In seiner Idee vom organischen Bauen versucht er, ein ganzheitliches Konzept zu realisieren. Architektur und Natur sollen eine Einheit bilden, eine Einheit mit den Menschen und auch mit der Kultur der Menschen. Natur und Kultur, bei ihm kein Gegensatz, sollen sich gegenseitig in ihrer Wirkung steigern. | »For Wright, the purpose of art and architecture was not slavishly to copy external nature, but to use it in the way Emerson recommended, as the occasion for exploring inner nature and thereby expressing universal spirit. For the artist, nature was a raw material awaiting transformation into some greater vision of a still more divine ideal. ›Nature is not fixed, but fluid‹, Emerson had declared. ›Spirit alters, moulds, makes it. The immobility or brutness of nature is the absence of spirit; to pure spirit, it is fluid, it is volatile, it is obedient. Every spirit builds itself a house; and beyond its world, a heaven. Know

1 F. L.Wright: *Ausgeführte Bauten und Entwürfe.* Wasmuth-Verlag, 1910, S. 14

"Along the wayside some blossom, with usually glowing colour of prettiness of form, attracts us: held by it, we accept gratefully its perfect loveliness; but, seeking to discover the secret of its charm, we find the blossom, whose more obvious claim first arrests our attention, intimately related to the texture and shape of its foliage; we discover a strange sympathy between the form of the flower and the system upon which the leaves are arranged about the stalk. From this we are led to observe a characteristic habit of growth, and resultant nature of structure, having its first direction and form in the roots hidden in the warm earth, kept most by the conservative covering of leaf mould. This structure proceeds from the general to the particular in a most inevitable way, arriving at the blossom to proclaim in its lines and form the nature of the structure that bore it. It is an organic thing. Law and order are the basis of its finished grace and beauty: its beauty is the expression of fundamental conditions in line, form and color, true to them, and existing to fulfill them according to design."[1] | (Fig.1) Wright was one of the first to use the term *organic architecture*. For him it meant an organic or grown coherence of architecture with the different elements of art, nature and human aspects of life creating a holistic idea of architecture. Everything should be considered and integrated. Through history, organic architecture has indicated an architecture that seeks a harmony between buildings and landscape; an architecture based on the proper use of building materials, with form developed organically from the function, as a biological, psychological and social utility. | F. L. Wright has a deep, almost religious understanding of nature. In his idea of organic architecture he tries to realize a holistic concept. Architecture and nature should form a single entity, a unity with human beings and with the culture of human beings. Nature and culture, not a contradiction to him, should mutual enhance each other's impact. | "For Wright, the purpose of art and architecture was not slavishly to copy external nature, but to use it in the way Emerson recommended, as the occasion for exploring inner nature and thereby expressing universal spirit. For the artist, nature was a raw material awaiting transformation into some greater vision of a still more divine ideal. 'Nature is not fixed, but fluid', Emerson had declared. 'Spirit alters, moulds, makes it. The immobility or brutness of nature is the absence of spirit; to pure spirit, it is fluid, it is volatile, it is obedient. Every spirit builds itself a house; and beyond its world, a heaven. Know then, that the world exists for you. Build therefore, your own world.'"[2] | This was Wrights ideal concept at the beginning of his career; that everything should form a single organism, everything should derive from nature and harmonize with nature, with an aim to create an organically grown whole, a holistic view of the world. But it never was an imitation of nature, instead a fine interpretation of materials, forms and structures of nature. | "Conceive now that a whole building might grow up out of conditions as a plant grows

1 F. L. Wright: *Studies and executed buildings*. Wasmuth-Verlag, 1910, p. 33

2 William Cronon: *inconstant unity: the passion of Frank Lloyd Wright*, in Terence Riley (ed.), *Frank Lloyd Wright," Architecture: Nature*, Museum of Modern Art, New York, 1994, p. 36

1

then, that the world exists for you. Build therefore, your own world.‹«[2] | Das war Wrights ideale Vorstellung zu Beginn seines Schaffens, dass alles eine Einheit bilden sollte, alles von der Natur ausgehen sollte und mit ihr harmonieren sollte, ein organisch gewachsenes Ganzes war das Ziel, ein holistisches Weltbild. Es war jedoch nie ein Nachahmen der Natur, sondern eine feine Interpretation der Materialien, Formen und Strukturen von Natur. | »Stellen Sie sich nun vor, dass ein ganzes Gebäude aufwachsen und sich nach gewissen Bedingungen entwickeln kann, wie eine Pflanze aufwächst und sich vom Boden her entwickelt und dennoch frei ist, sie selbst zu sein, das eigene Leben zu führen, indem es dem Wesen des Menschen folgt; es besitzt die ganze Würde eines Baumes in der Natur, ist aber ein Kind des menschlichen Geistes. Ich schlage nun ein Ideal für die Architektur des Maschinenzeitalters vor, für das ideale amerikanische Gebäude. Lassen Sie es auf dieses Bild zuwachsen: den Baum. Ich möchte damit jedoch keineswegs die Imitation des Baumes suggerieren.«[3] | Materialien benutzte er ihrer Natur entprechend, z. B. schwere rohe Steine als Lager für einen leichteren Aufbau, Holzbalken für größere Spannweiten und den oberen Teil von Gebäuden. Der Boden im Inneren wurde in den Wohnbereichen gerne aus rohen Steinplatten gebaut, teilweise wurde vorhandener Felsen sichtbar gelassen, in den Schlafräumen hingegen kam wärmerer und feinerer Holzboden zum Einsatz. In den Gebäuden ist diese *Natürlichkeit* zu spüren, sie schafft eine sehr intensive, behagliche Atmosphäre. Die

2 William Cronon: *inconstant unity: the passion of Frank Lloyd Wright*, in Terence Riley (ed.): *»Frank Lloyd Wright« Architecture: Nature*; Museum of Modern Art, New York, 1994, S. 36

3 F. L. Wright: *An Autobiography*. 1943, S. 171

Abb. 1 Prairie-Haus, Illinois

Fig. 1 Prairie House, Illinois

2

up out of soil and yet be free to be itself, to 'live its own life according to Man's nature.' Dignified as a tree in the midst of nature but a child of the spirit of man. I now propose an ideal for the architecture of the machine age, for the ideal American building. Let it grow up in this image: The tree. But I do not mean to suggest the imitation of the tree with."[3] | Materials were used by him in a way that correlated with their nature, for example heavy rough stones as foundations for a lighter construction, wooden beams for larger spans and the upper part of buildings. The floor in the inner living areas was often covered with stone slabs, sometimes an existing rock stayed visible, however in the sleeping rooms a warmer and finer wooden floor was used. One can feel this connection with nature in his buildings; it created a very intense and cozy atmosphere. People should feel comfortable in his buildings and together with them become part of the whole ensemble. Even in a spatial way the theme of nature and landscape are used. Flat wide rooms mirror the atmosphere of the American landscape. Horizontal windows show panoramas of the landscape, the flowing spaces give the feeling of a growing space, which constantly stays in motion. The border between outer and inner space dissolves. In the outer space cantilevered, flat roofs create a relationship to the landscape, while strong fixed foundation walls give a solid base that holds the house to the earth. Altogether it is an agglomeration of what was found in nature, a kind of super-nature. The expression is enhanced both in the whole and in the detail, a

3 F. L. Wright: *An Autobiography*. 1943, p. 171

Abb. 2 Taliesin, Wisconsin

Fig. 2 Taliesin, Wisconsin

Menschen sollten sich in seinen Gebäuden wohlfühlen und mit ihnen zusammen Teil des großen Ganzen sein. Auch räumlich wird das Thema der Natur, der Landschaft verarbeitet. Flache weite Räume spiegeln die Atmosphäre der Landschaft der USA wider. Horizontale Fensterbänder zeigen Ausschnitte der Landschaft, fließende Räume geben ein Gefühl von einem gewachsenen Raum, der in Bewegung ist. Die Grenze zwischen Außen- und Innenraum löst sich auf. Im Außenbereich sind es weit auskragende, flache Dächer, die den Bezug zur Landschaft herstellen, fest gefügte Grundmauern, die dem Haus einen festen Halt auf der Erde geben. Alles zusammen ist eine Verdichtung des in der Natur Vorgefundenen, eine Art Super-Natur. Der Ausdruck wird im Gesamten und im Detail gesteigert und zeugt so vom Schaffen des kreativen Menschen Frank Lloyd Wright. | Gesteigert werden diese horizontalen Linien in ihrer Wirkung durch vertikale Elemente, wie z. B. den Kaminen im Außen- wie im Innenraum. Auch das sind Themen, die in der Natur vorkommen, bei ihm jedoch bewusst eingesetzt werden, als künstlerische Überhöhung. Im Inneren trieb er die Idee vom weiten fliessenden Raum auf die Spitze. Innenwände wurden so weit als möglich reduziert, geschlossene Innenecken wurden konsequent vermieden, um den Raum fließen zu lassen. Er schuf Symphonien aus Licht und Raum. Die Grenzen des Raumes aufzulösen war sein Anliegen. Die Innenwände liefen nach außen und verzahnten die Landschaft mit dem Gebäude. Ein Motiv, das auch auf den Einfluss japanischer Kunst zurückzuführen ist. | »Taliesin sollte eine abstrakte Verbindung aus Stein und Holz sein, wie man sie in den umgebenden Hügeln natürlich wiederfindet. Und die Linien der Hügel würden die Linien der Dächer, die Abhänge der Hügel ihre Abhänge. Die mit Gips verputzten Oberflächen der leichten Holzmauern, die von den großen vorspringenden Dachflächen in Schatten getaucht werden, sollten sein, wie die flache Ausdehnung des Sandes unten im Fluss und von der gleichen Farbe. Denn von dort kommen die Materialien, die die Oberflächen bedecken.«[4] | (Abb. 2) Das ganze Haus war als organische Einheit gedacht, alles hatte einen Sinn und eine Funktion, alles Überflüssige wurde weggelassen, auch die Technik des Hauses bildete mit der Form und der Konstruktion eine Einheit. Das war seine Auffassung von der Aussage seines Lehrmeisters Sullivan »form follows function«. Nicht ein pragmatischer Funktionalismus war sein Ideal, sondern ein holistisches Verständnis für ein Gebäude als Werkzeug, das Gebäude als organischer Mechanismus, ähnlich einem lebenden Organismus. »Die Form folgt der Funktion? Jawohl, aber wichtiger ist jetzt, dass Form und Funktion eins sind.«[5] | Im Bezug auf die Funktion ist er oft missverstanden worden, er meinte, das Haus sollte eine Form entsprechend seiner Nutzung haben, d.h. eine Kirche sollte aussehen wie eine Kirche und ein Wohnhaus wie ein Wohnhaus. Das war es auch, was er in der Natur sah: »Und wie jeder sehen kann, ist eine Buche eben eine Buche. Sie versucht

4 F. L. Wright: *An Autobiography*. 1943, S. 171

5 F. L. Wright: *Eine organische Architektur*. aus *Bauweltfundament 25*, 1953, S. 190

testimony to the creative spirit of Frank Lloyd Wright. Horizontal lines are enhanced in their expression by vertical elements, such as the chimneys in the interior and exterior space. These are also themes that can be found in nature, but he uses them in a conscious way, as an artistic exaggeration. Inside, he pushed the idea of a wide flowing space to new levels. Interior walls have been reduced to a minimum, while closed interior corners were consistently avoided to let the space flow. He composed symphonies of light and space. His intention was to dissolve borders of space. The interior walls led outside and interlocked the landscape with the building. A motif derived from the influence of Japanese art. | "Taliesin was to be an abstract combination of stone and wood as they naturally met in the aspects of the hills around about. And the lines of the hills were the lines of the roofs, the slopes of the hills their slopes, the plastered surfaces of the light wood-walls, set back into shade beneath broad eaves, were like the flat stretches of sand in the river below and the same in color, for that is where the material that covered them came from."[4] | (Fig. 2) The whole house was considered as an organic entity, every single thing had a sense and a function, everything superfluous was avoided, and even the houses technique formed a unity together with its form and construction. That was an idea derived from his teacher, Sullivan's term "form follows function". Not a pragmatic functionalism, but a holistic understanding of a house as a tool, the building as an organic mechanism, similar to a living organism. "Form follows function? Yes, but the more important is, that form and function are one thing."[5] | "And as anyone might see, a beech tree is a beech tree. It isn't trying to be an oak.... there must be some kind of house that would belong to that hill, as trees and the ledges of rock did..."[6] | In relation to form he has often been misunderstood, he thought that the building should have a form relating to its function, which means a church should look like a church and a house like a house. An idea he also saw in nature. | Everything was integrated in the building's structure, even doors, furniture and tools, as well as technology, became part of the whole. This coincided with Wright's idea of life, for which he also had a holistic approach. This was also the case in his idea of social life. He lived in Taliesin together with his family and his colleagues. At the same time he was teaching some students that also lied there. Everything should form an entity. The students and colleagues not only used to work in his studio, but also in the connected farm. He was of the opinion that the students could develop a deeper understanding of nature and also of architecture this way. | In the later stage of his work he developed more and more an interest for non-organic themes; his architecture became more urban, and more and more distant from nature. In the vertical, he discovered an extension to his idea of space. Added to the movement in the horizontal, the relation to inside and outside came the vertical element as a further possibility. Overall in his

4 F. L. Wright: *An Autobiography*. 1943, p. 171

5 F. L. Wright: *Eine organische Architektur*. from *Bauweltfundament 25*, 1953, p. 190

6 F. L. Wright: *"Taliesin"* in *An Autobiography*. 1943, p. 168

nicht, eine Eiche zu sein.... es musste eine Art von Haus geben, das zu diesem Hügel gehörte, wie es die Bäume und Felsgesimse taten,«[6] | Alles wurde in die Struktur des Gebäudes mit einbezogen, auch Türen, Möbel und Geräte, wie auch die Technik, wurden zu einem Teil des Ganzen. | Dies entsprach Wrights Vorstellung vom Leben, an das er auch einen ganzheitlichen Anspruch hatte. Für ihn spielten alle Fragen des Lebens eine Rolle. Das ging bis zu seiner sehr eigenen Vorstellung von sozialem Leben. In Taliesin wohnte und arbeitete er mit seiner Familie und seinen Mitarbeitern. Gleichzeitig unterrichtete er auf seinem Anwesen eine Anzahl von Studenten, die auch dort untergebracht waren. Alles sollte eine Einheit bilden. Die Studenten und Mitarbeiter arbeiteten nicht nur im Büro, sondern auch im angeschlossenen landwirtschaftlichen Betrieb. Er war der Meinung, dass die Studenten so ein tieferes Verständnis von Natur und auch von Architektur entwickeln würden. | In seiner späteren Schaffensperiode interessierte sich Frank Lloyd Wright immer mehr für nicht landschaftliche Themen, seine Architektur wurde urbaner und entfernte sich weiter von den Bezügen zur Natur. Er entdeckte in der Vertikalen eine Erweiterung seines Raumideals. Zu einer Bewegung in der Horizontalen, der Beziehung von Außen und Innen, kam das vertikale Element als weitere Möglichkeit hinzu. Vor allem in seinen städtischen Gebäuden wurde das Vertikale bewusst eingesetzt. Z. B. das *Johnson Wax Gebäude* mit seinem Turm, eine Form, die er wegen ihrer vertikalen Ausrichtung Jahre zuvor kategorisch abgelehnt hatte. Seine Entwürfe für Hochhäuser waren dem seiner Meinung nach einzigen vertikalen Ding in der Natur, dem Baum, nachempfunden. Die Analogie zu einem Baum mit Stamm und Ästen ist unübersehbar. Sein Interesse für die Stadt führte konsequenterweise zu einer neuen Architektur, er wandte seine Herangehensweise an neue Aufgaben auch hier an und so entstanden zwangsläufig neue Formen und Räume. | Auf der Suche nach neuen Ausdrucksformen rückten immer mehr neuartige Themen in den Mittelpunkt. Er interessierte sich für futuristische Elemente, so sind z. B. in seinen Zeichnungen zu broadacre-city Ufos zu sehen und er entwickelte utopische Projekte, wie den *1 mile Wolkenkratzer.* | Er ist von der »Nautilius« des Autors Jules Verne fasziniert, dieser Mischung aus Höhle und Technik. Daran zeigt sich das Interesse an phantastischen Dingen, auch am Spektakulärem, das sich vom Naturbezug seiner Anfänge entfernt. In seinen letzten Jahren versuchte er, die Komplexität und Plastizität, die er in der Natur so bewunderte, mit seinen Bauten zu erreichen. Vor allem die Schneckenform von Muscheln hatte es ihm angetan. Er war regelrecht frustriert, dass ihm die technischen Möglichkeiten zur Realisierung einer solchen Form fehlten. Mit den aufkommenden Möglichkeiten des Stahlbetoms zeichnete sich die Möglichkeit zur Realisierung solcher Konstruktionen ab, er sollte es jedoch nicht mehr erleben. Ein Versuch, war die Schneckenform des Gugenheimmuseums, die

6 F. L. Wright: *»Taliesin«* aus *An Autobiography*. 1943, S. 168

3

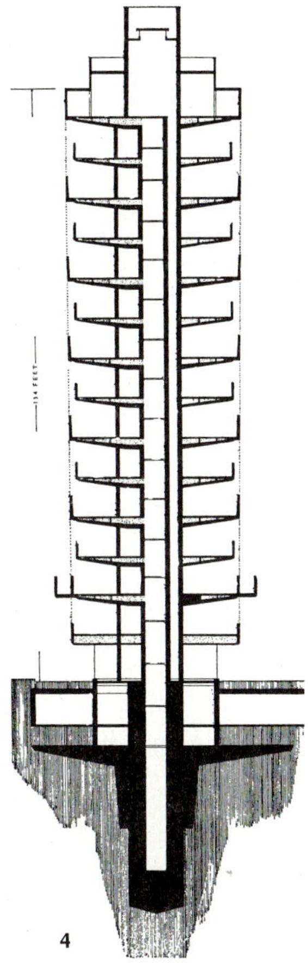

4

urban buildings the vertical was inserted as a decisive element. For example, the *Johnson Wax building*, with his tower, a form that he had decisively neglected because of its vertical dimension years before. His designs for high-rise buildings were associated with, in his opinion, the only vertical thing in nature, the tree. The analogy to a tree with its trunk and branches is obvious. His interests in the city consequently lead him to a new architecture; he applied his ideas and approach to new tasks and so inevitably new forms and spaces derived. | In the search for new ways of expression more and more new themes emerged. He was interested in futuristic elements, such as the UFO's seen in his sketches of broadacre-city, while also developing utopian projects like the *1 mile skyscraper*. He was fascinated by the "Nautilius" of the author Jules Verne, this mixture of cave and technique. Here one can see his interest for fantastic things, even in the spectacular, that moves far away from the reference to nature at the beginning of his career. In his last years he tried to reach the complexity and sculptural quality that he admired in nature with his buildings. Above all he was interested in the spiral form of seashells. He was really frustrated that there lacked the technical possibilities to realise such a form. With the emerging possibilities of reinforced concrete he saw the chance to realise such projects, but he would not live to see it happen. An attempt to use the form of the seashell can be seen in the Guggenheim Museum, which probably derived more from his wish to realize such a form as from its ability to function as

Abb.3 Johnson Wax, Innenraum
Fig.3 Johnson Wax, Interior

Abb.4 Johnson Wax, Schnitt
Fig.4 Johnson Wax, section

5

wohl eher seinem Wunsch nach dieser Form als ihrer Eignung für ein Museum entsprach. Architektur sollte organisch, flexibel und frei sein, wie die Natur selbst. (Abb. 5) | Edgar Kaufmann jr. sagte: »Für (Wright) war das menschliche Leben keine Ausnahme der Schöpfung, sondern ein Bestandteil natürlicher Vorgänge. Alles, was Menschen tun, geschieht in und mit einer Natur, die sie ihren Bedürfnissen anpassen, und in dieser Wechselwirkung tragen sie zum System der Natur bei. Entsprechend sah er auch die Architektur als eine natürliche Äußerung des menschlichen Lebens, die ihrerseits das System bereicherte, innerhalb dessen sie entstand. Architektur, Menschheit und Natur waren für Wright also in einem großen, dynamischen Kontinuum miteinander verbunden, und Kontinuität in der Natur war ein Zeichen dafür, dass die Menschen sich so, wie es seiner Auffassung nach sein sollte, in das Gefüge der Natur einordneten.«[7] | Wright plante seine Gebäude auf modularen, geometrischen Rastern auf, um alles in Einklang zu bringen mit dem, was er kosmische Grundmuster nannte. Für ihn war das das beste Mittel, um eine universale Kontinuität auszudrücken. Er gestaltete alles vom Möbel bis zum Städtebau nach diesem Prinzip, das er auch in der Natur sah, und wollte so alles miteinander in Einklang bringen. | Als Kind hatte er von seiner Mutter, die ihn schon vor seiner Geburt als großen Architekten sah, ein Spielzeug des deutschen Pädagogen Hermann Fröbel bekommen. Mit diesen geometrischen Formen übte sich Wright als Kind in der Komposition von Formen, die Klötze

7 Edgar Kaufmann jr.: *Frank Lloyd Wright: Plasicity, Continuity and Ornament.* Journal of the Society of Architectural Historians, S. 37

Abb. 5 Guggenheim Museum, New York **Fig. 5** Guggenheim Museum, New York

a museum. Architecture should have been, organic, flexible and free, like nature itself. (Fig.5) | Edgar Kaufmann said about Wright: "For (Wright) human life was not an exception of creation but a part of natural processes. Everything what men did happens in and with nature, that they adapt to their needs and in this interchange they are part of natures system. In this way he understood architecture as a natural comment of human life, that enriched the system out of that it derived. In his opinion, architecture, manhood and nature were connected in a big dynamic continuum and continuity in nature was a sign that people were integrated in the structure of nature, as he thought it should be."[7] | Wright planned his buildings on a modular geometric grid, so to bring everything in unison with what he called a cosmic master pattern. In his opinion it was the best way to express a universal continuity. He designed everything from the furniture to the urban design in accordance with this principle, which he saw in nature as a way to bring everything in unison. When he was a child he received a toy from his mother, who envisioned him as a leading architect before his birth, which was designed by the German pedagogue Hermann Fröbel. With this geometric forms Wright as a child trained himself in the composition of forms; the toy blocks were expandable by folded paper elements, balls, etc. Quite early on he developed a sense for the inner geometry of things. An idea that he developed while he was still working with Luis Sullivan, who himself had a preference for ornamental wall coverings. Wright understood these woven forms of nature as something structural, inherent in nature and from this he developed his idea of an organic architecture that went beyond the formal understanding of nature in the Art Nouveau. | "I knew well that no house should ever be *on* a hill or *on* anything. It should be *of* the hill. Belonging to it. Hill and house should live together each the happier for the other. That was the way everything found round about was naturally managed except when man did something. I wished to be part of my beloved southern Wisconsin, too. I did not want to put my small part of it out of countenance. Architecture, after all, I have learned – or before all, I should say – is no less a weaving and a fabric than the trees are."[8] | (Fig.6) He refused to think anything against "nature". His interest and understanding of nature went very far. He tried to discover the structure and the rules in nature and to understand it, which was beautiful to him. If there were rules behind things, beauty is the result. Wright despised all architectural styles. Architecture in his opinion had to grow out of the site and the specific task, and develop with it an organic entity. | He began his work with the Prairie houses, in which he saw something super American, the relation to the wideness of the American landscape, but also the unity with the family life. Walt Whitman, whom F. L. Wright admired, said: "I breath the space in big drags, east and west are mine, and north and south are mine … the earth is spreading left and right … o country

7 Edgar Kaufmann jr.: *Frank Lloyd Wright: Plasicity, Continuity and Ornament.* Journal of the Society of Architectural Historians, p. 37

8 F. L. Wright: *"Taliesin"* in *An Autobiography.* 1943, p. 168

6

waren erweiterbar um gefaltete Papierelemente, Kugeln usw. So konnte er schon früh einen Sinn für die den Dingen innewohnende Geometrie entwickeln. Eine Idee, die er entwickelte, als er noch bei Luis Sullivan arbeitete, der eine Vorliebe für ornamentale Wandflächen hatte. Wright begriff diese gewobenen Naturformen als etwas Strukturelles, was der Natur innewohnt, und begann daraus seine Vorstellung von einer organischen Architektur zu entwickeln, die über das formale Naturverständnis des Art Nouveau hinausging. | »Inzwischen wusste ich gut, dass kein Haus jemals *auf* einem Hügel oder *auf* sonst irgend etwas gesetzt werden durfte. Es sollte vom Hügel sein, zu ihm gehören, damit Hügel und Haus zusammen leben konnten, jedes wegen des anderen nur umso glücklicher. Alles, was man rund um das Haus findet, ist natürlich – außer wenn der Mensch etwas tut. [...] ich wünschte mir, ein Teil meines geliebten südlichen Wisconsin zu sein, und nicht, meinem kleinen Teil des Landes ein anderes Aussehen zu geben. Wie ich gelernt habe, ist die Architektur schließlich – oder vor allem anderen, wie ich sagen muss – nicht weniger ein Gewebe oder Gefüge als die Bäume..«[8] | (Abb. 6) Er lehnte es ab, etwas gegen die »Natur« zu denken. Sein Interesse und sein Verständnis für die Natur ging sehr weit. Er versuchte die Struktur und die Regelhaftigkeit in der Natur zu entdecken und zu begreifen, das machte für ihn Schönheit aus. Wenn diese Regelhaftigkeit hinter den Dingen steht, entsteht Schönheit. Wright verachtete alle architektonischen Stile, die Architektur musste bei ihm aus

8 F. L. Wright: *»Taliesin«* aus *An Autobiography*. 1943, S. 168

Abb. 6 Taliesin, Wisconsin

Fig. 6 Taliesin, Wisconsin

7

road ... you express myself better, than I can express myself. [...]" | This was and is the American dream of liberty and wideness, of democracy, a counter model to the European architecture of the city, which looked formally severe and unfree. He wanted to create something specific that was developed out of its surrounding and forming an entity with that. The center of his houses was the fireplace, the center of the family and the social life. From there, the house developed in the landscape, was not seen as a foreign object but became part of the landscape, through to the materiality. He used materials from the site, which condensed the expression of the site. The way he did that was very personal, so that he could give a personal note to his buildings. The signature of the architect should be quite visible. A holistic understanding of architecture, where inhabitants, nature, architecture and architect come together in a single building.

Abb.7 Fallingwater, Pennsylvania

Fig.7 Fallingwater, Pennsylvania

der Umgebung und der spezifischen Aufgabe wachsen und mit ihr eine organische Einheit entwickeln. | Begonnen hat er sein Schaffen mit den Prärie-Houses, in denen er etwas Uramerikanisches sah, den Zusammenhang mit der Weite der amerikanischen Landschaft, aber auch die Einheit mit dem Leben der Familie. Der von ihm verehrte Walt Whitman sagte: »Ich atme den Raum in großen Zügen ein, Ost und West sind mein, und Nord und Süd sind mein ... die Erde erstreckt sich links und rechts ... o Landstraße ... du drückst mich besser aus, als ich mich selber ausdrücken kann. [...]« | Das war und ist der amerikaniche Traum von Freiheit und Weite, von Demokratie. ein Gegenentwurf zur europäischen Architektur der Stadt, die ihm formal streng und unfrei vorkam. | Er wollte etwas Spezifisches schaffen, was aus der Umgebung heraus entstand und mit dieser eine Einheit einging. Mittelpunkt der Häuser war die Feuerstelle, der Mittelpunkt der Familie und des sozialen Lebens. Von dort aus entwickelte sich das Haus in die Landschaft, war nicht als ein Fremdkörper hingesetzt, sondern wurde Teil der Landschaft, bis hin zur Materialität. Er verwendete Materialien des Ortes, die den Ausdruck des Ortes verdichteten. Die Art und Weise, wie er das machte, war seine sehr persönliche, so dass er den Bauwerken einen sehr subjektiven Ausdruck gab. Die Handschrift des Architekten sollte durchaus sichtbar sein. Ein ganzheitliches Verständnis von Architektur, in dem Bewohner, Natur, Architektur und Architekt zu einem einzigen Bauwerk zusammenfanden.

20

Sternenschau und Schalenbau.

Das Zeiss-Planetarium Jena

Star Show and Gridshell Construction.

The Zeiss Planetarium Jena

Joachim Krausse

Das Objekt, das eine der größten Umwälzungen der Baugeschichte auslösen sollte, kommt ebenso wie sein Urheber in der Geschichte der Architektur nicht vor. Registriert wurde das Bauwerk, das Walter Bauersfeld bei der Firma Carl Zeiss in Jena nach dem Ersten Weltkrieg entwarf und das als Zeiss-Projektionsplanetarium weltbekannt wurde, lediglich durch eine Richtung im Bauingenieurwesen, die eng mit der Firmengeschichte der größten Betonfirmen verbunden ist.[1] | Hier, in der Geschichte der Betonkonstruktionen und der Hallenbauten, ist unser Objekt wieder zu finden, und zwar als ursprüngliche Anwendung der Schalenbauweise. Unter Schalen versteht man gekrümmte Flächen, die in einer äußerst dünnen Betonschicht gegossen oder gespritzt werden und eine hohe Tragfähigkeit besitzen. Ein Modell in der Natur stellt die Schale eines Eis dar. Mit der Schalenbauweise, so M. Sanchez-Arcas, wurde »auf Grund der umwälzenden Entwicklung von Naturwissenschaft und Technik im Bauwesen ein bisher in dieser Form nicht bekanntes konstruktiv-architektonisches Element in die Baukunst eingeführt.«[2] | Aber welches waren die umwälzenden Entwicklungen von Wissenschaft und Technik? Wo kamen sie her, wenn sie nicht auf dem Gebiet der Baukonstruktion selbst stattgefunden hatten? | Carl Zeiss Jena hatte 1913 den Auftrag bekommen, für das im Bau befindliche Deutsche Museum in München, das erste Museum für Naturwissenschaften und Technik in Deutschland, ein Planetarium zu bauen. Der Museumsgründer und -direktor, Oskar von Miller, war von der Idee besessen, die kaum vorstellbaren Vorgänge in Wissenschaft und Technik in geeigneten Modellen und Schaustücken einem großen Publikum nahe zu bringen. »Für die Erklärung besonders komplizierter Vorgänge, wie beispielsweise für die Bewegung der Gestirne, werden besonders kunstvolle Apparaturen wie das Planetarium geschaffen, die einen komplizierten Vorgang in seine einzelnen Phasen zerlegen und hierdurch begreiflich und fassbar machen.«[3] | Ausgehend von solchen Überlegungen wurde im Herbst 1913 der Auftrag an Carl Zeiss in Jena erteilt, zwei Himmelsmodelle zu bauen: Erstens ein so genanntes Kopernikanisches Planetarium – ein panoramaartiger Raum von 9 m Durchmesser mit einem mechanischen Modell unseres Planetensystems an der Decke, in dem die Sonne und die Planeten als Kugelleuchten verschiedener Größe und Helligkeit aus der Mechanik wie aus einem Schnürboden herabhängen, während der Fixsternhimmel als Panorama des Tierkreises auf der umschließenden Wand aufgemalt ist und das Sternenlicht von Glühlämpchen hinter kleinen Öffnungen der Wand in den Innenraum dringt. | Der Betrachter steht in einer auf Schienen laufenden, käfigartigen Kabine. Aus diesem ›Erdwagen‹ beobachtet er durch ein Periskop die Himmelskörper lediglich in einer horizontalen Richtung. Wie in den großen Panoramen und Dioramen des 19. Jahrhunderts verbinden sich plastische und flächige Elemente, wenn ein bestimmter Blickwinkel und eine bestimmte Augenhöhe eingehalten

1 Bauersfeld berücksichtigt und gewürdigt zu haben ist das Verdienst von: Jürgen Joedicke: *Schalenbau, Konstruktion und Gestaltung. Mit Beiträgen von Walter Bauersfeld und Herbert Kupfer.* Stuttgart: 1962.

2 M. Sanchez-Arcas: *Form und Bauweise der Schalen.* Berlin: 1961.

3 Oskar von Miller nach eigenen Aufzeichnungen, Reden und Briefen. München: 1932, S. 113

The object that was to spark off one of the greatest revolutions in building history, like its originator, does not make an appearance in the history of architecture. The structure that Walter Bauersfeld designed at the Carl Zeiss Company in Jena after the First World War and which became known throughout the world as the Zeiss projection planetarium has only been recorded as part of the history of civil engineering. It is here, in the history of concrete construction that our object is to be found, as the first application of monocoque construction or "shell building". By "shells" one understands curved surfaces which are cast or sprayed in a thin layer of concrete and possess a high load-bearing capacity. A model in nature is provided by the shell of an egg. With monocoque construction, an element was introduced into architecture which had not hitherto been known. Of interest to us in what follows, are the starting point and the conjunction of circumstances from which this innovation emerged. The point of contact linking, in this instance, optics and scientific instrument-making with building. | In 1913 Carl Zeiss Jena had received the contract to build a planetarium for the Deutsches Museum then under construction in Munich, the first museum of science and technology in Germany. The museum's founder and director Oskar von Miller was obsessed by the idea of getting the general public to understand the scarcely imaginable processes of science and technology through appropriate models and exhibits. | Miller had a feeling for the experimental dimension of machines, instruments and experimental arrays, and enjoyed such things as dioramas, mises-en-scène and special effects. For the planetarium, he wanted "a journey through the Copernican universe" – while of course retaining the greatest possible degree of scientific accuracy. The observer should be placed at the centre of cosmic events, in such a way that he can comfortably follow the simulated movements of the heavenly bodies with the naked eye. | On the basis of considerations like these, Carl Zeiss in Jena was commissioned to build two models of the heavens. The first was a so-called Copernican Planetarium, a panoramic room 9 metres in diameter with, on the ceiling, a mechanical model of our planetary system in which the Sun and the planets, represented by shining globes of varying size and brightness, hang down from the mechanism as from a rigging-loft in the theatre, with the fixed-star background painted as a panorama of the zodiac on the enclosing wall and starlight penetrating the interior from filament lamps behind tiny openings. | The second model that Oskar von Miller commissioned from Zeiss was to "place the viewer, in accordance with the ideas held by the astronomers of antiquity, on Earth, conceived as being at rest, on a fixed platform which was to be built inside a large, rotating tin globe, the sphere of the fixed stars and the planets of the ancients attached to special mechanisms were to be moved inside this globe in accordance with their apparent paths." | This second model was to be constructed at the

werden. | Man erkennt schnell, dass dieses Modell vollständig in der Tradition der Schaukünste des 19. Jahrhunderts steht, der Ära, in der die Ferne noch ohne Kino nahe gebracht wurde.[4] Die Wirkung war unbefriedigend; der Betrachter »fühlt nur zu deutlich, dass er sich in einem kleinen, der Natur künstlich nachgebildeten Raum befindet (...) Er muss feststellen, dass ihm in diesem Modell die Abmessungen der Körper des Sonnensystems und ihre Entfernungen voneinander gegenüber der Natur sehr gefälscht geboten werden, und wenn er nicht vorsichtig ist, bekommt er eine ganz falsche Vorstellung von der in Wirklichkeit so außerordentlich geringen Menge der leuchtenden Materie im Weltraum.«[5] | Die Reaktion auf diesen Fehlschlag war bezeichnenderweise zunächst die Rückkehr zum ptolemäischen Weltbild. Das zweite Modell, das Oskar von Miller bei Zeiss in Auftrag gab, sollte »nach den Anschauungen der Astronomen des Altertums den Beschauer auf der ruhend gedachten Erde auf einer festen Plattform aufstellen, die im Innern einer großen, drehbaren Blechkugel, der Fixsternsphäre, eingebaut sein sollte, und die Wandelsterne der Alten sollten innerhalb dieser Kugel an besonderen Mechanismen befestigt und ihrer scheinbaren Bahn entsprechend bewegt werden.«[6] | Das andere Modell, das ›ptolemäische‹, das Oskar von Miller in Auftrag gegeben hatte, sollte bei Zeiss Dr. Walter Bauersfeld konstruieren, ein Ingenieur, der auf dem Fachgebiet der Optik und Feinmechanik zu Hause war. Der Versuch – so Bauersfeld –, dem »Beschauer die Bewegungen der Gestirne in der ptolemäischen Anschauungsweise im Innern eines halbkugelförmigen Hohlraumes« nahe zu bringen, »genau so wie die Sterne von der Erde aus erscheinen«, zeigte schon bei den ersten Entwürfen, »dass hier mechanisch fast unlösbare Aufgaben vorlagen.«[7] | Wie Bauersfeld berichtet, blieben die Arbeiten an diesem Projekt während des Ersten Weltkrieges liegen, wurden aber nach Kriegsende wieder aufgenommen. In fünf Jahren entstand dann das Projektions-Planetarium als Typ eines modernen Weltraumsimulators. Es konnte freilich weder ›kopernikanisch‹ noch gar ›ptolemäisch‹ sein; wir werden sehen, dass es in einem bestimmten Sinne nur ›einsteinsch‹ sein konnte. | Die Fehlschläge der mechanischen Versuche führten Bauersfeld zu folgender Schlussfolgerung: »Eine getreue Nachbildung der Natur zu erhalten, erscheint so lange ausgeschlossen, als man daran festhält, das Ziel mit schwerfälligen Mechanismen zu erreichen, d. h. mit Einrichtungen, die niemals imstande sein werden, jenen geheimnisvollen, geräuschlosen Weltenlauf der Natur täuschend nachzubilden.«[8] Bauersfelds Konzept kehrte die Verhältnisse einfach um: »Der Grundgedanke der Lösung war, die Kugelschale fest anzuordnen und alle Gestirne auf ihrer Innenfläche durch ein System von Projektionsapparaten abzubilden, das in der Nähe des Kugelmittelpunktes anzuordnen war. Es war dazu nötig, die Kugelfläche oberhalb des Horizonts für die Aufnahme der Projektionsbilder weiß zu färben. Unterhalb des Ho-

4 Vgl. Stephan Oettermann: *Das Panorama. Die Geschichte eines Massenmediums.* Frankfurt a. M.: 1980.

5 Villiger: *Das Zeiss-Planetarium*, a.a.O., S. 8

6 Ebd.

7 Walter Bauersfeld: *Das Projektions-Planetarium des Deutschen Museums in München.* In: Zeitschrift VDI, Bd. 68, 1924, Nr. 31, S. 793

8 Villiger: *Das Zeiss-Planetarium*, a.a.O., S. 9

Zeiss factory by Dr. Walter Bauersfeld, an engineer who was at home in the specialist field of optics and precision engineering. Work on this project was interrupted during the First World War, but was taken up again after the end of the War. Then, over the next five years, the projection planetarium took shape as the exemplar of a modern Universe simulator. | The failure of the first mechanical experiments had let Bauersfeld to the following conclusion: "to obtain a faithful replica of nature seems to be out of the question so long as one insists on achieving the objective with heavy machinery, that is, with equipment that will never be equal to the task of convincingly reproducing that mysterious and soundless world-movement of nature."[1] Bauersfeld's plan simply stood the problem on its head: "The basic idea of the solution was to make the shell of the sphere fixed and produce an image of all the stars on its inner surface by means of a system of projectors set up close to the centre of the sphere. For this, it was necessary to colour the surface of the sphere above the horizon white in order for it to receive the projected images. Below the horizon, on the other hand, special measures had to be taken to render the images invisible."[2] We see here how the possibilities for omnidirectional photographic projection form the starting point for a revolutionary construction. It followed the projection idea that Bauersfeld not only had to develop a new and — because it had to be programmable — quite complicated type of photographic projector, but also a new, spherically curved projection screen. There was no immediate precedent for either. || The regular polyhedron which Bauersfeld chose as the shape of his projector could be replaced by another regular polyhedron, for example a tetrahedron, octahedron, cube, etc. without altering the principle. | This method of projecting a regular polyhedron onto the surrounding sphere creates on the latter's surface a network of lines which is known as geodesic, because the edges all lie on great circles, or 'geodesics'. Their radius is the radius of the sphere. But the geometric structure that Bauersfeld used in practice for the shape of his spherical projector, as well as that of the projection wall, was now also to become the structure for an architectural solution: the supporting structure for the dome of the planetarium. Bauersfeld thereby became the pioneer of a form of construction from which two revolutionary developments immediately followed: light-weight supporting structures, which were later developed so successfully by Konrad Wachsmann, Richard Buckminster Fuller, Max Mengeringhausen and many others; and monocoque construction, which opened up completely new possibilities for reinforced concrete in coping with large spans. It was the optician and precision engineer Bauersfeld who introduced this development, even though it belongs entirely in the field of building construction. | The novelty has very ordinary roots. Bauersfeld informs us about the circumstances: "In 1922 the first planetarium mechanism, which had been developed for the Deut-

1 Villiger: *Das Zeiss-Planetarium, loc. cit.*, p. 9

2 M. Sanchez-Arcas: *Form und Bauweise der Schalen.* Berlin: 1961.

1a

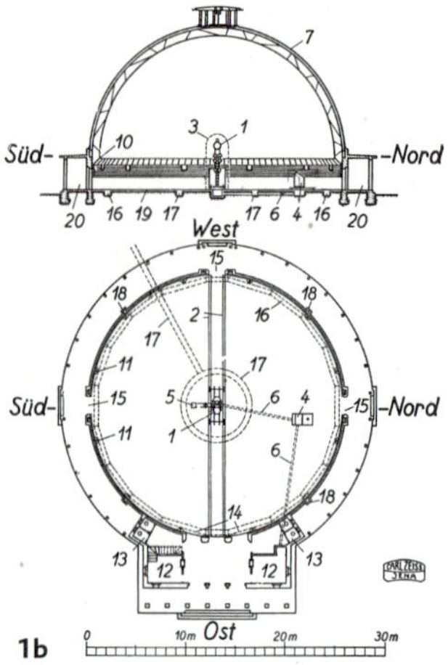

1b

rizonts dagegen mussten die Bilder durch besondere Maßnahmen unsichtbar gemacht werden.[9] Wir sehen, wie die Möglichkeiten einer allseitig beweglichen Lichtprojektion zum Ausgangspunkt einer revolutionären Konstruktion werden. Der tragende Gedanke des Wunders von Jena[10] ergibt sich aus den Methoden des Optikers, die Orte im Raum durch die Geometrie der Lichtstrahlen zu definieren, wobei in diesem Fall nicht das von realen Objekten im Raum ausgesendete Licht aufgefangen, sondern von einem Projektor gezielt auf eine Wand geworfen wird. Es ist bekannt, dass die Geschichte des Sehens von der prinzipiellen Umkehrbarkeit von Projektion und Rezeption handelt. | Aus der Projektionsidee folgte, dass Bauersfeld nicht nur einen neuartigen, ziemlich komplizierten – weil programmierbaren – Lichtbildwerfer zu entwickeln hatte, sondern auch noch eine neue, sphärisch gekrümmte Projektionswand. Für beide gab es keine unmittelbaren Vorläufer. || Der regelmäßige Vielflächner, den Bauersfeld als Projektionsfigur auswählt, könnte auch gegen einen anderen regelmäßigen Vielflächner wie z. B. Tetraeder, Oktaeder, Kubus etc. ausgetauscht werden, ohne am Prinzip etwas zu ändern. | Dieses Verfahren der Projektion regelmäßiger Polyeder auf die umschreibende Kugel erzeugt auf deren Oberfläche ein Liniennetzwerk, das geodätisch genannt wird, weil die Kanten sämtlich auf Großkreisen, sog. Geodäten, liegen. Ihr Radius ist der Radius der Kugel. Die geometrische Einteilung, die Bauersfeld für die Form seines Kugelprojektors ebenso wie für die der Projektions-

9 Bauersfeld, a.a.O.

10 So überschrieb der Direktor der Kopenhagener Sternwarte, Prof. Strömgren, einen enthusiastischen Bericht über das Zeiss-Planetarium. In: Politiken vom 19. 2. 1925

Abb. 1a / 1b Zeiss-Planetarium in Jena, 1925. Das Netzwerk der Kuppel wiegt nur 9 kg pro Quadratmeter.

Fig. 1a / 1b Zeiss planetarium at Jena, 1925. The network of the dome weighs only 9 kg per square meter.

2a

2b

2c

2d

Abb. 2a – 2e Die Kuppel des Planetariums ist eine geodätische Netzwerkkonstruktion. Die Geometrie des Netzwerks entspricht dem Kräftefeld der Lasten, so dass nur Druck- und Zug-, jedoch keine Biegespannungen auftreten.
Das selbsttragende Stabwerk wird in eine 3 cm dicke Betonschicht eingegossen. Die Konstruktion wiegt nur 1/30 herkömmlicher Kuppelkonstruktionen und ist der Ursprung von Raumtragwerken und Schalenbauweise (Versuchsbau auf dem Dach der Zeiss-Fabrik in Jena 1924).

Fig. 2a – 2e The planetarium dome is a geodesic network structure. The geometry of the network corresponds with the balance of forces so that there are only compression and tensile forces but no bending stress. The monocoque structure of the frame is cast into a 3 cm concrete shell.
The structure weighs only 1/30 of a conventional dome and marks the beginning of space frames and shell structures (experimental structure on the roof of the Zeiss factory in Jena 1924).

2e

sches Museum in Munich, was nearing completion. It was to be set up in Munich in a hemispherical room about 10 metres in diameter. For mounting and testing purposes, a similar spherical room was needed in Jena. Since no indoor space large enough for this purpose was available, a light-weight spherical building had to be erected outdoors, and needed to stand up to the effects of wind and weather at least for a few months. At first we thought of a construction along the lines of a circus tent. But that was ruled out because canvas, like all textiles at that time of very high inflation, was much too expensive. By contrast, steel, being a purely German product, was very low in price. So we ended up going for a steel construction. Since we attached great importance to creating the hemispherical shape very exactly, the construction of a hemispherical network out of steel rods seemed to us to be the most promising building technique."[3] | The Jena network dome was a model of radical light-weight construction. The structure had to be light, because there was no space for the construction except on the roof of the Zeiss factory. Bauersfeld describes the unique properties of this light-weight supporting structure thus: "Although it appears very fragile, it is strong enough for a number of people to be able to clamber about on it without any noticeable deformations occurring – and it is composed only of iron rods 80 x 20 mm in cross-section and about 60 cm in length. The essential feature is the configuration of the nodal points. The rods stand on end, they are grooved at the ends and are held firmly

3 Walter Bauersfeld: *lecture on gridshellconstruction at the celebration of the Siemens-Ring-Stiftung on the 12th of Dec.* 1942 in Berlin. Manuscript p. 1.
I thank the company Jenoptik Jena GmbH for the friendly surrender. A shortened and edited version of the text is printed in: Jürgen Joedicke: Schalenbau, loc. cit., p. 281–289

wand verwendete, sollte nun aber auch die Struktur für eine baukonstruktive Lösung werden: das Netzwerk als räumliches Tragwerk der Planetariumskuppel. Bauersfeld wird damit zum Pionier einer Baukonstruktion, von der gleich zwei Umwälzungen ausgehen: die leichten Raumtragwerke, die später so erfolgreich von Konrad Wachsmann, Richard Buckminster Fuller, Max Mengeringhausen und vielen anderen entwickelt worden sind; und andererseits die Schalenbauweise, die dem eisenbewehrten Beton und dem Beton ohne Bewehrung ganz neue Möglichkeiten beim Bewältigen großer Spannweiten geschaffen hat. Der Optiker und Feinmechaniker leitet diese Entwicklung ein, obwohl sie – so Bauersfeld – »ganz auf dem Gebiet des Hochbaus liegt, also völlig außerhalb meines eigentlichen Arbeitsgebietes, der Feinmechanik und Optik.«[11] | Das Neue entsteht aus Unscheinbarem. Ein Provisorium soll entstehen. Es soll nur stellvertretend für das ›eigentliche‹ Werk zum Zwecke des ›Ausprobierens‹, also als Simulator des ›wirklichen‹ Gebäudes, errichtet werden. Das nur befristet gebrauchte Objekt ist dadurch von vornherein näher an der Kategorie des Textilen als an der Kategorie des Gebäudes, eher ephemer als solide, eher ein Teil der ›fliegenden Bauten‹ als der ›Architektur‹. Bauersfeld gibt über die Umstände Auskunft: »Im Jahre 1922 näherte sich das erste Planetariumsgerät, welches für das Deutsche Museum in München entwickelt worden war, seiner Vollendung. Es sollte in München in einem halbkugelig gestalteten Raum von etwa 10 m Durchmesser aufgestellt werden. Für die Montage und Ausprobierung war ein ähnlicher Kuppelraum in Jena notwendig. Da ein hinreichend großer Innenraum hierfür nicht zur Verfügung stand, musste ein leichter Kuppelbau im Freien errichtet werden, der wenigstens für einige Monate den Einflüssen von Wind und Wetter gewachsen war. Zuerst dachten wir an eine Konstruktion nach Art eines Zirkuszeltes. Das schied aber aus, weil die Leinwand, wie alle Textilien damals, in der Zeit stärkster Inflation, viel zu teuer war. Im Gegensatz dazu war Eisen als rein deutsches Produkt sehr niedrig im Preise. So gelangten wir zu einer Eisenkonstruktion. Da wir großen Wert darauf legten, die Halbkugelform sehr genau herzustellen, so erschien uns der Aufbau eines halbkugeligen Netzwerkes aus Eisenstäben als die günstigste Bauweise.«[12] | Die Netzwerkkuppel von Jena ist beispielhaft für radikale Leichtbauweise. Leicht musste die Konstruktion sein, weil kein Platz für den Bau da war außer auf dem Dach der Zeiss-Fabrik. Bauersfeld beschrieb die einzigartigen Eigenschaften dieses Leichttragwerkes: »Obwohl es sehr zart erscheint, hat es doch solche Festigkeit, dass eine Anzahl Personen darauf herumklettern kann, ohne dass merkliche Deformationen eintreten. Dabei besteht es nur aus Eisenstäben von 80 x 20 mm Querschnitt und ca. 60 cm Länge. Wesentlich ist dabei die Gestaltung der Knotenpunkte ... Die Stäbe stehen hochkant, sie sind an den Enden eingekerbt und werden durch runde Scheiben fest zusammengehalten, die mit pas-

11 Walter Bauersfeld: *Vortrag über Schalenbau bei der Feier der Siemens-Ring-Stiftung am 12. Dez. 1942 in Berlin.* Manuskript S. 1. Für die freundliche Überlassung danke ich der Firma Jenoptik Jena GmbH. Eine gekürzte und redigierte Fassung des Textes ist abgedruckt in: Jürgen Joedicke: Schalenbau, a.a.O., S. 281–289

12 Ebd.

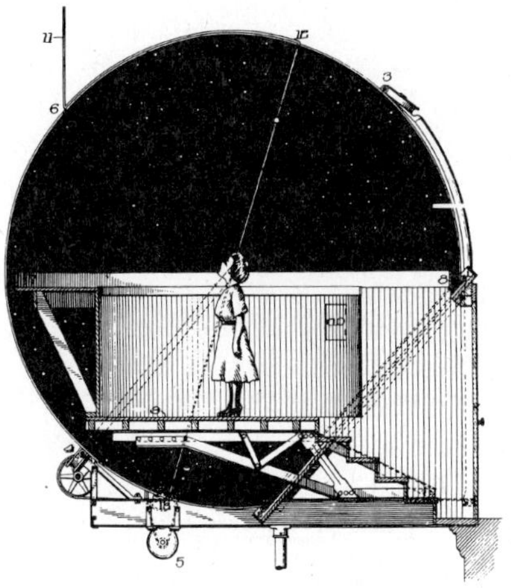

3

together by round plates fitted with appropriate necks. A high degree of rigidity of the nodes was thereby achieved. The dead weight amounted to only nine kg per square metre. Of course, the lengths of the rods had to be very exact, with a tolerance of some 1/20 mm, in order for the spherical shape to work out exactly. Some 50 different lengths of rod were required, and getting on for 4,000 rods in total. You will recognise in these details the involvement of the designer geared towards precision engineering."[4] | Zeiss obtained a patent for Bauersfeld's nodal construction.[5] It takes account of the fact that "in a network, the various nodes by no means present the same geometrical pattern at every point. The number of rods is not the same at all nodal points, and the same is true of the angles of inclination of the rods to each other and to the plates."[6] In order to guarantee this, the Zeiss nodes have notches running around the plates and the rods have ball pivots. The two angles in space are therefore not fixed in advance; the node can be used both for varying geometric figures and for differing sphere radii. | The time was not yet ripe for this structure to be covered with light alloy panels or fibre-glass, as was done experimentally by Richard Buckminster Fuller after the Second World War. On the other hand, there was already in Germany a developed concrete industry which was increasingly competing with iron and steel construction. | One of the leading concrete building firms, Dyckerhoff & Widmann AG (Dywidag), had already done various building work for Zeiss. It was natural for Bauersfeld to con-

4 Bauersfeld: lecture, loc. cit., p. 2

5 Node construction for iron networks. DRP 420 823, submitted in the 11-9-1922

6 at the same place

Abb.3 Bei der »Alwood Celestial Sphere« in Chicago von 1913 wurde die Sternenbewegung mit mechanisch bewegten Schalen simuliert.

Fig.3 In the "Aiwood Celestial Sphere" in Chicago (1913) mechanically moved shells were used to simulate the stellar motions.

senden Eindrehungen versehen sind. Damit wurde eine hohe Steifigkeit der Knotenpunkte erzielt. Das Eigengewicht betrug nur 9 kg pro Quadratmeter. Die Stablängen mussten natürlich sehr genau hergestellt werden, mit einer Toleranz von etwa 1/20 mm, damit die Kugelform genau herauskam. Es waren etwa 50 verschiedene nötig und nahezu 4.000 Stäbe im ganzen. Sie erkennen an diesen Details die Mitarbeit des nach der Feinmechanik ausgerichteten Ingenieurs.«[13] | Für Bauersfelds Knotenkonstruktion erhielt Zeiss ein Patent.[14] Es berücksichtigt, »dass bei einem Netzwerk an den verschiedenen Knotenpunkten keineswegs überall dasselbe geometrische Bild vorliegt. Es ist nicht an allen Knotenpunkten die Stabzahl die gleiche und ebenso der Neigungswinkel der Stäbe gegeneinander und gegen die Platten.«[15] Um dies zu gewährleisten, haben die Zeiss-Knoten an den Platten ringsum verlaufende Nuten und die Stäbe Kugelzapfen. Die beiden Winkel im Raum sind daher nicht festgelegt; der Knoten ist sowohl für variierende geometrische Figuren als auch für veränderliche Kuppelradien verwendbar. | Die Zeit war noch nicht reif, um diese Strukturen mit Paneelen aus Leichtmetall oder tiefgezogenen Kunststoffplatten zu decken, wie dies nach dem Zweiten Weltkrieg der amerikanische Universalarchitekt R. Buckminster Fuller experimentell erprobte und in den fünfziger und sechziger Jahren mit Industrie und Militär realisierte. Dagegen war in Deutschland bereits eine entwickelte Betonindustrie vorhanden, die zunehmend dem Eisen- und Stahlbau Konkurrenz machte. Für Zeiss hatte eine der führenden Betonbaufirmen, Dyckerhoff & Widmann AG, schon verschiedentlich gebaut. Es lag nahe, dass Bauersfeld sich in der Frage der Herstellung der Schale mit den Leuten von Dywidag besprach. Zunächst ging er von der Annahme aus, die Fläche mit Rabitzputz füllen zu können. Da es aber um eine möglichst glatte Oberfläche für die Projektion ging, entstand noch eine andere Idee: »nämlich das Aufspritzen einer Betonmischung nach dem Torkret-Verfahren. Dieses Betonspritzverfahren war erst kurz vorher entwickelt worden, hatte aber sehr gute Resultate ergeben. Die glatte Innenfläche sollte dadurch erzielt werden, dass von innen her eine Holzschalung von kugeliger Krümmung am Netzwerk angelegt wird, dass das Drahtgeflecht und das Stabwerk selbst von Beton ganz eingehüllt werden. Die Kugelform bietet den weiteren Vorteil, dass die Holzschalung, die etwa 3 x 3 m groß auszuführen war, nach dem Erhärten des Betons abgenommen und zum gleichen Zweck mehrfach verwendet werden konnte. Dazu kam als günstiger Umstand die Möglichkeit der Verwendung von sehr fein gemahlenem Zement, der nur eine sehr kurze Abbindezeit erforderte, und der auch erst kurz vorher im Betonbau eingeführt worden war.«[16] | Auf diesem Wege waren dünnwandige Kuppelschalen herstellbar, mit denen auch große Spannweiten überwölbt werden konnten. Das Münchener Planetarium hatte 10 m Durchmesser, die erste Zeiss-Kuppel auf dem Fabrikdach 16 m, die zweite für die be-

13 Bauersfeld: Vortrag, a.a.O., S. 2

14 Knotenpunktverbindung für eiserne Netzwerke. DRP 420 823, eingereicht am 9.11.1922

15 Ebd.

16 Bauersfeld: Vortrag, a.a.O.

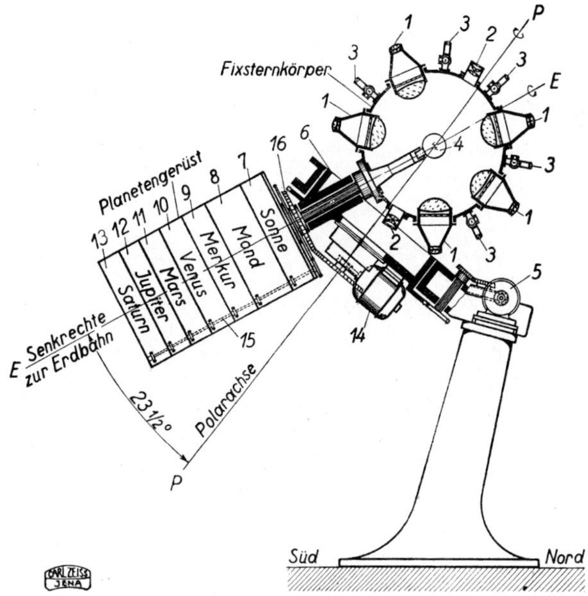

4

sult the people from Dywidag about manufacturing the shell. At first he assumed that the surface could be filled in with plaster. But since the smoothest possible surface was needed for the projection, another idea took shape: "namely the spraying on of a concrete mixture using the shotcrete technique. This concrete spraying technique had only been developed a short time before, but had yielded very good results. The smooth inner surface was to be achieved by fitting a wooden casing with spherical curvature onto the network from the inside, so that the wire mesh and the rod structure itself were completely encased in concrete. The spherical shape offered the further advantage that the wooden casing, which was to be made about 3 x 3 m in size, could be removed after the concrete had hardened and used several times for the same purpose. By happy circumstance, there was also the possibility of using very finely ground cement, which only required a very short time to set, and which had also only very recently been introduced into concrete Construction."[7] | In this way, it was possible to produce thin-walled dome shells with which even large spans could be vaulted. The Munich Planetarium had a diameter of 10 metres, the first Zeiss dome on the factory roof 16 m, the second for the neighbouring glass factory Schott and Gen 40 m; the later polygonal domes, for example that of the Market Hall in Leipzig, which created the effect of opened-out umbrellas, each had a diameter of 76 m, more than Max Berg's Centenary Hall in Breslau, with only a third of the weight. The Schott Dome, at 40 m, came close

7 Walter Bauersfeld: *Das Projektions-Planetarium des deutschen Museums in München*. In Magazin VDI, Bd.68, 1924, Nr.31, p.793.

Abb.4 Die patentierte Projektionsmaschine; die geodätisch unterteilte Kugel sieht Lichtstrahlen für örtlich festgelegte Sterne vor, der Zylinder enthält bewegliche Punktlampen für die Planeten.

Fig.4 Patented projecting machine; the geodesic divided sphere provides light beams for fixed stars, the cylinder contains moving spot lamps for the planets.

5

6

nachbarte Glasfabrik Schott & Gen. 40 m, die späteren Vieleck-Kuppeln z. B. der Großmarkthalle in Leipzig, die wie ausgespannte Regenschirme wirken, hatten je 76 m, mehr als Max Bergs Jahrhunderthalle in Breslau, dabei nur ein Drittel des Gewichts. Die Schott-Kuppel reichte mit ihren 40 m schon an die Spannweite des Petersdoms in Rom mit 42,6 m heran. Die Schale hatte eine Wandstärke von nur 6 cm, sie wog nur 330 Tonnen, ein Dreißigstel der 10 000 Tonnen schweren Domkuppel. Schon um 1930 gab es Projekte für Schalenkuppeln von 150 m Spannweite. Um die Schalenbauweise für Nutzbauten jeder Art verwendbar zu machen, war noch eine Reihe von Schritten notwendig, die im Wesentlichen aus der Zusammenarbeit zwischen Bauersfeld einerseits und den Bauingenieuren von Dywidag, vor allem Franz Dischinger und Ulrich Finsterwalder, andererseits resultierten. Erwähnt seien hier nur zwei wesentliche Neuerungen: ▪ Die Entdeckung, dass nicht nur doppelt gekrümmte Schalen – wie z. B. die Kugel – mit der Schalenbauweise herstellbar sind, sondern ebenso einfach gekrümmte, wie z. B. Zylinder, mit denen man bequem rechteckige Grundrisse überdachen kann. Damit öffnete sich der riesige Aufgabenbereich der Hallenbauten für diese Bauweise; Großmarkthallen, Ausstellungs- und Messehallen, Bahnhöfe und Flugzeughallen, Sporthallen und Fabriken – für alle diese Bauaufgaben wurde das Zeiss-Dywidag-Verfahren gebraucht. Und es wurden weitere Patente angemeldet. ▪ Die Entdeckung, die anlässlich der Gesolei-Ausstellung 1926 in Düsseldorf mit einer zylin-

Abb.5 Die Zeiss Fabrik, 1924. Die Kuppel des Planetariums oben auf dem Dach (rechts).

Fig.5 Zeiss factory 1924. On top of the roof (right) the dome of the planetarium.

Abb.6 Walter Bauersfeld (1879 – 1960), Direktor und leitender Ingenieur der optischen Arbeiten. Carl Zeiss Jena erfand das Projektionsplanetarium und eine neue Methode der Leichtbaukonstruktion für Kuppeln und Gewölbe. Foto: vor 1914.

Fig.6 Walter Bauersfeld (1879 – 1960), director and chief engineer of the Optical Works. Carl Zeiss Jena invented the projection planetarium and a new method of lightweight construction for domes and vaults. Photo: before 1914.

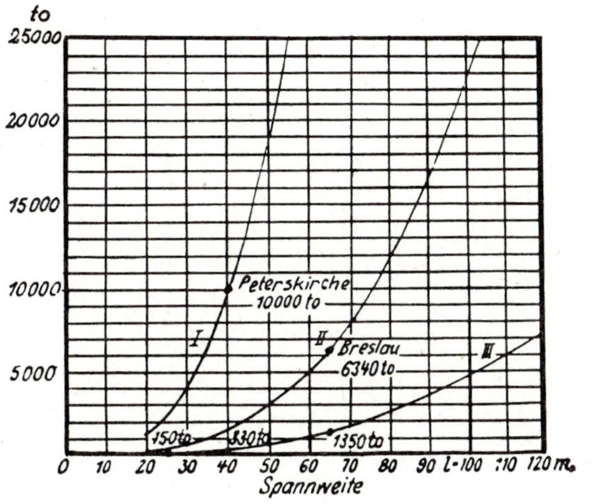

7

to the span of St Peter's in Rome, which is 42.6 m. The shell had a wall width of only 6 cm, and weighed only 330 tonnes, a thirtieth of the weight of St Peter's dome at 10,000 tonnes. Around 1930 there were already plans for shell domes with a span of 150 m. And it is likely that Speer's design for a Congress Hall for the capital city 'Germania', crowned with a dome of 250 m in diameter, could only have been built using the technology of monocoque construction. In any event, there was a proposal in existence by the Dywidag monocoque builder Franz Dischinger for a concrete dome with double shells, which even made allowance for bomb strikes. | A number of further steps were still necessary in order to make monocoique construction practicable for usable building of every kind, and these steps were made as a result of the cooperation between Bauersfeld on the one hand and Dywidag's construction engineers. ▪ A major step was made with the discovery that not only double-curved shells — such as the sphere — can be made using monocoque construction techniques, but also simple curved shells, such as cylinders, which can easily be erected over rectangular ground-plans. This opened up the enormous field of contracts for hall buildings to this construction technique; market halls, exhibition and trade halls, railway stations and hangars — for all these building tasks the Zeiss-Dywidag method was used. And more patents were filed. ▪ Another major discovery was made on the occasion of the 1926 Gesolei Exhibition in Düsseldorf: that the network no longer needed to be concreted

Abb.7 Diagramm für den Vergleich von Kuppelbauten:
1. Dom-Kuppel der Peterskirche in Rom.
2. »Jahrhunderthalle« von Max Berg in Breslau.
3. Schott-Kuppel als Beispiel für die Schalenbauweise.

Das Diagramm wurde von Bauersfelds Mitarbeiter, Ingenieur F. Dischinger von Dyckerhoff & Widmann 1925 veröffentlicht.

Fig.7 Diagramm comparing dome construction:
1. Dome of St Peter's in Rome.
2. Max Berg's "Jahrhunderthalle" in Breslau.
3. Schott dome as example for "Schalenbauweise".

The diagram was published by the collaborator of Bauersfeld the civil engineer F. Dischinger of Dyckerhoff & Widmann in 1925.

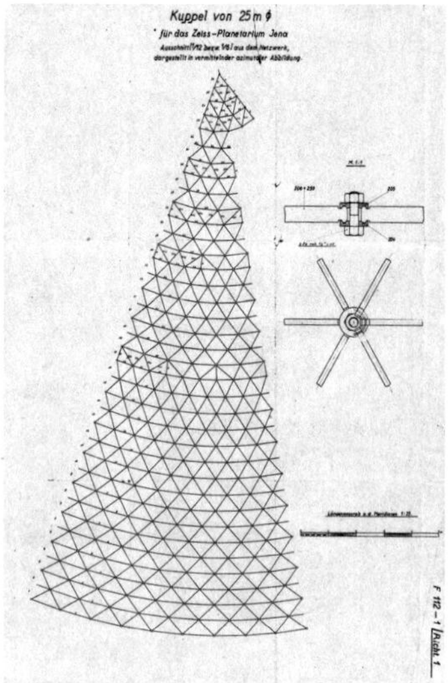

8

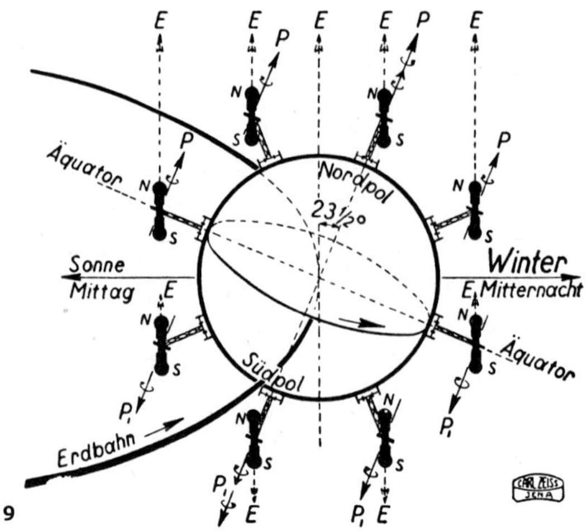

9

drischen Schale gemacht wurde: dass das Netzwerk nicht mehr in die Schale einbetoniert werden musste und damit wieder verwendbar wurde. Das tetraedrisch aufgebaute Stabtragwerk dient nun lediglich als Träger der Holzverschalung, auf die die eigentliche Betonschale mit normaler Bewehrung aufgebracht wird.[17] | Diese Entwicklungen trieb neben Dischinger maßgeblich Ulrich Finsterwalder voran. Er hatte parallel zu Bauersfeld eine Berechnungsgrundlage für die Statik der Netzwerke bzw. der Schalen ausgearbeitet. Sie ist als Membrantheorie bekannt. Die Statik eines Netzwerkes aus eisernen Stäben wird ihr zufolge rechnerisch wie die eines entsprechenden homogenen Blechs behandelt. Daraus lassen sich Spannungen im Stabwerk genau genug abschätzen. Die Membrantheorie entwickelte Walter Bauersfeld parallel zu Ulrich Finsterwalder, der die Berechnung der Stabkräfte zum Thema seiner Diplomarbeit an der TH München machte. Danach fing er bei Dyckerhoff & Widmann als Bauingenieur an und arbeitete seit Herbst 1923 einige Monate im Zeiss-Werk in Jena, während die Schott-Kuppel gebaut wurde. Er war nicht nur Schüler des bedeutendsten Statikers der Zeit, August Föppel, sondern auch Sohn des Mathematikers Prof. Finsterwalder, eines Pioniers der Ballonphotographie und -photogrammetrie und Experten der Geometrie der Polyederprojektion. Der Sohn berichtet über den Vater: »Er hat die ersten photogrammetrischen terrestrischen Aufnahmen gemacht. Er hat sich von Rodenstock dazu einen Apparat bauen lassen und hat in einer Spezialvorlesung diese Netzwerke

17 Vgl. Karl-Heinz Hüter: *Schalenbau. In: Form + Zweck*, 1980, 6, S. 38

Abb. 8 Konstruktion des dreieckigen Netzwerkes aus Eisenstangen für die Kuppel des Zeissplanetariums. Konstrukteure: Bauersfeld und Dischinger. Patentinhaber waren Zeiss und die Firma Dyckerhoff & Widmann, Fachmann für die Stahlbetonberatung der deutschen Bauindustrie.

Fig. 8 Design of the triangulated network of iron rods for the dome of Zeiss planetarium. Designers: Bauersfeld and Dischinger. Patent holders were Zeiss and the Dyckerhoff & Widmann company, specialist reinforced concrete consultants for the German construction industry.

Abb. 9 Mit Apparaturen, wie man sie auch für Spezialeffekte in der Theaterbeleuchtung verwendet, können Sternschnuppen, Polarlichter, Kometen etc. simuliert werden.

into the shell, and could therefore be re-used. The tetrahedral supporting rod structure now serves only to hold the formwork on which the actual concrete shell is applied with normal reinforcement.[8] | These developments were driven alongside Dischinger, significantly by Ulrich Finsterwalder. Parallel to Bauersfeld, he had elaborated a calculation basis for the statics of mesh-works respectively gridshells. It is known as the membrane-theory. The structural analysis of a network of iron rods is calculated according to this mathematical theory, treated as an appropriate homogeneous sheet. Following this, tensions in a truss can be estimated almost exactly. Walter Bauersfeld developed the membrane-theory parallel to Ulrich Finsterwalder, who had chosen the calculation of stresses as the major theme of his thesis at the TH Munich. He then joined Dyckerhoff & Widmann as a construction engineer, and from autumn 1923, worked for a few months at the Zeiss-factory in Jena, while the Schott-dome was constructed. Not only had he been a student of the most important structural engineer of the time, August Föppel, but also the son of the mathematician Prof. Finsterwalder, a pioneer of the balloon-photography and photogrammetry and an expert of the geometry of polyeder-projection. The son comments about his father: “He took the first photogrammetric terrestrial pictures. He let Rodenstock build an apparatus for this and developed in a special lecture these canvases of survey, which Bauersfeld had constructed for his Planetarium”.[9] If we add that Prof. Finsterwalder, the father of the construction engineer, had been together with Oskar von Miller in the foundation board of the German museum, we can see how the network of personal constellations and intense relations are repeated, that exist between geometry, optics and architecture. | The first architect to understand these revolutionary developments was Adolf Meyer, the Bauhaus architect and for many years the associate of Walter Gropius. Meyer, whose importance for the development of modern architecture has up to now been totally undervalued – a result in part of his early death in 1929 – was deeply involved in the new construction method, and in the work he did in Frankfurt for the city, he erected some remarkable buildings using monocoque techniques. When a competition was advertised for designs for the public planetarium in the Prinzessinnengarten in Jena, Adolf Meyer took part. But in 1926 a design was picked for execution which was not Meyer’s but that of Schreiter & Schlag. While Meyer parabolically superelevates the shell slightly and thereby emphasizes it as a shape (association: egg in eggcup), the more conventional design of the building actually erected, with its ring of colonnades and wide entrance hall on supporting columns, makes allusion to the Roman model of the Pantheon. It had not been understood that the dome had removed all justification for the existence of the column as a support. | Adolf Meyer had recognised the relationship with the natural sciences, and in a short article for the first volume of the

8 cf. Karl-Heinz Hüter: *Schalenbau*. In: Form + Zweck, 1980, 6, p.38

9 Interview by the author with Dr. Ulrich Finsterwalder, Berlin. 14th november 1987; cf. also Bauersfeld: Lecture, l.c., p.7

Fig. 9 The same mechanisms used for spezial lighting effects in theaters can be employed to simulate shooting stars, polar lights, comets etc.

10

entwickelt, die Bauersfeld für sein Planetarium konstruiert hatte.«[18] Wenn wir hinzufügen, dass Prof. Finsterwalder, der Vater des Bauingenieurs, zusammen mit Oskar von Miller im Gründungsausschuss des Deutschen Museums saß, sehen wir, wie sich auch im Netz der persönlichen Konstellationen die intensiven Beziehungen wiederholen, die es in der Sache zwischen Geometrie, Optik und Architektur gibt. | Der erste Architekt, der die Revolution der Architektur durch die geometrische und optische Projektion verstand, war wahrscheinlich Adolf Meyer, der Bauhausmeister und langjährige Büropartner von Walter Gropius. Meyer hat sich sofort intensiv mit der neuen Bauweise beschäftigt und in seiner Frankfurter Arbeit für die Stadt bemerkenswerte Bauten in Schalenbauweise errichtet. Er hatte den Bezug zum Weltbild der Naturwissenschaften erkannt und in einem kurzen Artikel für das erste Heft der Werkbund-Zeitschrift *Die Form* 1925 deutlich gemacht: »Die Kuppel-Bauten der Zeiss-Planetarien gehören wegen der Kühnheit und Leichtigkeit ihrer Konstruktion zu den beachtenswertesten Erscheinungen auf dem Gebiet der Bau- und Ingenieurkunst der Zeit, deren Einwirkung auf die gesamte Baukunst heute noch nicht abzusehen ist (...) Hier decken sich Konstruktion und Form im reinsten Maße, die Konstruktion würde, auf die äußerste Kuppelhaut projiziert, ein Bild von kristallischer Klarheit und Eindeutigkeit der Form ergeben. Die Beziehungswerte zwischen den tragenden und füllenden Teilen sind in ein klar ablesbares Verhältnis gebracht, wobei eine richtungslose

18 Interview des Autors mit Dr. Ulrich Finsterwalder. Berlin, 14. November 1987; vgl. ferner: Bauersfeld: Vortrag, a.a.O., S. 7

Abb.10 Das Jenaer Planetarium vollzieht bereits 1924 den Übergang vom mechanischen zum Informationszeitalter. Mit seinem Analogcomputer zur Steuerung der Raum-Zeit-Reisen ist es das erste »Intelligent Building«.

Fig.10 As early as 1924 the Jena planetarium marks the transition from the mechanical age to the age of information. With ist analogue computer for controlling space-time travel it represents the first "intelligent building".

Werkbund journal *Die Form* in 1925 stated: "The dome buildings of the Zeiss planetariums are, because of the audacity and grace of their construction, among the most remarkable phenomena in the field of the architecture and engineering of the age, and their influence on architecture as a whole cannot yet be foreseen."[10] | It took nearly another 30 years for this knowledge to be converted into a series of constructional and aesthetic experiences for architecture. This took place principally in the USA, through the work of R. Buckminster Fuller, who — like Bauersfeld — had developed his geodesic dome constructions out of the ideal of projection. During the Second World War he was employed at the Board of Economic Warfare working on logistic studies; in 1943 he published a polyhedral map projection of the Earth, which, when folded and stuck together, formed a folding globe. In 1951 he filed a patent for domes; like Bauersfeld's his Geodesic Dome was once again created through the projection of an icosahedron onto the circumscribed sphere. In the following 30 years more than 300,000 geodesic domes were built: from the radar stations of the DEW Line to the US Pavilion in Montreal in 1967. They have become the symbols of the most diverse currents and movements all over the world.

10 Today known under the name 'De Stijl'.

11

Abb.11 Verbesserte Projektionsmaschine, entworfen durch Bauersfeld und Villinger. Diese wurde seit 1926 in einigen Reihen produziert.

Fig.11 Improved projecting machine, designed by Bauersfeld and Villinger. From 1926 on this type was produced in several series.

12

Gleichwertigkeit der Ausdehnungsgrenzen in Parallele gestellt werden kann zu der unter dem Namen ›horizontal-vertikal‹ bekannten gleichlautenden Bewegung in der bildenden Kunst[19], diese zu einer kosmischen Totalität ergänzend. In dieser neuen Bausynthese lassen sich Vergleichsmomente aufdecken, die auf Zusammenhänge mit den wissenschaftlichen Erkenntnissen hinweisen, die bereits mit Leibniz begonnen haben, sich aber jetzt erst allmählich in alle Gestaltungskomplexe der Zeit durchsetzen wollen. Auf dem Gebiet des mathematisch-wissenschaftlichen Niederschlags seiner Philosophie beginnt auch die Kunst eine neue Deutung und Formung zu finden und in ihrer optischen Erscheinungswelt ein neues Weltbild darzustellen.«[20] | Es dauerte noch fast dreißig Jahre, bis diese Erkenntnis in eine Reihe von konstruktiven und ästhetischen Erfahrungen für die Architektur umgesetzt werden konnte. Dies geschah vor allem in den USA, und zwar durch R. Buckminster Fuller, der seine geodätischen Kuppelkonstruktionen – wie Bauersfeld – aus der Projektionsidee heraus entwickelte. Obwohl seine ›domes‹ sehr populär waren, hat kein Bau es zu der Popularität bringen können wie der 32-Flächner, in dem weiße Sechsecke mit schwarzen Fünfecken abwechseln, so dass sich gute Wahrnehmbarkeit mit exzellenter Festigkeit verbindet, welche er für seine schnell wechselnden Bewegungen nötiger hat als Bauersfelds Kugelprojektor: der Fernsehfußball.

19 Heute bekannt unter dem Namen ›De Stijl‹

20 Adolf Meyer: *Das Zeiss-Planetarium in Jena: Der Bau.* In: Die Form, 1. 1925/26, Heft 1, S. 17

Abb. 12 Das Jenaer Planetarium; das Baugerüst befindet sich innerhalb der Haube. Der Beton wurde auf die Verschalung von der Außensteite aufgesprüht.

Fig. 12 Jena planetarium; scaffolding inside the dome. Concrete sprayed onton formwork from outside.

21

Hermann Finsterlin.

Ein phantastischer Architekt

Hermann Finsterlin.

A fantastic architect

Rudolf Finsterwalder

Hermann Finsterlin, geboren 1887 in München, gestorben 1973 in Stuttgart, war ein utopischer Architekt, Maler, Dichter, Essayist und Komponist. | Seine phantastischen Bilder und Modelle zur Architektur waren ein Gegenentwurf zur nüchternen Architektur der Moderne vor allem des Bauhauses, ohne dass er jemals eine seiner Visionen gebaut hätte. Seine biomorphen Formen, die sich wie Gewächse organisch aus der Landschaft entwickeln, waren das expressive Ergebnis einer Gesellschaftskritik, an »den Rieseneierschachtelhaufen unserer Städte und den primitiven Würfeln unserer Wohnstätten«.[1] Seine utopischen Formen, die jeglicher Vorstellung der Zeit vom üblichen Bauen widersprachen, waren ein Affront gegen die zeitgenössische Architektur. | Ausgangspunkt für sein Schaffen war ein Erweckungserlebnis während einer Besteigung des Watzmanns: »Und dann kam in höchster innerer Not die Schicksalsstunde. ... Ein Verzweiflungsaufstieg auf den Watzmann bei Berchtesgaden in einer zauberhaften Frühlingsvollmondnacht. ... Die Erkenntnis meines Lebens unter der einen Firmenkuppel, statt einer Katastrophe. ... Wo der unabsehbare Teppich des Seins als Wissenschaft nicht erreichbar ist, bleibt nur der eigene Webstuhl. Und dieser Webstuhl unendlicher Combination, der Gottheit lebendiges Kleid, unendlicher Complexverbindungen, unendlichen Werdens, Vergehens, Veränderns ist allein menschlicher Kunst wert. Sinn und Ziel dieser Kunst, die in so seltenen Fällen, wenn überhaupt in historischer Zeit gefasst und geübt ward, abgesehen von altem fernöstlichem Weistum

1 Reinhard Döhl: *Hermann Finsterlin. Eine Annäherung*. S. 33, Anm. 81 und Anm. 3; *Am Beginn meines bewussten Lebens*. aus einem dreiseitigen Typoskript Finsterlins

Abb. 1 Siedlung
Fig. 1 Settlement

Born 1887 in Munich, died 1973 in Stuttgart. | He was a utopian architect, painter, poet, essay-writer and composer. | His fantastic paintings and architectural models constituted an alternative to sober Modernist architecture, above all that of the Bauhaus, even though none of his projects were realized. | His biomorphic shapes grew organically out of the landscape like plants and were the expressive outcome of a critique of society that attacked "the enormous egg carton conglomerations of our cities and the primitive cubes of our habitats".[1] His futuristic shapes that contradicted every prevailing architectural concept were an affront to the architecture of the time. | The starting point for his work as an artist was an experience that jolted him into awareness while climbing the Watzmann, a myth-shrouded mountain in Bavaria: "And then in a state of extreme emotional misery came the hour of destiny. Driven by desperation I ascended the Watzmann near Berchtesgaden one enchanting springtime night with a full moon. Here, under the dome of the firmament I experienced not a catastrophe but the revelation of my life. Where the unpredictable tapestry of existence is inaccessible as a science all that remains is one's own weaving loom. And this loom of infinite combinations, living garment of divinity, of infinite, complex linkages, infinite becoming, transience and change – that is the only worthy subject of human art. The meaning and purpose of this art, which in such rare cases, if at all was expressed and practiced in historical time, apart from the old Far Eastern wisdom and old mythical creations;

1 Reinhard Döhl: *Hermann Finsterlin. Eine Annäherung.* p. 33, annot. 81 and annot. 3; *the start of my conscious life* from a three-page typescript by H. Finsterlin

Abb. 2 Kathedrale des Lichts (1920–24)
Fig. 2 Cathedral of light (1920–24)

Abb. 3 Wolkenkuckucksheim (1920–24)
Fig. 3 Wolkenkuckucksheim (1920–24)

Abb. 4 Architekturen — Das Haus am Meer (1919)

Fig. 4 Architecture — House by the Sea (1919)

Abb.5 Grundrisse 1924
Fig.5 Floor plans 1924

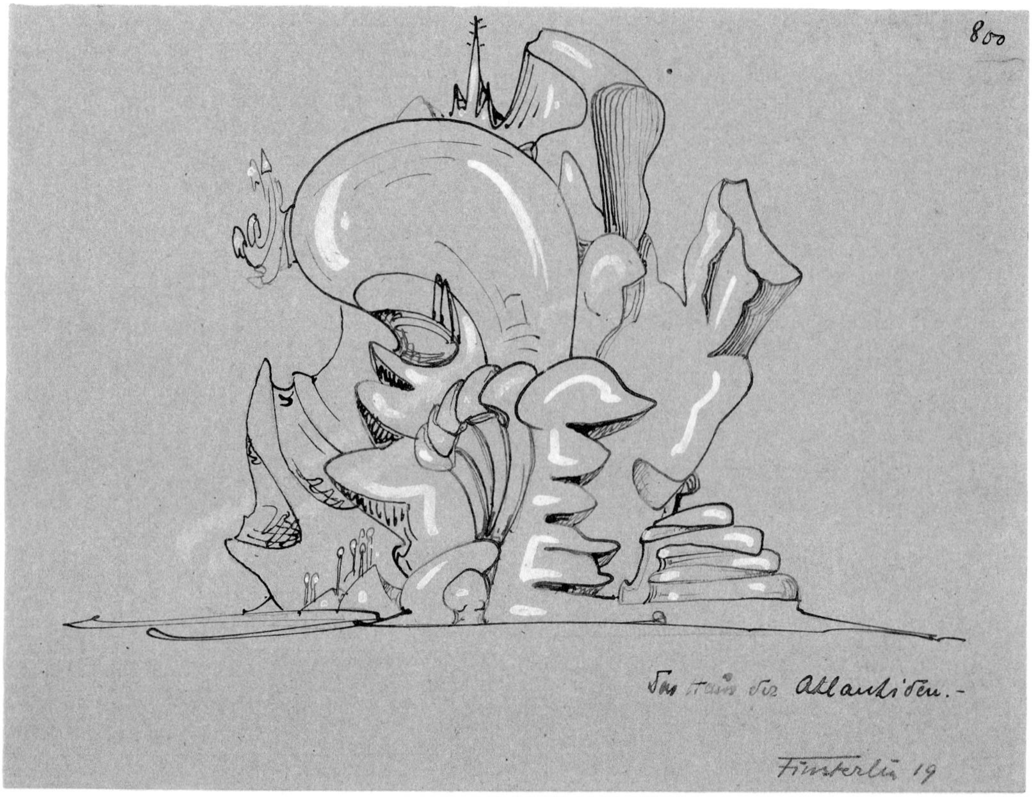

Abb.6 Die Häuser der Atlantiden. Montsalvat (1919)

Fig.6 Houses of the Atlantiden. Montsalvat (1919)

art as magic. | From this time onwards all that has existed for me was to create for myself, be it in word, image, sound or architecture." [2] | In his paintings and models we discern a blend of fantasies, mythologies and the strong desire to create an alternative model to the prevailing architecture. | The very lonesome artist found friends amongst the *gläserene Kette* group around Bruno Taut and Paul Scheerbart, and he corresponded with both. He was also familiar with the nature studies of Ernst Haeckel, which impressed him. | "The omnipotence of being able to infinitely compose and vary the primary phenomena, to create events untrammeled by time or space using images, sounds or words in an unlimited, boundless game subject solely to the image, sound or word composition being viable, in other words subject to the aesthetics ruled by the laws of nature sine qua non, which even characterizes the grotesque creations of nature, as here all dissonance, all exaggeration moves into a viable balance, into the overarching harmony; this was the most important thing in my life." [3] | What makes Hermann Finsterlin's work so interesting is the way he melds various fantasies and topics to form his architectural visions, which appeared utopian yet could feasibly be built in principle, maybe not during his lifetime, but for sure our times. A kind of super-organic architecture in which technology and nature truly merge into a single organism. What makes it so fascinating is the combination of biomorphic shapes with elements from architecture, such as design features or components. This approach is still pursued today, for example in the field of film production, one example being the creations of artist and designer H. Giger for the movie "Alien". Giger called the resulting shapes "biomechanoid"; they were a merging of biomorphic shapes with technical elements. | Finsterlin's architecture still appears futuristic today; his creations have lost none of their visionary appeal.

2 Reinhard Döhl: *Hermann Finsterlin. Eine Annäherung*. p. 33, annot. 81 and annot. 3; *the start of my conscious life* from a three-page typescript by H. Finsterlin, Hatje 1988

3 Reinhard Döhl: *Hermann Finsterlin. Eine Annäherung*. p. 33, annot. 81 and annot. 3; from the *lecture to neighbors and friends* by H. Finsterlin, Hatje 1988

Abb. 7 Villayette 1952
Fig. 7 Villayette 1952

und alten Mythenschöpfungen – Kunst als Magie. [...] Von diesem Zeitpunkt an gab es für mich nur mehr eigenes Schaffen in Wort, Bild, Ton und Bau. ... «[2] | In seinen Bildern und Modellen mischen sich Phantasien, Mythologien und der starke Wunsch, ein Gegenmodell zur vorherrschenden Architektur zu schaffen. | Der zurückgezogen lebende Künstler fand Anschluss an die Gruppe der *gläserenen Kette* um Bruno Taut und Paul Scheerbart, mit denen er korrespondierte und sich austauschte. Bekannt waren ihm auch die Naturstudien Ernst Haeckels, die ihn beeindruckten. | »Die Allmacht, die Urphänomene ins Unendliche, komponieren und variieren zu können, zeit- und raumlos gebundene Ereignisse bildhaft, klanglich oder wörtlich erstehen zu lassen in einem unbeschränkten, grenzenlosen Spiel unter dem einzigen Gesetz einer lebensfähigen Bild-, Klang- oder Wortorganisation eben der naturgesetzlichen Ästhetik sine qua non, die selbst den groteskesten Naturgebilden eignet, da hier jede Dissonanz, jede Übertreibung in die lebensfähige Balance rückt, in die übergeordnete Harmonie; war mir höchster Sinn des Lebens.«[3] | Was die Arbeit Hermann Finsterlins so interessant macht, ist das Verschmelzen verschiedener Phantasien und Themen zu seinen architektonischen Visionen, die utopisch wirken, prinzipiell aber baubar wären, vieleicht noch nicht zu seiner Zeit, aber bestimmt mit den heutigen Mitteln. Eine Art von super-organischer Architektur, in der tatsächlich Technik und Natur zu einem einzigen Organismus verschmelzen. Die Verbindung von biomorphen Formen mit Versatzstücken von Architektur, wie konstruktiven Elementen oder Bauteilen, macht ihren Reiz aus. Dieser Ansatz wird bis heute weiterverfolgt, z. B. im Bereich der Filmarchitektur, wie in den Gestaltungen des Künstlers und Designers H. Giger für den Film *Alien*. Die enstehenden Formen nannte Giger »biomechanoid«, sie waren eine Verschmelzung von biomorphen Formen mit technischen Elementen. | Auch heute noch wirken die Architekturen Finsterlins futuristisch, ihren visionären Reiz haben sie bis heute nicht verloren.

2 Reinhard Döhl: *Hermann Finsterlin. Eine Annäherung*. S. 33, Anm. 81 und Anm. 3; *Am Beginn meines bewussten Lebens*. aus einem dreiseitigen Typoskript Finsterlins, Hatje 1988

3 Reinhard Döhl: *Hermann Finsterlin. Eine Annäherung*. S. 36, Anm. 98 und Anm. 2. aus dem *Vortrag vor Nachbarn und Freunden* von H. Finsterlin, Hatje 1988

Abb. 8 Gaststätte (um 1965)
Fig. 8 Restaurant (ca. 1965)

22

Über Spiculae und Spicularskelette

On spicules and spicular Skeletons

D'Arcy W. Thompson

Die Skelettbildungen bei Radiolarien Wenn ich die anorganischen Skelettbildungen bei Schwämmen verhältnismäßig kurz behandelt habe, trotz des Interesses, das dem Gegenstand vom physikalischen Standpunkt aus zukommt, so ist das auf verschiedene Gründe zurückzuführen. Obgleich das Phänomen in seinen allgemeinen Zügen klar ist, muss man zunächst einmal zugeben, dass viele Punkte noch ungeklärt sind und nur im Verlauf einer langen Diskussion behandelt werden könnten. Zweitens steht die physikalische Theorie (wie ich gezeigt habe) zu oft im Widerspruch mit den Aussagen der Embryologen über die Entwicklung der Spiculae und mit den jeweiligen biologischen Theorien, die sich aus ihren Beschreibungen ergeben. Es würde zu weit führen, sich mit solchen Beschreibungen im Einzelnen zu befassen. Außerdem sind wir in der Lage, die gleichen physikalischen Prinzipien klarer und zuverlässiger bei einer anderen Tiergruppe zu veranschaulichen, nämlich bei den Radiolarien. | Die als Radiolaria bekannte Gruppe mikroskopisch kleiner Organismen ist außerordentlich reich an verschiedenen Formen oder »Arten«. Ich weiß nicht, wie viele solcher Arten von den Naturforschern beschrieben und bestimmt worden sind; doch sollen es vor etwa fünfzig Jahren mehr als viertausend gewesen sein, die in mehr als siebenhundert Gattungen angeordnet waren[1]; in der letzten Zeit war man bestrebt, die Zahl zu verringern. Aber abgesehen von der außerordentlichen Vielfalt der Radiolarienformen, gibt es innerhalb dieser Mannigfaltigkeit noch ganz bestimmte Merkmale, die unsere Aufmerksamkeit erregen. Die räumliche Verteilung der Radiolarien ist merkwürdig und unbestimmt; viele Arten findet man über die ganze Erde verbreitet oder wenigstens jede von ihnen hier und dort, ohne Anzeichen einer besonderen Begrenzung ihres geographischen Fundortes. Einige kommen in der Nachbarschaft der beiden Pole vor, andere sind auf warme und wieder andere auf kalte ozeanische Strömungen beschränkt. Nicht weniger unbestimmt ist ihre zeitliche Verteilung: Dies geht so weit, dass von ihnen behauptet wurde, »vom Kambrium an erschienen die Familien und selbst die Gattungen als identisch mit den jetzt lebenden«. | Schließlich finden wir, außer vielleicht im Fall einiger größerer »Kolonialformen«, selten (wenn überhaupt) eine lokale Vorherrschaft einer bestimmten Art, wie es sonst bei den meisten Tieren üblich ist. Wir können im Gegenteil in einem Krümel Tiefseeschlamm oder in etwas fossilem »Radiolariensand« wahrscheinlich Dutzende, ja Hunderte verschiedener Formen finden. Ferner sind die Skelette der Radiolarien außerordentlich zart und kompliziert gebaut trotz ihrer Winzigkeit und der relativen Einfachheit der »einzelligen« Organismen, in denen sie sich entwickeln; und diese komplexen Gebilde zeigen eine wunderbare und ungewöhnliche geometrische Regelmäßigkeit.

1 Haeckel schätzte in seiner *Challenger Monogr.* (1887), p. CLXXXVIII, die Zahl der bekannten Formen auf 4314 Arten, die in 739 Gattungen enthalten sind. Von ihnen wurden 3508 Arten zum ersten Mal in diesem Werk beschrieben.

The Skeletons of Radiolaria If I have dealt comparatively briefly with the inorganic skeletons of sponges, in spite of the interest of the subject from the physical point of view, it has been owing to several reasons. In the first place, though the general trend of the phenomena is clear, it must be admitted that many points are obscure, and could only be discussed at the cost of a long argument. In the second place, the physical theory is too often (as I have shown) in conflict with the accounts given by embryologists of the development of the spicules, and with the current biological theories which their descriptions embody; it is beyond our scope to deal with such descriptions in detail. Lastly, we find ourselves able to illustrate the same physical principles with greater clearness and greater certitude in another group of animals, namely the Radiolaria. | The group of microscopic organisms known as the Radiolaria is extraordinarily rich in diverse forms or "species". I do not know how many of such species have been described and defined by naturalists, but some fifty years ago the number was said to he over four thousand, arranged in more than seven hundred genera[1]; of late years there has been a tendency to reduce the number. But apart from the extraordinary multiplicity of forms among the Radiolaria, there are certain features in this multiplicity which arrest our attention. Their distribution in space is curious and vague; many species are found all over the world, or at least every here and there, with no evidence of specific limitations of geographical habitat; some occur in the neighbourhood of the two poles, some are confined to warm and others to cold currents of the ocean. In time their distribution is not less vague: so much so that it has been asserted of them that "from the Cambrian age downwards, the families and even genera appear identical with those now living". Lastly, except perhaps in the case of a few large "colonial forms", we seldom if ever find, as is usual in most animals, a local predominance of one particular species. On the contrary, in a little pinch of deep-sea mud or of some fossil "radiolarian earth", we shall probably find scores, and it may be even hundreds, of different forms. Moreover, the radiolarian skeletons are of quite extraordinary delicacy and complexity, in spite of their minuteness and the comparative simplicity of the "unicellular" organisms within which they grow; and these complex conformations have a wonderful and unusual appearance of geometric regularity.

1 Haeckel, in his *Challenger Monogr.* (1887), p. clxxxviii, estimated the number of known forms at 4314 species, included in 739 genera. Of these, 3508 species were described for the first time in that work.

Analogy with Snow-Crystals All these general considerations seem such as to prepare us for some physical hypothesis of causation. The little skeletons remind us of such things as snow-crystals (themselves almost endless in their diversity), rather than of a collection of animals, constructed in accordance with functional needs and distributed in accordance with their fitness for particular situations. Nevertheless, great efforts have been made to attach "a biological meaning" to

1

Analogie mit Schneekristallen Alle diese allgemeinen Betrachtungen scheinen uns auf irgendeine physikalische Hypothese über ursächliche Zusammenhänge vorzubereiten. Die kleinen Skelette erinnern uns eher an Gebilde wie Schneekristalle (die selbst fast endlose Varianten aufweisen) als an eine Sammlung von Tieren, die gemäß ihren funktionellen Bedürfnissen gebaut sind und entsprechend ihrer Anpassung an bestimmte Situationen eine besondere räumliche Verteilung zeigen. Gleichwohl hat man sich sehr bemüht, für diese komplizierten Strukturen eine (biologische) Bedeutung zu finden und »die Hoffnung zu rechtfertigen, dass mit der Zeit ihr zweckbestimmter Charakter noch besser erkannt werden wird«.[2] | So wie Ernst Haeckel viele hundert Arten von Radiolarienskeletten beschrieben und gezeichnet hat, so haben die Physiker mehrere tausend verschiedene Formen von Schneekristallen geschildert.[3] Diese verdanken ihre unzähligen Variationen den symmetrischen Wiederholungen einer einfachen kristallinen Form – eine wunderschöne Darstellung von Platos *Einer unter vielen*, το εν παρα τα πολλα [das Eine unter den Vielen]. Andererseits erlangt das Radiolarienskelett seine unendlichen Variationen zum gleichen Thema durch Kombination bestimmter Flächen, Ecken und Kanten innerhalb einer häutigen und blasigen Masse. Der große Unterschied zwischen den beiden Erscheinungen ist sehr einleuchtend und lehrreich. | Kepler untersuchte die Schneeflocke mit bedächtiger Sorgfalt, obwohl er meinte, die Beschäftigung mit einer solchen Kleinigkeit käme der Messung ei-

2 Vgl. Gamble: *Radiolaria* (Lankester's *Treatise on Zoology*) I (1909), 131. – Vgl. auch die Arbeiten von V. Häcker, in Jena Z. Naturw. 39 (1905), 581; Z. wiss. Zool. 83 (1905), 336; Aren. Protistenk. 9 (1907), 139; usw.

3 Siehe Kepler: *De nive sexangula* (1611), Opera, Ed. Fritsch, VII, 715–730; Erasmus Bartholin: *De figura nivis*, Diss. (Hafniae 1661); Dom. Cassini: *Obs. de la figure de la neige* (auszugsweise), Mém. Acad. R. Sci. (1666–1699) X, 1730; J. C. Wilcke: *Om de naturliga snö-figurers*, K. V. Akad. Handl. 22 (1761). Auch zahlreiche Hinweise aus neuerer Zeit.

Abb. 1 Schneekristalle oder »Schneeblumen«, aus Dominic Cassini (ca. 1600).

Fig. 1 Snow-crystals, or "snow flowers". From Dominic Cassini (c. 1600).

2

these elaborate structures, and "to justify the hope that in time their utilitarian character will be more completely recognised".[2] | As Ernst Haeckel described and figured many hundred "species" of radiolarian skeletons, so have the physicists depicted snow-crystals in several thousand different forms.[3] These owe their multitudinous variety to symmetrical repetitions of one simple crystalline form – a beautiful illustration of Plato's *One among the Many*, το εν παρα τα πολλα. On the other hand, the radiolarian skeleton rings its endless changes on combinations of certain facets, corners and edges within a filmy and bubbly mass. The broad difference between the two is very plain and instructive. | Kepler studied the snowflake with care and insight, though he said that to care for such a trifle was like Socrates measuring the hop of a flea. The first drawings I know are by Dominic Cassini; and if that great astronomer was content with them they show how the physical sciences lagged behind astronomy. They date from the time when Maraldi, Cassini's nephew, was studying the bee's cell; and they show once more how very rough his measurements of the honeycomb are bound to have been. | Crystals lie outside the province of this book; yet snow-crystals, and all the rest besides, have much to teach us about the variety, the beauty and the very nature of form. To begin with, the snowcrystal is a regular hexagonal plate or thin prism; that is to say, it shows hexagonal faces above and below, with edges set at co-equal angles of 120°. Ringing her changes on this fundamental form, Nature superadds

2 Cf. Gamble: *Radiolaria* (Lankester's *Treatise on Zoology*), I (1909), 131. Cf. also papers by V. Häcker, in *Jena Z. Naturw.* 39 (1905), 581; *Z. wiss. Zool.* 83 (1905), 336; *Arch. Protistenk.* 9 (1907), 139; etc.

3 See Kepler: *De nive sexangula* (1611), Opera, ed. Fritsch, VII, 715–31; Erasmus Bartholin, *De figura nivis*, Diss. (Hafniae, 1661); Dom. Cassini: *Obs. de la figure de la neige* (Abstr.), Mém. Acad. R. Sci. (1666–99), x, 1730; J. C. Wilcke: *Om de naturliga snö-figurers*, K. V. Akad. Handl. 22 (1761). Also numerous modern references.

Abb. 2 Schneekristalle. Aus Bentley und Humphreys, 1931.

Fig. 2 Snow-crystals. From Bentley and Humphreys, 1931.

nes Flohsprungs durch Sokrates gleich. | Die ersten mir bekannten Zeichnungen stammen von Dominic Cassini, und wenn sie dem großen Astronomen genügten, so beweist dies, wie sehr die Physik hinter der Astronomie zurückgeblieben war. Sie entstanden zur gleichen Zeit, in der Maraldi, Cassinis Neffe, die Zelle der Biene untersuchte, und sie veranschaulichen nochmals, wie grob seine Messungen an der Honigwabe notgedrungen gewesen sind. Die Betrachtung der Kristalle liegt nicht mehr im Bereich dieses Buches; jedoch können wir von den Schneekristallen und allen übrigen Kristallbildungen manches über die Vielfalt, die Schönheit und die wahre Natur der Form lernen. Zunächst ist der Schneekristall eine regelmäßige hexagonale Platte oder ein dünnes Prisma, das heißt, er zeigt oben und unten sechseckige Flächen mit Kanten, die gleich große Winkel von 120° bilden. Alle Variationen zu dieser grundlegenden Form abspielend, fügt die Natur zu den primären Hexagonen zahllose Kombinationen ähnlicher Platten oder Prismen hinzu, die alle gleiche Winkel, aber verschieden lange Seiten haben; und sie wiederholt mit großartiger Symmetrie in Bezug auf alle drei Achsen des Hexagons, was immer sie zur Ausschmückung und Vollendung einer einzigen getan haben mag. Diese Schneekristalle scheinen (wie Tutton behauptet) einen sichtbaren Beweis für das Raumgitter zu erbringen, in das ihre Struktur eingefügt ist. | Die Schönheit eines Schneekristalls beruht auf seiner mathematischen Regelmäßigkeit und Symmetrie. Unsere Freude und Bewunderung wird aber noch außerordentlich erhöht durch die Verbindung von vielen Varianten eines einzigen Typus, die alle verwandt, von denen aber nie zwei gleich sind. Von derselben Art ist die auserlesene Schönheit, die ein japanischer Künstler im Binsenwuchs oder in einer vom Wind berührten Bambusgruppe sieht; und so ist auch (wie wir schon früher gesehen haben) die mannigfaltige Schönheit eines blühenden Zweiges, wenn er alle Stadien von der sich öffnenden Knospe bis zur welkenden Blüte zeigt. | Der Schneekristall wird noch komplizierter und schöner durch Einschluss winziger Luftblasen oder Wassertropfen, deren symmetrische Form sehr merkwürdig und nicht immer leicht zu erklären ist.[4] Schließlich können wir unsere Schneekristalle noch beobachten, nachdem ein leichtes Auftauen ihre scharfen Kanten abgerundet und ihre Anmut durch Verschleiern ihrer Konturen noch erhöht hat.

Der Bau der Radiolarien In den meisten Fällen ist das Skelett der Radiolarien wie das so vieler Schwämme aus Kieselsäure aufgebaut; bei der großen Familie der Acantharia und vielleicht noch bei einigen anderen besteht es aus einem sehr ungewöhnlichen Grundstoff, nämlich aus Strontiumsulfat.[5] Die aus diesen beiden Substanzen aufgebauten Schalen unterscheiden sich nicht wesentlich in ihrer Morphologie; keinesfalls kann den chemischen Eigenschaften dieser anorganischen Stoffe ein Einfluss auf die Form des komplexen Skelettes oder der Schale

4 In Emil Hatscheks Abhandlung: *Gestalt und Orientierung von Gasblasen in Gelen*, Kolloid. Z. 20 (1914), 226–234, können wir einige naheliegende Analogien zu diesen Einschlüssen finden.

5 Bütschli: *Über die chemische Natur der Skelettsubstanz der Acantharia*, Zool. Anz. 30 (1906), 784.

to the primary hexagon endless combinations of similar plates or prisms, all with identical angles but varying lengths of side; and she repeats, with an exquisite symmetry, about all three axes of the hexagon, whatsoever she may have done for the adornment and elaboration of one. These snow-crystals seem (as Tutton says) to give visible proof of the space-lattice on which their structure is framed. | The beauty of a snow-crystal depends on its mathematical regularity and symmetry; but somehow the association of many variants of a single type, all related but no two the same, vastly increases our pleasure and admiration. Such is the peculiar beauty which a Japanese artist sees in a bed of rushes or a clump of bamboos, especially when the wind's ablowing; and such (as we saw before) is the phasebeauty of a flowering spray when it shows every gradation from opening bud to fading flower. | The snow-crystal is further complicated, and its beauty is notably enhanced, by minute occluded bubbles of air or drops of water, whose symmetrical form and arrangement are very curious and not always easy to explain.[4] Lastly, we are apt to see our snow-crystals after a slight thaw has rounded their sharp edges, and has heightened their beauty by softening their contours.

Construction of Radiolaria In the majority of cases, the skeleton of the Radiolaria is composed, like that of so many sponges, of silica; in one large family, the Acantharia, and perhaps in some others, it is made of a very unusual constituent, namely strontium sulphate.[5] There is no important morphological character in which the shells made of these two constituents differ from one another; and in no case can the chemical properties of these inorganic materials be said to influence the form of the complex skeleton or shell, save only in this general way that, by their hardness, toughness and rigidity, they give rise to a fabric more slender and delicate than we find among calcareous organisms. | A slight exception to this rule is found in the presence of true crystals, which occur within the central capsules of certain Radiolaria, for instance the genus *Collosphaera*.[6] Johannes Müller (whose knowledge and insight never fail to astonish us)[7] remarked that these were identical in form with crystals of celestine, a sulphate of strontium and barium; and Bütschli's discovery of sulphates of strontium and of barium in kindred forms renders it all but certain that they are actually true crystals of celestine.[8] | In its typical form, the radiolarian body consists of a spherical mass of protoplasm, around which, and separated from it by some sort of porous "capsule", lies a frothy protoplasm, bubbled up into a multitude of alveoli or vacuoles, filled with a fluid which can scarcely differ much from sea-water.[9] According to their surface-tension conditions, these vacuoles may appear more or less isolated and spherical, or joined together in a "froth" of polyhedral cells; and in the latter, which is the commoner condition, the cells tend to be of equal size, and the

4 We may find some suggestive analogies to these occlusions in Emil Hatschek's paper: *Gestalt and Orientierung von Gasblasen in Gelen*, Kolloid. Z. 20 (1914), 226–34.

5 Bütschli: *Über die chemische Natur der Skelettsubstanz der Acantharia*, Zool. Anz. 30 (1906), 784.

6 For figures of these crystals see Brandt: *F. u. Fl. d. Golfes von Neapel*, 13, Radiolaria (1885), pl. v. Cf. Johannes Müller: *Über die Thalassicollen, etc.*, Abh. K. Akad. Wiss. Berl. (1858).

7 It is interesting to think of the lesser discoveries or inventions, due to men famous for greater things. Johannes Müller first used the tow-net, and Edward Forbes first burrowed the oyster-man's dredge. When we watch a living polyp under the microscope in its tiny aquarium of a glass-cell, we are doing what John Goodsir was the first to do; and the microtome itself was the invention of that best of laboratory-servants, "Old Stirling", Goodsir's right-hand man.

8 Cetestine, or celestite, is $SrSO_4$, with some BaO replacing SrO.

zugeschrieben werden – es sei denn, dass sie durch ihre Härte, Zähigkeit und Starrheit ein schlankeres und zarteres Gebilde erzeugen, als wir es bei den kalkhaltigen Organismen finden. | Eine geringfügige Ausnahme von dieser Regel ist in der Gegenwart echter Kristalle anzutreffen, die innerhalb der Zentralkapseln von bestimmten Radiolarien, zum Beispiel der Gattung *Collosphaera*, vorkommen.[6] Johannes Müller (dessen Wissen und Einsicht uns immer wieder in Erstaunen versetzen)[7] bemerkte, dass diese Gebilde in ihrer Form identisch seien mit Kristallen von Zölestin, einem Strontium- und Bariumsulfat, und Bütschlis Entdeckung von Strontium- und Bariumsulfaten in verwandten Formen macht es fast zur Gewissheit, dass es sich tatsächlich um echte Zölestinkristalle handelt.[8] | In seiner typischen Form besteht der Körper der Radiolarien aus einer kugeligen Protoplasmamasse, um die herum und durch eine Art poröser »Kapsel« von ihr getrennt, schaumiges Protoplasma liegt; es ist zu zahlreichen Alveolen oder Vakuolen aufgeblasen, die mit einer sich wohl kaum vom Meerwasser unterscheidenden Flüssigkeit gefüllt sind.[9] Entsprechend den Verhältnissen ihrer Oberflächenspannung können diese Vakuolen mehr oder weniger isoliert und rund erscheinen oder in einem »Schaum« polyedrischer Zellen miteinander verbunden sein; im letzteren, dem häufigsten Fall sind die Zellen annähernd gleich groß und das aus ihnen gebildete polygonale Netz ist wunderbar regelmäßig. In einigen Fällen sind solche einfachen Einzelorganismen in großer Zahl vereinigt, wobei sie eine schwimmende Kolonie bilden, und wahrscheinlich gehörten viele andere, von denen uns nur die verstreuten Skelette bekannt sind, ebenfalls zu einem solchen koloniebildenden Organismus. | Anders ausgedrückt, ist ein großer Teil des Radiolarienkörpers und insbesondere jener äußere Teil, den Haeckel als »Calymma« bezeichnet hat, aus einer Masse von »Bläschen« aufgebaut, die eine Art steifen Schaumes bilden und im physikalischen, wenn auch nicht unbedingt im biologischen Sinn mit »Zellen« gleichzusetzen sind; die kleinen Bläschen besitzen nämlich ihre genau festgelegten Grenzen und ihre eigenen Oberflächenphänomene. Alles, was wir über Zelloberflächen und Zellstrukturen in unserer Besprechung der Zellen und Gewebe gesagt haben, ist in gleicher Weise und unter entsprechenden Bedingungen auf diese Bläschen anwendbar. In gewissen Fällen, selbst in einem so häufigen und einfachen wie bei der blasigen Substanz von *Actinosphaerium*, können wir eine große Ähnlichkeit oder Formanalogie mit einem zellulären oder parenchymatischen Gewebe beobachten in der dichtgedrängten Anordnung und der sich daraus ergebenden Form dieser Bläschen, sogar gelegentlich auch in einer leichten, hautartigen Verstärkung ihrer Wände. Leidy hat einige merkwürdige kleine Körper, ähnlich kleinen Massen erhärteten Schaumes, dargestellt, die anscheinend nichts anderes sind als die toten und leeren Schalen oder die häutigen Skelette von *Actinosphaerium*[10]; und Carnoy hat in gewissen Zellkernen ein

6 Bezüglich Abbildungen dieser Kristalle siehe Brandt: *F. u. Fl. des Golfes von Neapel.* 13, Radiolaria (1885), Tafel V. Vgl. Johannes Müller: *Über die Thalassicollen, usw.*, Abh. K. Akad. Wiss. Berl. (1858).

7 Es ist interessant, sich der bescheideneren Entdeckungen oder Erfindungen von Männern zu erinnern, die durch größere Dinge berühmt geworden sind. Johannes Müller benutzte als erster das Schleppnetz und Edward Forbes verwendete zum ersten Mal ein Netz, wie es Austernhändler gebrauchen. Wenn wir einen lebenden Polypen in seinem aus einer Glaszelle bestehenden winzigen Aquarium unter dem Mikroskop beobachten, tun wir dasselbe, was John Goodsir als Erster getan hat; und das Mikrotom war die Erfindung des besten aller Laboratoriumsdiener, von »Old Stirling«, Goodsirs rechter Hand.

8 Zölestin oder Zölestit ist $SrSO_4$, in dem SrO durch etwas BaO ersetzt wird.

9 Im Einvernehmen mit den Kolloidchemikern können wir (wie Rhumbler es tat) die Ausdrücke *Spumoid* oder *Emulsoid* auf eine Ansammlung mit Flüssigkeit gefüllter Bläschen anwenden und die Bezeichnung *Schaum* auf solche Bläschen beschränken, die mit Luft oder irgendeinem anderen Gas gefüllt sind.

10 J. Leidy: *Fresh-water Rhizopods of North America* (1879), p. 262, Tafel XLI, Abb. 11, 12.

resulting polygonal meshwork beautifully regular. In some cases a large number of such simple individual organisms are associated together, forming a floating colony; and it is probable that many others, with whose scattered skeletons we are alone acquainted, had likewise formed part of a colonial organism. | In other words a great part of the body of the Radiolarian, and especially that outer portion to which Haeckel has given the name of the "calymma", is built up of a mass of "vesicles", forming a sort of stiff froth, and equivalent in the physical though not necessarily in the biological sense to "cells", inasmuch as the little vesicles have their own well-defined boundaries, and their own surface phenomena. All that we have said of cell-surfaces and cell-conformations in our discussion of cells and of tissues will apply in like manner, and under appropriate conditions, to these. In certain eases, even in so common and so simple a one as the vacuolated substance of an *Actinosphaerium*, we may see a close resemblance, or formal analogy, to a cellular or parenchymatous tissue in the close-packed arrangement and consequent configuration of these vesicles, and even at times in a slight membranous hardening of their walls. Leidy has figured some curious little bodies like small masses of consolidated froth, which seem to be nothing else than the dead and empty husks, or filmy skeletons, of *Actinosphaerium*[10]; and Carnoy has demonstrated in certain cell-nuclei an all but precisely similar framework, of extreme minuteness and tenuity, formed by adsorption or partial solidification of interstitial matter in a close-packed system of alveoli (Fig. 3).[11] In short, we are again dealing or about to deal with a network or basketwork, whose meshes correspond to the boundary lines between associated cells or vesieles. It is just in those boundary walls or films, still more in their edges or at their corners, that surface-energy will be concentrated and adsorption will be hard at work; and the whole arrangement will follow, or tend to follow, the rules of areae minimae — the partition-walls meeting at co-equal angles, three by three in an edge, and their edges meeting four by four in a corner.

9 With the colloid chemists, we may adopt (as Rhumbler has done) the terms *spumoid or emulsoid* to denote an agglomeration of fluid-filled vesicles, restricting the name *froth* to such vesicles when filled with air or some other gas.

10 J. Leidy: *Fresh-water Rhizopods of North America* (1579), p. 262, pl. xli. figs. 11, 12.

11 Carnoy: *Biologie Cellulaire* p. 244, fig. 108; cf. Dreyer: *op. cit.* (1892), fig. 185.

Hexagonal Skeletons Let us suppose the outer surface of our radiolarian to be covered by a layer of froth-like vesicles, uniform in size or nearly so. We know that their mutual tensions will *tend* to conform them into the fashion of a honeycomb, or regular meshwork of hexagons, and that the free end of each hexagonal prism will be a little spherical cap. Suppose now that it be at the outer surface of the protoplasm (in contact with the surrounding sea-water) that the siliceous particles have a tendency to be secreted or adsorbed; the distribution of surface-energy will lead them to accumulate in the grooves which separate the vesicles, and the result will be the development of a delicate sphere composed of tiny rods arranged, or apparently arranged, in a hexagonal network after the fashion of Carnoy's *reticu-*

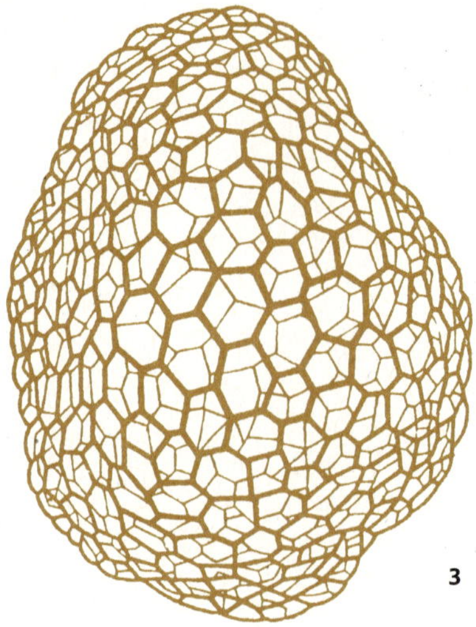

3

fast genau gleiches, außerordentlich kleines und zartes Gerüst nachgewiesen, das durch Adsorption oder teilweise Verfestigung von interstitieller Substanz in einem dicht zusammengeballten System von Alveolen gebildet worden war (Abb. 3).[11] Kurz, wir haben es wiederum mit einem Netz- oder Flechtwerk zu tun, dessen Maschen den Grenzlinien zwischen zwei miteinander in Verbindung stehenden Zellen oder Bläschen entsprechen. Gerade in diesen Grenzwänden oder -häutchen und noch mehr in ihren Kanten oder an ihren Ecken konzentriert sich die Oberflächenenergie und ist die Adsorption besonders wirksam. Die ganze Anordnung folgt offensichtlich den Regeln der areae minimae, wobei die Trennungswände sich in gleichen Winkeln zu dritt in einer Kante treffen und sich je vier Kanten in einer Ecke begegnen.

11 Carnoy: *Biologie Cellulaire*, p. 244, Abb. 108; vgl. Dreyer. op. cit. (1892), Abb. 185.

Hexagonale Skelette Nehmen wir an, die äußere Oberfläche unserer Radiolarien sei mit einer Schicht gleich oder fast gleich großer schaumartiger Bläschen bedeckt. Wir wissen, dass ihre gegenseitigen Spannungen danach streben, ihnen die Form einer Bienenwabe oder eines regelmäßigen Netzwerks von Hexagonen zu geben und dass das freie Ende jedes hexagonalen Prismas eine kleine kugelige Kappe bildet. Nehmen wir nun an, dass die Kieselsäureteilchen eine Tendenz zeigen, an der äußeren Oberfläche des Protoplasmas (im Kontakt mit dem die Radiolarien umgebenden Meerwasser) abgesondert oder angelagert zu werden;

Abb. 3 *Reticulum plasmatique.* Nach Carnoy.

Fig. 3 *Reticulum plasmatique.* After Carnoy.

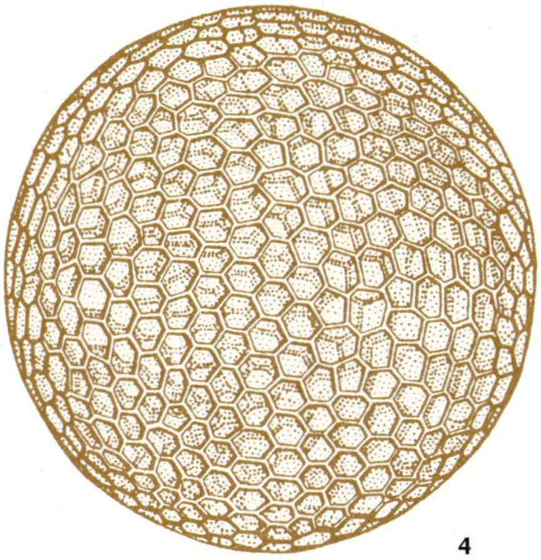

4

lum plasmatique — only more solid — and still more neat and regular. Just such a spherical basket, looking like the finest imaginable Chinese ivory ball, is found in the siliceous skeleton of *Aulonia*, another of Haeckel's Radiolaria from the "Challenger". | But here a strange thing comes to light. *No system of hexagons can enclose space*; whether the hexagons be equal or unequal, regular or irregular, it is still under all circumstances mathematically impossible. So we learn from Euler: the array of hexagons may be extended as far as you please, and over a surface either plane or curved, but *it never closes in*. Neither our *reticulum plasmatique* nor what seems the very perfection of hexagonal symmetry in *Aulonia* are as we are wont to conceive them; hexagons indeed predominate in both, but a certain number of facets are and must be other than hexagonal. If we look carefully at Carnoy's careful drawing we see that both pentagons and heptagons are shown in his reticulum, and Haeckel actually states, in his brief description of his *Aulonia hexagona*, that a few square or pentagonal facets are to be found among the hexagons. | Such skeletal conformations are common: and Nature, as in all her handiwork, is quick to ring the changes on the theme. Among its many variants may be found cases (e.g. *Actinomma*) where the vesicles have been less regular in size; and others in which the mesh-work has been developed not on an outer surface only but at successive levels, producing a system of concentric spheres. If the siliceous material be not limited to the linear junctions of the cells but spread over a portion of the outer

Abb. 4 *Aulonia hexagona* Hkl.

Fig. 4 *Aulonia hexagona* Hkl.

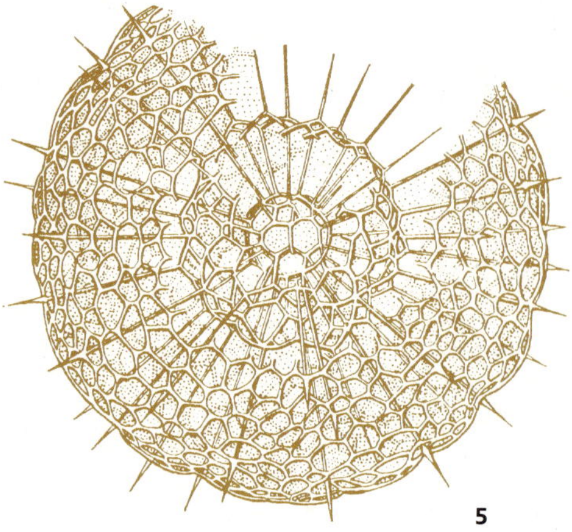

5

dann werden sie sich durch die Verteilung der Oberflächenenergie in den Furchen ansammeln, welche die Bläschen voneinander trennen; das Resultat wird die Entwicklung einer zarten Kugel sein, wobei die winzigen Stäbchen, aus denen sie zusammengesetzt ist, zu einem hexagonalen Netzwerk nach der Art von Carnoys *reticulum plasmatique* — nur fester und noch klarer und regelmäßiger — angeordnet sind oder zu sein scheinen. Genau solch ein kugelförmiges Körbchen, das der allerzierlichsten chinesischen Elfenbeinkugel gleicht, stellt das Kieselskelett von *Aulonia* dar, einer anderen Vertreterin von Haeckels Radiolarien der »Challenger«-Expedition. (Abb. 5) Doch hier zeigt sich etwas Merkwürdiges. *Kein hexagonales System kann einen Raum einschließen*; seien die Hexagone nun gleich oder ungleich, regelmäßig oder unregelmäßig — es ist unter allen Umständen mathematisch unmöglich. So erfahren wir von Euler: Wie beliebig die Anordnung der Hexagone auch erweitert und über eine ebene oder gekrümmte Oberfläche ausgedehnt werden mag, *so schließt sie nie ein*. Weder unser *reticulum plasmatique* noch die anscheinend vollkommene Symmetrie von *Aulonia* stellt etwas dar, was wir zu erwarten gewohnt sind; zwar herrschen in beiden Fällen Hexagone vor, aber eine bestimmte Anzahl von Flächen sind anders ausgebildet und müssen es auch sein. Wenn wir Carnoys sorgfältige Zeichnung genau betrachten, erkennen wir sowohl Pentagone wie Heptagone in seinem Netzwerk; und tatsächlich behauptet auch Haeckel in der kurzen Beschreibung seiner *Aulonia hexagona*, dass ein paar qua-

Abb. 5 *Actinomma arcadophorum* Hkl.

Fig. 5 *Actinomma arcadophorum* Hkl.

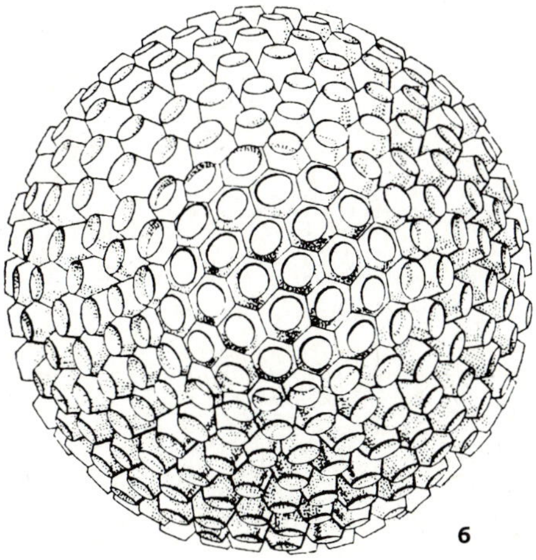

6

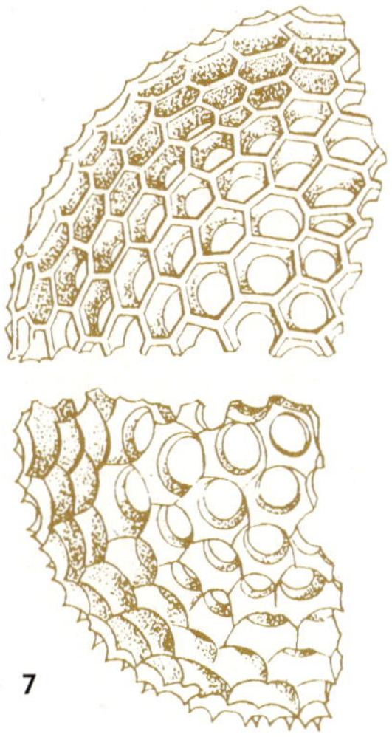

7

spherical surfaces or caps, then we shall have the condition represented in Fig. 6 (*Ethmosphaera*), where the shell appears perforated by circular instead of hexagonal apertures and the circular pores are set on slight spheroidal eminences; and, interconnected with such types as this, we have others in which the accumulating pellicles of skeletal matter have extended from the edges into the substances of the boundary walls and have so produced a system of films, normal to the surface of the sphere, constituting a very perfect honeycomb, as in *Genosphaera favosa* and *vesparia*.[12] | In one or two simple forms, such as the fresh-water *Clathrulina*, just such a sphèrical perforated shell is produced out of some organic, acanthin-like substance; and in some examples of *Clathrulina* the chitinous lattice-work of the shell is just as regular and delicate, with the meshes for the most part as beautifully hexagonal as in the sili- ceous shells of the oceanic Radiolaria. This is only another proof (if proof be needed) that the peculiar form and character of these little skeletons are due not to the material of which they are composed, but to the moulding of that material upon an underlying vesicular structure.

12 In all these latter cases we recognise a relation to, or extension of, the principle of Plateaus's *bourrelet*, or van der Mensbrugghe's *masse annulaire*, or Gibb's ring, of which we have had much to say.

Abb. 6 *Ethmosphaera conosiphonia* Hkl.
Fig. 6 *Ethmosphaera conosiphonia* Hkl.

Abb. 7 Schalenteile zweier »Arten« von *Cenosphaera*: oben *C. favosa*; unten *C. vesparia* Hkl.

Fig. 7 Portions of shells of two "species" of *Cenosphaera*: upper figure *C. favosa*; lower *C. vesparia* Hkl.

dratische oder fünfeckige Flächen zwischen den Sechsecken zu finden seien. | Solche Skelettbildungen kommen häufig vor: und die Natur ist, wie überall in ihrem Wirken, schnell dabei, das Thema in immer neuer Weise abzuwandeln. Unter den vielen Varianten findet man Fälle (zum Beispiel *Actinomma*), bei denen die Bläschen in ihrer Größe weniger regelmäßig sind, und andere, wo sich das Netzwerk nicht nur an der äußeren Oberfläche, sondern in aufeinander folgenden Ebenen entwickelt hat, so dass ein System konzentrischer Kugeln entstanden ist. | Wenn die Kieselsäuresubstanz nicht auf die Verbindungslinien der Zellen beschränkt, sondern über einen Teil der äußeren kugelförmigen Oberflächen oder Kappen ausgebreitet ist, finden wir den in Abb. 6 (*Ethmosphaera*) dargestellten Zustand; hier scheint die Schale durch runde statt durch hexagonale Öffnungen durchlöchert zu sein und diese runden Poren liegen auf schwach kugeligen Erhebungen. In enger Verbindung mit solchen Typen finden wir andere, bei denen sich die Schar der skelettbildenden Häutchen von den Kanten bis in die Substanzen der Grenzwände ausgedehnt und auf diese Weise ein senkrecht zur Kugeloberfläche angeordnetes System von Membranen erzeugt hat; dadurch entsteht die Form einer äußerst vollkommenen Honigwabe, wie bei *Cenosphaera favosa* und *vesparia*.[12] | Bei einer oder zwei einfachen Formen, wie zum Beispiel bei der Süßwasserform *Clathrulina*, ist genau eine solche kugelförmige, durchlöcherte Schale aus einer organischen, akanthinartigen Substanz aufgebaut worden; und bei einigen Vertretern der Gattung *Clathrulina* finden wir das chitinige Gitterwerk der Schale genauso regelmäßig und zart und mit zum größten Teil ebenso wunderschön hexagonalen Maschen ausgebildet wie bei den Kieselsäureschalen der marinen Radiolarien. Das ist (falls es nötig ist) nur ein weiterer Beweis dafür, dass die besondere Form und die Eigenart dieser kleinen Skelettgebilde nicht vom Material abhängen, aus dem sie aufgebaut sind, sondern von der Gestaltung dieses Stoffes auf der Grundlage einer Bläschenstruktur.

12 In allen diesen letzteren Fällen sehen wir das Prinzip von Plateaus *bourrelet* oder van der Mensbrugghes *masse annulaire* oder dem Gibbschen Ring angedeutet oder erweitert, von dem wir schon so viel gesprochen haben.

23

Kunstformen der Natur

Art forms of nature

Ernst Haeckel

Vorwort Die Natur erzeugt in ihrem Schoße eine unerschöpfliche Fülle von wunderbaren Gestalten, durch deren Schönheit und Mannigfaltigkeit alle vom Menschen geschaffenen Kunstformen weitaus übertroffen werden. Die Naturprodukte, aus deren Nachahmung und Modellierung die bildende Kunst des Menschen hervorgegangen ist, gehören begreiflicherweise solchen höheren Gruppen des Pflanzenreichs und des Tierreichs an, mit denen der Mensch in beständiger Berührung lebt, vor allem den Blütenpflanzen und Wirbeltieren. Dagegen ist den meisten Menschen größtenteils oder ganz unbekannt jenes unermessliche Gebiet der niederen Lebensformen, die versteckt in den Tiefen des Meeres wohnen oder wegen ihrer geringen Größe dem unbewaffneten Auge verschlossen bleiben. Der größte Teil dieser verborgenen Schönheiten der Natur ist erst durch die ausgedehnten Forschungen des 19. Jahrhunderts aufgedeckt worden. | Besonders ergiebig an eigenartigen und wundervollen Gestalten ist das weite Reich der Protisten oder Zellinge, jener einfachsten Organismen, deren ganzer lebendiger Körper nur aus einer einzigen Zelle besteht: Radiolarien, Thalamophoren und Infusorien unter den Urtieren (Protozoen); Diatomeen, Kosmarieen und Peridineen unter den Urpflanzen (Protophyten). Die erstaunliche Fülle von zierlichen und phantastischen Formen, die diese einzelligen Protisten hervorbringen, ist uns erst durch das verbesserte Mikroskop, die verfeinerten Beobachtungsmethoden und die planmäßige Meeresforschung der Neuzeit zugänglich geworden. Diesen verdanken wir aber auch einen überraschenden Reichtum an Entdeckungen auf den benachbarten Gebieten, auf denen größere Organismen niederen Ranges ihre bewundernswürdige Gestaltungskraft entfalten: Algen, Pilze und Moose unter den niederen Pflanzen; Polypen, Korallen und Medusen unter den Nesseltieren. Die Mehrzahl der vorhandenen Abbildungen dieser formenschönen Organismen ist in teuren und seltenen Werken versteckt und dem Laien schwer erreichbar. Die vorliegenden »Kunstformen der Natur« dagegen verfolgen den Zweck, jene verborgenen Schätze ans Licht zu ziehen und einem größeren Kreise von Freunden der Kunst und der Natur zugänglich zu machen. Seit frühester Jugend von dem Formenreize der lebendigen Wesen gefesselt und seit einem halben Jahrhundert mit Vorliebe morphologische Studien pflegend, war ich nicht nur bemüht, die Gesetze ihrer Gestaltung und Entwickelung zu erkennen, sondern auch zeichnend und malend tiefer in das Geheimnis ihrer Schönheit einzudringen. Auf zahlreichen Reisen, die sich über einen Zeitraum von fünfundvierzig Jahren erstrecken, habe ich alle Länder und Küsten Europas kennen gelernt und auch an den interessantesten Gestaden des nördlichen Afrika und des südlichen Asien längere Zeit gearbeitet. Tausende von Figuren, die ich auf diesen wissenschaftlichen Reisen nach der Natur gezeichnet habe, sind bereits in meinen größeren Monographien publiziert; einen anderen Teil will ich bei dieser Gelegenheit veröffentlichen. Außerdem werde

Preface Nature has produced in its womb an inexhaustible wealth of wonderful creatures, which in their beauty and diversity far exceed all the art forms ever created by mankind. Understandably, these products of nature whose imitation and simulation gave rise to fine art belong to those higher groups of fauna and flora with which man lives in constant contact, above all the flowering plants and vertebrates. By contrast, the majority of people are largely or entirely ignorant of that immense field of lower life forms that lie hidden in the depths of the ocean or remain invisible to the naked eye owing to their minute size. The lion's share of these hidden beauties of nature was only revealed by the extensive research conducted in the 19th century. | Particularly rich in unusual and wonderful creatures is the broad realm of protista or single-cell organisms. Those highly simple organisms whose entire living bodies consist of just a single cell: radiolarians, thalamophora and infusorians (amongst the protozoa); diatoms, cosmaria and peridiniums (amongst the protophytes). The amazing wealth of delicate and fantastic shapes, which these one-cell protista produce has only become accessible to us through improvements to the microscope, refined observation methods and the systematic marine research conducted in modern times. But we also owe them [these improved methods] a startling abundance of discoveries in associated fields in which larger, lower-ranking organisms unfold their remarkable creative power: algae, fungi and mosses amongst the lower plants; polyps, corals and medusa amongst the cnidarians. The majority of the existing illustrations of these elegantly proportioned organisms is hidden in expensive and rare works and poorly accessible to the amateur. By way of contrast, this publication "Art forms in Nature" aims to bring to light these hidden treasures and make them accessible to a larger circle of friends of art and nature. From my very earliest youth I have been captivated by the attractive designs of living creatures and for half a century I have cultivated a special liking for morphological studies; I was not only keen to recognize the laws underlying their design and development but sought through drawing and painting to penetrate deeper into the secret of their beauty. On numerous journeys extending over a period of forty-five years I have acquainted myself with all the nations and coasts of Europe and have also worked for longer periods at the most interesting shores of Northern Africa and Southern Asia. Thousands of figures that I drew on these scientific expeditions have already been published in my larger monographs; I wish to publish others on this occasion. Moreover, I will be at pains to select and compile from the extensive existing literature the most beautiful and most aesthetically valuable forms. Should the first journals be received favorably then more seldom and less well known beauties from the realms of the higher animal and plant world will also be featured. Initially, 50 plates will be published from "Art Forms in Nature" (five informal journals each containing

ich bemüht sein, aus der umfangreichen Literatur die schönsten und ästhetisch wertvollsten Formen auszulesen und zusammenzustellen. Wenn die ersten Hefte beifällig aufgenommen werden, so sollen später auch die selteneren und weniger bekannten Schönheiten aus dem Gebiete der höheren Tier- und Pflanzenwelt eine entsprechende Darstellung finden. Zunächst werden von den »Kunstformen der Natur« 50 Tafeln erscheinen (fünf zwanglose Hefte zu je zehn Tafeln), jede Tafel von einem erläuternden Textblatt begleitet. Im Falle einer günstigen Aufnahme ist eine größere Zahl von Heften in Aussicht genommen; ich hoffe dann, nach Vollendung von zehn Heften (100 Tafeln), eine allgemeine Einleitung zu dem Werke geben zu können, welche die systematische Ordnung sämtlicher Formengruppen enthält, ferner eine ästhetische Erörterung ihrer künstlerischen Gestaltung sowie Angaben über die wichtigsten Quellen der betreffenden Literatur. | Die moderne bildende Kunst und das moderne, mächtig emporgeblühte Kunstgewerbe werden in diesen wahren »Kunstformen der Natur« eine reiche Fülle neuer und schöner Motive finden. Bei ihrer Zusammenstellung habe ich mich auf die naturgetreue Wiedergabe der wirklich vorhandenen Naturerzeugnisse beschränkt, dagegen von einer stilistischen Modellierung und dekorativen Verwertung abgesehen; diese überlasse ich den bildenden Künstlern selbst. | Für die künstlerische Ausführung der Figuren und ihre naturwahre Lithographie bin ich meinem treuen, bewährten Mitarbeiter, Herrn Adolf Giltsch in Jena, zu aufrichtigem Danke verpflichtet. Seinem lebhaften Interesse für die gestellte Aufgabe, seinem morphologischen Verständnis und künstlerischen Talente ist es zu verdanken, dass ich den schon vor langer Zeit entworfenen Plan dieses Werkes endlich in der gewünschten Form ausführen konnte. | Lebhaften Dank schulde ich außerdem für materielle und intellektuelle Förderung meines Unternehmens Herrn Dr. Paul von Ritter in Basel, dem begeisterten Freunde und opferwilligen Gönner der Naturwissenschaft. Als er im Jahre 1886 die »Paul von Ritter'sche Stiftung für phylogenetische Zoologie« an der Universität Jena gründete, sprach er den Wunsch aus, dass deren Mittel nicht nur zur Förderung akademischer Studien und Forschungsreisen verwendet würden, sondern auch zur Erweckung des Interesse an den Wunderwerken und Schönheiten der Natur in weiteren Volkskreisen. Die Quellen ästhetischen Genusses und veredelnder Erkenntnis, die überall in der Natur verborgen sind, sollen mehr und mehr erschlossen und Gemeingut weitester Bildungskreise werden. | Diesen Anschauungen kam auch das Bibliographische Institut in Leipzig entgegen, das die kostspielige Ausführung und die Veröffentlichung der Tafeln bereitwillig übernahm; auch ihm statte ich für seine Opfer und Mühen meinen besten Dank ab. Möge unsere gemeinsame Absicht erreicht werden, durch die Bekanntschaft mit den »Kunstformen der Natur« gleichzeitig das künstlerische und das wissenschaftliche Interesse an der herrlichen uns umgebenden Gestaltenwelt zu fördern!

Jena, am 16. Februar 1899. *Ernst Haeckel*

ten plates), and each plate accompanied by an explanatory text sheet. In the event of a favorable reception a larger number of journals is planned; I then hope after completing ten journals (100 plates), to be able to write a general introduction to the work, which would contain the systematic classification of all the formal groups, and also an aesthetic discussion of their artistic design but also information on the most important sources of relevant literature. | In these true "Art Forms in Nature" modern fine art and the modern, greatly flourishing arts and crafts will find a wealth of new and attractive motifs. In compiling them I have confined myself to a true-to-life depiction of products of nature that actually exist, and have desisted from a stylistic modeling or decorative exploitation; I leave this to the artists themselves. | I should like to extend sincere thanks to my true and trusted employee, Adolf Giltsch in Jena to whom I owe the artistic execution of the figures and their true-to-life lithography. I owe it to his lively interest in the task at hand, his morphological understanding and artistic talents that I could finally realize this work in the desired work that has existed as a plan for so long. | In addition, for his material and intellectual support I owe great thanks to Dr. Paul von Ritter in Basle, an enthusiastic friend, enthusiastic friend and self-sacrificing patron of the sciences. When he founded the "Paul Ritter Foundation for Phylogenetic Zoology" at Jena University 1886 he expressed the desire that its funds should not only be employed to promote academic studies and research expeditions but also to arouse interest in the wondrous works and beauties of nature in broad sections of the public. The sources of aesthetic enjoyment and improving knowledge that are an intrinsic part of nature should be increasingly exploited and become the common property of broad sections of the educated public. | The Bibliography Institute in Leipzig also found these views very congenial and willingly took on the costly execution and publication of the tables; my best thanks also go to this Institute for the sacrifices and efforts this involved. May our common intention be achieved, namely by familiarizing people with the "Art Forms in Nature" to promote both artistic and scientific interest in the magnificent morphology of the world surrounding us!

Jena, February 16, 1899. *Ernst Haeckel*

Diatomea. Schachtellinge.

Stamm der Urpflanzen (Protophyta) / Hauptklasse der Algarien
Klasse der Diatomeen (Schachtel- oder Kiesel-Algarien).

Die Diatomeen oder Schachtellinge bilden eine formenreiche Klasse von einzelligen Urpflanzen, welche massenhaft sowohl im Süßwasser als im Meere leben; über 2000 Arten sind bekannt. Sie zeichnen sich vor anderen Protophyten durch die Bildung einer zierlichen, zweiklappigen Kieselschale aus; die beiden Hälften oder Klappen derselben verhalten sich wie eine Schachtel und ihr Deckel. Die obere, etwas größere Hälfte, die Deckelklappe, greift mit einem breiten Rande, dem Gürtelbande, über den Rand der unteren größeren Hälfte, der Schachtelklappe, hinüber. Daher hat jede Schale zwei sehr verschiedene Ansichten, die parallele (horizontale) Boden- oder Hauptseite (Fig. 1, 4 etc.), und die ringförmige (vertikale) Gürtel- oder Nebenseite (Fig. 20, 21 etc.). Die erstere ist meistens durch sehr zierliche Skulptur ausgezeichnet: Rippen, Leisten, Felder, Körner etc. Sie ist von sehr feinen Poren durchbrochen. Die meisten Diatomeen sind sehr klein, schweben frei im Wasser und bilden einen wichtigen Bestandteil des Planktons; andere Arten sind durch Gallertstiele am Boden befestigt. Viele Arten bilden Cönobien oder Zellvereine, indem die durch Teilung entstehenden Tochterzellen in Zusammenhang bleiben. Alle auf dieser Tafel abgebildeten Arten gehören zu den einsam lebenden (Monobien) und frei schwimmenden. Ihre Schalen sind meist durch eine sehr regelmäßige geometrische Grundform ausgezeichnet: zweiseitig (Fig. 2, 3, 10), breitstrahlig (Fig. 1, 4, 22), vierstrahlig (Fig. 7, 9, 11), fünfstrahlig (Fig. 5), vielstrahlig (Fig. 16). Der lebendige, weiche Zellenkörper, welcher in der Schale eingeschlossen ist (Fig. 15), enthält in der Mitte einen Zellenkern; von der feinen Plasmaschicht, die ihn umgibt, strahlen verzweigte Plasmafäden aus, welche die strömende Bewegung der lebendigen Zellsubstanz zeigen. Im Plasmanetz zerstreut liegen viele Chromatellen oder Farbkörner; ihre grüne Farbe (Chlorophyll) wird meistens durch einen gelben oder braunen Farbstoff verdeckt (Diatomin).

Fig. 1 Triceratium digitale *(Brun)*
Fig. 2 Navicula lyra *(Ehrenberg)*
Fig. 3 Navicula excavata *(Greville)*
Fig. 4 Triceratium mirificum *(Brun)*
Fig. 5 Triceratium pentacrinus *(Wallich)* vgl. Fig. 21.
Fig. 6 Actinoptychus constellatus *(Brun)*
Fig. 7 Aulacodiscus mammosus *(Greville)*
Fig. 8 Navicula Wrightii *(Meara)*
Fig. 9 Auliscus crucifer *(Brun)*
Fig. 10 Biddulphia pulchella *(Gray)*
Fig. 11 Auliscus craterifer *(Brun)*
Fig. 12 Auliscus mirabilis *(Greville)*
Fig. 13 Aulacodiscus Grevilleanus *(Norman)*
Fig. 14 Surirella Macraeana *(Greville)*
Fig. 15 Denticella regia *(Max Schultze)*
Fig. 16 Asterolampra eximia *(Greville)*
Fig. 17 Actinoptychus heliopelta *(Brun)*
Fig. 18 Plagiogramma barbadense *(Brun)*
Fig. 19 Pinnularia Mülleri *(Haeckel)*
Fig. 20 Biddulphia granulata *(Smith)*
Fig. 21 Triceratium pentacrinus *(Wallich)* vgl. Fig. 5.
Fig. 22 Triceratium moronense *(Greville)*

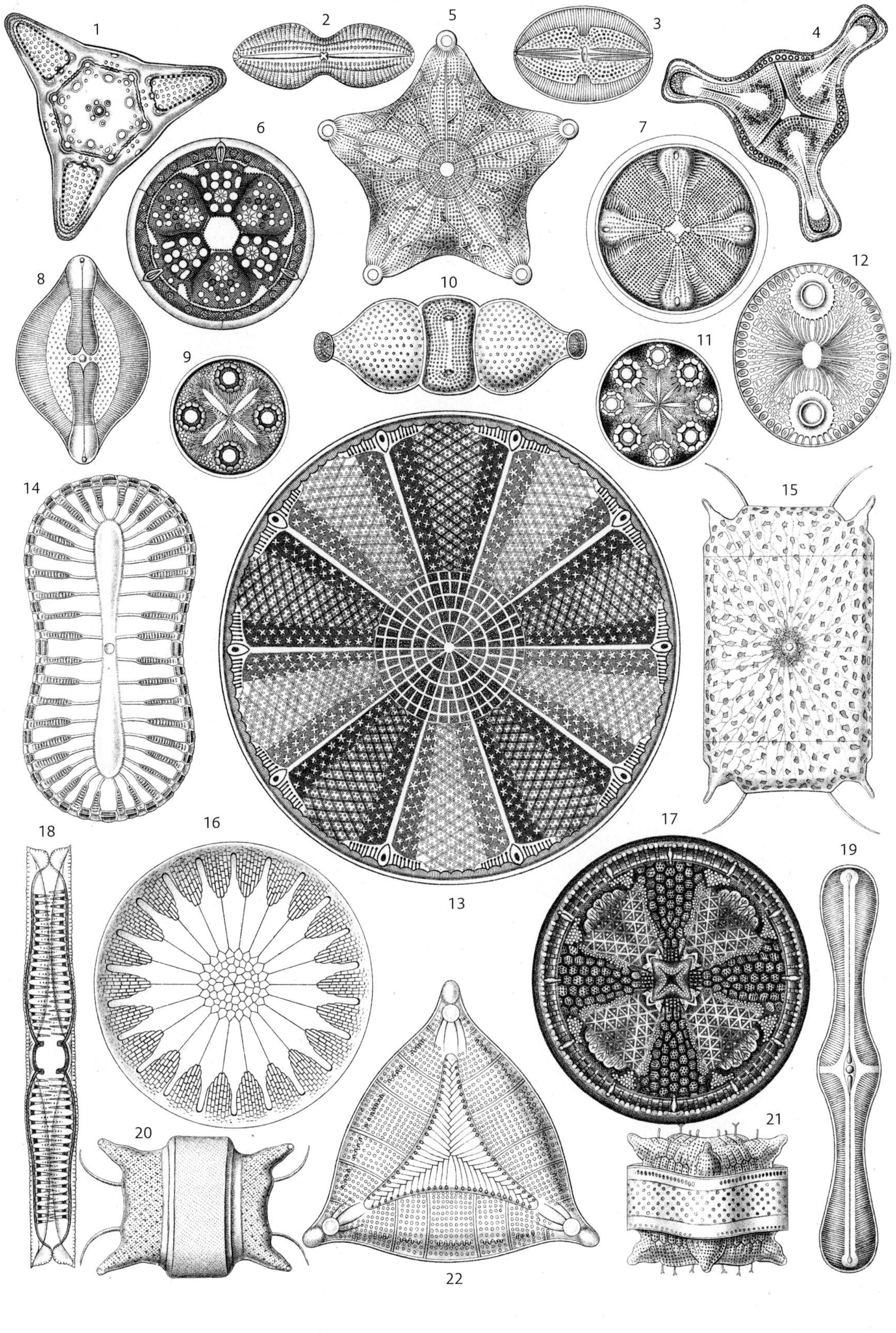
1
2
5
3
4
6
7
8
10
12
9
11
14
13
15
18
16
17
19
20
22
21

Spyroidea. Nüßchenstrahlinge.

Stamm der Urtiere (Protozoa) / Hauptklasse der Wurzelfüßer (Rhizopoda)
Klasse der Strahlinge (Radiolaria) / Legion der Korbstrahlinge oder Monopyleen (Nassellaria)
Ordnung der Nüßchenstrahlinge (Spyroidea).

Die Figuren dieser Tafel stellen die gereinigten Kieselskelette von Spyroideen dar, einer besonderen Ordnung aus der formenreichen Legion der Korbstrahlinge oder Nassellarien. Das Skelett dieser kleinen, dem bloßen Auge nur als ein Pünktchen erscheinenden Radiolarien bildet eine zierliche Gitterschale von zweiseitiger Grundform, zusammengesetzt aus zwei Seitenhälften, welche durch eine mittlere Einschnürung (Fig. 1, 9, 13) oder durch einen senkrecht stehenden Ring (Fig. 6, 8, 11) geschieden erscheinen, ähnlich den beiden Hälften einer Walnuß. Der lebendige weiche Körper, welcher innerhalb dieser Schale liegt und meistens eine rundliche, kegelförmige oder nußförmige Zentralkapsel enthält, ist auf dieser Tafel nicht dargestellt, ebenso auch nicht die zahlreichen feinen Plasmafäden (Scheinfüßchen oder Pseudopodien), welche von demselben ausstrahlen. Die gegitterte Kieselschale der meisten Spyroideen ist mit Stacheln oder flügelförmigen Anhängen versehen, welche teils als Schutzwaffen, teils als Schwebeapparate dienen, sowie als Stützen für die Scheinfüßchen. Oben auf dem Gipfel vieler Schalen steht ein Horn oder Scheitelstachel (Fig. 6, 8, 11), daneben oft noch zwei Seitenstacheln (Fig. 1, 7, 12). Unten ist die Mündung des Gehäuses, aus welcher die Scheinfüßchen hauptsächlich vortreten, oft mit zwei langen Seitenstacheln oder Füßen versehen (Fig. 1, 5, 11) oder mit einem Kranze von Blättern oder Stacheln umgeben (Fig. 4, 6, 7, 8).

Fig. 1 Triceraspyris gazella
Schale oben mit drei Hörnern, unten mit drei Füßen.

Fig. 2 Clathrospyris pyramidalis
Schale oben mit einem Scheitelhorn, unten mit fünf Füßen.

Fig. 3 Pylospyris canariensis
Schale oben mit Helmaufsatz, unten ohne Füße.

Fig. 4 Anthospyris mammillata
Schale oben mit drei Hörnern, unten mit einem Kranz von blattförmigen Füßen.

Fig. 5 Dendrospyris polyrrhiza
Schale oben mit einem Scheitelhorn, unten mit zwei gekrümmten Füßen.

Fig. 6 Sepalospyris pagoda
Schale oben mit Helm und Scheitelhorn, unten mit einem Kranz von blattförmigen Füßen.

Fig. 7 Elaphospyris cervicornis
Schale oben mit drei Hörnern, unten mit zwei Paar ästigen geweihartigen Füßen.

Fig. 8 Tholospyris cupola
Schale oben mit einem Scheitelhorn, unten mit drei geweihartigen Füßen.

Fig. 9 Dictyospyris stalactites
Schale ohne Hörner und Füße, mit kleinen, stalaktitenähnlichen Knorren bedeckt.

Fig. 10 Dictyospyris anthophora
Schale ohne Hörner und Füße, mit dicken, teilweise gespaltenen Knorren bedeckt.

Fig. 11 Dorcadospyris dinoceras
Schale oben mit einem Scheitelhorn, unten mit zwei mächtigen gekrümmten Füßen, die eine Reihe von dornigen Stacheln tragen.

Fig. 12 Triceraspyris damaecornis
Schale oben mit drei Hörnern, unten mit zwei Paar ästigen Füßen (von oben gesehen).

Fig. 13 Ceratospyris Strasburgeri Schale stachelig, mit zahlreichen geflügelten Hörnern und Füßen.

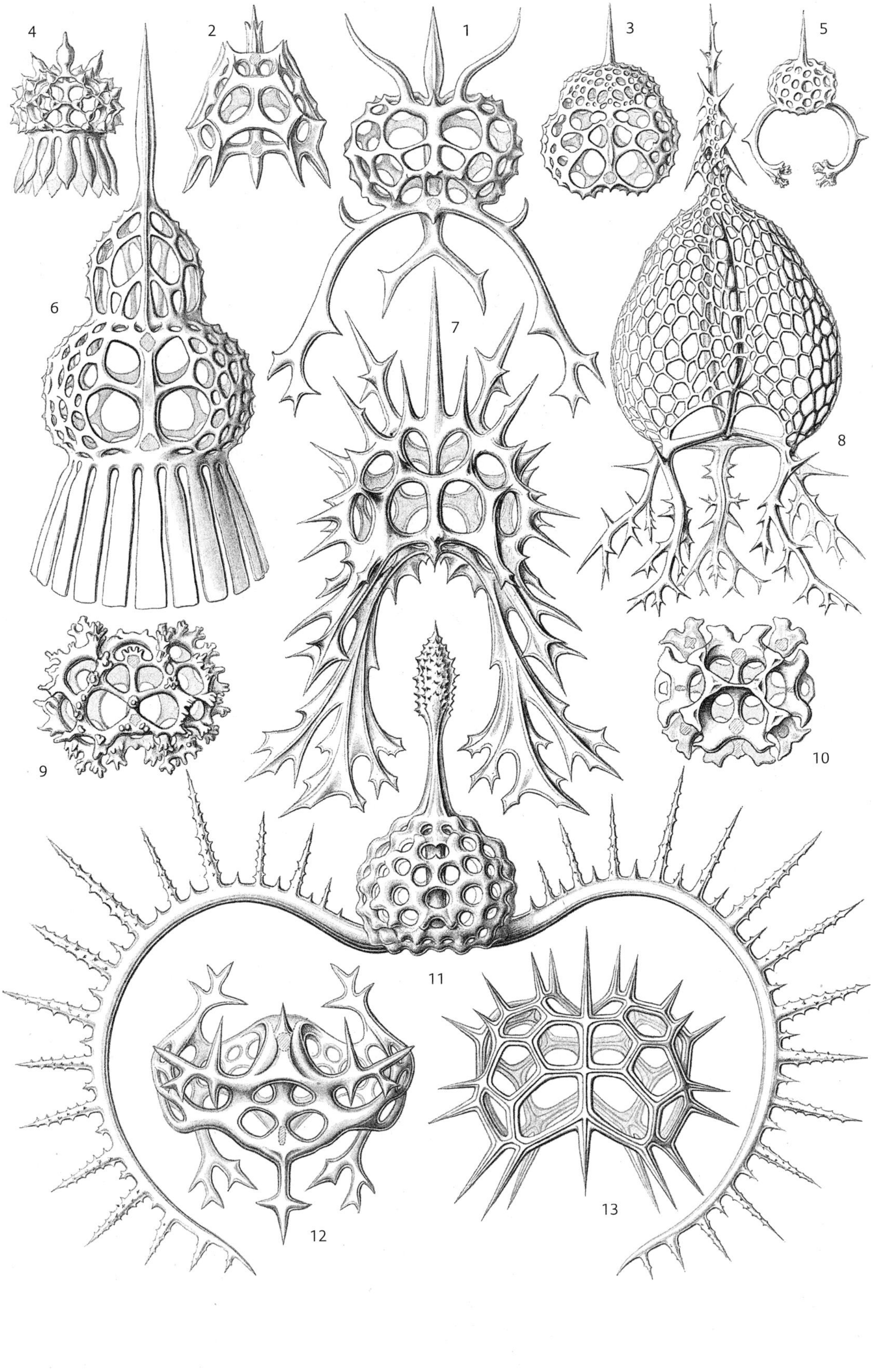
4
2
1
3
5
6
7
8
9
10
11
12
13

Melethallia. Gesellige Algetten.

Stamm der Urpflanzen (Protophyta) / Hauptklasse der Algetten (Zoosporata)
Klasse der Melethallien (Coenobiotica) / Familie der Wassernetzchen (Hydrodictyea).

Die Melethallien (Coenobiotica) sind kleine, das Süßwasser bewohnende Urpflanzen aus der Hauptklasse der Algetten (der sogenannten »einzelligen Algen mit Schwärmsporen«, Zoosporata); sie unterscheiden sich von den übrigen Algetten dadurch, daß ihre grünen Zellen nicht einzeln leben (Monobia, die naheverwandten Protokokken), sondern beständige Zellvereine bilden (Coenobia). Bei der Gattung Pediastrum, mit zahlreichen, im Süßwasser lebenden Arten, sind die geselligen Zellen stets in Form einer flachen Scheibe von zierlicher Form geordnet, in einer einzigen Schicht. In dem grünen Zellinhalt findet sich stets ein glänzender Eiweißkristall (Pyrenoid) sowie mehrere (hier rötlich gefärbte) kleine Zellkerne. Die Fortpflanzung erfolgt gewöhnlich durch Schwärmsporen, welche in bestimmter Zahl (4, 8, 16, 32) in einer Zelle entstehen; dann springt an einer Stelle die Zellwand auf, und aus dieser Geburtsspalte der Mutterzelle tritt eine gallertige Blase hervor, in welche die beweglichen Tochterzellen übertreten (Fig. 8); schon innerhalb der Blase ordnen sich letztere zu einer neuen Scheibe. Bei den meisten Arten ist die Scheibe aus 8 oder 16 Zellen zusammengesetzt; seltener aus 4, 32 oder 64. Die Randzellen unterscheiden sich gewöhnlich von den Mittelzellen durch Bildung von Lappen, Zacken oder Stacheln, welche in den einzelnen Arten verschieden geformt sind.

Fig. 1 Pediastrum tetras *(Ehrenberg)*.
Die Scheibe besteht aus vier gleichen, im Kreuz stehenden, dreieckigen Zellen, deren Außenrand zweilappig und vierspitzig ist. Jede Zelle enthält ein Pyrenoid und zwei Kerne.

Fig. 2 Pediastrum rotula *(Kützing)*.
Die Scheibe ist aus acht zweilappigen Zellen zusammengesetzt; die Mittelzelle enthält fünf Kerne, die sieben Randzellen je sechs Kerne.

Fig. 3 Pediastrum granulatum *(Kützing)*.
Die Scheibe besteht aus acht Zellen, von denen jede ein zentrales Pyrenoid und zahlreiche kleine Kerne einschließt. Die beiden symmetrischen Mittelzellen sind trapezoid, ohne Fortsätze. Die sechs Randzellen sind zweilappig, jede mit zwei radialen, körnig rauhen Kolben bewaffnet.

Fig. 4 Pediastrum octonum *(Haeckel)*.
Die Scheibe besteht aus acht gleichen, flaschenförmigen Zellen, die einen regulären achtstrahligen Stern bilden. Jede Zelle besitzt einen Kern (mitten), ein Pyrenoid (innen) und einen radialen Fortsatz, gleich einem Flaschenhals (außen).

Fig. 5 Pediastrum cruciatum *(Haeckel)*.
Die Scheibe besteht aus vier gleichen, im Kreuz stehenden zweilappigen Zellen; jede Zelle enthält ein Pyrenoid und sechs kleine Kerne. Jeder der acht Lappen trägt außen zwei feine Spitzen.

Fig. 6 Pediastrum selenaea *(Kützing)*.
Die Scheibe besteht aus 16 Zellen, welche die gewöhnliche Anordnung zeigen (wie auch in Fig. 7); eine zentrale Mittelzelle ist von fünf ähnlichen umgeben, und diese von einem Kranze von zehn zweilappigen Randzellen. Jede Zelle enthält ein zentrales Pyrenoid und 4–6 Kerne.

Fig. 7 Pediastrum pertusum *(Kützing)*.
Die Scheibe besteht aus 16 Zellen, in derselben Anordnung wie Fig. 6, nur durch größere Zwischenräume getrennt. Die spitzen, dreieckigen Lappen der zehn Randzellen sind gezähnt. Die kleinen Kerne dieser Art sind zahlreicher.

Fig. 8 Pediastrum elegans
Die Scheibe besteht aus 32 Zellen; eine zentrale Mittelzelle ist von drei Zellenringen zu fünf, zehn und 16 Zellen umgeben.
Von den sechs Mittelzellen enthält jede ein zentrales Pyrenoid und 5–6 Kerne. Die zehn Zellen des zweiten Ringes sind in Vermehrung begriffen; jede zerfällt durch Teilung in 8–16 Zellen. Die 16 Zellen des äußeren Ringes oder die Randzellen (mit je vier spitzen, gezähnten Randlappen) zeigen den Geburtsakt der Tochterscheiben, die sich in jeder einzelnen Zelle der Mutterscheibe durch Teilung gebildet haben. In vier diagonalen Randzellen beginnt die Geburt, indem aus einem Sprung oder Geburtsspalt der Zellenwand eine gewölbte Blase vortritt. In vier anderen, zwischen jenen liegenden Randzellen ist die Geburt weiter vorgeschritten; die vier jungen Tochterscheiben (deren 16 bewegliche Zellen sich noch nicht regelmäßig geordnet haben) sind aus ihrer Mutterzelle in die Blase übergetreten.
Die acht Randzellen zwischen den gebärenden acht Zellen sind bereits entleert; jede zeigt noch den schrägen Sprung der Zellenwand, aus welchem die geborenen Tochterscheiben ausgetreten sind.

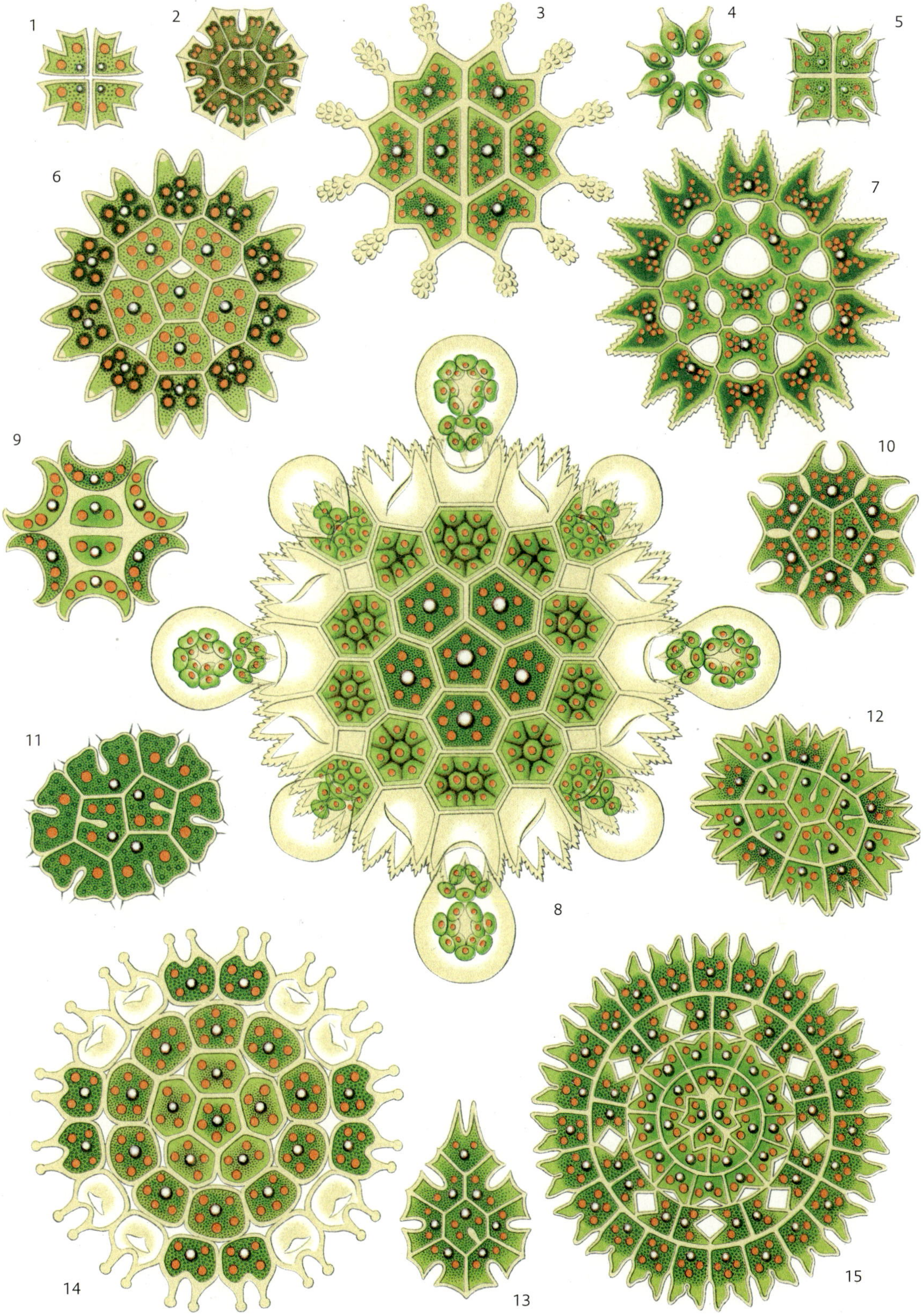
1
2
3
4
5
6
7
8
9
10
11
12
13
14
15

Fig. 9 Pediastrum lunatum
Die Scheibe besteht aus acht Zellen, ähnlich geordnet wie Fig. 3 und 10. Die beiden Mittelzellen sind halbkreisförmig, mit je zwei Kernen; die sechs Randzellen sind halbmondförmig, mit je vier Kernen. Jede Zelle enthält ein zentrales Pyrenoid.

Fig. 10 Pediastrum furcatum
Die Scheibe besteht aus acht Zellen, ähnlich geordnet wie Fig. 3 und 9. Die beiden Mittelzellen sind fünfeckig, die sechs Randzellen sind sechseckig, außen gabelteilig.

Fig. 11 Pediastrum Braunii
Die elliptische Scheibe besteht aus acht pentagonalen Zellen, deren jede einen tiefen Einschnitt zeigt. Die sechs Randzellen sind mit je vier zarten Spitzen bewaffnet. Diese Art ist nach dem feinsinnigen Botaniker Alexander Braun benannt.

Fig. 12 Pediastrum ellipticum *(Ehrenberg)*
Die elliptische Scheibe besteht aus 16 Zellen, von denen jede ein Pyrenoid und vier Kerne enthält. Die fünf Mittelzellen sind zweiteilig, die elf Randzellen am Rande vierlappig.

Fig. 13 Pediastrum Darwinii
Die birnförmige Scheibe ist bilateral und besteht aus acht symmetrisch geordneten Zellen; eine Mittelzelle ist von sechs zweiteiligen Randzellen umgeben; die achte Zelle liegt exzentrisch zwischen zwei Randzellen.

Fig. 14 Pediastrum trochiscus
Die Scheibe, einem Zahnrad ähnlich, besteht aus 32 Zellen; eine Mittelzelle ist von drei Zellenringen umgeben; der erste (innere) Ring ist aus fünf, der zweite (mittlere) aus zehn und der dritte (äußere) aus 16 Zellen zusammengesetzt. Von letzteren sind vier Paar leer und zeigen die Geburtsspalte, aus welcher die Tochterscheiben ausgetreten sind.

Fig. 15 Pediastrum solare
Die Scheibe besteht aus 64 Zellen; drei Zentralzellen sind von vier Ringen umgeben; der erste Ring enthält 7, der zweite 13, der dritte 18 und der vierte 23 Zellen. Jede Zelle enthält ein Pyrenoid und mehrere Kerne.

Leptomedusae. Faltenquallen.

Stamm der Nesseltiere (Cnidaria) / Klasse der Schleierquallen (Craspedotae)
Ordnung der Faltenquallen (Leptomedusae) / Familie der Sonnenquallen (Aequoridae).

Die Sonnenquallen (Aequoridae), welche auf dieser Tafel dargestellt sind, zeichnen sich vor den übrigen Faltenquallen (Leptomedusae) durch die ansehnliche Größe und die ungewöhnliche Zahl der Strahlkanäle aus, die an der unteren Fläche des Gallertschirmes verlaufen; von unten gesehen (Fig. 1, 3) gleicht dieser einer Sonnenblume mit vielen Strahlen. Während bei den meisten übrigen Medusen die Zahl der Radialkanäle vier oder acht beträgt, steigt sie hier auf zweiunddreißig bis vierundsechzig oder selbst über hundert. Der kreisrunde Gallertschirm dieser Sonnenquallen (Umbrella) ist meistens flach scheibenförmig (Fig. 1–3), seltener glockenförmig hoch gewölbt (Fig. 4–6). Die gewölbte obere Fläche (Exumbrella) ist bisweilen von radialen Rippen durchzogen, wie ein Kristallteller (Fig. 2, 5). An der ausgehöhlten unteren Fläche (Subumbrella) liegen die Ringmuskeln, welche den Schirm zusammenziehen und durch Ausstoßen des Wassers aus der Schirmhöhle dessen Schwimmbewegung vermitteln. In der Mitte der unteren Schirmfläche liegt eine flache, kreisrunde Magenhöhle; diese öffnet sich durch einen sehr dehnbaren Mund, der von einem Kranze dünner, beweglicher Mundlappen umgeben ist (Fig. 1, 3). Bisweilen sitzen letztere am unteren Ende eines umgekehrt kegelförmigen Gallertstieles, der unten weit aus der Schirmhöhle hervorragt (Fig. 5, 6). Die Strahlkanäle, die aus dem Umkreise des Magens entspringen, steigen dann erst am Magenstiel empor, biegen oben nach außen um und laufen an der Subumbrella zum Schirmrande; hier vereinigen sie sich zu einem Ringkanal. An diesem liegt auch der Nervenring sowie ein Kranz von Gehörbläschen; nach innen davon ein horizontaler Schwimmring (Velum, Fig. 1, 3). Die zahlreichen Tentakeln oder Fangfäden, die vom Schirmrande abgehen, geraten beim Schwimmen in die anmutigsten wellenförmigen Bewegungen. Die Äquoriden sind getrennten Geschlechtes wie die meisten anderen Medusen; die Geschlechtsdrüsen sind hier kleine, wurstförmige Säckchen, die beim Weibchen Eier, beim Männchen Sperma erzeugen; sie liegen bald am Anfange der Radialkanäle (Fig. 6), bald im Verlaufe oder am Ende derselben (Fig. 1, 3, 5). Die Farbe der zarten, durchsichtigen Äquoriden ist meistens bläulich oder licht rötlich.

Fig.1 Aequorea discus
Aus dem Mittelmeer, in natürlicher Größe, von unten gesehen. Der geöffnete zentrale Mund ist von einem Kranze kurzer Lippenfransen umgeben und führt in eine flache Magenhöhle, von deren Umfang 32 Radialkanäle ausstrahlen; in der Mitte ihres Verlaufes liegen ebensoviele Geschlechtsdrüsen, am Rande kurze Fangfäden. Nach innen springt von dem einwärts gekrümmten Rande ein horizontaler Muskelring vor (Velum).

Fig.2 u. 3 Zygocanna diploconus
Aus Neuguinea, in natürlicher Größe; Fig.2 von der Seite, Fig.3 von unten gesehen. Der durchsichtige Gallertschirm bildet eine kristallartige Scheibe, deren flach gewölbte obere Fläche von 32 Furchen eingeschnitten ist. Vom Schirmrande strahlen 16 lange, sehr bewegliche Fangfäden aus. In der Mitte der hohlen unteren Fläche liegt der kreisrunde Mund, von 16 gekräuselten Mundlappen umgeben; er führt in die flache Magenhöhle, von deren Rande 16 gabelspaltige Radialkanäle ausstrahlen. An jedem Gabelaste der letzteren sitzt eine wurstförmige Geschlechtsdrüse. An dem Nervenringe des Schirmrandes, von welchem die 16 Tentakeln entspringen, sind sehr zahlreiche kleine Körnchen sichtbar (Gehörbläschen); nach innen davon ein kreisrunder, horizontaler Muskelring (Velum).

Fig.4 Polycanna germanica
Von Helgoland, in natürlicher Größe, von der Seite gesehen, in lebhafter Schwimmbewegung begriffen. Der flach gewölbte Gallertschirm ist fast halbkugelig zusammengezogen und stößt Wasser aus der unteren Schirmhöhle aus. Dadurch wird der Kranz von langen Fangfäden, der vom Schirmrande herabhängt, in eine zierliche Wellenbewegung versetzt. In der Mitte der Schirmhöhle hängt oben der Magen herab, von dessen Umkreise 50–70 Radialkanäle ausstrahlen. Die Ringmuskeln an der unteren Fläche des Schirmes sind an drei Stellen besonders stark zusammengezogen.

Fig.5 Zygocannula diploconus
Aus dem Indischen Ozean, in natürlicher Größe, von der Seite gesehen. Der hochgewölbte Gallertschirm gleicht einer Kristallglasglocke, deren Oberfläche in strahlige Rippen geteilt ist. Vom Schirmrande hängen unten sehr zahlreiche lange Fangfäden herab. In der Mitte der Schirmhöhle sitzt der umgekehrt kegelförmige Gallertstiel des Magens, der in 16 lange, gekräuselte, wellenförmig bewegte Mundlappen gespalten ist. Zahlreiche Radialkanäle gehen vom Magen aufwärts, biegen oben in der Schirmhöhle um und laufen abwärts gegen den freien Rand des Schirmes; hier sind sie gabelförmig gespalten, und jeder Gabelast trägt eine faltige Geschlechtsdrüse. Die zahlreichen Körnchen am Schirmrande sind kleine Gehörbläschen.

Fig.6 Orchistoma elegans
Halb von unten, halb von der Seite gesehen, in natürlicher Größe. Aus der unteren Fläche des halbkugeligen Schirmes hängt ein dikker, gallertiger Magenstiel herab, dessen unteres, dünnes Ende den Magen trägt; dieser ist fast bis zum Grunde in 32 lange, dünne, bandförmige Mundlappen gespalten, die sich kräuselnd bewegen. Gleich oberhalb derselben liegt ein Kranz von 32 Geschlechtsdrüsen, am Beginne der aufsteigenden Radialkanäle; oben biegen letztere nach außen um und laufen zum Schirmrande, wo sich jeder Kanal in einen langen, beweglichen Fangfaden fortsetzt. Diese elegante neue Art, in der Nähe der Azoren-Inseln gefangen, unterscheidet sich von der verwandten Orchistoma Steenstrupii der Antillen durch den schlankeren Magenstiel und die viel längeren Mundlappen und Tentakeln.

1
2
5
4
3
6

Acanthophracta. Wunderstrahlinge.

Stamm der Urtiere (Protozoa) / Hauptklasse der Wurzelfüßer (Rhizopoda)
Klasse der Strahlinge (Radiolaria) / Legion der Aktipyleen (Acantharia)
Ordnung der Wunderstrahlinge (Acanthophracta).

Die Wunderstrahlinge oder Akanthophrakten gehören zu den wunderbarsten und interessantesten Bildungen, die der einzellige Organismus der Protozoen hervorzubringen im Stande ist. Diese zierlichen Radiolarien haben sich aus der Ordnung der Stachelstrahlinge oder Akanthometren entwickelt, die auf Tafel 21* dargestellt sind; sie gehören gleich diesen zur Legion der Akantharien. Das eigentümliche Acanthin-Skelett der Akanthophrakten umhüllt den einfachen Zellenkörper in Gestalt einer Gitterschale, die sich durch ihre höchst zierliche und regelmäßige Bildung auszeichnet. Die Grundlage des Skeletts bilden zwanzig Stacheln, die vom Mittelpunkte des einzelligen Körpers ausstrahlen und nach jenem merkwürdigen Ikosakanthen-Gesetze verteilt sind, das bereits bei den Akanthometren beschrieben wurde. Während aber bei diesen letzteren die zwanzig Stacheln einfach bleiben oder ihre Querfortsätze höchstens einfache, freie Gittertafeln bilden, treten sie hier zur Bildung einer vollständigen Gitterschale zusammen. Die Scheinfüßchen oder Pseudopodien, die von der Zentralkapsel des lebenden Zellenkörpers ausstrahlen, durchsetzen die Gallerthülle (Calymma), welche sie von der Schale trennt, und treten durch deren Gitterlöcher hervor. (Auf unserer Tafel 41 sind nur die gereinigten Skelette dargestellt, nicht der Weichkörper.) Die Gitterbildung der Schale ist in zwei Hauptgruppen der Akanthophrakten dergestalt verschieden, daß in der einen Gruppe von jedem Radialstachel (an der Oberfläche des Calymma) je zwei gegenständige Querfortsätze auswachsen (Diporaspida), in der anderen Gruppe dagegen je vier kreuzständige Querfortsätze (Tessaraspida). Im ersten Falle, bei den Diporaspiden (Fig. 1 und 2) entstehen durch Verwachsung der Querfortsätze am Abgang vom Stachel zwei gegenständige Aspinalporen, im zweiten Falle dagegen, bei den Tessaraspiden (Fig. 3 und 4), vier kreuzständige Aspinalporen. Außen auf der Gitterschale bilden sich später oft zarte Beistacheln, deren Richtung derjenigen der zwanzig radialen Hauptstacheln parallel ist (Fig. 2, 3 und 4). Die ursprüngliche Kugelform der Schale (Fig. 1–3) geht später oft in die linsenförmige über (Fig. 9 und 10) und zuletzt in die Doppelkegelform (Fig. 6 und 7).

* Anmerkung:
Tafel Nr. 21 wurde in diesem Buch nicht abgebildet.

Fig. 1 Dorataspis typica
Polaransicht der kugeligen Schale. In der Mitte der Figur ist der Polarporus sichtbar, umgeben von den Schildern der vier Polarstacheln, deren jeder einen Kragen mit zwei Aspinalporen trägt. In denselben beiden, sich kreuzförmig schneidenden Meridianebenen wie die Polarstacheln liegen die vier Äquatorialstacheln, die am Rande der Figur (im Äquator der Schale) vortreten. In zwei anderen Meridianebenen, welche die ersteren unter Winkeln von 45 Grad diagonal schneiden, liegen die acht Tropenstacheln, vier nach vorn, vier nach hinten gerichtet. In jeder Naht, die zwei aneinanderstoßende Tafeln verbindet, liegt ein Koronalporus.

Fig. 2 Diporaspis nephropora
Äquatorialansicht der kugeligen Schale. In der Mitte ist einer der vier Äquatorialstacheln sichtbar, umgeben von zwei nierenförmigen Aspinalporen und sechs kleinen, runden Koronalporen. In der Horizontalebene treten rechts und links zwei Äquatorstacheln vor. Oben sieht man die vier Polarstacheln der nördlichen, unten die der südlichen Hemisphäre. Von den acht Tropenstacheln sind nur die vier vorderen, hell leuchtenden, sichtbar. Die Oberfläche der Schale ist mit gabelförmigen Beistacheln bedeckt.

Fig. 3 Lychnaspis miranda
Polaransicht der kugeligen Schale. In der Mitte ist der vierlappige Polarporus sichtbar, und in diesem das Zentrum der Kugel, in dem die 20 Radialstacheln zusammenstoßen. Von diesen sieht man 16: die vier äquatorialen und die vier vorderen polaren, zwischen diesen die acht Tropenstacheln (diagonal, vier vordere und vier hintere). Zahlreiche, zickzackförmig gebogene Beistacheln laufen parallel den 20 Hauptstacheln, auf deren Gittertafeln sie sich erheben. Diese neue Art, im September 1899 in Ajaccio auf Korsika beobachtet, unterscheidet sich von den verwandten Arten der Gattung durch die vier starken kreuzständigen Flügel, die von dem pyramidalen Außenteil jedes Hauptstachels abgehen.

4
1
2
5
6
3
7
9
8
10

Fig. 4 Lychnaspis polyancistra
Ein einzelner von den zwanzig Hauptstacheln, welche die kugelige Gitterschale zusammensetzen. Die vier kreuzständigen Querfortsätze, die von der Mitte des Stachels abgehen, umschließen durch ihre Verbindung vier runde Aspinalporen und tragen auf den Enden ihrer Äste dünne, gezähnte Beistacheln.

Fig. 5 Echinaspis echinoides
Ein einzelner von den zwanzig Hauptstacheln, welche die kugelige Gitterschale zusammensetzen. Die vier kreuzständigen Querfortsätze, die von der Mitte des Stachels abgehen, tragen auf den Enden ihrer gabelteiligen Äste dünne, gezähnte Beistacheln.

Fig. 6 Diplocolpus costatatus
Die Schale dieser Gattung weicht am meisten von der ursprünglichen Stammform ab; von den 20 Radialstacheln, welche die eigentliche Gitterschale zusammensetzen (in der Mitte der Figur), sind 18 rudimentär: nur zwei gegenständige Stacheln (senkrecht) sind sehr stark entwickelt und von einem glockenförmigen Kragen mit gezähntem Rande umgeben.

Fig. 7 Diploconus hexaphyllus
Die Schale dieser Gattung unterscheidet sich von der vorigen (Fig. 6) dadurch, daß die 18 rudimentären Radialstacheln noch äußerlich vortreten. Die beiden großen, senkrecht stehenden Stacheln sind durch sechs radiale Flügel mit dem kegelförmigen Kragen verbunden, der ihre Basis umgibt.

Fig. 8 Icosaspis elegans
Eine isolierte (Polar-)Platte, getrennt von dem Verbande der 20 Gitterplatten (ähnlich wie Fig. 3).

Fig. 9 Hexaconus serratus
Die linsenförmige Gitterschale ist aus 20 Radialstacheln zusammengesetzt, von denen 14 kleinere nur wenig über die Außenfläche vortreten, 6 größere von starken, gezähnten Scheiden umgeben sind.

Fig. 10 Hexacolpus nivalis
Die linsenförmige Gitterschale ist aus 20 Radialstacheln zusammengesetzt, von denen 14 kleinere gar nicht über die Außenfläche der Schale vortreten, sechs größere (am Linsenrande) von starken, gerippten Stachelscheiden umschlossen sind.

Polycyttaria. Vereins-Strahlinge.

Stamm der Urtiere (Protozoa) / Hauptklasse der Wurzelfüßer (Rhizopoda)
Klasse der Strahlinge (Radiolaria) / Legion der Peripyleen oder Schaumsternchen (Spumellaria)
Ordnung der Vereins-Strahlinge (Polycyttaria).

Die kleine Ordnung der »Vereins-Strahlinge« oder »Sozialen Radiolarien«, welche im System unter dem Namen Polycyttaria zusammengefaßt werden, zeichnet sich vor den anderen Urtieren dieser Klasse durch ihre Neigung zur Associon aus. Während die große Mehrzahl der Radiolarien durch isoliert lebende »Einsiedlerzellen« vertreten wird (Eremobia, Tafel 41), bilden dagegen die Polycyttarien permanente Zellvereine (Coenobia). Die zahlreichen geselligen Zellen, welche einen solchen Verein oder eine »Zellkolonie« darstellen, besitzen jede ihre besondere, von einer festen Membran umschlossene Zentralkapsel, in deren Mitte der Zellkern liegt (im Alter durch eine Fettkugel ersetzt). Dagegen ist das weiche Calymma, die Gallerthülle, in welche die Zentralkapseln eingebettet liegen, allen gemeinsam; oft ist die Gallerte von Wasserblasen oder Vakuolen erfüllt, und bisweilen zeichnet sich eine größere kugelige Zentralblase durch besondere Beschaffenheit aus (Fig. 12). Die unzähligen Plasmafäden, welche von den einzelnen Zentralkapseln ausstrahlen, verästeln sich und verbinden sich innerhalb des Calymma zu einem dichten Netzwerk; an der Oberfläche dagegen strahlen sie in Form feiner radialer Fühler aus. Zahlreiche gelbe Zellen, welche im Calymma zerstreut zwischen den Kapseln liegen, gehören nicht zur Kolonie selbst, sondern sind einzellige Pflanzen (Algarien) aus der Gattung Xanthella; sie leben mit den Radiolarien in Symbiose (Genossenschaft zum gegenseitigen Vorteil; Fig. 2, 10, 11, 12). Die Polycyttarien gehören (zufolge der Struktur ihrer Zentralkapsel) zu der Legion der Schaumsternchen (Spumellaria) und bilden innerhalb dieser Legion eine besondere Ordnung. Diese umfaßt drei Familien: I. die Collozoida, die kein Kieselskelett bilden; II. Die Sphaerozoida mit einem Kieselskelett, das aus vielen einzelnen, locker und unverbunden im Calymma zerstreuten Kieselstücken von Nadelform besteht (Fig. 2-5); III. die Collosphaerida, bei denen jede einzelne Zentralkapsel von einer gegitterten Kieselschale umschlossen ist (Fig. 6-12).

Fig. 1 Collosphaera primordialis
Ein ringförmiges Cönobium in natürlicher Größe; die feinen Punkte in der Gallertmasse sind die einzelnen Zentralkapseln.

Fig. 2 Thalassoxanthium medusinum
Ein einzelnes, einzelliges Tier, zusammengesetzt aus der blauen Zentralkapsel und deren gelblicher Gallerthülle. In der Mitte der Zentralkapsel der kugelige Zellkern, mit vielen Kernkörperchen; die glänzenden Kugeln in der Peripherie sind Fettkörner. Zahlreiche gelbe Körner im Calymma sind symbiotische Algarien: einzellige Pflanzen aus der Gattung Xanthella. Eine äußere schützende Dornenkrone wird durch vierstrahlige Kieselnadeln gebildet, deren gekrümmte Schenkel dornig sind.

Fig. 3 Sphaerozoum ovodimare
Ein kugeliges Cönobium, schwach vergrößert. Die blauen Zentralkapseln der Einzeltiere, welche an der Oberfläche des gemeinsamen Calymma verteilt liegen, sind linsenförmig und enthalten eine zentrale Fettkugel. Die Kieselkörper, die in großer Zahl im Calymma zerstreut sind, tragen an beiden Polen eines Stabes je drei Schenkel.

Fig. 4 Thalassoxanthium cervicorne
Ein einzelner, dreistrahliger Kieselkörper, dessen drei Schenkel gleiche Winkel bilden und wiederholt gabelförmig verästelt sind.

Fig. 5 Sphaerozoum spinosissimum
Ein einzelnes Kieselstück (Spiculum) mit drei divergenten Schenkeln an jedem Pole des Mittelstabes. Diese neue Art (aus dem Indischen Ozean) unterscheidet sich von den nächstverwandten Arten der Gattung (S. armatum und S. punctatum) dadurch, daß die zahlreichen, in Wirteln stehenden Seitenäste der sechs Strahlen selbst wieder verästelt und mit Dornen besetzt sind.

Fig. 6 Coronosphaera diadema
Ein einzelnes, einzelliges Tier, dessen blaue Zentralkapsel von einer gegitterten Kieselschale mit kronenförmigen Aufsätzen umgeben ist.

Fig. 7 Trypanosphaera trepanata
Eine einzelne Zelle, umgeben von der kugeligen Kieselschale, deren Öffnungen die Form von Trepankronen tragen.

Fig. 8 Aerosphaera inflata
Eine einzelne Zelle, umgeben von der kugeligen Kieselschale, aus deren Oberfläche sich mehrere pyramidenförmige Fortsätze erheben, jeder mit einem radialen Gipfelstachel.

Fig. 9 Mazosphaera lagotis
Eine einzelne Zelle, umgeben von einer kugeligen Kieselschale, aus deren Oberfläche sich viele radiale Röhren erheben, jede mit einer seitlichen Öffnung und einem gekrümmten Spitzenfortsatz, ähnlich einem Hasenohr.

Fig. 10 Caminosphaera dendrophora
Eine einzelne Zelle, umgeben von einer kugeligen Kieselschale, deren Oberfläche zahlreiche radiale Röhren mit baumförmig verzweigten Ästen trägt; die erweiterte Mündung der Röhrenäste ist trichterförmig, mit gezacktem Rande. Außerhalb gelbe Zellen.

Fig. 11 Coronosphaera calycina
Eine einzelne Zelle, deren kugelige Kieselschale eine Anzahl von großen trichterförmigen Aufsätzen trägt; die Achse derselben ist radial, die Wand gegittert, die äußere Mündung mit gezähntem Rande.

Fig. 12 Solenosphaera familiaris
Ein kugeliges Cönobium, im Sunda-Meere lebend beobachtet, schwach vergrößert. Im gelblichen Calymma sind zahlreiche einzellige Einzeltiere eingeschlossen, deren Zentralkapsel blau gefärbt ist und eine zentrale Fettkugel einschließt. Unter den zahlreichen Wasserblasen oder Vakuolen, welche die Gallerte zwischen den einzelnen Zellen erfüllen, zeichnet sich die zentrale kugelige Blase (oder »Zentral-Alveole«) durch eine verdickte Wand und einen netzförmigen Überzug von Sarkode aus. Die kleineren und jüngeren Zellen (im Inneren des Cönobiums) sind noch nackt und vermehren sich lebhaft durch Teilung. Die größeren und älteren Zellen (an der Oberfläche) haben als Schutzhülle eine poröse Kieselschale ausgeschieden, welche mehrere radiale, trichterförmig erweiterte Aufsätze trägt. Von der nächstverwandten Solenosphaera cornucopiae (Haeckel) unterscheidet sich diese neue Art durch die kleineren und regelmäßigen Poren der Gitterschale.

1
2
3
4
5
6
7
8
9
10
11
12

Stephoidea. Ringel-Strahlinge.

Stamm der Urtiere (Protozoa / Hauptklasse der Wurzelfüßer (Rhizopoda)
Klasse der Strahlinge (Radiolaria) / Legion der Korbstrahlinge oder Monopyleen (Nassellaria)
Ordnung der Ringelstrahlinge (Stephoidea).

Die formenreiche Ordnung der Ringelstrahlinge (Stephoidea) ist nächstverwandt derjenigen der Nüßchenstrahlinge (Spyroidea, Tafel 22) und der Flaschenstrahlinge. Wie bei allen Nassellarien, ist ihre Zentralkapsel durch einen besonderen Fußkegel (Podoconus) ausgezeichnet, welcher unterhalb des Zellkernes liegt (Fig. 1, 10 und 11) und ein Porenfeld besitzt zum Austritte der unzähligen feinen Plasmafäden (Scheinfüßchen oder Pseudopodien). Diese letzteren bauen ein Kieselskelett von sehr zierlicher und mannigfaltiger Form auf. Seine Grundlage bildet das Calymma, eine helle Gallerthülle, welche die Zentralkapsel umschließt und von den Scheinfüßchen durchsetzt wird. Im Calymma zerstreut liegen bei den meisten Radiolarien zahlreiche gelbe Zellen (Fig. 1 und 10); diese gehören nicht zum Organismus selbst, sondern sind einzellige Pflanzen (Algarien) aus der Gattung Xanthella; sie vermehren sich selbständig durch Teilung und bilden mit den Radiolarien eine Genossenschaft zum gegenseitigen Vorteil (Symbiose; vgl. Tafel 51, Fig. 2, 10, 11 und 12). Die Ordnung der Stephoideen enthält vier verschiedene Familien. Bei den Stephaniden oder »Kränzelradiolarien« (Fig. 1) bildet den wesentlichsten Teil des Kieselskeletts ein einfacher, vertikal stehender Ring (Sagittalring). Bei den Semantiden oder »Siegelstrahlingen« (Fig. 2) tritt dazu ein zweiter, horizontaler Ring, welcher mit der Basis des ersteren zusammenhängt (ähnlich der Platte eines Siegelringes). Die Coroniden hingegen oder die »Kronenradiolarien« (Fig. 3-5) besitzen drei Ringe, die in drei senkrecht aufeinanderstehenden Ebenen liegen. Die Tympaniden endlich oder die »Trommelstrahlinge« (Fig. 6-13) zeichnen sich durch den Besitz von zwei Horizontalringen aus, die durch einen oder mehrere Vertikalringe verbunden sind.

Fig. 1 Lithocircus magnificus
Familie der Stephaniden.
Die eiförmige Zentralkapsel (rot) enthält unten den Podoconus, oben den wurstförmigen Zellkern, seitlich ein paar Ölkugeln. Die Pseudopodien, die davon ausstrahlen, sind stark verästelt. In dem Calymma zerstreut liegen zahlreiche gelbe Zellen, die symbiotischen Xanthellen. Von dem vertikalen elliptischen Kieselring, der das Calymma umschließt, gehen viele geweihförmig verzweigte Stacheln ab.

Fig. 2 Semantis sigillum
Familie der Semantiden.
Das Skelett besteht aus zwei Kieselringen, die senkrecht aufeinanderstehen, einem vertikalen Ring (gleich Fig. 1) und einem horizontalen Ring (unten).

Fig. 3 Acanthodesmia corona
Familie der Coroniden.
Das kronenähnliche Skelett besteht aus einem horizontalen Basalring und zwei vertikalen bogenförmigen Spangen, die sich oben kreuzen.

Fig. 4 Tristephanium dimensivum
Familie der Coroniden.
Das Skelett besteht aus drei Kieselringen, die in drei aufeinander senkrechten Ebenen liegen. Der horizontale Ring liegt tiefer als die Mitte der beiden vertikalen Ringe; daher sind die oberen vier Tore größer als die vier unteren.

Fig. 5 Trissocyclus sphaeridium
Familie der Coroniden.
Die innere Zentralkapsel (rot) ist kugelförmig und von einer konzentrischen, ebenfalls kugeligen Gallerthülle (Calymma) umgeben. Auf deren Oberfläche sind drei kreisrunde Kieselringe abgelagert, die in drei aufeinander senkrechten Ebenen liegen. Daher sind die acht Tore, die zwischen den Ringabschnitten offen bleiben, gleich groß und gleichseitig dreieckig.

Fig. 6-13 Familie der Tympaniden.
Die Familie der Trommelstrahlinge (Tympanida) unterscheidet sich von den drei anderen Familien der Stephoideen dadurch, daß zwei parallele horizontale Kieselringe das Skelett bilden (wie bei einer Trommel); beide sind durch einen oder mehrere vertikale Ringe verbunden. Der obere (Mitralring) ist meistens kleiner als der untere (Basalring).

Fig. 6 Octotympanum cervicorne
Die beiden Horizontalringe sind von gleicher Größe und so gegeneinander gebogen, daß sie sich an den beiden Polen der horizontalen Transversalachse berühren und verwachsen. Daher bleiben zwischen ihnen und dem kleineren Vertikalring acht große Tore offen.

Fig. 7 Microcubus zonarius
Zwischen dem oberen (mitralen) und unteren (basalen) Kieselring findet sich hier noch ein dritter (äquatorialer) Horizontalring. Alle drei werden von dem vertikalen (Sagittal-) Ring in der Mitte halbiert.

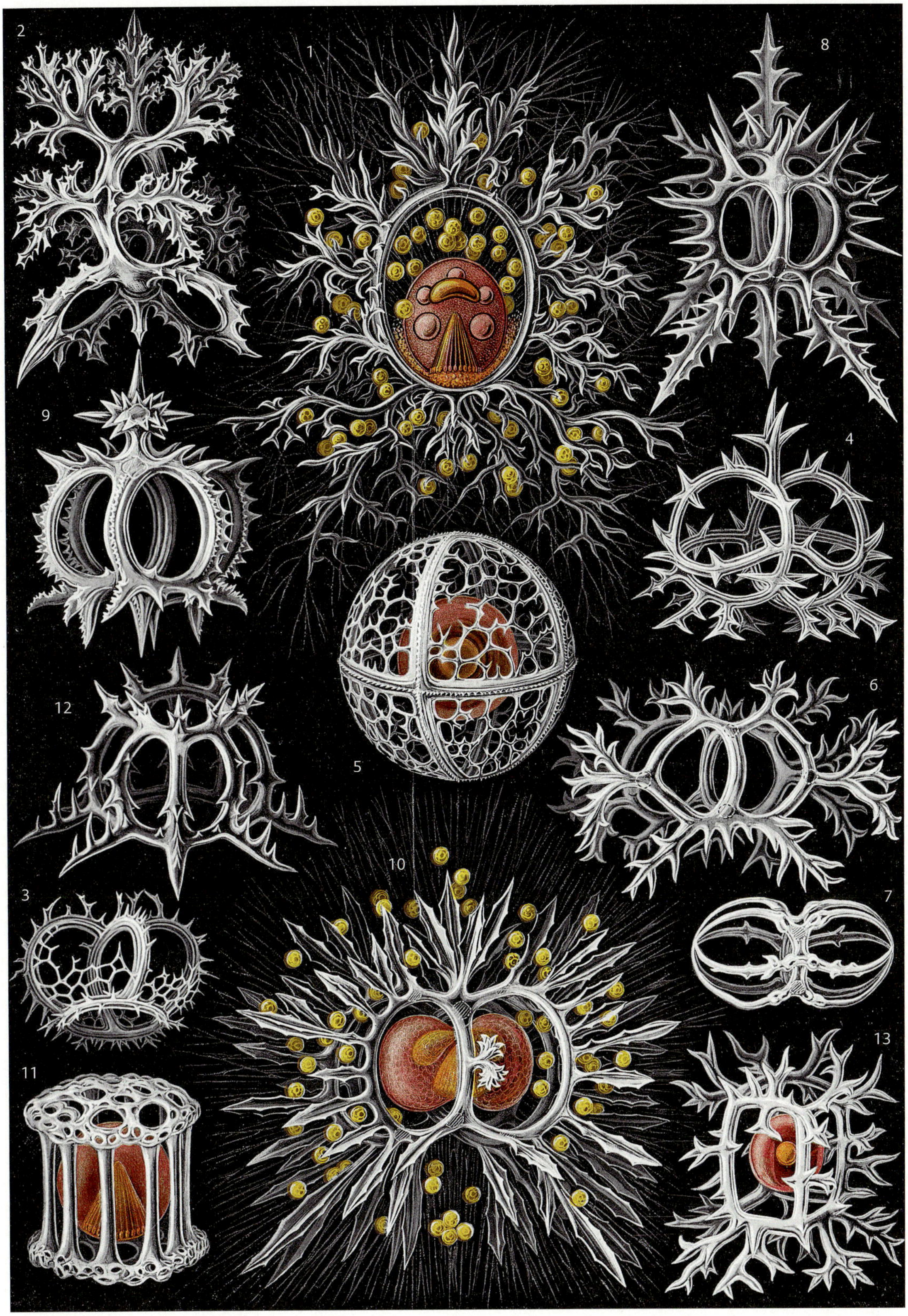
2
1
8
9
4
12
5
6
10
3
7
13
11

Fig. 8 Tympaniscus tripodiscus
Die beiden Horizontalringe sind durch sechs vertikale Säulen (die Hälften von drei Meridianringen) verbunden. Der Scheitel trägt oben einen Gipfelstachel. Unten stehen drei divergente stachelige Füße.

Fig. 9 Tympaniscus quadrupes
Die Kieselschale dieser Art ist einem Diadem ähnlich; der Scheitelstachel oben ist sehr stark und dornig. An der Basis vier kurze senkrechte Füße.

Fig. 10 Tympanidium foliosum
Das Kieselskelett ist aus zwei horizontalen und vier vertikalen Ringen zusammengesetzt, die sich in der senkrechten Mittelebene schneiden. Die beiden Horizontalringe werden durch den Medianring halbiert. Zwischen den blattartig verbreiterten Stacheln zahlreiche gelbe Zellen (Xanthellen) und feine Pseudopodien, die von der (roten) Zentralkapsel ausstrahlen.

Fig. 11 Lithotympanum tuberosum
Die Kieselschale gleicht einer Trommel, deren beide Trommelfelle durch eine gewölbte Gitterplatte ersetzt sind. Die beiden horizontalen Ringe sind durch 10 bis 12 senkrechte parallele Stäbe verbunden.

Fig. 12 Circotympanum octogonium
Das kronenähnliche Kieselskelett besteht aus zwei parallelen Horizontalringen, die durch acht Stäbe verbunden sind. Da der untere (basale) Ring doppelt so groß ist als der obere (mitrale), steigen die Stäbe schräg nach oben und innen auf. Von beiden Enden jedes Stabes springt ein gezähnter Dorn vor.

Fig. 13 Lithocubus astragalus
Die beiden horizontalen Kieselringe sind viereckig und von gleicher Größe, an den vier Ecken durch vier parallele, senkrecht aufsteigende Stäbe verbunden.
So entsteht ein Gerüst mit sechs viereckigen Toren, von der mathematischen Grundform des Würfels.

Spumellaria. Schaumstrahlinge.

Stamm der Urtiere (Protozoa); — Hauptklasse der Wurzelfüßer (Rhizopoda);
— Klasse der Strahlinge (Radiolaria); — Legion der Peripyleen oder Schaumstrahlinge (Spumellaria).

Die formenreiche Legion der Schaumstrahlinge (Spumellaria), von denen über 2000 Arten bekannt sind, zeichnet sich durch die große Regelmäßigkeit und Mannigfaltigkeit im Aufbau ihrer zierlichen Kieselschalen aus. Diese bilden meistens ein äußerst feines Gitterwerk, das von Tausenden feiner Plasmafäden ausgeschieden wird, Ausstrahlungen des einzelligen Weichkörpers. Während einfache oder mehrfache Gitterschalen den letzteren schützend umgeben, dienen ihm stachelige oder flügelförmige Fortsätze als Schwebeapparate, die das Untersinken des kleinen, im Meere schwimmenden Körpers verhindern. Als vier Ordnungen dieser großen Legion werden die kugeligen (Sphaeroïdea), pflaumenförmigen (Prunoïdea), scheibenförmigen (Discoïdea) und lentelliptischen (Larcoïdea) unterschieden. | Die Zentralkapsel, der innere Teil des einzelligen Körpers (der den Zellkern einschließt), und ebenso die äußere Gallerthülle (Calymma), welche sie schützend umgibt, sind ursprünglich kugelig, bei den Sphaeroïdea, Fig. 1, 2. Indem sich eine Achse der Zentralkapsel verlängert, wird sie ellipsoïd oder pflaumenförmig, bei den Prunoïdea, Fig. 3-10. Wenn sich dagegen eine Achse verkürzt, wird sie linsenförmig oder scheibenförmig, bei den Discoïdea, Fig. 11-14. Endlich wird die Zentralkapsel und das Calymma, entsprechend auch die auf letzterem abgelagerte Gitterschale, ein abgeplattetes Ellipsoid (lentelliptisch) bei den Larcoïdea, Fig. 15; hier sind die drei aufeinander senkrechten Achsen des Körpers von ungleicher Länge. || Alle Figuren dieser Tafel sind stark vergrößert; es sind nur die Kieselskelette dargestellt. || Über den eingeschlossenen Weichkörper und die von ihm ausstrahlenden Pseudopodien sind Tafel 11* und 51 zu vergleichen.

* Anmerkung:
Tafel Nr. 11 wurde in diesem Buch nicht abgebildet.

Fig. 1 Astrosphaera stellata *(Haeckel).*
Ordnung: Sphaeroïdea; Familie: Astrosphaerida.
Die kugelige Gitterschale, von 0,2 mm Durchmesser, besteht aus regelmäßig sechseckigen Maschen, von deren Knotenpunkten feine Radialnadeln ausstrahlen. Die Gitterbalken werden gekreuzt durch Tangentialstäbchen. Die zwanzig starken Radialstacheln zeigen drei gezähnte Kanten.

Fig. 2 Hexancistra quadricuspis *(Haeckel).*
Ordnung: Sphaeroïdea; Familie: Cubosphaerida.
Die Gitterschale, von 0,17 mm Durchmesser, ist aus zwei konzentrischen Kugeln zusammengesetzt, die durch sechs starke, gleichgroße Radialstacheln zusammenhängen. Diese liegen paarweise in drei aufeinander senkrechten Durchmessern. Ihre drei Kanten sind etwas spiralig gedreht und laufen in drei gekrümmte Seitenäste aus.

Fig. 3 Cannartidium mammiferum
Ordnung: Prunoïdea; Familie: Cyphinida.
Äußere Gitterschale zweikammerig, 0,13 mm lang, durch eine ringförmige Einschnürung in zwei gleiche, fast kugelige Kammern geschieden, die warzenförmige Höcker tragen. An beiden Polen der Längsachse eine gegitterte Röhre.

Fig. 4 Cannartidium mastophorum
Ordnung: Prunoïdea; Familie: Cyphinida.
Äußere Gitterschale ähnlich der vorigen (Fig. 3), 0,14 mm lang, von ihr verschieden durch die scharfen Spitzen der kegelförmigen Polarröhren und Warzen. In der Mitte zwei kleine konzentrische Markschalen.

Fig. 5 Cannartiscus amphiconiscus
Ordnung: Prunoïdea; Familie: Cyphinida.
Äußere Gitterschale, 0,14 mm lang, ähnlich den beiden vorigen (Fig. 3, 4); aber die innere Markschale einfach.

Fig. 6 Cyphinus amphilophus
Ordnung: Prunoïdea; Familie: Cyphinida.
Äußere Gitterschale, 0,2 mm lang, ähnlich den drei vorigen; aber statt der hohlen gegitterten Röhren an beiden Polen der Längsachse ein Schopf von starken kegelförmigen Stacheln. Markschale einfach.

Fig. 7 Panartus diploconus
Ordnung: Prunoïdea; Familie: Panartida.
Die äußere Gitterschale, 0,34 mm lang, besteht aus zwei Paar Kammern, die in der verlängerten Achse einander gegenüberliegen und durch drei ringförmige Einschnürungen getrennt werden; die beiden inneren Kammern sind sphäroïdal, die beiden äußeren kegelförmig. In der Mitte liegen zwei kleine konzentrische Markschalen.

Fig. 8 Peripanartus amphiconus
Ordnung: Prunoïdea; Familie: Panartida.
Gitterschale vierkammerig, 0,3 mm lang, ähnlich der vorigen (Fig. 7); aber von einer zweiten, äußeren Rindenschale von gleicher Form umgeben.

Fig. 9 Panicium coronatum
Ordnung: Prunoïdea; Familie: Panartida.
Gitterschale vierkammerig, 0,22 mm lang, ähnlich Fig. 7; aber die beiden polaren Kammern nicht kegelförmig, sondern flach kappenförmig, mit einem starken Stachelkranz.

11
2
1
3
5
7
9
12
6
4
8
10
14
15
13

Fig.10 Peripanicium amphicorona
Ordnung: Prunoïdea; Familie: Panartida.
Gitterschale vierkammerig, 0,26 mm lang, ähnlich der vorigen (Fig.9); aber von einer zweiten, äußeren Rindenschale umgeben.

Fig.11 Trochodiscus stellaris
Ordnung: Discoïdea; Familie: Cenodiscida.
Gitterschale linsenförmig, einfach, von 0,25 mm Durchmesser; in der Äquatorebene mit einem breiten Strahlengürtel, der in zwölf dreieckige Zacken ausläuft.

Fig.12 Dicranastrum bifurcatum
Ordnung: Discoïdea; Familie: Porodiscida.
Gitterschale kreuzförmig, mit vier schlanken, doppelt gabelspaltigen Armen, die in einer Ebene aufeinander senkrecht liegen.

Fig.13 Archidiscus pyloniscus
Ordnung: Discoïdea; Familie: Porodiscida.
Gitterschale scheibenförmig, gleichseitig dreieckig, von 0,05 mm Durchmesser. Die zentrale linsenförmige Markschale ist von einem Gitterring umgeben, der in sechs Kammern geteilt ist; drei größere abwechselnd mit drei kleineren.

Fig.14 Pylodiseus triangularis
Ordnung: Discoïdea; Familie: Pylodiscida.
Gitterschale scheibenförmig, gleichseitig dreieckig, von 0,15 mm Durchmesser. Um eine innere Markschale, die der vorigen (Fig.13) gleicht, ist ein äußerer Gürtel angewachsen, der wieder in sechs Kammern geteilt ist.

Fig.15 Tholoma metallasson
Ordnung: Larcoïdea; Familie: Tholonida.
Gitterschale kreuzförmig, lentelliptisch, neunkammerig, von 0,26 mm Durchmesser. Um eine einfache elliptische Zentralkammer sind vier innere Kammern angewachsen, die kreuzförmig in einer Ebene liegen, und über diesen vier äußere Kammern von gleicher Form, je zwei gegenständige größer als die zwei anderen Kammern.

24

Strena.

Vom sechseckigen Schnee

Strena.

About tne hexagonal snow

Johannes Kepler

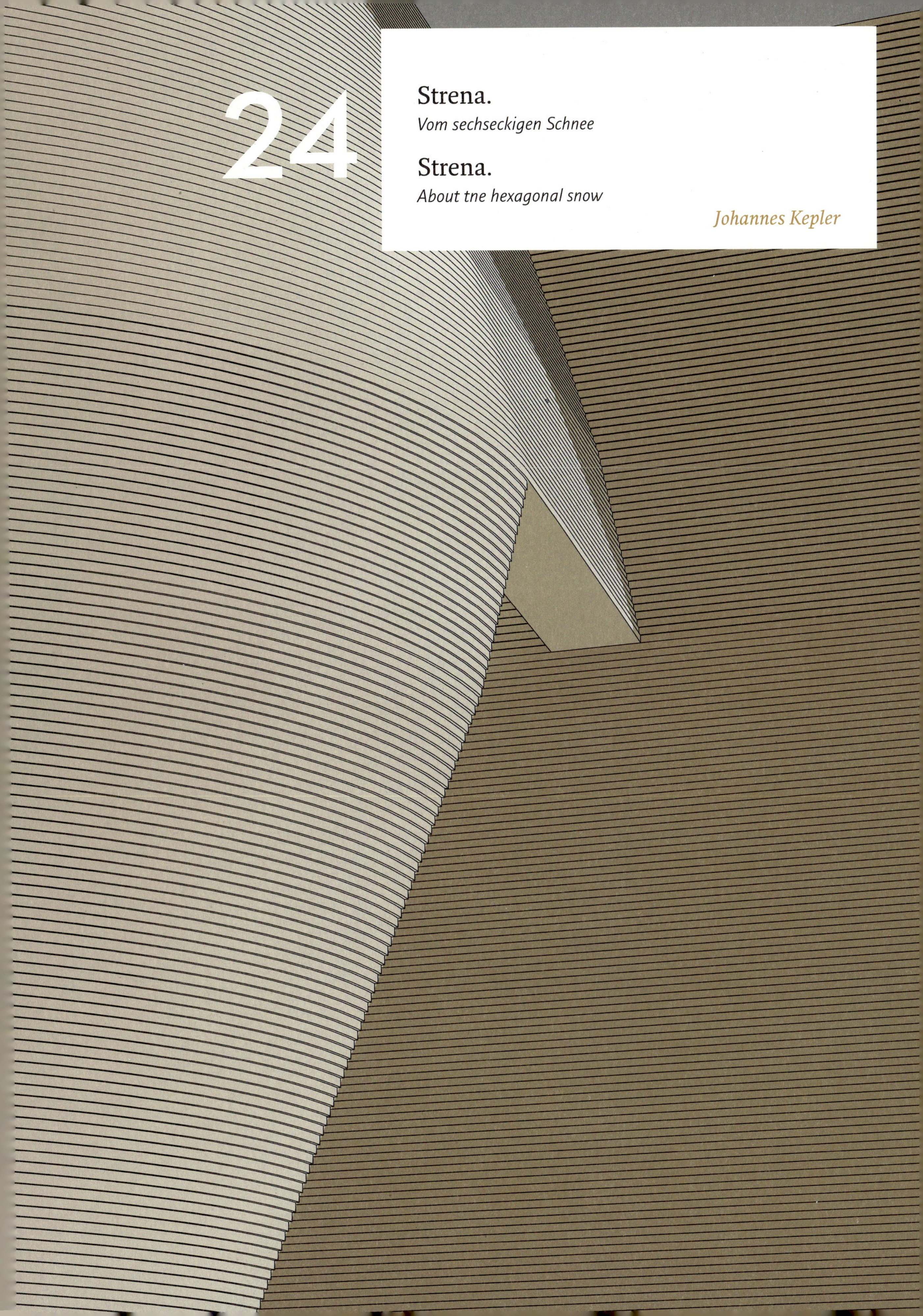

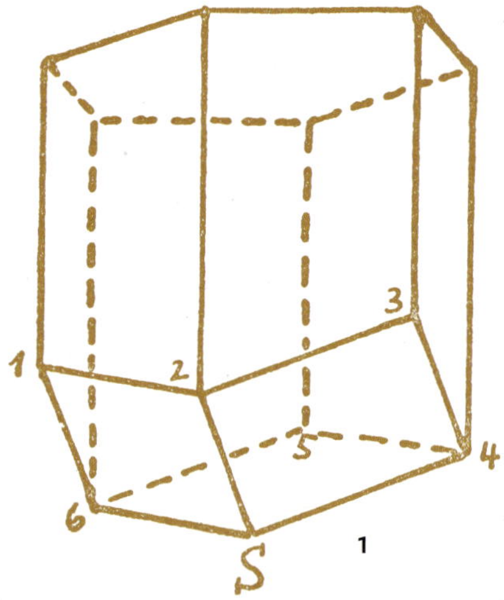

1

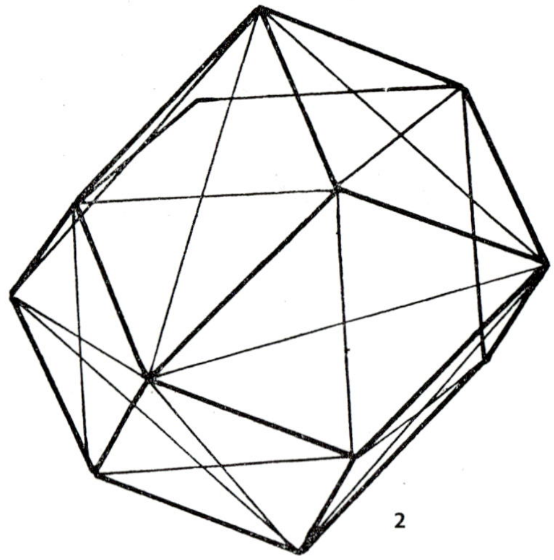

2

[...] Wenn du die Mathematiker fragst, in welcher Weise die Bienenzellen gebaut sind, sagen sie: sechseckig. Diese glatte Antwort ergibt sich schon bei bloßer Betrachtung der Öffnungen oder Eingänge und der Seitenflächen, durch die die Zellen gebildet werden. Jede einzelne Zelle ist von sechs anderen umgeben, die alle mit je einer Seitenfläche mit der mittleren zusammenhängen. Betrachtest du aber den Boden der Zellen, so siehst du ihn von drei in stumpfen Winkeln zusammenstoßenden Ebenen gebildet. Dieser Boden (den man besser als Zeltdach bezeichnet) stößt mit den sechs Seitenflächen der Zelle in je sechs Raumecken zusammen: Die drei oberen sind dreikantig und ganz ähnlich sogar der Zeltspitze, die drei unteren vierkantig und zwischen jenen angeordnet.[1] Außerdem ist zu beobachten, dass die Zellen in doppelter Lage angeordnet sind, wobei die Öffnungen jeweils nach entgegengesetzten Seiten zeigen, während sie hinten dicht und lückenlos aneinander grenzen. Dabei liegt jede Zeltspitze der einen Lage zwischen drei Dachflächen dreier Zeltdächer der anderen Lage so eingebettet, dass jede Zelle sowohl auf sechs Seiten mit sechs Zellen derselben Lage verbunden ist, wie auch durch die drei Zeltdachebenen mit drei Zellen der gegenüberliegenden Lage. So kommt es, dass jede Biene neun Nachbarinnen hat und von jeder nur durch eine einzige gemeinsame Wand geschieden ist. Die drei ebenen Dachflächen sind einander ähnlich und haben eine Form, die die Mathematiker Rhombus nennen. Durch diese Rhomben angeregt, habe ich geometrisch untersucht, ob ein Körper,

Die Bienenzellen

1 Um diese genaue Beschreibung der Bienenzellen anschaulich zu machen, soll die perspektivische Ansicht einer Bienenzelle hinzugefügt werden (Tafel). Die Zeltspitze ist bezeichnet mit S, die sechs Raumecken, in denen das Zeltdach mit den sechs Seitenflächen zusammenstößt, mit 1, 2, 3, 4, 5 und 6, wobei die drei unteren vierkantigen Ecken gerade, die oberen dreikantigen Ecken ungerade Zahlen haben. Aus der Zeichnung geht auch klar hervor, dass diese der Zeltspitze ähnlich sind, denn hier stoßen die Rhomben mit ihren stumpfen Winkeln zusammen, in den vierkantigen Ecken aber mit den spitzen.

Abb. 1 Bienenzelle (Anm. 1)
Fig. 1 Honeycomb cell (remark 1)

Abb. 2 Rhombenzwölfflächner als Hüllkörper von Würfel und Oktaeder

Fig. 2 Twelve-sided rhombus as enveloping body of a cube and octahedron

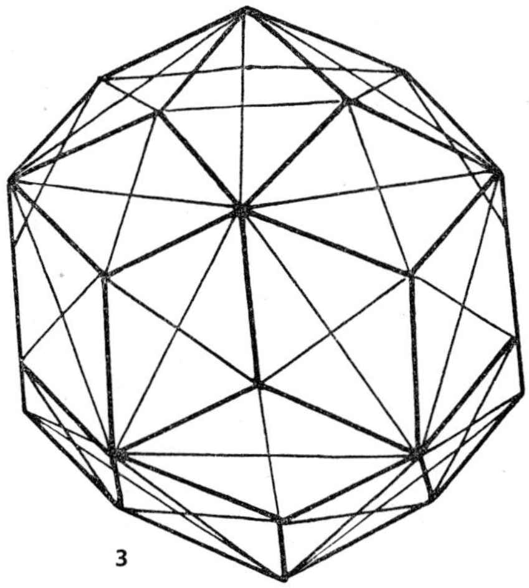

3

[...] If you ask mathematicians about how honeycomb cells are structured, they say they are hexagonal. This ready answer comes from mere observation of the openings or entrances and the sides, which make up the cells. Each individual cell is surrounded by six others, each of which is connected to the middle one on one side. However, if you take a look at the base of the cells you will see that it is made up of three levels that adjoin at obtuse angles. This base (which is best referred to as the tent roof) collides with the six sides of the cell in six corners: the three upper ones are triangular and very similar to the tip of the tent, while the three lower ones are tetragonal and configured between them.[1] Furthermore it can be seen that the cells are configured in a double layer, with the openings in each case pointing to opposite sides, whereas at the rear they border closely on each other without any gap. Each tent tip of the one layer lies embedded between three roof surfaces of three tent tips of the other layer in such a way that each cell is connected on six sides to six cells of the same layer, as well as the three tent roof levels with three cells of the opposite layer. As such each bee has nine neighbors and is only separated from each by a single joint wall. The three flat roof surfaces are similar to each other and are what mathematicians call a rhombus in shape. Inspired by these rhombuses I geometrically investigated whether, similar to the five regular and fourteen *Archimedean*, a body can be made up of rhombuses alone. And I found two such bodies, one of which is related to the cube and octahedron,

The honeycomb cells

1 In order to make this precise description of the honeycomb cell graphic the perspective view of a honeycomb cell should be added (panel). The tip of the tent is indicated by S, the six corners, in which the tent roof of the six sides meets by 1, 2, 3, 4, 5 and 6, where the four lower tetragonal corners have an even and the upper triangular corners uneven numbers. It is also clear from the diagram that these are similar to the tip of the tent, as this is where the rhombuses with their obtuse, in the tetragonal corners with their acute angles come together.

Abb. 3 Rhombendreißigflächner als Hüllenkörper von Dodekaeder und Ikosaeder

Fig. 3 Thirty-sided rhombus as enveloping body of dodecahedron and icosahedron

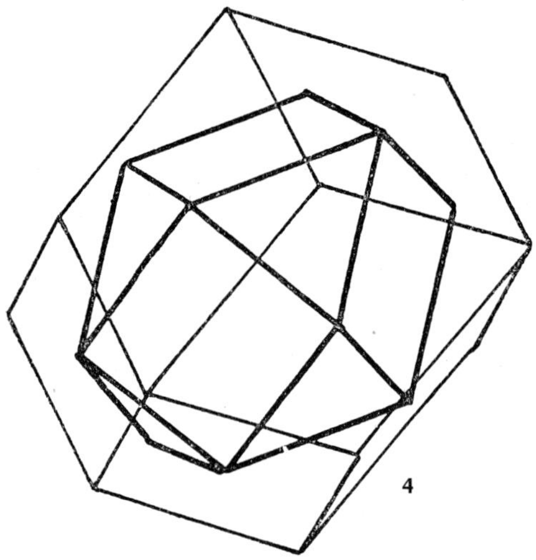

4

ähnlich den fünf regelmäßigen und vierzehn *Archimedischen*, nur aus Rhomben aufgebaut werden kann. Und ich fand zwei solche Körper, von denen der eine dem Würfel und Oktaeder, der andere dem Dodekaeder und Ikosaeder verwandt ist (der Würfel selbst ist ein dritter solcher Körper und ist verwandt mit einem Körper, der aus zwei entgegengesetzt liegenden Tetraedern zusammengesetzt ist). Der erste wird von zwölf, der zweite von dreißig Rhomben umschlossen. Der erste aber teilt folgende Eigenschaft mit dem Würfel. Wie die acht Ecken von acht Würfeln, die an einem Punkte zusammenstoßen, den Raum vollständig und lückenlos ausfüllen, so leistet der erste Rhombenkörper dasselbe mit vier stumpfwinkligen dreikantigen Ecken und entsprechend mit sechs vierkantigen Ecken. Man kann daher den Raum aus lauter solchen rhombischen Körpern aufbauen, so dass immer vier dreikantige oder auch sechs vierkantige Ecken an einem einzigen Punkte zusammenstoßen. Und um das Ergebnis zu ziehen: wenn der ganze Raum durch gleiche Würfel in regelmäßiger Anordnung ausgefüllt wird, dann berühren einen Würfel die anderen mit zweiunddreißig einzelnen Ecken und außerdem sechs mit je vier Ecken. Die Zahl der Berührungen ist also achtunddreißig. Wenn aber der Raum ebenso durch gleiche rhombische Körper ausgefüllt wird, dann berühren einen solchen sechs andere mit einzelnen vierkantigen Ecken, und außerdem zwölf mit je vier Ecken. Die Zahl dieser Berührungen ist daher achtzehn. | Dies ist also der geometrische Körper, der nächst dem regelmäßigen den körperlichen Raum völlig

Die regelmäßigen rhombischen Körper

Abb. 4 Kuboktaeder und polar-inverser Rhombenzwölfflächner

Fig. 4 Cuboctahedron and inverse polar twelve-sided rhombus

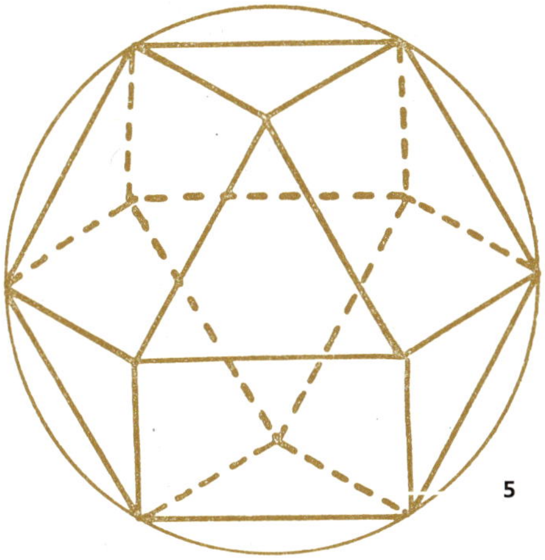
5

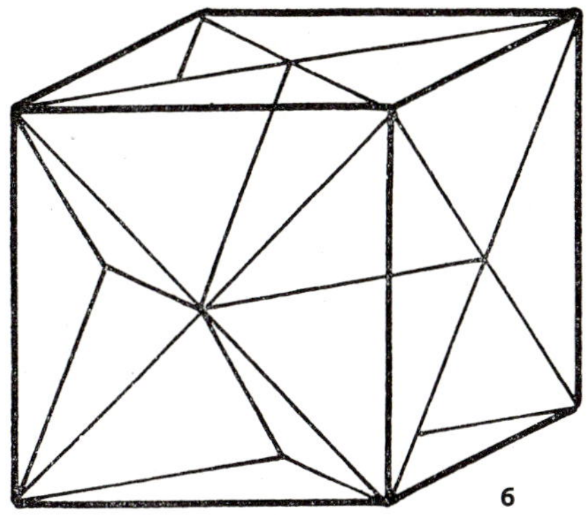
6

The regular rhombic bodies

and the other to the dodecahedron and icosahedron (the cube itself is a third such body and is related to a body composed of two opposing tetrahedrons). The first is surrounded by twelve rhombuses, the second by thirty. The first, however, shares the following characteristic with the cube. As the eight corners of eight cubes that meet at one point fill the space completely and without any gap, so the first rhombus body achieves the same with four obtuse-angle triangular corners and correspondingly with six tetragonal corners. As such one can structure the space purely with rhombic bodies such as these, such that at all times four triangular or six tetragonal corners meet at a single point. And taking the result further: if the entire space is filled with identical cubes in regular configuration, then the other cubes touch another at thirty-two individual corners, and furthermore six at four corners each. The number of points of contact is thus thirty-eight. However, if the space is likewise filled with identical rhombic bodies, one of these touches six others with individual tetragonal corners, and twelve of them with four corners each. The number of points of contact is thus eighteen. | So this is the geometric body that alongside the regular totally fills the bodily space; just as the regular hexagon, the square and the equilateral triangle entirely cover the level. As stated, this is the shape the bees use when building their cells, if one disregards the fact, that is, that the cells have no cover similar to a roof. | If they were to add this and each bee were then to be positioned between twelve or eighteen others, it would then no

Abb. 5 Kuboktaeder aus Richtung einer sechszähligen Symmetrieachse

Fig. 5 Cuboctahedron from the direction of a six-symmetry axis

Abb. 6 Rhombensechsflächner als Hüllkörper zweier entgegengesetzt liegender Tetraeder

Fig. 6 Six-sided rhombus as enveloping body of two opposing tetrahedrons

erfüllt; ähnlich wie das regelmäßige Sechseck, das Quadrat und das gleichseitige Dreieck die Ebene lückenlos bedecken. Das ist auch, wie gesagt, die Figur, die die Bienen beim Bau ihrer Zellen anwenden, wenn man davon absieht, dass die Zellen keine dachähnlichen Deckel haben. | Wenn sie auch diesen anfügten und so jede Biene zwischen zwölf oder achtzehn anderen liegen würde, so stünde ihr kein Ausweg mehr offen und sie wäre völlig eingeschlossen. Daher können sie Zellen mit Deckel nicht gebrauchen, aber nichts hindert sie, an die rhombischen Flächen des Zeltdaches sechs Wände, entsprechend dem Maß ihres Körperchens, anzubauen und diese in entgegengesetzter Richtung jenen unähnlich auszubilden. | Ferner wenn jemand einen ziemlich großen Granatapfel durchschneidet, sieht er, dass die meisten Kerne dieselbe Figur angenommen haben, wenn dies nicht durch die Reihe Fasern, die ihnen die Nahrungssäfte zuführen, gestört wird. | Nun fragt man sich bei diesen beiden Beispielen, wer der Urheber für die rhombische Figur der Bienenzellen und der Granatapfelkerne ist. In der Materie liegt die Ursache nicht. Denn nirgends finden die Bienen solche rhombischen Blättchen fertig vor, die sie sammeln und für den Bau ihrer Behausungen verwenden könnten. Es ist auch nicht wahrscheinlich, dass allein bei den Granatäpfeln die Kerne von sich aus eckige Bildung bekommen, während sie in allen anderen Früchten rund werden, sofern sie nicht behindert werden, wenn die zugeführte Feuchtigkeit die zähen Schalen durchdringt und aufweicht, dass sie schwellen und gegebenenfalls keimen. | Bei den Granatapfelkernen liegt daher die Ursache für die Figur in einer Seele der Pflanze, die das Wachstum der Frucht bewirkt. Aber diese ist als Ursache für die Figur nicht ausreichend, denn sie leistet dies bei der Frucht nicht auf Grund einer formalen Beschaffenheit, sie wird vielmehr unterstützt durch materielle Notwendigkeit. Denn die Kerne sind anfangs, solange sie noch klein sind und ihnen der Raum innerhalb der Schale genügt, rund und werden, während die Schale erhärtet und sie beständig weiterwachsen, zusammengedrängt und gepresst, wie es auch bei den Erbsen in ihren länglichen Schoten geschieht. Aber die Erbsen haben keinen Raum, wohin sie ausweichen könnten, sind sie doch in ihren Schoten in einer langen Reihe festgewachsen; sie werden daher nur von zwei Seiten gepresst. Die runden Kerne im Granatapfel dagegen finden zunächst einen freieren Raum vor, sie zwängen sich daher, von ihrer runden Gestalt begünstigt, leicht einzeln zwischen je drei Kerne der gegenüberliegenden Schicht, wobei die Flüssigkeit von der Druckstelle weg in den freieren Raum gedrängt wird. Wenn man eine Menge runder gleich großer Kügelchen aus weichem Material in ein rundes Gefäß tut und dieses durch Metallreifen von allen Richtungen gleichmäßig zusammendrückt, dann werden die meisten Kügelchen nach Art rhombischer Körper umgepresst, besonders wenn man sie vorher durch geeignetes Schütteln des Gefäßes in freiem Umschwung auf einen möglichst engen Raum gebracht hat.

Welches ist die Figur der Granatapfelkerne?

Welche Figur haben die Erbsen?

longer have any way out and would be totally enclosed. For this reason they have no use for cells with covers, though nothing prevents them from adding six walls, corresponding to the size of their body, to the rhombic surfaces of the tent roof and arranging them in opposite directions in a dissimilar way. | Furthermore, if you dissect a fairly large pomegranate you will see that most of the pips have assumed the same shape, as long as this is not disturbed by the fibers that transport the nutritious juices to them. | With both these examples the question arises as to the creator of the rhombic shape of the honeycomb cells and pomegranate pips. The cause is not to be found in the material itself. As nowhere do bees come across rhombic leaves they can collect and use to construct their abode. Nor is it probable that in pomegranates alone the pips take on a cornered shape, while in all other fruits they are round, provided they are not hindered by the moisture transported to them penetrating and softening the tough shells such that these swell and possibly germinate. | For this reason in the case of pomegranate pips the reason for the shape is to be found in a soul of the plant that causes the fruit to grow. This, however, is not sufficient with regard to the cause of the shape, as with the fruit it assumes this not on account of a formal property, far more it is supported by material necessity. In the beginning the pips, as long as they are still small and have sufficient space in the shell, are round and, as the shell hardens and they continue growing, are pushed together and pressed, as is the case with peas in their long pods. The peas, however, have no room to which they can move, they are fixed in a long row in their pods; as such they are only pressed from two sides. The round pips in a pomegranate, on the other hand, initially encounter more free space, and for this reason, favored by the round shape, they can easily squeeze themselves individually between three pips of the opposite layer, with the liquid being forced from the pressure point into the more open space. If one puts a number of identical-size round pellets made of a soft material in a round vessel, and squeezes it together evenly from all directions with metal strips, most of the pellets will be squeezed into rhombic bodies, particularly if by appropriately shaking the vessel beforehand they were put in as tight a space as possible. Distributing the pellets in a straight line that cannot be disturbed and applying pressure also produces cubes. | In general, identical pellets in a vessel configure themselves in two ways, corresponding to two possible configurations on a level plane. | If one pushes identical-size pellets together on a horizontal plane such that they are touching, they will be arranged either in a triangular or rectangular shape; in some places six pellets surround one in the middle, in other places four, and in both cases the manner in which they touch is the same for all pellets, except those on the outside. A pentagon shape cannot provide even cover, and the hexagon can be traced back to the triangle. As previously stated, therefore, there are only two configura-

What is the shape figure of the pomegranate pips?

What shape are the peas?

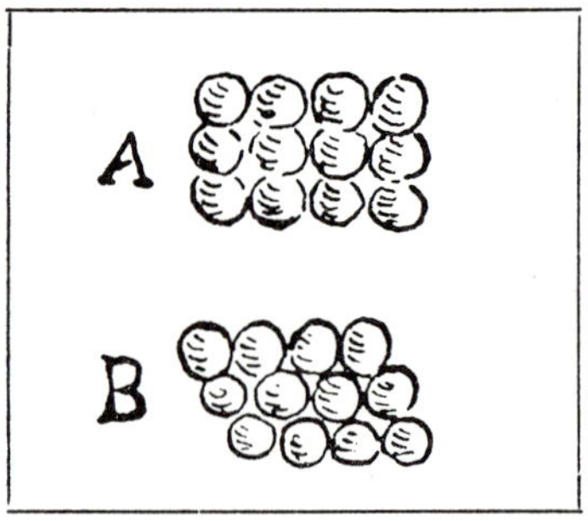

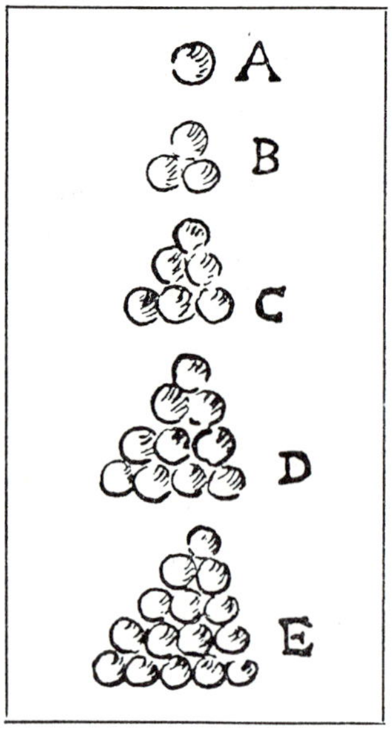

Bei einer Verteilung der Kugeln in geradliniger Ordnung, die nicht durcheinander gebracht werden kann, bringt man durch Druck ebenso Würfel hervor. | Allgemein ordnen sich gleiche Kugeln in einem Gefäß auf zweierlei Art an, entsprechend zwei möglichen Anordnungen in einer Ebene. | Denn wenn man in einer waagerechten Ebene gleich große Kugeln zusammenschiebt, so dass sie sich gegenseitig berühren, legen sie sich entweder in Dreiecks- oder in Vierecksform zusammen; dort umgeben sechs, hier vier Kugeln eine mittlere, und je in beiden Fällen ist die Art der Berührung bei allen Kugeln dieselbe, ausgenommen bei den äußeren. Die Fünfecksform kann eine gleichmäßige Bedeckung nicht ergeben und die Sechsecksform lässt sich auf die Dreiecksform zurückführen. Es gibt also, wie gesagt, nur zwei Anordnungen. | Wenn du nun an den Bau möglichst fest gefügter Körper gehst, indem du Schicht auf Schicht von ebenen Kugelanordnungen aufeinander setzt, so werden diese quadratisch (A) oder dreieckig (B). Bei der Quadratform stehen entweder die einzelnen Kugeln der oberen Schicht auf den einzelnen der unteren Schicht oder aber jene sitzen immer zwischen je vier Kugeln der unteren Schicht. Im ersten Falle wird jede Kugel von vier benachbarten in derselben Ebene berührt und von je einer darüber und darunter, im Ganzen also von sechs. Dies ist eine kubische Anordnung und beim Zusammendrucken ergeben sich Würfel; es ist aber nicht die dichteste Packung. | Bei der zweiten Anordnung wird jede Kugel außer von den vier benachbarten in derselben Ebene auch noch

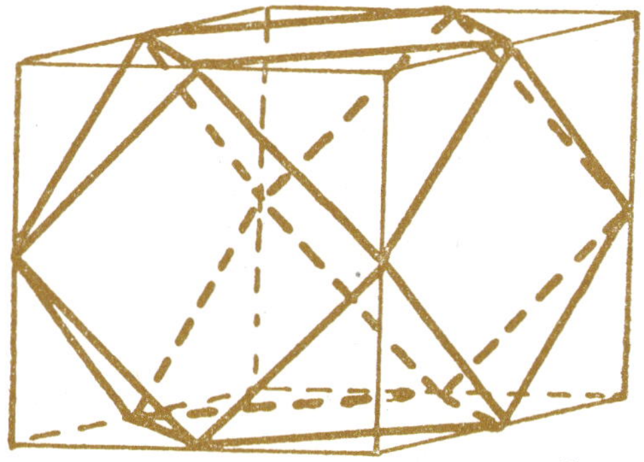

7

tions. | If you now set about building as firmly joined a body as possible by placing layer upon layer of level spherical configurations one on top of the other, these turn out square (A) or triangular (B). With regard to the square shape either the individual spheres of the upper lower level are on top of the individual ones of the lower level, or these are always each positioned between four pellets of the lower layer. In the first case each pellet is touched by four neighboring ones on the same plane and by one above and below, so in total by six. This is a cubic configuration, and when pressed together cubes emerge; it is not, however the most concentrated packaging. With the second configuration each sphere, apart from the four neighboring ones on the same plane, is also touched by four on top and four below, so in total by twelve sphere; and when pressed together rhombic bodies emerge from the spheres. This configuration follows more from the octahedron and the pyramid. The packaging becomes as concentrated as possible; in no other configuration can more spheres be accommodated in the same vessel. On the other hand, if the spheres are configured in a triangle on the plane, then in the spatial configuration either the individual spheres of the upper layer are positioned on the individual spheres of the lower layer, and once again in the far packaging, or the individual spheres are located between three of the upper and lower layer. In the first case each sphere is touched by six neighboring ones on the same plane, as well as by one above and one below, so in total by eight others. This configuration follows on

Abb. 7 Kuboktaeder durch Entecken eines Würfels

Fig. 7 Cuboctahedron through removal of the corners of a cube

von vier darüber und von vier darunter, im Ganzen also von zwölf Kugeln berührt; und beim Zusammenpressen entstehen aus den Kugeln rhombische Körper. Diese Anordnung schließt sich mehr an das Oktaeder und die Pyramide an. Die Packung wird so dicht wie möglich; in keiner anderen Anordnung können mehr Kugeln im selben Gefäß untergebracht werden. | Wiederum, wenn die Kugeln in der Ebene dreieckig angeordnet wurden, dann stehen bei der Raumanordnung entweder die einzelnen Kugeln der oberen Schicht auf den einzelnen Kugeln der unteren Schicht, und zwar wieder bei der weiten Packung, oder die einzelnen sitzen zwischen drei der oberen und unteren Schicht. Im ersten Falle wird jede Kugel von sechs benachbarten in derselben Ebene berührt, außerdem von einer darüber und einer darunter, im Ganzen also von acht anderen. Diese Anordnung schließt sich dem Prisma an und beim Zusammendrücken werden aus den Kugeln Säulen mit sechs rechteckigen Seitenflächen und zwei sechseckigen Grundflächen. Bei der zweiten Anordnung ergibt sich dasselbe, wie früher im zweiten Falle der quadratischen Anordnung. Es sei nämlich (B) eine Gruppe von drei Kugeln. Auf diese lege Kugel (A) als Spitze. Es sei weiter (C) eine Gruppe von sechs Kugeln, entsprechend (D) eine solche von zehn und endlich (E) eine von fünfzehn Kugeln. Nun lege immer die kleinere Gruppe auf die größere, so dass sich eine Pyramidenfigur ergibt. Und wenn so durch diese Schichtung je eine der oberen zwischen je drei darunter liegenden Kugeln sitzt, dann werden, wenn du die Figur umlegst, so dass nicht die Spitze, sondern eine Seitenfläche der Pyramide nach oben zeigt, und wenn du eine Kugel aus dem Ganzen herausnimmst, die vier darunter liegenden eine quadratische Ordnung bilden. Und wieder wird wie vorher eine Kugel von zwölf anderen berührt, nämlich von sechs benachbarten in derselben Ebene und von je dreien oben und unten. So kann bei einer möglichst dichten Packung eine dreieckige Anordnung nicht ohne eine viereckige vorkommen und umgekehrt. Daraus geht klar hervor, dass die Granatapfelkerne durch materielle Notwendigkeit, die sich aus ihren Wachstumsverhältnissen ergibt, in die Figur des rhombischen Körpers gepresst werden. Die runden Kerne können nicht dauernd unversehrt gegeneinander drängen, sondern weichen dem Druck aus, indem sie sich in die Zwischenräume zwischen je drei oder vier gegenüberliegenden Kernen hinein schieben. | In den Bienenstöcken aber herrscht ein anderes Gesetz. Denn die Bienen bilden keinen ungeordneten Haufen wie die Apfelkerne, sondern ordnen sich nach ihrem freien Willen in Reihen an, wobei sie sämtlich die Köpfe nach der einen oder nach der entgegengesetzten Seite richten und sich mit dem Hinterleib gegeneinander stemmen. Wenn jedoch aus einem derartigen Zusammendrängen die Figur entstünde, dann müssten die Bienen den Baustoff der Zellen ausschwitzen und gleichsam von ihm umhüllt werden, ähnlich wie wir es von den Schnecken wissen, deren Häuser um sie herum wachsen. Aber es ist sicher, dass die Bienen sich ihre

Die Ursache der Figur der Granatapfelkerne

Woher stammt die Figur der Schneckenhäuser?

from the prism, and when pressed together the spheres become columns with six rectangular sides and two hexagonal bases. In the second configuration the same emerges, as earlier in the second case of the square configuration. Let there be namely (B) a group of three spheres. On this lay sphere (A) as the tip. It is still (C) a group of six spheres, correspondingly (D) one of ten and finally (E) one of fifteen spheres. Now always lay the smaller group on the larger, so that a pyramid emerges. And if, as a result of this layering one of the upper spheres is positioned between three spheres beneath and you then turn the figure over so that a side and not the tip of the pyramid is pointing upward, if you remove a sphere from the whole structure the four beneath form a square configuration. And once again, as previously, one sphere is touched by twelve others, namely six neighboring ones on the same plane and from three each above and below. In this way, with as concentrated packaging as possible, there cannot be a triangular configuration without a square one, and vice versa. This is clear evidence of the fact that the pomegranate pips are pressed into the shape of a rhombic body through material necessity that emerges from the circumstances of their growth. The round pips cannot permanently push against each other unscathed, but avoid the pressure by forcing themselves into the spaces between three or four opposite pips. | In the beehives, however, a different law applies. As the bees do not form a disordered pile like the pomegranate pips but align themselves of their own accord in rows, all their heads pointing towards one or the opposing side and bracing themselves against each other with their rear bodies. If however, the shape were to emerge from a crush of this nature, the bees would have to sweat out the building material for the cells and be cloaked by it, as it were, as we are familiar with in the case of snails, whose shells grow around them. It is for certain, however, that the bees shape their cells freely and as such construct the entire honeycomb from scratch. | Bees, then, have a natural instinct that enables them to build things best of all in this shape. This form of construction was given to them by their creator, and neither the wax, the bees' tiny bodies nor growth can do anything against it. | On the back of this observation the question also emerges as to the ultimate purpose, not to the one the bee itself pursues through its own consideration, but to the one God, the creator of the bees, had given himself, when he prescribed these building regulations for them. | And it is precisely here, to determine the ultimate purpose again, that the observation of the bodies and the material begins. There are, namely, three things that can be said about this ultimate purpose. The first is familiar to any nature observer, who only observes a hexagonal structure, as represented here with the openings from the outside. Without any gap whatever, the plane is only filled by the following figures, the triangle, the square, and the hexagon. Of these the hexagon has the greatest surface area, but the bees want to

The reason for the pomegranate pips shape

Where does the shape of the snail shell come from?

The cause of the shape of the honeycomb shells

The hexagon has a large capacity

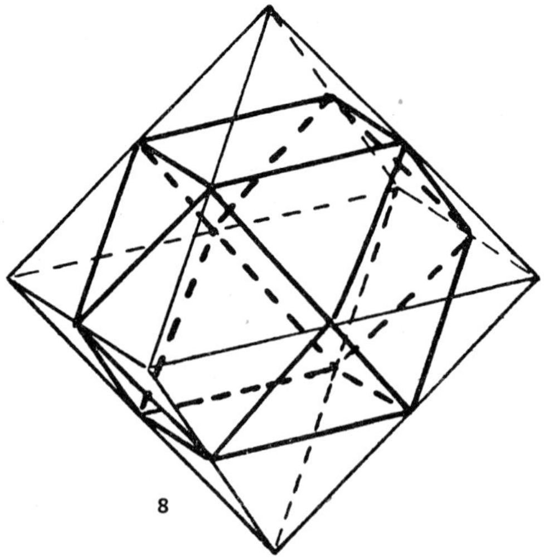

8

Zellen frei formen und so die ganze Wabe von Grund aus errichten. | Die Biene hat also von Natur aus einen ihr eigentümlichen Instinkt, der sie gerade in dieser Figur am besten bauen lässt. Diese Bauform ist ihr vom Schöpfer eingepflanzt und nichts vermag hier der Stoff des Wachses, noch des Bienenkörperchens, nichts das Wachstum. | Nach dieser Beobachtung entsteht weiter die Frage nach dem Endzweck, nicht nach dem, den die Biene selbst durch eigene Überlegung verfolgt, sondern nach dem, den Gott, der Schöpfer der Bienen, sich gesteckt hatte, als er ihnen diese Baugesetze vorschrieb. | Und gerade da beginnt, wieder um den Endzweck zu bestimmen, die Betrachtung der Körper und der Materie. Dreierlei nämlich lässt sich über diesen Endzweck sagen. Das Erste ist jedem Naturbeobachter geläufig, der nur eine sechseckige Struktur betrachtet, wie sie sich hier mit den Öffnungen von außen her darstellt. Lückenlos wird nämlich die Ebene nur von folgenden Figuren ausgefüllt, vom Dreieck, vom Quadrat und vom Sechseck. Von diesen kommt dem Sechseck der größte Flächeninhalt zu. Höchstes Fassungsvermögen aber wollen sich die Bienen schaffen zur Unterbringung des Honigs. | Diese Überlegung kann man auch auf räumliche Verhältnisse ausdehnen in der Weise, dass der ganze Raum nur durch Würfel oder Rhombenflächner lückenlos erfüllt wird; dabei haben die Rhombenflächner größeres Fassungsvermögen als die Würfel. Aber diese Überlegung genügt nicht; denn wenn die Bienen höchstes Fassungsvermögen anstreben, warum baut sich nicht jede ein rundes Nest, war-

Die Ursache der Figur bei den Bienenzellen

Die Sechseckfigur hat größtes Fassungsvermögen

Abb. 8 Kuboktaeder durch Entecken eines Oktaeders

Fig. 8 Cuboctahedron through removal of the corners of an octahedron

create the greatest capacity in which to store their honey. | One can also extend this theory to spatial relationships in a manner in which the entire space is totally filled only with cubes or rhombahedrons, with the latter having a greater capacity than cubes. But this theory is not sufficient, because if the bees are striving for maximum capacity, why does each individual bee not build a round nest, why is it necessary to make use of the smallest of spaces, as if there were no room left in the entire hive? The following second reason is more probable, though according to the theories mentioned it too is not sufficient, namely that the small, tender bodies of the young bees are more comfortable in a nest shape with several obtuse corners and which is related to the sphere, than in cube, which has fewer, acute corners, not to mention a level base, which is unacceptable to a rounded body. | As such there has to be a third reason as well: the workload is reduced if two always build a wall together. In addition the level paneling of the entire honeycomb is more robust and has greater stability than if the individual abodes were round and thus more liable to be crushed. Ultimately there are gaps between round bodies, no matter how closely they are connected, through which cold air could penetrate. All this has been taken care of, as "only bees have a city with common abodes", as can be read in *Virgil's* poems. | So these theories, which take material necessity into account are sufficient, I believe, such that there is no need for me here to mention philosophical observations about the perfection and beauty or the superb figure of the rhombohedron. Nor need I worry about how to recognize the being of the soul hidden in bees from observation of the figure it makes. Admittedly this would have had to happen were there no obvious purpose for the figure. | The same applies to the pomegranate. Material necessity appears to be what forces the pips into rhombic bodies in their subsequent growth. For this reason, reflecting on the being of the soul that lives in this tree, and which produces rhombohedrons in particular, would be effort made in vain. | If, on the other hand, one asks why all trees and bushes (or at least the majority) develop a blossom based on the five-fold system, in other words one that has five leaves, — this is also the case with apples and pears, with the same figure five or the related ten, as on the inside the seeds are kept in five compartments and ten sheaths, which also occurs with cucumbers and other such fruits — I would say that here consideration of the beauty or characteristic of the shape, which has given these plants' soul its character is appropriate. And while I am at it I would also like to relate my thoughts about the matter. | There are two regular bodies, the dodecahedron and the icosahedron, which are structured on the basis of the pentagon. The latter is verily surrounded by pentagons, the former by equilateral triangles, though in each case five are aligned in a pentagon shape. Construction of both bodies, first and foremost though the pentagon itself, cannot occur without the proportion today's mathematicians refer to

The actual purpose of the rhombic shape of honeycomb cells

The cause of the five blossom leaves

The five-part regular bodies and their origin in divine proportion

um ist es nötig, den kleinsten Platz auszunutzen, als ob im ganzen Bienenstock kein Raum übrig bliebe? Wahrscheinlicher wäre der folgende zweite Grund, obgleich auch er nach den genannten Überlegungen nicht ausreicht, nämlich dass die zarten Körperchen der jungen Bienen bequemer in einer Nestform mit mehreren und stumpferen Ecken, die der Kugel verwandter ist, liegen als in einem Würfel, der weniger und spitzere Ecken hat, außerdem einen ebenen Boden, der mit einem rundlichen Körper ganz unverträglich ist. | Daher muss ein dritter Grund hinzukommen: Es vermindert sich die Arbeit, wenn immer zwei an einer gemeinsamen Wand bauen. Ferner geben die ebenen Vertäfelungen der ganzen Honigwabe mehr Halt und größere Festigkeit, als wenn die einzelnen Häuschen rund und so dem Zerdrücken leichter ausgesetzt wären. Endlich bleiben zwischen runden Körpern, wenn sie noch so eng verbunden sind, Lücken, durch die Kälte eindringen könnte. Für alles dies ist Vorsorge getroffen, denn: »Nur die Biene besitzt eine Stadt mit gemeinsamen Häusern«, wie's in Vergils Gedichten heißt. | Diese Überlegungen also, die die materielle Notwendigkeit berücksichtigen, genügen, glaub' ich, so dass ich hier nicht noch philosophische Betrachtungen über die Vollkommenheit und Schönheit oder die edle Figur der Rhombenflächner anzustellen brauche. Auch braucht's mich nicht zu kümmern, wie das Wesen des Seelchens, das in der Biene steckt, aus der Betrachtung der Figur, die sie hervorbringt, erkannt werden könne. Dies hätte freilich geschehen müssen, wenn ein Zweck der Figur nicht klar zutage getreten wäre. | Dasselbe gilt für den Granatapfel. Die materielle Notwendigkeit erscheint als das, was die Kerne bei ihrem folgenden Wachstum in rhombische Körper zwingt. Daher wäre es vergebene Liebesmühe, über das Wesen der Seele nachzudenken, die in diesem Baume wohnt und die etwa den Rhombenflächner besonders gern hervorbringt. | Wenn man dagegen fragt, warum alle Bäume und Sträucher (oder wenigstens die Mehrzahl) gerade eine Blüte nach dem Fünfersystem entwickeln, d.h. eine, die fünf Blütenblätter hat, — bei den Äpfeln und Birnen entspricht dieser Blütenform auch die Fruchtanlage mit derselben Zahl fünf oder der verwandten zehn, denn im Innern sind die Samen in fünf Fächern und zehn Hüllen untergebracht, was auch bei Gurken und anderen solchen Früchten vorkommt, so sage ich, hier ist eine Betrachtung über die Schönheit oder die Eigentümlichkeit der Figur am Platze, die der Seele dieser Pflanzen ihren Charakter gegeben hat. Und beiläufig will ich auch hierüber meine Gedanken sagen. | Es gibt zwei regelmäßige Körper, das Dodekaeder und das Ikosaeder, die nach dem Grundplan des Fünfecks gebaut sind. Jenes wird geradezu von Fünfecken umschlossen, dieses zwar von gleichseitigen Dreiecken, aber je fünf sind stets in Fünfeckform angeordnet. Die Konstruktion beider Körper, vor allem aber die des Fünfecks selbst, ist nicht durchführbar ohne jene Proportion, die die heutigen Mathematiker die *göttliche* nennen. Diese aber ist so beschaffen,

Der eigentliche Zweck der rhombischen Figur bei den Bienenzellen

Die Ursache der Fünfzahl bei den Blütenblättern

Die fünfzähligen regelmäßigen Körper und ihr Ursprung aus der göttlichen Proportion

as the divine. This however, is composed such that together the two smaller members of a proportion produce the third, and this way the sum of two neighboring members is always the one immediately following, with the same proportion continuing ad infinitum. Totally expressing the result in figures is impossible. However, the further we get away from the unit, the better the convergence becomes. You have to imagine the two smallest figures 1 and 1 as being of unequal value. Added together they make 2. Adding the larger 1 makes 3, adding another 2 makes 5. If you add another 3 you get 8, plus 5 makes 13, plus another 8, 21. 5 to 8 is now approximate to 8 to 13 and 8 to 13 almost the same as 13 to 21. | The analogy of this self-perpetuating proportion illustrates, I believe, the ability to procreate, and as such an unadulterated five-finger pennon is held up as the ability to procreate. I am passing over much that could be added to back up this point of view. But I would need my own space of my own for that. We only considered this example first in order to have more practice in and better prepared for investigating the hexagonal shape of snow.

dass die beiden kleineren Glieder einer stetigen Proportion zusammen das dritte ergeben, und in dieser Weise bildet immer die Summe zweier benachbarter Glieder das unmittelbar folgende, wobei die gleiche Proportion bis ins Unbegrenzte fortgeht. In Zahlen das Ergebnis vollkommen auszudrücken ist unmöglich. Je weiter wir uns jedoch von der Einheit entfernen, desto besser wird die Annäherung. Es seien die beiden kleinsten Zahlen 1 und 1, die du dir als ungleichwertig vorstellen musst. Addiert geben sie 2. Die größere 1 hinzu addiert gibt 3, ferner 2 zugezählt ergibt 5. Addiert man weiter 3 dazu, so kommt 8, dazu 5 geschlagen macht 13, 8 dazu addiert ergibt 21. Es verhält sich nun 5 zu 8 angenähert wie 8 zu 13 und 8 zu 13 nahezu wie 13 zu 21. | Nach Analogie dieser sich selbst fortsetzenden Proportion ist, wie ich glaube, das Zeugungsvermögen versinnbildet, und so wird in der Blüte dem Zeugungsvermögen ein unverfälschtes fünfzackiges Fähnlein vorangetragen. Ich übergehe manches, was zur Bekräftigung dieser Ansicht in ergötzlicher Betrachtung hinzugefügt werden könnte. Aber dafür brauchte ich schon einen eigenen Platz. Hier haben wir dies nur als Beispiel vorausgeschickt, um zur Erforschung der sechseckigen Figur des Schnees besser geübt und gerüstet zu sein.

Anhang / Annex

AUTOREN | AUTHORS

Kristin Feireiss Kristin Feireiss stammt aus Berlin / Deutschland, ist Architekturkuratorin, Autorin und Redakteurin. Sie studierte Kunstgeschichte und Philosophie an der Johann Wolfgang Goethe-Universität in Frankfurt. 1980 gründete sie (zusammen mit Helga Retzer, †1984) das unabhängige Architekturforum Aedes in Berlin. Sie leitet diese internationale Plattform zusammen mit Hans-Jürgen Commerell seit 1994. Feireiss' Arbeit erweitert das Verständnis für international Architektur und Stadtplanung einschließlich der dazugehörigen kulturellen, sozialen und wirtschaftlichen Faktoren. Als Leiterin des Niederländischen Architekturinstituts (NAi) machte sie von 1996 bis 2001 auf städtische Veränderungsprozesse aufmerksam und führte bahnbrechende Forschungen auf diesem Gebiet durch. Feireiss erhielt 2001 das Deutsche Bundesverdienstkreuz am Bande für ihre internationalen Verdienste in Architektur. 2007 verlieh die Technische Universität Carolo-Wilhelmina zu Braunschweig Feireiss den Ehrendoktortitel. 1996 und 2000 gab sie den niederländischen Pavillon bei der Architekturbiennale in Venedig in Auftrag; 2012 war sie Mitglied der internationalen Jury bei der Architekturbiennale in Venedig. Seit 2007 arbeitet sie für das Europäische Kulturparlament. 2013 erhielt sie den Orden vom Niederländischen Löwen, einen der wichtigsten Auszeichnungen, die von der niederländischen Regierung verliehen werden. Seit 2013 ist Kristin Feireiss Jurymitglied für den Pritzker-Preis. || Feireiss hat mehrere Monografien und Themenbände zu Architektur und Stadtplanung herausgegeben. Kürzlich erschienene Werke umfassen unter anderem *Informal City — Caracas Case*, welches das Potential von informellen Siedlungen beleuchtet und *Architecture in Times of Need* zum Thema innovative, nachhaltige und erschwingliche Wohnmöglichkeiten, die in New Orleans' Lower Ninth Ward nach Hurrikan Katrina errichtet wurden. Feireiss war auch Mitherausgeberin von zwei Bänden in der Reihe *Architecture of Change: Sustainability and Humanity in the Built Environment*. Diese Bücher präsentieren außergewöhnliche Architektur und soziale Initiativen und zeigen, wie Architekten Themen der Nachhaltigkeit und sozialen Gerechtigkeit in öffentlichen Gebäuden, Wohnhäusern und im Bereich Stadtplanung ansprechen sollen.

Kristin Feireiss Kristin Feireiss of Berlin, Germany, is an architecture curator, writer, and editor. She studied art history and philosophy at Johann Wolfgang Goethe University in Frankfurt. In 1980, she founded (with Helga Retzer, †1984) the independent forum for architecture Aedes in Berlin. Since 1994 she has led this international platform in partnership with Hans-Jürgen Commerell. Feireiss's work deepens understanding of international architecture and urban development including the cultural, social and economic factors involved. As director of the Netherlands Architecture Institute (NAi) from 1996 to 2001, she brought greater attention to the transformative processes affecting cities and has carried out groundbreaking research in this area. Feireiss received the Deutsche Bundesverdienstkreuz am Band (Cross of the Order of Merit of the Federal Republic of Germany) in 2001 for her international engagement with architecture. 2007 Wilhelmina University Braunschweig honored Feireiss with the Doctor honoris causa. She was commissioner of the Dutch Pavilion at the Architecture Biennale in Venice in 1996 and 2000; and a 2012 member of the International Jury for the Architecture Biennale in Venice. Since 2007, she has served in the European Cultural Parliament. 2013 was announced as Knight in the Order of the Netherlands Lion, one of the most prestigious merit awarded by the Dutch Government. Since 2013 Kristin Feireiss is jury member of the Pritzker Prize. || Feireiss has edited numerous monographs and thematic volumes on international architecture and urban context. Her recent books include, *Informal City — Caracas Case*, focussing the potentials of informal settlements and *Architecture in Times of Need*, which highlights innovative, sustainable, affordable housing to redevelop New Orleans' Lower Ninth Ward, after Hurricane Katrina. Feireiss also co-edited two volumes of *Architecture of Change: Sustainability and Humanity in the Built Environment*. These books feature outstanding sustainable architecture and social initiatives and underscore how architects must address issues of sustainability and social justice in public buildings, housing and city planning.

Rudolf Finsterwalder wurde 1966 in Rosenheim geboren. Er arbeitet als Architekt in Stephanskirchen und Berlin. Während des Studiums an der TU-Berlin hat er unter anderem in Berlin mit Benedict Tonon und dem Büro Hufnagel, Pütz, Rafaelian gearbeitet, sowie in Porto mit Àlvaro Siza. Nach dem Studium hat er als Projektleiter im Büro Ortner und Ortner die Sächsische Landesbibliothek - Staats- und Universitätsbibliothek Dresden geplant. Im Jahr 2000

Rudolf Finsterwalder was born 1966 in Rosenheim. He is a practicing architect in Stephanskirchen and Berlin. During his studies at the Technical University in Berlin he worked with Benedict Tonon and Hufnagel, Pütz, Rafaelian, and Àlvaro Siza in Porto. After he finished university, he was project-manager of the Sächsische Landesbibliothek-Staats- und Universitätsbibliothek (library) in Dresden. In 2000 he founded Finsterwalder Architekten together with

hat er sich mit seiner Frau Maria-José da Silva Áraujo Finsterwalder selbständig gemacht. Sie betreiben gemeinsam ein Büro in Stephanskirchen. | Rudolf Finsterwalder erhielt zahlreiche Preise und Stipendien, u.a. Brick Award 2010 (zusammen mit Álvaro Siza), Nominierung zum DAM-Preis (2009), Villa Massimo, Rom (2007). Seine Arbeiten wurden in vielen nationalen und internationalen Ausstellungen präsentiert, u.a. Gropius-Bau, Berlin (2011), 12. Architekturbiennale, Venedig (2010), Tefaf, Maastricht (2009), DAM, Frankfurt (2009), Museum Ludwig, Köln (2008), Gropius-Bau, Berlin (2008), Villa Massimo, Rom (2007), AIA, New York (2005), 9. Architekturbiennale, Venedig (2004). | Seine Bauten und Projekte wurden in vielen renommierten nationalen und internationalen Fachzeitschriften und Büchern veröffentlicht, u.a. *architecture now / museums, L'architecture d'aujourd'hui, a + u, Bauwelt, Casabella, Area.*

Dieter Dolezel wurde 1977 in München geboren. Nach anfänglichem Philosophie-Studium an der LMU in München studierte Dolezel zunächst Gitarre am Richard-Strauss-Konservatorium. Es folgten Kompositions-Studien bei Wilfried Hiller sowie ein Aufbaustudium bei Louis Andriessen am Königlichen Konservatorium in Den Haag. Ergänzt wurden seine Studien durch diverse Meisterkurse unter anderem bei Tristan Murail, Alvin Curran und Richard Ayres. | Im Jahre 2002 gründete Dolezel die doArte-Kulturstiftung, für die er nach wie vor konsultativ tätig ist. | Aufführungen seiner Werke fanden u.a. bei der Gaudeamus Music Week, Novembermusic, den Hamburger Klangwerktagen, im Münchner Prinzregententheater sowie im Auditorium Santa Cecilia in Rom statt. | Zu den Interpreten seiner Musik zählen u.a. das Ensemble Modern, die King's Singers, Mike Svoboda, John Snijders, BL!NDMAN [drums], Holland Symfonia, sowie das ensemble unitedberlin. | Zu seinen Auszeichnungen gehören unter anderem das Stipendium im Internationalen Künstlerhaus Villa Concordia in Bamberg, das Musikstipendium der Landeshauptstadt München, der NOG Jonge Componistenprijs, sowie der Gaudeamus Publikumspreis 2007. 2006 war er Preisträger beim 1. Internationalen Kompositionswettbewerb der Hamburger Klangwerktage, 2007 - 2008 Stipendiat der Villa Massimo in Rom.

Prof. em. Dr. rer. nat. Werner Nachtigall wurde 1934 in Saaz, Tschechoslowakei, geboren, studierte an der Ludwigs-Maximilian-Universität und an der Technischen Universität München unter anderem Biologie mit Schwerpunkt Zoologie, Chemie und Technische Physik mit Schwerpunkt Strömungsmechanik. Er promovierte mit einem technisch-biologischen Thema und habilitierte sich über Biomechanik des Tierflugs. Auf seine Tätigkeit als Wissenschaftlicher Assistent und Stipendiat der DFG am Zoologischen und Strahlenbiologischen Institut der Universität München folgte ein Aufenthalt als Associate Professor an der University of California, Berkeley.

his wife Maria-José da Silva Áraujo Finsterwalder. | Rudolf Finsterwalder has received numerous prizes and fellowships such as Brick Award 2010 (together with Álvaro Siza), Nomination to DAM-Prize (2009), Villa Massimo, Rome (2007). His works have been shown at major national and international exhibitions, e.g. Gropius-Bau, Berlin (2011), 12th Biennale at Venice (2010), Tefaf, Maastricht (2009), DAM, Frankfurt (2009), Museum Ludwig, Cologne (2008), Gropius-Bau, Berlin (2008), Villa Massimo, Rome (2007), AIA, New York (2005), 9th Biennale at Venice (2004). | His buildings and projects have been published in many prestigious books and professional journals, such as, *architecture now / museums, L'architecture d'aujourd'hui, a + u, Bauwelt, Casabella, Area.*

Dieter Dolezel was born 1977 in Munich. After he had studied at the LMU in Munich, Dolezel studied guitar in the Richard-Strauss-Konservatorium. After this he studied composition with Wilfried Hiller and a postgraduate study with Louis Andriessen at the Royal Conservatoire in The Hague. He complemented his studies with various master classes with including Tristan Murail, Alvin Curran and Richard Ayres. | In 2002 Dolezel founded the doArte-cultural Foundation. He is still an active consultant. | Performances of his works took place at the Gaudeamus Music Week, Novembermusic, Hamburger Klangwerktagen in Hamburg, Prinzregententheater in Munich, and at the auditorium Santa Cecilia in Rome, among other. | Performers of his music include the Ensemble Modern, the King's Singers, Mike Svoboda, John Snijders, BL!NDMANN [drums], Holland Symfonia and the ensemble unitedberlin. His honours include a fellowship at the Internationales Künstlerhaus Villa Concordia in Bamberg, a music-fellowship from the state capital Munich, the NOG Jonge Componistenprijs, and the Gaudeamus Publikumspreis 2007. In 2006 he was a award winner at 1st International Kompositionswettbewerb der Hamburger Klangwerktagen, 2007 - 2008 fellowship of the Villa Massimo in Rome.

Prof. em. Dr. rer. nat. Werner Nachtigall was born 1934 in Saaz, Czechoslovakia. He studied at the Ludwigs-Maximilian-Universität Munich and the Technische Universität Munich among others biology with specialization in zoology, chemistry and technical physics with specialization in fluid mechanics. He received a doctorate in a technical-biological subject and qualified as a professor of biomechanics in animal flight. After his activity as a scientific assistant and Fellow of the DFG at the Zoological and Rays-biological Institute of the University of Munich, he was an associate professor at the University of California, Berkeley. Following his appoint-

Nach seiner Berufung an die Universität des Saarlandes baute er in Saarbrücken eine bewegungsphysiologische Arbeitsgruppe auf und arbeitete auf den Gebieten der Technischen Biologie und Bionik. Dazu gehörten seit der Kooperation im SFB 231 mit Frei Otto in Stuttgart auch Aspekte der Bau-Bionik, die ihren Niederschlag in einem gleichnamigen Buch gefunden haben. Der Ideentransfer zwischen Biologie und Technik gehörte schon früh zu seinen Anliegen. In diesem Zusammenhang hat er eine Ausbildungsrichtung »Technische Biologie und Bionik« und eine Gesellschaft gleichen Namens gegründet und war massgeblich beteiligt am Aufbau des Bionik-Kompetenznetzes BioKon. Er hat mehrere Hundert Publikationen und mehrere Dutzend Bücher verfasst und ist gewähltes Mitglied von Akademien und Organisationen sowie Träger einer Reihe von Preisen und Auszeichnungen.

ment to the University of Saarland in Saarbrücken he founded a study group of physiological motions and worked in the fields of technical biology and bionics. He also worked on aspects of the construction-bionics in cooperation with Frei Otto in the SFB 231 in Stuttgart. He has written a book with same name. The transfer of ideas between biology and technology was one of his earliest concerns. In this context, he founded a school and association under the name "Technical Biology and Bionics" where he was involved at the construction of the Bionics Competence Network BioKon. He has written several hundred publications and several dozen books and is an elected member of academies and organizations. His work is honoured with a series of prizes and awards.

Cornelius Thywissen wurde 1948 geboren, Studium der Biologie und Architektur in Freiburg, Aachen, Stuttgart und Hohenheim. Abgeschluß als Dipl.-Ing. mit der Diplomarbeit über Methoden zur Rekultivierung von Städten mit Pflanzen. Stipendium der INRA (Institut National de Recherches Agronomiques) in Paris für praktische Experimente mit Pflanzen; 76/78 Arbeit als Architekt. Seit 1978 Mitglied des IL; Hauptgebiete: Minimalstrukturen, Biologie und Architektur.

Cornelius Thywissen was born 1948. He studied biology and architecture in Freiburg, Aachen, Stuttgart and Hohenheim. He graduated as Dipl.-Ing. with his thesis about methods to re-cultivate cities with plants. He was awarded a fellowship of the INRA (Institut National de Recherches Agronomiques) in Paris for practical experiments with plants. Between 76/78 worked as architect. He is a member of the IL since 1978; main tasks: minimal structures, biology and architecture.

Günther Hartmann wurde 1924 in Berlin geboren. Studium der Ethnologie, Amerikanistik und Geographie. 1959 Dr. phil. Lehrbeauftragter der Freien Universität. Museum für Völkerkunde seit 1960, Oberkustos seit 1971. Hauptschriftleiter der Zeitschrift für Ethnologie seit 1965. Vorstandsmitglied der Berliner Gesellschaft für Anthropologie, Ethnologie und Urgeschichte. Mehrere Studienaufenthalte in Nord-, Mittel- und Südamerika. Forschungsreisen noch Nord-, Zentral- und Südamerika. Zahlreiche Buch- und Zeitschriftenpublikationen.

Günther Hartmann was born in 1924 in Berlin where he studied cultural anthropology, American studies and geography. In 1959 he received a PhD in philosophy and lectured at the Freie Universität. He worked at the Museum of Ethnology since 1960, and as chief curator since 1971. He was Chief editor of the Journal of Ethnology since 1965. He was a board member of the Berliner Gesellschaft für Anthropologie, Ethnologie und Urgeschichte. He conducted several studies and explorations in North, Central and South America. He has numerous books and magazine publications.

Frei Otto wurde 1925 in Siegmar, Sachsen geboren, er lebt in Leonberg-Warmbronn bei Stuttgart. 1943-45 im Kriegseinsatz. 1948 Studium in Berlin. Studium der Soziologie, des Städtebaus an der University of Virginia. 1952 Dipl. Ing. freier Architekt in Berlin. 1954 Promotion (Dr.-Ing.) über »Das hängende Dach«. 1958-69 Atelier in Berlin mit Lohs, Bubner, Frank, Werhahn, Medlin u.a. 1964-91 Honorarprofessor, seit 1975 ordentlicher Professor und Leiter des Instituts für Leichte Flächentragwerke (IL) an der Universität Stuttgart. Seit 1958 Gastprofessor an verschiedenen Universitäten in den USA und Deutschland. | Hauptprojekte u.a.: Zeltpavillons — Bundesgartenschau Kassel 1955, Interbau Berlin 1957, Versuchsbauten Deubau 1962, Deutscher Pavillon auf der Expo Montreal 1967 (mit Gutbrod), Olympiadach Mün-

Frei Otto was born in 1925 in Siegmar, Saxony. He lives in Leonberg near Stuttgart-Warmbrohn. Part of the war effort in 1943-1945. Studied in Berlin in 1948. He studied sociology, urban planning at the university of Virginia. Received Dipl.-Ing. in 1952, and worked as a freelance architect in Berlin. In 1954, he received a PhD (Dr.-Ing.) for "Das hängende Dach". 1958-69 founded an atelier in Berlin with Lohs, Bubner, Frank, Werhahn, Medlin. 1964-91 Honorary Professor, Associate Professor since 1975 and Director of the "Institut für leichte Flächentragwerke" (IL). Since 1958, a visiting professor at several universities in the USA and Germany. | Main projects include: 1955 tent pavilions — National Garden Show in Kassel, Berlin Interbau 1957, experimental buildings Deubau 1962, German pavilion at the Montreal

chen (mit Behnisch) 1972, Government Center Riyadh (mit Gutbrod, Kendel u.a.), 1975 Bundesgartenschau Mannheim Multihalle (mit Mutschler, Langner), Bühnendach Pink Floyd (mit Happold) 1977. 1986 Diplomatic Club Riyadh (mit Omrania, Happold), 1989 Ökohaus Berlin-Tiergarten (mit Kendel). 2000 Projekt Stuttgarter Bahnhof (mit Ingenhoven). | Preis und Auszeichnungen (Auswahl): 1967 Kunstpreis der Stadt Berlin. Mitglied der Akademie der Künste Berlin 1970. 1971 Paul-Bonatz-Preis der Stadt Stuttgart. 1978 Hugo-Häring-Preis des BDA Baden-Württemberg. 1990 Honda Preis für Ökotechnologie. 1992 Preis des Deutschen Werkbundes Bayern. 1998 Aga-Kahn-Award for architecture. 2006 Praemium Imperiale. | Buchveröffentlichungen (Auswahl): *Das hängende Dach* 1954. *Zugbeanspruchte Konstruktionen* (2 Bde.) 1962/1965. *Natürliche Konstruktionen* 1982. *Gestaltwerdung* 1988. Herausgeber der IL-Schriftenreihe Universität Stuttgart.

Expo 1967 (with Gutbrod), Munich Olympia roofing (with Behnisch) 1972, Government Center Riyadh (with Gutbrod, Kendel i.a.), 1975 National Garden Show Multihalle Mannheim (with Mutschler, Langner), stage roof Pink Floyd (with Happold) 1977. 1986 Riyadh Diplomatic Club (with Omrania, Happold), 1989 Ökohaus Berlin-Tiergarten (with Kendel). Project Stuttgart main station in 2000 (with Ingenhoven). | Awards (selection): 1967 Kunstpreis of the city of Berlin). Member of the Akademie der Künste Berlin 1970. 1971 Paul-Bonatz-Preis of the city of Stuttgart. 1978 Hugo-Häring-Preis of BDA Baden-Württemberg. 1990 Honda Award for ecotechnology. 1992 Prize of the Deutschen Werkbund Bayern, exhibition, 1998 Aga-Kahn-Award for architecture. 2006 Praemium Imperiale. | Book Publications (selection): *Das hängende Dach* 1954. *Zugbeanspruchte Konstruktionen* (2 volumes) 1962/1965 *Natürliche Konstruktionen* 1982. *Gestaltwerdung* 1988. Editor of IL-Series of Publications, university of Stuttgart.

Eda Schaur 1945 in Ljubljana, Jugoslawien, geboren. 1964-71 Architekturstudium an der Technischen Universität Graz, Österreich, abgeschlossen als Dipl.-Ing. 1991 Dissertation »Ungeplante Siedlungen« bei Frei Otto, Universität Stuttgart. Seit der Gründung 1972 bis 1995 wissenschaftliche Mitarbeiterin am Institut für leichte Flächentragwerke (IL), Universität Stuttgart. Von 1973 bis 1985 Tätigkeit im Sonderforschungsbereich 64 »Weitgespannte Flächentragwerke« der Deutschen Forschungsgemeinschaft. Von 1984 bis 1995 Mitglied und Teilprojektleitung im SFB 230 »Natürliche Konstruktionen — Leichtbau in Architektur und Natur« an den Universitäten Stuttgart und Tübingen. | Von 1976 bis 1987 Lehr- und Forschungstätigkeit in Indien an der School of Architecture (SA), Ahmedabad 1981-1987 Forschungsprojekt »Leichtbau mit traditionellen Baumaterialien«. Von 1983 bis 1994 Zusammenarbeit mit Yona Friedman bei dem Aufbau des »Communication Centre of Scientific Knowledge for Self-Reliance« in Paris an der Universität der Vereinten Nationen (UNU). 1993. Seit 1974 Buchveröffentlichungen und zahlreiche Aufsätze, wie: *Ungeplante Siedlungen* IL 39, Stuttgart, 1991. *Intelligent Bauen — Aspekte einer anderen Baukultur* IL 41, 1995. | Seit März 1995 o. Univ.-Prof. an der Universität Innsbruck, Leitung des Instituts für Konstruktion und Gestaltung.

Eda Schaur was born 1945 in Ljubljana, Yugoslavia. Between 1964-71 studied architecture at the Technische Universität Graz, Austria, graduated with Dipl.-Ing. In 1991 dissertation, "Ungeplante Siedlungen" under Frei Otto, Universität Stuttgart. Research Assistant, since its foundation, from 1972 to 1995, at the Institut für leichte Flächentragwerke (IL). From 1973 to 1985 worked in the Collaborative Research Centre (SFB) 64 "Weitgespannte Flächentragwerke" of the German Research Foundation (DFG). From 1984 to 1995 member and partly project-manager in the SFB 230 "natural constructions — lightweight constructions in architecture and nature" at the Universität Stuttgart and Universität Tübingen. | From 1976 to 1987 teaching and research in India at the School of Architecture (SA), Ahmedabad and during 1981 to 1987 research project "lightweight constructions with traditional materials". From 1983 to 1994 together with Yona Friedman in the construction of the "Communication Centre of Scientific Knowledge for Self-Reliance" in Paris at the United Nations University (UNU). 1993 to 1994 management at the Institut für leichte Flächentragwerke, Universität Stuttgart. Since 1974, numerous articles and book publications, such as: *Ungeplante Siedlungen* (Non-Plannend Settlements) IL 39, Stuttgart, 1991. *Intelligent Bauen — Aspekte einer anderen Baukultur* (Building with Intelligence — Different Aspects of Building a Culture) IL 41, 1995. | Since March 1995 Professor at the University of Innsbruck, head of the Institute for Construction and Design.

Carsten Nicolai ist bildender Künstler, Komponist und Musiker und wurde 1965 in Karl-Marx-Stadt geboren. Er lebt und arbeitet in Berlin und Chemnitz. Nachdem er 1984 als Gärtner beschäftigt war, absolvierte er zwischen 1985 und 1990 ein Studium der Landschaftsgestaltung in Dresden. 1992 war er Mitbegründer des VOXXX - Kultur- und Kommunikations-

Carsten Nicolai born 1965 in Karl-Marx-Stadt, is a visual artist, composer and musician who lives and works in Berlin and Chemnitz. After he worked as a gardener in 1984, he studied landscape architecture in Dresden from 1985 to 1990. In 1992 he was a co-founder of VOXXX - Kultur- und Kommunikationszentrums in Chemnitz. In 1994 he started his own

zentrums in Chemnitz. 1994 rief er das Label *noton.archiv für ton und nichtton* ins Leben, welches 1999 mit *rastermusic* zu *raster-noton* fusionierte. | Bekannt geworden ist er vor allem als bildender Künstler der Objekt- und Installationskunst im Grenzbereich zwischen Kunst und Wissenschaft, wobei er die Grenzen zwischen den verschiedenen menschlichen Sinneswahrnehmungen aufzulösen sucht, etwa indem er physikalische Phänomene wie Sound- und Lichtfrequenzen sowohl für das Auge als auch für das Ohr wahrnehmbar macht. Bei Nicolais Auseinandersetzung mit mathematischen, physikalischen und natürlichen Phänomenen und Theorien sind das kreative Potential von Fehlern und des Zufalls sowie selbstorganisierende Prozesse (z. B. Schneekristalle) weitere Aspekte seiner Arbeit. | Nicolai erhielt zahlreiche Preise und Stipendien, u. a. Villa Massimo, Rom (2007), Zurich Prize, Basel (2007), Villa Aurora, Los Angeles (2003), Goldene Nica, Ars Electronica, Linz (2000), F6-Philip Morris Graphik-Preis, Dresden (2000). Seine Arbeiten wurden in vielen nationalen und internationalen Ausstellungen renommierter Galerien und Museen präsentiert und fanden Aufnahme in bedeutende private und öffentliche Sammlungen. | Unter den Pseudonymen *noto* und *alva noto* veröffentlichte Nicolai eine umfangreiche Anzahl von Tonträgern, welche er bei zahlreichen Live-Performances in nationalen und internationalen Konzerthäusern, Clubs und Museen präsentiert hat. Dabei kooperierte er mit bedeutenden Komponisten und Musikern, u. a. Ryuichi Sakamoto, Blixa Bargeld, Ryoji Ikeda, Michael Nyman, Mika Vainio, Olaf Bender + Frank Bretschneider (signal), sowie mit diversen Orchestern für moderne Musik, u. a. Ensemble Modern, BCN 216 und zeitkratzer.

label *noton.archiv für ton und nichtton*, which fused with *rastermusic* to *raster-noton* in 1999. | Nicolai is part of an artist generation who works intensively in the transitional area between art and science. As a visual artist mainly creating objects and installations, Nicolai seeks to overcome the separation of the art forms and genres by trying to endeavor a holistic artistic approach. Thus he seeks to overcome the separation of the sensual perceptions of man, e.g. by making sound and light frequencies perceivable for both eyes and ears to sensitise the viewer to the connection of the different sensory levels. Further aspects of his works consider the integration of the creative potential of error and chance as well as the implementation of mathematical, physical or natural phenomena and theories. Special interest he also puts on self-organizing processes, like the growing of snow crystals. | Nicolai received numerous prizes and stipends, such as Villa Massimo, Rome (2007), Zurich Prize, Basle (2007), Villa Aurora, Los Angeles (2003), Golden Nica, Ars Electronica, Linz (2000 und 2001), F6-Philip Morris Graphik-Preis, Dresden (2000). His works have been shown at national and international major exhibitions (documenta, Venice Biennial), as well as at numerous solo exhibitions in prestigious galleries and museums. Many of his works are part of prominent private and public collections worldwide. | For several years now Nicolai has released albums under the pseudonyms *noto* and *alva noto* which he has presented at live shows in national and international concert halls, clubs and museums. He has cooperated with renowned composers and musician including Ryuichi Sakamoto, Blixa Bargeld, Ryoji Ikeda, Michael Nyman, Mika Vainio, Olaf Bender + Frank Bretschneider (signal), as well as with diverse orchestras for modern music, such as Ensemble Modern, BCN 216 and zeitkratzer.

Paul Oskar Prochnow wurde 1884 in Wendisch-Buchholz, Brandenburg, geboren. Von 1890 bis 1896 besuchte er die Gemeindeschule seiner Heimatstadt und wurde danach durch Privatunterricht für die Quarta eines Gymnasiums vorbereitet. Sieben Jahre lang besuchte er das Kgl. Friedrich-Wilhelm-Gymnasium in Cottbus, das er im Jahr 1903 mit dem Zeugnis der Reife verließ. Daraufhin studierte er sieben Semester in Berlin hautpsächlich Zoologie, Botanik, Physik, Mathematik und Philosophie unter anderen bei Geh. Reg.-Rat H. A. Schwarz, Prof. Frobenius, Prof. Knoblauch, Prof. M. Planck, Prof. Blasius, Geh. Reg.-Rat Riehl, Geh. Reg.-Rat F. E. Schulze, Geh. Reg.-Rat Möbius, Prof. Brauer, Geh. Reg.-Rat Schwendener, Dr. Deegener und Dr. Baur. | Am 5. März 1908 schloss er seine Promotions-Prüfung *cum laude* ab.

Paul Oskar Prochnow was born 1884 in Wendisch-Buchholz, Brandenburg, Germany. From 1890 to 1896 he attended the local school of his hometown and afterwards was prepared by private tuition for the third year of a high school. For seven years he attended the Königliche Friedrich-Wilhelm-Gymnasium in Cottbus. He graduated in 1903 with a General Certificate of Secondary Education. He then spent seven semesters in Berlin studing zoology, botany, physics, mathematics and philosophy under many professors included Geh. Reg.-Rat H. A. Schwarz, Prof. Frobenius, Prof. Knoblauch, Prof. M. Planck, Prof. Blasius, Geh. Reg.-Rat Riehl, Geh. Reg.-Rat F. E. Schulze, Geh. Reg.-Rat Möbius, Prof. Brauer, Geh. Reg.-Rat Schwendener, Dr. Deegener and Dr. Baur. | On the 5th March, 1908 he completed his doctorate *cum laude*.

Ian Stewart wurde 1945 in England geboren. Er studierte in Cambridge und an der Universität Warwick, wo er 1969 bei Brian Hartley promovierte. Er ist Professor und Direktor des »Mathematics Awareness Centre at Warwick« an der Universität Warwick und hatte unter anderem Gastpro-

Ian Stewart was born 1945 in England. He studied in Cambridge and at the University of Warwick. In 1969 he received his doctorate under Brian Hartley. He is now Professor of Mathematics at the University of Warwick and director of "Mathematics Awareness Centre at Warwick".

fessuren in Deutschland (1974), Neuseeland (1976) und den USA (Universität von Connecticut 1977 - 78, Universität von Houston 1983-84). Außerdem ist er Professor an der Universität von Houston. | Stewart schrieb wissenschaftliche Lehrbücher und Monographien, populärwissenschaftlichen Arbeiten zur Mathematik unter anderem für New Scientist und Scientific American und Discover. Stewart hielt 1997 die berühmten Christmas Lectures der Royal Institution und 1995 erhielt er die Faraday Medaille der Royal Society für seine Beiträge zur Popularisierung der Wissenschaften. Er ist Ehrendoktor von Westminster, Löwen, der Open University und von Kingston. 2001 wurde er zum »Fellow of the Royal Society« gewählt.

Stewart has held visiting academic positions in Germany (1974), New Zealand (1976), and the U.S. (University of Connecticut 1977 - 78, University of Houston 1983 - 84) and he is Professor at the University of Houston. | Stewart has published educational books and monographs, popular expositions of mathematics, including New Scientist, Scientific American and Discover. In 1995 Stewart received the Michael Faraday Medal and in 1997 he gave the Royal Institution Christmas Lectures. He is honorary doctor of Westminster, Lions, Open University and of Kingston. He was elected as a Fellow of the Royal Society in 2001.

Paul Bühler wurde 1936 in Freudenstadt im Schwarzwald geboren. 1957 - 65 studierte er Biologie, Chemie und Geologie an der Technischen Hochschule Stuttgart und an der Universität Kopenhagen. Im Jahre 1966 wurde er Assistent am Zoologischen Institut der Universität Hohenheim. Seine Promotion erfolgte 1969. | Arbeiten: Systematik und Ökologie von Fledermäusen und Spitzmäusen, Verhalten von Vögeln, Funktionelle Morphologie von Wirbeltieren.

Paul Bühler was born 1936 in Freudenstadt, Black Forest. Between 1957 - 65 he studied biology, chemistry, geology at the Technische Hochschule Stuttgart and the University of Kopenhagen. In 1966 he was assistant at the Zoologisches Institut at the University Hohenheim. In 1969 he received his doctorate. | Works: Systematics and ecology of field mice and shrews, behavior of birds, functional morphology of vertebrates.

Frank Lloyd Wright wurde 1867 in Richland Center, Wisconsin, geboren. Er studierte von 1885 bis 1887 an der University of Wisconsin Ingenieurwesen, sein Studium brach er ab, um in Chicago im Architekturbüro von Louis Sullivan zu arbeiten. 1893 kam es zum Bruch mit Sullivan, woraufhin er sich selbstständig machte. | Wright entwickelte sein erstes Wohnkonzept der »Prairie Houses«. Seine Architektur wurde als wichtiges Element des »Arts and Crafts Movement« in Amerika angesehen. Wright arbeitete im Einklang mit der Natur, er versuchte Abstand von traditioneller Gestaltung zu gewinnen und verband organische Formen mit geometrischen Elementen. | 1909 reiste er für zwei Jahre nach Europa. Auch dort wurde er zu einem wichtigen Impulsgerber der Architektur,so beeinflussten seine Werke u. a. Walter Gropius und De Stijl. Wieder in Amerika begann Wright 1911 mit dem Bau von Taliesin, einem Wohnbau mit Ateliers bei Spring Green, Wisconsin. 1913 reiste er mit Antonin Raymond nach Japan, um unter Anwendung eines neuartigen, erdbebensicheren Konstruktionsprinzips das Imperial Hotel in Tokio zu realisieren. 1917 bis 1920 baute er das Barnsdall House in Los Angeles. | Nach dem Börsencrash 1929 hielt Wright an der Princton Universität Vorlesungen, welche er auch veröffentlichte. Außerdem schrieb er mehrere Bücher, z. B. über seine Ideen zur Stadtplanung und gab auch Zeitschriften heraus. 1935 bis 1939 entstand eine seiner genialsten Arbeiten, das Haus »Fallingwater« in Pennsylvania. Im Rahmen der Weltwirtschaftskrise entwickelte er einen neuen Haustypus, das so genannte »Usonian House«. Außergewöhnliche organische Formen verwendete Wright 1956 bis 1959 bei der Spirale des »Guggenheim Museums« in New York. | Frank Lloyd Wright starb am 9. April 1959 in Phoenix, Arizona.

Frank Lloyd Wright was born 1867 in Richland Center, Wisconsin. He studied engineering from 1885 till 1887 at the University of Wisconsin, which he quit in order to work in Chicago at the architecture office of Louis Sullivan. In 1893 he broke apart with Sullivan and started his own business. | Wright developed his first living concept for the "Prairie Houses". His architecture was considered the most important element of the "Arts and Crafts Movement" in America. He worked in harmony with nature and tried to gain distance to the traditional designs and united organic forms with geometric ones. | In 1909, he travelled to Europe for two years where he too, became one of the most important sources of inspiration regarding architecture, for example his work influenced Walter Gropius and De Stijl. Back in America, Wright began the construction of "Taliesin", Wisconsin in 1911. Together with Antonin Raymond he travelled to Japan in 1913 to realise the Imperial Hotel in Tokyo, applying a new earthquake proof construction principle. From 1917 to 1920 he built the Barnsdall House in Los Angeles. | After the stock market crash in 1929 Wright gave lectures at the Princeton University, which he also published. Moreover did he write several books and magazines about his idea of urban planning for instance and issued magazines. In the period between 1935 and 1939 he created "Fallingwater", one of his most ingenious works, in Pennsylvania. In the context of the global economic crisis he developed a new type of house, the so called "Usonian House". From 1956 to 1959 he used extraordinary organic shapes for the spiral of the "Guggenheim Museum" in New York. Frank Lloyd Wright died on April 9, 1959 in Phoenix, Arizona.

Joachim Krausse Joachim Krausse wurde mit seinen Arbeiten über den amerikanischen Architekten, Designer und Erfinder R. Buckminster Fuller bekannt, zuletzt mit Ausstellung und Buch *Your Private Sky: R. Buckminster Fuller — The Art of Design Science* 1999 - 2002. Im Zentrum seiner Forschungen und Veröffentlichungen, Ausstellungen und Fernsehdokumentationen stehen die Fragen nach einer technischen Kultur und die Bedeutung von Konstruktion als Verkörperung von Intelligenz. | Parallel zu einem geisteswissenschaftlichen Studium (Philosophie, Literaturwissenschaften und Publizistik) erschienen Beiträge zur visuellen Poesie, Beteiligung an Ausstellungen wie *Visuelle Poesie* (1964) oder *Cybernetic Serendipity* (1967). Redakteur des Verlags Edition Voltaire. Lehrtätigkeit seit 1970 u. a. an der Hochschule der Künste, der TU Berlin, HfG Schwäbisch Gmünd. Seit 1999 Professor für Designtheorie am Fachbereich Design der Hochschule Anhalt (FH) in Dessau. Seit 2002 Leiter des internationalen Masterstudiengangs Integrated Design (MAID). Seit 1991 ständiger Mitarbeiter der Architekturzeitschrift Archplus. Seit 2008 emeritierter Professor an der Hochschule Anhalt (FH). Seit März 2009 Fellow an der Bauhaus-Universität Weimar. | Weitere Artikel zum Planetarium: *Architektur aus dem Geist der Projektion. Das Zeiss Planetarium*. In: *Wissen in Bewegung. 80 Jahre Zeiss-Planetarium Jena*. Hrsg. Ernst-Abbe-Stiftung / Hans-Christian von Herrmann, Jena 2006, S. 51 - 84. *Mechanischer Affe und Quantum Machine. Bau- und Bühnenlaboratorium vom Bauhaus zum Black Mountain College*. In: *Spuren der Avantgarde. Theatrum Machinarum*. Hrsg. Helmar Schramm, Berlin, New York: de Gruyter 2008, S. 407 - 444.

Hermann Finsterlin wurde 1887 in München geboren. 1905 - 08 privater Kunstunterricht in München, u. a. bei Walter Thor und Hermann Groeber. Mit Freunden Gründung eines freien Ateliers. 1907 erste Gedichte. 1914/15 Studium (ohne Abschluss): Naturwissenschaften (Chemie, Physik), Medizin, Indologie an der Universität München. Ab 1917 Architekturentwürfe. 1918 regt das Erlebnis der Bergwelt am Watzmann Finsterlins Hinwendung zum Naturexpressionismus an. 1919 Kontakt zu Walter Gropius; Finsterlin etabliert sich als »Architekt« durch Teilnahme an einer Ausstellung des Arbeitsrats für Kunst in Berlin und mit Entwürfen utopischer »Traumhäuser«. 1920 - 22 Beteiligung an dem von Bruno Taut initiierten Architekten-Briefwechsel der »Gläsernen Kette« und Artikel in der Zeitschrift *Frühlicht* von B. Taut. 1921 - 28 Arbeit am Baukasten Stilspiel, einer »Genesis der Weltarchitektur«. 1922 Gründungsmitglied des Berchtesgadener Künstlerbunds. 1924 Finsterlin-Sonderheft der Zeitschrift *Wendingen*. 1924 Briefwechsel mit Antoni Gaudí. Um 1963 neuer Kontakt zu den Mitglied der »Gläsernen Kette«, ab 1969 mit Wijdeveld. Hermann Finsterlin stirbt am 16. September 1973 in Stuttgart.

Joachim Krausse Joachim Krausse became known with his work concerning the American architect, designer and inventor R. Buckminster Fuller, most recently with his exhibition and book *Your Private Sky: R. Buckminster Fuller — The Art of Design Science* 1999 - 2002. In his research and publications, exhibitions and TV documentaries the focus lays on the demand for a technical culture and the importance of the construction as an embodiment of intelligence. Parallel to a study of human sciences (philosophy, literature and journalism) essays about visual poetry ands participation in exhibitions like *Visuelle Poesie* (1964) or *Cybernetic Serendipity* (1967) were published. Editor of Edition Voltaire publishers. Teaching activity since 1970 at, amongst others, the University of Arts, the TU Berlin, HfG Schwäbisch Gmünd. Since 1999 professor for design theory at the design department at the University of Anhalt (FH) in Dessau. Since 2002 head of the international master's programme Integrated Design (MAID). Since 1991 regular contributor for the architectural magazine Archplus. Since 2008 Emeritus professor at the University of Anhalt (FH) Since march 2009 fellow at the Bauhaus University Weimar. | Further articles about the planetarium: *Architektur aus dem Geist der Projektion. Das Zeiss Planetarium*. In: *Wissen in Bewegung. 80 Jahre Zeiss-Planetarium Jena*. Publisher Ernst-Abbe-Stiftung / Hans-Christian von Herrmann, Jena 2006, p. 51 - 84. *Mechanischer Affe und Quantum Machine. Bau- und Bühnenlaboratorium vom Bauhaus zum Black Mountain College*. In: *Spuren der Avantgarde. Theatrum Machinarum*. Publisher Helmar Schramm, Berlin, New York: de Gruyter 2008, p. 407 - 444.

Hermann Finsterlin was born 1887 in Munich. 1905 - 08 took private art lessons in Munich, including Walter Thor and Hermann Groeber. He founded with friends a free atelier. Wrote his first poems in 1907. 1914/15 study (no degree): Natural Sciences (chemistry, physics), medicine, and Indian studies at Universität München. In 1917 architectural designs. 1918 the very intense experience of the mountain "Watzmann" stimulates him to turn to expressionism of nature. 1919 contact with Walter Gropius; Finsterlin established himself as "architect" by participating in an exhibition of the Workers' Art in Berlin and with designs of utopian "dream houses". 1920 - 22 initial correspondences with Bruno Taut led to participation in the "Gläsernen Kette" (pseudonym Prometh) and articles in the Journal *Frühlicht* of B. Taut. 1921 - 28 work on the construction kit Stilspiel, a "genesis of the world's architecture". 1922 founding member of the Berchtesgadener Künstlerbundes. In 1924 Finsterlin special issue of the journal *Wendingen*. In 1924 began correspondence with Antoni Gaudí. Around 1963 made contact with a member of the "Gläsernen Kette", Wijdeveld, until 1969. Hermann Finsterlin died on September 16, 1973 in Stuttgart.

Sir D' Arcy Wentworth Thompson geboren 1860 in Edinburgh. Er war Biologe, Mathematiker und Altphilologe. Bekannt wurde er hauptsächlich als der Autor des Buches *Über Wachstum und Form* (1917), eine einflussreiche Arbeit mit auffallender Originalität und Eleganz. | 1878 schrieb er sich an der Universität von Edinburgh ein, um Medizin zu studieren. Zwei Jahre später, wechselte er an der Cambridge University zu Zoologie und erhielt 1883 den BA in der Naturwissenschaft. 1884 wurde er Professor für Biologie bzw. Naturgeschichte in Dundee, ein Posten, den er 64 Jahre hielt. 1917 wurde er zum Professor der Naturgeschichte an der St. Andrews University berufen. 1896 führte er eine Expedition durch die Bering-Straße. Ab 1916 war er Fellow of the Royal Society (London) und 1937 wurde er geadelt. 1946 erhielt er die Darwin-Medaille. Er verstarb am 21. Juni 1948 in St. Andrews, Schottland

Sir D'Arcy Wentworth Thompson was born 1860 in Edinburgh. He was a biologist, mathematician, and classics scholar. He is mainly remembered as the author of the 1917 book *On Growth and Form*, an influential work of striking originality and elegance. | In 1878, he matriculated at University of Edinburgh to study medicine. Two years later, he shifted his studies to Cambridge University, obtaining the BA in Natural Science in 1883. In 1884, he was appointed Professor of Biology at University College, Dundee, a post he held for a record 64 years. In 1917, he was appointed to the Chair of Natural History at St. Andrews University. In 1896, he carried out an expedition to the Bering Straits. Elected a Fellow of the Royal Society in 1916, he was knighted in 1937 and awarded the Darwin Medal in 1946. He died June 21, 1948 in St. Andrews, Scotland.

Ernst Haeckel wurde 1834 in Potsdam, Preußen, geboren. Er war ein deutscher Zoologe und Evolutionist. Nach seinem Hochschulabschluss in Medizin 1857 erhielt er ein Doktorat in Zoologie an der Universität von Jena, und von 1862 bis 1909 lehrte er Zoologie in Jena. Zwischen 1872 und '76 nahm er an der Challenger-Expedition teil, auf der er ca. 3500 neue Radiolarien erfasste und zeichnerisch darstellte. Seine Arbeit konzentrierte sich auf verschiedene wirbellose Meerestiere. Beeinflusst durch Charles Darwin, sah Haeckel Entwicklung als die Basis für eine Erklärung aller Arten und des Grundprinzips einer philosophischen Annäherung. Er versuchte, den ersten Abstammungsbaum des gesamten Tierreiches herzustellen.
Er glaubte, dass jede Art ihre Evolutionsgeschichte in ihrer embryologischen Entwicklung veranschaulicht (»Ontogenese rekapituliert Phylogenese«). Mit seinen Theorien über die Entwicklung der Menschen wurde die Aufmerksamkeit zu den wichtigen biologischen Fragen auf ihn gelenkt. Durch seine zahlreichen Bücher war er ein einflussreicher Popularisator der Evolutionstheorie. Er verstarb am 9. August 1919 in Jena, Deutschland.

Ernst Haeckel was born 1834 in Potsdam, Prussia. He was a German zoologist and evolutionist. After receiving a degree in medicine in 1857, he obtained a doctorate in zoology from the University of Jena, and from 1862 to 1909 he taught zoology at Jena. From 1872 - 76 he travelled with the challenger-expedition and discovered approximately 3500 new types of radiolarians which he also documented with drawings. His work concentrated on diverse marine invertebrates. Influenced by Charles Darwin, Haeckel understood evolution as the basis for an explanation of all nature and the basic principle of a philosophical approach. He attempted to create the first genealogical tree of the entire animal kingdom. He proposed that each species illustrates its evolutionary history in its embryological development ("Ontogeny recapitulates phylogeny"). Through his theories of the evolution of humans, he brought attention to important biological questions. Through his numerous books, he was an influential popularizer of evolutionary theory.
He died at August 9, 1919 in Jena, Germany.

Johannes Kepler Der Astronom, Naturphilosoph, Theologe, Astrologe, Optiker und Mathematiker Johannes Kepler wurde 1571 in Weil, Württemberg, geboren. Im Jahr 1589 begann er sein Studium an der Schwäbischen Hochschule Tübingen. Im August 1591 erwarb Kepler die Magisterwürde. Anschließend widmete er sich mehr der Theologie. Im dritten Studienjahr der Theologie erhielt Kepler einen Ruf an die Universität Graz, wo er ab März 1594 lehrte. Durch seine treffenden Voraussagen für das Jahr 1594 (kalter Winter, Türkeneinfall) wurde Kepler auch als Astrologe schnell berühmt. 1596 erschien sein *Mysterium cosmographicum* (das Weltgeheimnis). Im Herbst 1604 erschienen die *Silvae chronologicae*, 1605 die *Betrachtungen über den neuen Stern im Ophiuchus* sowie 1607 die Schrift über den in diesem Jahr erschienenen

Johannes Kepler The astrologer, natural philosopher, theologian, astrologer, optician and mathematician was born 1571 in Weil, Württtemberg. In 1589 he started his study at the University of Thübingen. After he received his master's degree in August 1591, he dedicated himself to theology. In his third year of theology studies, Kepler was appointed professor at the University of Graz, where he started teaching in March 1594. Due to his correct forecasts for the year 1594 (cold winter, entry of the Turks) Kepler quickly became also known as an astrologer. In 1596 his *Mysterium cosmographicum* (secret of the world) was published. | In fall 1604 the *Silvae chronologicae* was released, in 1605 the *observations about the new star in the Ophiuchus* and in 1607 the essay concerning the comet that appeared that year. In 1618

Kometen. 1618 erschien Keplers drittes Gesetz der Planetenbewegung. Dies Gesetz findet sich in *Harmonices mundi* (1619). 1609 erschien sein Werk *De Stella Martis* (1609). 1611 erfindet Kepler das astronomische Fernrohr, genannt Kepler'sches Fernrohr. Kepler bestimmt die astronomische Strahlenbrechung und stellt Formeln dazu auf. Seine wichtigste Schrift zur Mathematik *Nova stereometria doliorum vinariorum* erschien 1615. Er verstarb am 15. November 1630 in Regensburg.

Keplers third law over planetary motion was published in the *Harmonices mundi* (1619). In 1609 he released his work *De Stella Martis* (1609). Kepler also invented the astronomical telescope in 1611, called the Kepler telescope. He defines the astronomical refraction and sets up formulas for it. His most important mathematical work for the *Nova stereometria doliorum vinariorum* appeared in book form in 1615. He died on November 15, 1630 in Regensburg.

PERSONENREGISTER | INDEX OF PERSONS

REGISTER

INDEX

BIBLIOGRAPHIE | BIBLIOGRAPHY

Krausse, Joachim: *Das Wunder von Jena.* In Arch+, Nr. 116, März 1993

Blake, Peter: *Drei Meisterarchitekten.* R. Piper & Co Verlag, 1962

Blüchel, Kurt G.; Malik, Fredmund: *Faszination Bionik.* Malik Management Zentrum, 2006

Bühler, Paul: *Sandwichstrukturen der Schädelkapsel verschiedener Vögel — Zum Leichtbauprinzip bei Organismen.* Institut für leichte Flächentragwerke, IL4, 1972

De Long, David: *Frank Lloyd Wright. Die lebendige Stadt.* Vitra Design Museum, 1998

Döhl, Reinhard: *Finsterlin.* Verlag Gerd Hatje, 1988

Geiger, Annette; Hennecke, Stefanie; Kempf, Christin: *Spielarten des Organischen in Architektur, Design und Kunst.* Dietrich Reimer Verlag, 2005

Haeckel, Ernst: *Kunstformen der Natur.* Marix Verlag, 2004

Hartmann, Günter: *Domestic architecture of the primitive people.* Institut für leichte Flächentragwerke, IL27, 1980

Hildebrand, Grant: *The Wright space: pattern and meaning in Frank Lloyd Wright's houses.* University of Washington Press, 1994

Hitchcook, Henry-Russell: *The Building of Frank Lloyd Wright 1887-1941.* In the Nature of Materials. Duell, Sloan and Pearce, 1942

Hoffmann, Donald: *Frank Lloyd Wright Architecture and Nature.* Dover Publications, 1986

Jodidio, Philip: *Architecture: Nature.* Prestel Verlag, 2006

Kepler, Johannes: *Strena vom Sechseckigen Schnee.* W. Keiper Berlin, 1943

Le Corbusier: *Ausblick auf eine Architektur.* Birkhäuser Verlag, 2001

Le Corbusier: *Der Modulor.* J. G. Cotta'sche Buchhandlung, 1953

Mandelbrot, Benoît B.: *Die fraktale Geometrie der Natur.* Birkhäuser Verlag, 1991

Müller, Rudolf: *Der umgekehrte Weg.* Verlagsgesellschaft Rudolf Müller, 1990

Nachtigall, Werner: *Bau-Bionik.* Springer-Verlag, 2003

Nachtigall, Werner: *Vorbild Natur.* Springer-Verlag, 1997

Nachtigall, Werner; Blüchel, Kurt G.: *Das große Buch der Bionik.* Deutsche Verlags-Anstalt, 2000

Nicolai, Carsten: *Static Fades.* JRP Ringier, 2007

Otto, Frei: *Wachsende und sich teilende Pneus.* Institut für leichte Flächentragwerke, IL19, 1979

Otto, Frei: *Natürliche Konstruktionen.* Deutsche Verlags-Anstalt, MIT Press, 1982

Otto, Frei: *Schriften und Reden, 1951–1983.* Friedr. Vieweg & Sohn, Verlagsgesellschaft, 1984

Otto, Frei: *Schalentiere und Seeigel.* Institut für leichte Flächentragwerke, IL35, 1995

Otto, Frei: *Gestaltwerdung. Zur Formentstehung in Natur, Technik und Baukunst.* Rudolf Müller Verlag, 1998 Otto, Frei: *Das Netz der lebenden Gestalt.* Verlag Ulrich Keicher, 2006

Otto, Frei; Rasch, Bodo: *Gestalt finden.* Edition Axel Menges, 2001

Pearce, Peter: *Structure in Natur is a Strategy for design.* 1990

Prochnow, Oskar: *Formenkunst der Natur.* Ernst Wasmuth Verlag, 1934

Schaur, Eda: *Bauweise der lebenden Natur.* Institut für leichte Flächentragwerke, IL27, 1980

Schneider, Ulrich: *Hermann Finsterlin und die Architektur des Expressionismus.* Ernst Wasmuth Verlag, 1999

Stewart, Ian: *Das Rätsel der Schneeflocke.* Spektrum Akademischer Verlag, 2001

Stewart, Ian: (text © 2001 Ian Stewart, originally published in) *What Shape is a Snowflake?* The Ivy Group, 2001

Stewart, Ian: *Die Zahlen der Natur.* Spektrum Akademischer Verlag, 2001

Stewart, Ian; Golubitsky, Martin: *Denkt Gott symmetrisch?* Birkhäuser Verlag, 1993

Thompson, D'Arcy Wentworth: *On Growth and Form.* Cambridge University Press, 2006

Thompson, D'Arcy Wentworth: *Über Wachstum und Form.* Eichborn Verlag, 2006

Thywissen, Cornelius: *Tierbauten zwischen Spezialisierung und Anpassung.* Institut für leichte Flächentragwerke, IL27, 1980

Thywissen, Cornelius: *Konstruktionen der nicht lebenden Natur,* Institut für leichte Flächentragwerke, IL27, 1980

Treiber, Daniel: *Frank Lloyd Wright.* Birkhäuser Verlag, 1988

Van der Ree, Pieter: *Organische Architektur.* Verlag Freies Geistesleben, 2001

Wyss, Beat: Vitruv, *Baukunst.* Band I–V und VI–X. Artemis Verlag, 1987

Wille, Rudolf: *Symmetrie in Geistes- und Naturwissenschaft.* Springer-Verlag, 1988

Wolfram, Stephen: *A new kind of science.* Wolfram Media, 2002

Wright, Frank Lloyd: *Modern Architecture.* Princeton University Press, 1931

Wright, Frank Lloyd: *An autobiography.* Duell, Sloan and Pearce, 1943

Wright, Frank Lloyd: *Usonien when democracy builds.* Verlag Gebr. Mann, 1950

Wright, Frank Lloyd: *Das Natürliche Haus.* Albert Langen - Georg Müller Verlag, 1954

Wright, Frank Lloyd: *Humane Architektur.* Bertelsmann Fachverlag, 1969

Wright, Frank Lloyd: *Ausgeführte Bauten und Entwürfe von Frank Lloyd Wright.* Ernst Wasmuth Verlag, 1998

BILDNACHWEIS | PICTURE CREDITS

Vorwort / Preface (8–12)
Abbildung / Image by ©Thomas Mayer

Einleitung / Introduction (13–34)
Abb. / Fig.1 Vitruv, Abb. / Fig.2, 3 ©2013 Finsterwalder Architekten, Abb. / Fig.4, 6, 11, 15–18 ©IL, Uni Stuttgart, Abb. / Fig.5 Modulor, Le Corbusier, Abb. / Fig.7 Álvaro Siza, Abb. / Fig.8 Benoit B. Mandelbrot, Abb. / Fig.9 Frei Otto, Abb. / Fig.10 J. Krausse, Abb. / Fig.12 Eichborn-Verlag, Abb. / Fig.13 Marix Verlag, Abb. / Fig.14 Speedo

Kapitel / Chapter 1 (35–46)
Abbildung / Image by ©2013 Finsterwalder Architekten

Kapitel / Chapter 2 (47–56)
Abbildung / Image by ©2011 Finsterwalder Architekten

Kapitel / Chapter 3 (57–92)
Abbildungen von Dieter Dolezel
Images by Dieter Dolezel

Kapitel / Chapter 4 (93–126)
Abb. / Fig.1, 13 ©2011 Finsterwalder Architekten, Abb. / Fig.2, 11, 14, 15 Frei Otto, Abb. / Fig.3a–d J. Krausse, Abb. / Fig.4, 5, 7, 8, 10, 12 ©IL, Uni Stuttgart, Abb. / Fig.6 Eichborn-Verlag, Abb. / Fig.9 www.nps.qov

Kapitel / Chapter 5 (127–144)
©MCB Verlag Publishing House GmbH München für den Beitrag von Werner Nachtigall
©MCB Verlag Publishing House GmbH München for the contribution of Werner Nachtigall

Kapitel / Chapter 6 (145–164)
Abbildungen ©2011 Finsterwalder Architekten
Images by ©2011 Finsterwalder Architekten, S. / p.150 Alberto Novelli

Kapitel / Chapter 7 (165–196)
Abbildungen ©2011 Finsterwalder Architekten
Images by ©2011 Finsterwalder Architekten, Abb. / Fig.7 Peter Pohl, Abb. / Fig.20–65 Josefine Unterhauser

Kapitel / Chapter 8 (197–208)
Abbildungen ©2000 Carsten Nicolai, VG Bild-Kunst, Bonn und Courtesy Galerie EIGEN + ART Leipzig / Berlin und The Pace Gallery
Images by © 2000 Carsten Nicolai, VG Bild-Kunst, Bonn and Courtesy Galerie EIGEN + ART Leipzig / Berlin and The Pace Gallery

Kapitel / Chapter 9 (209–238)
Abb. / Fig.1 ©moni.sertel, »Drillinge« *
Abb. / Fig.5 ©Hansjoerg Morandell, »Sand Dune« *
Abb. / Fig.6 ©juerjen.by, »Dünenwanderung« *
S. / p.195 © PolluxTS, »Trio« *
* Quelle / Source: www.piqs.de
Abb. / Fig.17 ©NASA/JPL/Space Science Institute
Abb. / Fig.18 ©NASA/JPL-Caltech/T. Pyle (SSC)
Abb. / Fig.4 Imolcho, Abb. / Fig.2, 3, 7–11, 13–16 ©2011 Finsterwalder Architekten

Kapitel / Chapter 10 (239–258)
Kapitel / Chapter 11 (259–294)
Kapitel / Chapter 12 (295–308)
Kapitel / Chapter 13 (309–320)
Kapitel / Chapter 14 (321–330)
Kapitel / Chapter 15 (331–338)
Kapitel / Chapter 16 (339–354)
Kapitel / Chapter 17 (355–368)
Abbildungen ©IL, Uni Stuttgart
Images by ©IL, Uni Stuttgart

Kapitel / Chapter 18 (369–386)
Abbildungen ©Wasmuth Verlag
Images by ©Wasmuth Verlag

Kapitel / Chapter 19 (387–400)
Abbildungen ©2011 Rudolf Finsterwalder
Images by ©2011 Rudolf Finsterwalder
Abb. / Fig.3 Nachzeichnung

Kapitel / Chapter 20 (401–420)
Abbildungen von J. Krausse
Images by J. Krausse

Kapitel / Chapter 21 (421–428)
Abb. / Fig.2, 3, 4, 7, 8 ©Staatsgalerie Stuttgart, Abb. / Fig.1, 5, 6 ©Westfälisches Landesmuseum Münster

Kapitel / Chapter 22 (429–442)
Abbildungen ©Eichborn-Verlag
Images by ©Eichborn-Verlag

Kapitel / Chapter 23 (443–470)
Abbildungen ©Marix-Verlag
Images by ©Marix-Verlag

Kapitel / Chapter 24 (471–486)
Abbildungen von Johannes Kepler
Images by Johannes Kepler

DANKSAGUNGEN | ACKNOWLEDGMENTS

Ingrid und Frei Otto
für die wichtigen Gespräche

Joachim Blüher, Direktor der Villa Massimo Rom
für seine unermüdliche und einzigartige Unterstützung

Dr. Karl Eisele & Elisabeth Eisele Stiftung
für die finanzielle Unterstützung

Mercedes-Benz Italien und Volker Wiedmeyer
für die finanzielle Unterstützung

Ingrid and Frei Otto
for the importatant discussions

Joachim Blüher, Director Villa Massimo Rome
for his never ending and exceptional support

Dr. Karl Eisele & Elisabeth Eisele Stiftung
for the financial aid

Mercedes-Benz Italien and Volker Wiedmeyer
for the financial aid

DR. KARL EISELE
& ELISABETH EISELE
STIFTUNG

Mercedes-Benz

IMPRINT

Editor Rudolf Finsterwalder
www.finsterwalderarchitekten.com

in collaboration with:
Birgit Ostrzecha, Linda Walter, Anne Schmidt,
Monika Grandl, Marco Ravanelli, Johnathan Buckley

Translation German-English, English-German:
Jeremy Gains, Rudolf Finsterwalder, Johnathan Buckley,
Andrea Lyman and Brandon Johnson (2nd edition)

Layout and Coverdesign:
Anne Schmidt (Calysto), Erfurt, Germany

Proof reading:
Michael Walch, Vienna, Austria (1st edition)
Karin Huck, Allschwil, Switzerland (2nd edition)

Image processing:
Carsten Humme, Leipzig, Germany

Printing:
Holzhausen Druck GmbH, Wolkersdorf, Austria

Library of Congress Cataloging-in-Publication data A CIP catalog record for this book has been applied for at the Library of Congress.

Bibliographic information published by the German National Library.The German National Library lists this publication in the Deutsche Nationalbibliografie; detailed bibliographic data are available on the Internet at http://dnb.dnb.de.

P.O. Box 44, 4009 Basel, Switzerland
Part of Walter de Gruyter GmbH, Berlin/Boston

Printed on acid-free paper produced from chlorine-free pulp. TCF ∞

Printed in Austria

ISBN 978-3-0356-0790-1

9 8 7 6 5 4 3 2
www.birkhauser.com